Grundlehren der mathematischen Wissenschaften 219

A Series of Comprehensive Studies in Mathematics

G. Duvaut J. L. Lions

Inequalities in Mechanics and Physics

Translated from the French by C. W. John

With 28 Figures

Springer-Verlag
Berlin Heidelberg New York 1976

Georges Duvaut
Université de Paris VI, Mécanique Théoretique, 75230 Paris, France

Jacques Louis Lions
Collège de France, 75015 Paris, France

Translator: Charlotte W. John
New Rochelle, New York 10804, U.S.A.

Translation of the French Original Edition
"Les inéquations en mécanique et en physique", Paris: Dunod 1972

AMS Subject Classification (1970): 35A15, 35B45, 35Gxx, 35Jxx, 35Kxx, 35Lxx, 49H05, 73B99, 76A99, 90A10

ISBN 3-540-07327-2 Springer-Verlag Berlin Heidelberg New York
ISBN 0-387-07327-2 Springer-Verlag New York Heidelberg Berlin

ISBN 2-04-004962-2 (French edition) Dunod Paris

Library of Congress Cataloging in Publication Data. Duvaut, G. Inequalities in mechanics and physics. (Grundlehren der mathematischen Wissenschaften in Einzeldarstellungen; 219). Includes bibliographies and index. 1. Mechanics. 2. Physics. 3. Inequalities (Mathematics) I. Lions, Jacques Louis, joint author. II. Title. III. Series. QA808.D8813. 531. 75-26891.

Printed in Germany.

Typesetting and printing: Zechnersche Buchdruckerei, Speyer. Bookbindung: Konrad Triltsch, Würzburg.

Table of Contents

Introduction

1. We begin by giving a simple example of a *partial differential inequality* that occurs in an elementary physics problem.

We consider a fluid with pressure $u(x,t)$ at the point x at the instant t that occupies a region Ω of $\mathbb{R}^3$ bounded by a membrane Γ of negligible thickness that, however, is *semi-permeable*, i.e., a membrane that permits the fluid to enter Ω freely but that prevents all outflow of fluid.

One can prove then (cf. the details in Chapter I, Section 2.2.1) that

$$\frac{\partial u}{\partial t} - \Delta u = g \left(\Delta u = \frac{\partial^2 u}{\partial x_1^2} + \frac{\partial^2 u}{\partial x_2^2} + \frac{\partial^2 u}{\partial x_3^2} \right) \quad \text{in} \quad \Omega,\, t>0, \tag{1}$$

g a given function, *with boundary conditions in the form of inequalities*[1]

$$\begin{aligned} &u(x,t)>0 \Rightarrow \partial u(x,t)/\partial n=0, \quad x\in\Gamma,\\ &u(x,t)=0 \Rightarrow \partial u(x,t)/\partial n\geqslant 0, \quad x\in\Gamma, \end{aligned} \tag{2}$$

to which is added the initial condition

$$u(x,0)=u_0(x). \tag{3}$$

We note that conditions (2) are *non linear*; they imply that, at each fixed instant t, there exist on Γ two regions Γ_0^t and Γ_1^t where $u(x,t)=0$ and $\partial u(x,t)/\partial n=0$, respectively. These regions are not prescribed; thus we deal with a "free boundary" problem.

We can restate (1), (2) in the (equivalent) form of *inequalities*. For that purpose, we introduce the set K of "test functions" v:

$$K = \{v| \quad v=\text{function defined in } \Omega\,^2,\ v\geqslant 0 \text{ on } \Gamma\}; \tag{4}$$

then (1), (2) are *equivalent* to

$$\begin{aligned} &u(.,t)\in K \quad \forall t\geqslant 0,\\ &\int_\Omega \left[\frac{\partial u}{\partial t}(v-u)+\operatorname{grad}_x u.\operatorname{grad}_x(v-u)-g(v-u)\right]\mathrm{d}x\geqslant 0 \qquad \forall v\in K. \end{aligned} \tag{5}$$

---

[1] $\partial/\partial n$ denotes the derivative in the direction of the normal to Γ directed towards the exterior of Ω.

[2] We must take v in the Sobolev space $H^1(\Omega)$; this will beformulated more precisely in Chapter 1.

The problem to find a solution u of (5) with the initial condition (3) is what we call *an inequality of evolution* (of parabolic type).

2. The preceding example has features of a general character: we will encounter problems *that can be expressed in terms of inequalities* in situations where the constraints, the equations of state, the physical laws change when certain thresholds are crossed or attained.

The aim of the present work is to discuss examples of such situations in Mechanics and Physics.

3. The "program" indicated in the above paragraph covers an immense field that we have not studied exhaustively; we limited ourselves in this volume to the simplest classic laws. We have treated the following subjects:

1) problems of semi-permeable walls, of diffusion, applications to thermodynamics and hydrodynamics;
2) problems of control, particularly in theromodynamics;
3) problems in (linearized) elasticity involving friction and unilateral conditions;
4) problems of bending of plane plates;
5) phenomena of elastic-visco-plasticity, perfect elasticity, plasticity, rigid-visco-plasticity, rigid-perfect plasticity, and locking materials;
6) flows of Bingham fluids;
7) problems of inequalities connected with the system of Maxwell operators.

4. In order to avoid ambiguity in the formulation of the problems enumerated above, it was necessary to give a concise but precise review of the mechanical or physical bases for the situations envisioned. This is done at the beginning of each chapter. We now give a short description of the contents of the chapters.

5. In Section 1. above, we gave an example of the problems treated in *Chapter* 1; other problems concern temperature control.

In *Chapter* 2, control problems are discussed that lead to inequalities of the type (compare with (5))

$$\partial u/\partial t(.,t)\in K,$$

$$(6)\qquad \int_\Omega \left[\frac{\partial u}{\partial t}\left(v-\frac{\partial u}{\partial t}\right)+\operatorname{grad}_x u.\operatorname{grad}_x\left(v-\frac{\partial u}{\partial t}\right)-g\left(v-\frac{\partial u}{\partial t}\right)\right]\mathrm{d}x\geqslant 0 \quad \forall v\in K,$$

with the initial condition (3).

Chapter 3 treats the classic linear theory of elasticity rather completely (in particular, we give a proof of Korn's inequality, the indispensable mathematical basis for the theory); then we go on to problems of friction that lead to inequalities; we adopt Coulomb's law and indicate some modifications.

Chapter 4 deals with problems of friction connected with the mechanics of thin plates.

Chapter 5 is devoted to phenomena of elasto-visco plasticity from which we derive, by various passages to the limit, the elastic-perfectly plastic case, the rigid-visco-plastic case and the rigid-perfectly plastic case, all of these problems being stated in the form of inequalities. In this chapter, we also investigate Hencky's law and locking materials.

Chapter 6 treats the flow of a certain type of non-newtonian fluid: Bingham fluids. Here, we are led to inqualities of evolution containing, as a special case, the classic system of Navier-Stokes equations.

Chapter 7 is concerned with the problems of inequalities connected with the system of Maxwell operators. We will first study conducting media where the relation between the electric field and the current density is expressed by the classic Ohm's law, i. e., media with constant resistivity (we call such a medium "stable"). Subsequently, we treat the case of media susceptible to ionization under the influence of the electric field. The resistivity then abruptly becomes infinite: these are the phenomena that occur in connection with breakdown of condensors or antennas.

"Hybrid" problems simultaneously involving two of the situations described in the outline for the preceding chapters are treated in separate articles by the authors (see: Duvaut-Lions [7], [8]).

6. Throughout this book, we made use of the most direct methods possible, generally representing *inequalities* (the absolutely indispensable tool, especially for *problems of evolution*) as *limiting cases of non linear equations* (which, moreover, usually have a mechanical or physical interpretation).

In addition, in order to facilitate the reading of the book, we presented each chapter as independent as possible (at the price of some repetition).

7. There are numerous earlier works on stationary inequalities in Mechanics. The classic approach (see P. Germain [1], G. Mandel [1], E. Tonti [1] and the bibliographies of these works) consists in studying stationary elasticity in relation to minimization of quadratic functionals on *vector spaces*. The minimization of analogous functionals on convex sets that *are not* vector spaces made its appearance in *perfect plasticity* (where the stress tensor remains in a closed *bounded* convex set) (cf. W.I. Koiter [1], G. Mandel [2], W. Prager [1] and the bibliographies of these works), subsequently in *unilateral elasticity* in the problem of Signorini, solved in G. Fichera [1], then in J.L. Lions-G. Stampacchia [1].

Similarly, the phenomena of cavitation studied by J. Moreau [3] and the investigation of minimal surfaces with constraints (J.C. Nitsche [1]) also lead to problems in variational inequalities.

The *inequalities of evolution* were introduced in Lions-Stampacchia for the parabolic case, in Lions [4] for the hyperbolic case and have been investigated particularly by H. Brézis [2][3] (cf. also the book Lions [1] and the bibliography of this work). It seems that the applications of the inequalities of evolution to Mechanics and Physics are being investigated here for the first time. As might be expected, these applications lead to many new problems, some of them still open; we mention specifically:

- the problem of *regularity* of solutions (the methods of Brézis-Stampacchia [1], Brézis [2] are not applicable to numerous situations in this book);
- the problem of inequalities of evolution in connection with convex sets or with functions *depending on* t (they occur particularly in the theory of dynamic elastic-visco plasticity.

[3] where one will find, in particular, the use of the theory of *non linear semigroups*, a theory that has not been used in this book.

8. There are other situations in physics leading to inequalities, either stationary or of evolution. We will return to this subject, e.g. in the discussion of thermo-elastic-visco plasticity and of optimal control of systems governed by inequalities. We also would like to point out that a free boundary problem, occurring in hydrodynamics, was solved with inequality methods by C. Baiocchi [1].

We did not treat two subjects related to this book:

i) *singular perturbations* related to inequalities (theory of singular layers); we refer to J.L. Lions [5], [6];

ii) methods of *numerical approximation* of solutions of inequalities of evolution, methods that will be treated in the book by R. Glowinski, J.L. Lions and R. Trémolières [1]. We refer to the works on this subject by D. Bégis [1], J.F. Bourgat [1], H. Brézis et M. Sibony [1], J. Céa et R. Glowinski [1], J. Céa, R. Glowinski et J. Nédelec [1], R. Commincioli [1], [2], [3], B. Courjaret [1], M. Frémond [1], A. Fusciardi, U. Mosco, F. Scarpini et A. Schiaffino [1], M. Goursat [1], Y. Haugazeau [1], P.G. Hodge [1], A. Marrocco [1], M. Sibony [1], D. Viaud [1].

9. The authors wish to express their sincere gratitude to M. Alais with whom they had fruitful discussions, to M.A. Lichnérowicz who graciously accepted the French edition in the series which he edits, and to C.W. John for her excellent work done in translating this book.

G. Duvaut, J.L. Lions

Chapter I

Problems of Semi-Permeable Media and of Temperature Control

1. Review of Continuum Mechanics

In this review, we do not intend to develop the complete theory of continuous media. For such an exposition, we refer the reader to the works of P. Germain [1, 2], G. Mandel [1], W. Noll and C. Truesdell [1], Sedov [1].

We would, however, like to review the principles of this theory and the essential results that we will subsequently need. We also introduce the notation which will be used. This review will deal with the stress tensor, conservation laws, tensor of deformations and constituent equations.

1.1. Stress Tensor

Let there be given a continuous medium that occupies an open region Ω of $\mathbb{R}^3$ referred to a system of orthonormal axes $Ox_1x_2x_3$. This medium is in a state of equilibrium under the influence of exterior forces, generally consisting of a volume distribution of forces in Ω and a surface distribution of forces on the boundary of Ω.

Inside the continuous medium, these forces induce a field of stresses that can be described in the following way: let M be a point of Ω, and $\mathscr{V}$ a two-dimensional continuously differentiable manifold passing through M and dividing the continuous medium, at least in the neighborhood of M, into two regions Ω_1 and Ω_2. Let $\mathbf{n}$ be the unit normal to $\mathscr{V}$ at M directed toward Ω_2. Under certain assumptions[1], generally well satisfied in practice, one can prove that the action of Ω_2 on Ω_1 is equivalent to a force density $\mathbf{F}$ on $\mathscr{V}$ and that $\mathbf{F}$ depends on the point M and on the normal vector $\mathbf{n}$ according to the formula

$$F_i = \sigma_{ij} n_j \qquad i=1,2,3; \quad j=1,2,3, \tag{1.1}$$

where the coefficients σ_{ij} depend on the point M.

[1] These assumptions amount to considering as negligible a possibly existing distribution of couples on the manifold $\mathscr{V}$. If one gives up these simplifying assumptions, one is led to a theory of so-called oriented media in which the stress tensor is no longer symmetric. On this subject, see P. Casal [1], G. Duvaut [1,2], A.E. Green and R.S. Rivlin [1], R. Hayart [1], R.D. Mindlin and H.F. Tiersten [1], A.C. Eringen and Suhubi [1], R.A. Toupin [1,2].

The quantities F_i and n_j are the components of the vectors $\mathbf{F}$ and $\mathbf{n}$. In Equation (1.1), we made use of the summation convention concerning repeated indices, i.e., that (1.1) stands for

$$F_i = \sum_{j=1}^{3} \sigma_{ij} n_j .$$

Throughout this book, we will make use of this convention, unless the contrary is expressly stated. With a change of base in $\mathbb{R}^3$, the quantities F_i and n_j transform like the components of vectors and, consequently, the quantities σ_{ij} are components of a tensor of second order, called the *stress tensor*. ☐

The introduction of this tensor permits the formulation of *boundary conditions*. Indeed, let Γ be the boundary of Ω, which we assume to be regular, and $\mathbf{n}$ the exterior unit normal of Γ. If $\mathbf{F}$ is the surface density of exterior forces acting on Ω in the points of Γ, we obtain the relation

$$F_i = \sigma_{ij} n_j$$

in every point of Γ, since it is the exterior medium which then plays the role of Ω_2.

1.2. Conservation Laws

The axioms or basic principles of classical mechanics of continua are the three laws of conservation: conservation of mass, conservation of momentum and conservation of energy.

i) *Conservation of mass.* Let $\mathbf{v}(M,t)$ be the field of velocity vectors at the instant t of the points of a continuous medium that is in motion relative to the system $Ox_1x_2x_3$. By $\mathcal{S}$ we denote an arbitrary domain of $\mathbb{R}^3$ that is contained in the region occupied by the continuous medium; this domain $\mathcal{S}$ contains a certain portion of matter, i.e., a certain number of material particles; we consider $\mathcal{S}$ as the region of $\mathbb{R}^3$ that contains these material particles at each instant; this implies that $\mathcal{S}$ is a variable domain that moves along with the fluid. The principle of conservation of mass then says: *The mass of material contained in the arbitrary domain $\mathcal{S}$, that is followed in its motion, is independent of time.* It follows that

$$\text{(1.2)} \qquad d/dt \int\int\int_{\mathcal{S}} \rho \, dx = 0, \qquad \forall \mathcal{S}$$

where the following notation has been used:

- $\rho = \rho(M,t)$ is the density at the point M at the instant t,
- dx is the element of volume $dx_1 dx_2 dx_3$.

If the field of velocities is continuous in the interior of $\mathcal{S}$, the integral equation (1.2) is equivalent to the pointwise equation

$$\text{(1.3)} \qquad \partial\rho/\partial t + \operatorname{div}(\rho \mathbf{v}) = 0,$$

where div$(\rho\mathbf{v})$ naturally denotes the divergence of the vector field $\rho\mathbf{v}$, i.e.,

$$\partial(\rho v_i)/\partial x_i .$$

ii) *Conservation of momentum.* This conservation law, also known as the fundamental principle of dynamics, can be expressed in the following way: *There exists a coordinate system and a time scale t, called Galilean, such that for any material system and at any instant, the "wrench"** (see definition at the end of this section) *of the exterior forces, applied to the system, is equal to the time derivative of the wrench of the momenta.*

Let us again consider the portion $\mathscr{S}$ of matter introduced in i); it constitutes a material system to which we can apply the principle stated above. The system of axes $Ox_1x_2x_3$ is assumed to be Galilean as is the time scale t under consideration.

We express the equality of the two wrenches in terms of their elements of reduction at the point O (cf. Remark 1.1 at the end of this section),

$$\int_{\mathscr{S}} f_i \mathrm{d}x + \int_{\partial\mathscr{S}} \sigma_{ij} n_j \mathrm{d}S = \mathrm{d}/\mathrm{d}t \int_{\mathscr{S}} \rho v_i \mathrm{d}x , \quad \forall \mathscr{S} , \tag{1.4}$$

$$\int_{\mathscr{S}} \varepsilon_{ijk} x_j f_k \mathrm{d}x + \int_{\partial\mathscr{S}} \varepsilon_{ijk} x_j \sigma_{kl} n_l \mathrm{d}S = \mathrm{d}/\mathrm{d}t \int_{\mathscr{S}} \varepsilon_{ijk} x_j \rho v_k \mathrm{d}x , \quad \forall \mathscr{S}. \tag{1.5}$$

We now explain the notation and the significance of the different terms.

The vector $\mathbf{f}$ with components f_i $(i=1,2,3)$ represents a volume distribution of exterior forces (example: the forces of gravity). Accordingly, the term $\int_{\mathscr{S}} f_i \mathrm{d}x$ is the i-th component of the resultant of the volume forces.

At its boundary $\partial\mathscr{S}$, the system $\mathscr{S}$ is subjected to a surface density of forces that, according to i), is given by $\sigma_{il}n_l$. Thus, the term $\int_{\partial\mathscr{S}} \sigma_{il} n_l \mathrm{d}S$ represents the i-th component of the resultant of the surface forces applied to the system $\mathscr{S}$. The surface element on $\partial\mathscr{S}$ is denoted by $\mathrm{d}S$.

The term $\mathrm{d}/\mathrm{d}t \int_{\mathscr{S}} \rho v_i \mathrm{d}x$ represents the i-th component of the time derivative of the kinetic resultant.

The quantities ε_{ijk} are the components of the third order tensor (for direct orthonormal changes of base) that is completely antisymmetric and is such that $\varepsilon_{123}=+1$ (cf. Remark 1.2 at the end of this section).

The two left-hand terms of Equation (1.5) are the i-th components of the moments that at the point O, result from the volume and surface forces acting on the system $\mathscr{S}$.

The right-hand side of (1.5) is the t-derivative of the i-th component of the kinetic moment at O.

By transforming the surface integral (1.4) into a volume integral, making use of conservation of mass and assuming the various quantities that occur under the integral sign to be sufficiently regular, we see that (1.4) is equivalent to

$$\sigma_{ij,j} + f_i = \rho\gamma_i , \quad i=1,2,3 , \tag{1.6}$$

---

* Fr. "torseur" (tanslator).

where we set

(1.7) $$\gamma_i = \frac{dv_i}{dt} = \frac{\partial v_i}{\partial t} + \frac{\partial v_i}{\partial x_j} v_j,$$

(1.8) $$X_{,j} = \partial X / \partial x_j.$$

The quantity γ_i is the i-th component of acceleration of the particle situated at the point with the coordinates (x_i) at the instant t. We observe that the expression γ_i, and therefore Equation (1.6), contains a *term that is nonlinear* with respect to the velocity components.

Equations (1.6) are known as the *equations of motion*.

If the problem under consideration is a *static* one, ($\mathbf{v} \equiv 0$), the righthand sides of Equations (1.6) are identically zero; we then speak of *equations of equilibrium*; in that case, the equations are *linear* with respect to the components σ_{ij} of the stress tensor.

Equations (1.5) can be changed by transforming the surface integral into a volume integral and by making use of conservation of mass and the equations of motion. Thus, they become

(1.9) $$\int_{\mathscr{S}} \varepsilon_{ijk} \sigma_{jk} dx = 0 \quad \forall \mathscr{S}$$

which is equivalent to

$$\sigma_{ijk} = 0$$

that is, to

(1.10) $$\sigma_{kj} = \sigma_{jk}.$$

Hence, the stress tensor is symmetric. □

The following two remarks are actually comments on wrenches and on the tensor with components ε_{ijk}.

Remark 1.1. We call *wrench* the entity consisting of

a) a free vector $\mathbf{R}$, called the resultant of the wrench,
b) a field of vectors $\mathbf{M}(P)$ defined at every point P and such that

(1.11) $$\mathbf{M}(Q) = \mathbf{M}(P) + \mathbf{QP} \wedge \mathbf{R}.$$

The vector $\mathbf{M}(P)$ is called the resultant moment at P of the wrench. The pair of vectors $\mathbf{R}$ and $\mathbf{M}(A)$, originating from a point A, is referred to as the elements of *reduction at A* of the wrench. Clearly, a wrench is completely determined if its elements of reduction at one point are known. □

Remark 1.2. The quantities ε_{ijk}, defined in the above section, are convenient for the performance of certain calculations due to the following basic relations

$$\varepsilon_{ijk}\varepsilon_{pqr} = \mathrm{Det}\begin{vmatrix} \delta_{ip} & \delta_{iq} & \delta_{ir} \\ \delta_{jp} & \delta_{jq} & \delta_{jr} \\ \delta_{kp} & \delta_{kq} & \delta_{kr} \end{vmatrix}, \tag{1.12}$$

$$\varepsilon_{ijk}\varepsilon_{pqk} = \delta_{ip}\delta_{jq} - \delta_{iq}\delta_{jp}, \tag{1.13}$$

$$\varepsilon_{ijk}\varepsilon_{pjk} = 2\delta_{ip}, \tag{1.14}$$

$$\varepsilon_{ijk}\varepsilon_{ijk} = 6. \tag{1.15}$$

The quantities δ_{ip} are the components of the Kronecker tensor, that is

$$\delta_{ip} = 0 \quad \text{if} \quad i \neq p,$$

$$\delta_{ip} = 1 \quad \text{if} \quad i = p.$$

Formula (1.12) is proved by utilizing the antisymmetric properties of the ε_{ijk} which permit reduction to the case where $(i,j,k) = (1,2,3)$ and $(p,q,r) = (1,2,3)$. Formulas (1.13), (1.14), (1.15) are proved successively, starting with (1.12).

The quantities ε_{ijk} occur in writing the i-th component of the vector product of two vectors or of the curl of a field of vectors and also in the expansion of the determinant of a 3×3 matrix. Indeed, one can easily verify that

$$(a \wedge b)_i = \varepsilon_{ijk} a_j b_k, \tag{1.16}$$

where a and b are vectors with the components (a_i) and (b_i), respectively, and that

$$(\mathrm{Curl}\, v)_i = \varepsilon_{ijk} v_{k,j}, \tag{1.17}$$

where $\mathbf{v}$ is a vector field with components (v_k).

Furthermore,

$$\varepsilon_{ijk}\, \mathrm{Det}\, M = \varepsilon_{pqr} M_{ip} M_{jq} M_{kr}, \tag{1.18}$$

$$\mathrm{Det}\, M = \tfrac{1}{6} \varepsilon_{ijk}\varepsilon_{pqr} M_{ip} M_{jq} M_{kr}, \tag{1.19}$$

where M is a 3×3 matrix with elements (M_{ij}).

Relation (1.18) shows, in the case where M is the matrix of a direct ($\mathrm{Det}\, M = +1$) change of orthonormal bases, that the ε_{ijk} are components of a third-order tensor with respect to such changes of bases.

Finally, if M^{-1} is the reciprocal of the matrix M—in the case $\mathrm{Det}\, M \neq 0$—, the elements of M^{-1} are given by

$$(M^{-1})_{ij} = (2\,\mathrm{Det}\, M)^{-1} \varepsilon_{jpq}\varepsilon_{irs} M_{pr} M_{qs}. \tag{1.20}$$

Due to these different relations, it is in particular easy to establish the following classic formulas

$$\mathrm{Det}(A\,.\,B)=(\mathrm{Det}\,A)\,.\,(\mathrm{Det}\,B)\,,$$

$$\mathbf{a}\wedge(\mathbf{b}\wedge\mathbf{c})=(\mathbf{a}\,.\,\mathbf{c})\,\mathbf{b}-(\mathbf{a}\,.\,\mathbf{b})\mathbf{c}\,,$$

$$\mathrm{Curl}\,\mathrm{Curl}\,\mathbf{v}=\mathrm{grad}\,\mathrm{div}\,\mathbf{v}-\Delta\mathbf{v}\,,\quad \text{etc.}\quad \square$$

iii) *Conservation of energy.* This law is also known as the *first principle of thermodynamics.* It says that *the time derivative of the total energy of a system* (internal energy + kinetic energy) *is equal to the power of the exterior forces applied plus the influx of energy per unit time.*

Applying this law to the previously introduced material system $\mathscr{S}$, we obtain

$$\text{(1.21)}\qquad \mathrm{d}/\mathrm{d}t\int_{\mathscr{S}}\rho(\tfrac{1}{2}\mathbf{v}^2+e)\,\mathrm{d}x=\int_{\mathscr{S}}f_iv_i\,\mathrm{d}x+\int_{\partial\mathscr{S}}\sigma_{ij}n_jv_i\,\mathrm{d}S+\int_{\mathscr{S}}\rho w\,\mathrm{d}x-\int_{\partial\mathscr{S}}q_in_i\,\mathrm{d}S\,.$$

The scalar e denotes the specific internal energy of the continuous medium and, consequently, the term $\int_{\mathscr{S}}\rho(\frac{1}{2}\mathbf{v}^2+e)\mathrm{d}x$ represents the total energy of the system under consideration.

The terms $\int_{\mathscr{S}}f_iv_i\,\mathrm{d}x$ and $\int_{\partial\mathscr{S}}\sigma_{ij}n_jv_i\,\mathrm{d}S$ are equal, respectively, to the power of the volume and surface forces.

The scalar w denotes the quantity of energy entering per unit of mass and time and, consequently, $\int_{\mathscr{S}}\rho w\,\mathrm{d}x$ represents the volume influx of energy to the system $\mathscr{S}$ per unit of time.

The vector $\mathbf{q}$ with components q_i is the transport of energy vector and the term $\int_{\partial\mathscr{S}}q_in_i\,\mathrm{d}S$ represents the surface influx of energy per unit of time.

Equation (1.21) can be reduced by making use of Equations (1.3), (1.6), (1.10), after transforming the surface integrals into volume integrals. Thus, we obtain

$$\text{(1.22)}\qquad \int_{\mathscr{S}}\rho(\mathrm{d}e/\mathrm{d}t)\,\mathrm{d}x=\int_{\mathscr{S}}\sigma_{ij}v_{i,j}\,\mathrm{d}x+\int_{\mathscr{S}}(\rho w-q_{i,i})\,\mathrm{d}x\,.$$

We introduce the tensor D of strain velocities with the components

$$\text{(1.23)}\qquad D_{ij}=\tfrac{1}{2}(v_{i,j}+v_{j,i})\,.$$

Since Equation (1.22) holds for an arbitrary domain $\mathscr{S}$, it is equivalent to (1.24)

$$\text{(1.24)}\qquad \rho\,\mathrm{d}e/\mathrm{d}t=\sigma_{ij}D_{ij}+\rho w-q_{i,i}\,.\quad\square$$

Summary. Altogether, the conservation laws furnished us three equations, or groups of equations:

1) *The equation of continuity*

$$\text{(1.25)}\qquad \partial\rho/\partial t+\mathrm{div}\,(\rho\mathbf{v})=0\,.$$

2) *The equations of motion*

$$\rho\gamma_i = \sigma_{ij,j} + f_i \,. \tag{1.26}$$

3) *The equation of energy*

$$\rho\, de/dt = \sigma_{ij} D_{ij} + \rho w - q_{i,i}\,, \tag{1.27}$$

where the *stress tensor* with components σ_{ij} is *symmetric*. ▯

Remark 1.3. Starting out from the conservation laws, we derived the pointwise Equations (1.25)–(1.27) under assumptions of continuity. If, on the contrary, there exists a line of discontinuity of velocities or of the components of the stress tensor inside the continuous medium, we can show (P. Germain [2]) that on this line the partial differential Equations (1.25)–(1.27) must be replaced by discontinuity relations.

These discontinuity conditions can be included in the relations (1.25)–(1.27), provided the derivatives that occur there are to be interpreted in the sense of distributions. ▯

Remark 1.4. Equations (1.25)–(1.27) constitute a total of five scalar relations. There are fourteen unknown functions:

i) the six components σ_{ij} of the (symmetric) stress tensor;
ii) the three components v_i of the velocity;
iii) the density ρ, the internal energy e, the components q_i of the transport of energy vector.

From this enumeration it is clear that, simply from the mathematical point of view, it is quite unlikely that, with five equations, one can determine fourteen unknown functions!

Moreover, from the point of view of physics, one has to observe that the stated conservation laws are universal laws, valid for all continuous media, liquid, solid, or gaseous. If therefore Equations (1.25)–(1.27), which were obtained from those laws, were sufficient to determine all the parameters, this would mean that the different continuous media would have identical behavior if subjected to identical conditions. This is obviously absurd.

The stated conservation laws therefore are insufficient by themselves to describe the motions of continous media; they must be augmented by other relations that are described by the general term of *constituent relations*. To introduce these, we first have to define the strain tensor.

1.3. Strain Tensor

i) *Kinematik description*. Let there be a continuous medium in motion that at the instant t occupies an open set Ω of $\mathbb{R}^3$, referred to a direct orthonormal base $Ox_1x_2x_3$.

Let (a_α) $(\alpha = 1,2,3)$ be the coordinates at the time 0 of the material particle located in the point with coordinates (x_i) $(i = 1,2,3)$ at the time t. We assume, in accordance with the physical reality of every particle of matter, that there exists a one-to-one correspondence between the (a_α) and the (x_i) of one and the same particle, say

$$(1.28) \qquad \begin{aligned} a_\alpha &= g_\alpha(x,t), \quad & \alpha &= 1,2,3, \quad & x &= (x_1, x_2, x_3), \\ x_i &= f_i(a,t), \quad & i &= 1,2,3, \quad & a &= (a_1, a_2, a_3). \end{aligned}$$

The relations (1.28), assumed to be continuous, also establish a correspondence between the open region Ω and an open region Ω_0 consisting of the set of points occupied, at the instant $t=0$, by the particles which are contained in Ω at the instant t. The components (a_α) are the *Lagrange coordinates* of the material particle under consideration, while the (x_i) are its *Euler coordinates* at the instant t.

ii) *Deformation gradient.* In transformation (1.28), an infinitesimal vector $\mathrm{d}M_0$ with components $(\mathrm{d}a_\alpha)$ becomes, at the instant t, an infinitesimal vector $\mathrm{d}M$ with components $\mathrm{d}x_i$ given by

$$(1.29) \qquad \mathrm{d}x_i = x_{i,\alpha} \mathrm{d}a_\alpha$$

or also by

$$(1.30) \qquad \mathrm{d}\mathbf{M} = \mathbb{F}\,\mathrm{d}\mathbf{M}_0$$

where $\mathbb{F}$ denotes the tensor with components $F_{i\alpha}$ given by

$$(1.31) \qquad F_{i\alpha} = f_{i,\alpha} = x_{i,\alpha}.$$

The tensor F is the *deformation gradient* tensor.

iii) *Tensor of dilatations.* Let $\mathrm{d}\mathbf{M}_0$ and $\delta\mathbf{M}_0$ be two infinitesimal material vectors with initial point $M_0 \in \Omega_0$; at the instant t, they become two infinitesimal vectors $\mathrm{d}\mathbf{M}$ and $\delta\mathbf{M}$ with initial point M and such that

$$\mathrm{d}\mathbf{M} = \mathbb{F}\,\mathrm{d}\mathbf{M}_0, \qquad \delta\mathbf{M} = \mathbb{F}\,\delta\mathbf{M}_0.$$

If, at any time t, we know the scalar products $\mathrm{d}\mathbf{M}^2$, $\mathrm{d}\mathbf{M}.\delta\mathbf{M}, \delta\mathbf{M}^2$, we also know the lengths and angles of the material elements $\mathrm{d}\mathbf{M}$ and $\delta\mathbf{M}$; now,

$$(1.32) \qquad \mathrm{d}\mathbf{M}.\delta\mathbf{M} = x_{i,\alpha} x_{i,\beta} \mathrm{d}a_\alpha \delta a_\beta,$$

therefore, these lengths and angles can be derived from the quantities $C_{\alpha\beta}$ given by

$$(1.33) \qquad C_{\alpha\beta} = x_{i,\alpha} x_{i,\beta}.$$

These are the components of a symmetric second order tensor, called the *dilatations tensor* and denoted by $\mathbb{C}$. We also have

$$(1.34) \qquad \mathbb{C} = \mathbb{F}^T \mathbb{F}$$

where $\mathbb{F}^T$ is the tensor that is the transpose of the tensor $\mathbb{F}$ and where the product $\mathbb{F}^T\mathbb{F}$ is carried out like a product of matrices.

If the continuous medium moves like a *rigid* body, that means without deformation, we have

$$(1.35)\qquad \mathrm{d}\mathbf{M}.\delta\mathbf{M}=\mathrm{d}\mathbf{M}_0.\delta\mathbf{M}_0\,,\qquad \forall \mathrm{d}\mathbf{M}_0\,,\qquad \forall\delta\mathbf{M}_0\,,$$

which implies that

$$(1.36)\qquad \mathbb{C}=\mathbb{1}$$

where $\mathbb{1}$ denotes the unit tensor, i.e., the tensor with components $(\delta_{\alpha\beta})$. Conversely, (1.36) implies (1.35); consequently: the necessary and sufficient condition that a continuous medium moves without deformation is that (1.36) holds.

iv) *Strain tensor.* This is the tensor defined by

$$(1.37)\qquad \mathbb{X}=\tfrac{1}{2}(\mathbb{C}-\mathbb{1})$$

or, in terms of components, by

$$(1.38)\qquad X_{\alpha\beta}=\tfrac{1}{2}(C_{\alpha\beta}-\delta_{\alpha\beta})\,.$$

We have then: *the necessary and sufficient condition for a continuous medium to move without deformation is that the strain tensor is zero.*

Let us introduce the displacement vector $\mathbf{u}$ by

$$(1.39)\qquad u_i=x_i-a_i\,,\qquad i=1,2,3\,.$$

We obtain

$$F_{i\alpha}=u_{i\alpha}+\delta_{i\alpha}$$

whence

$$(1.40)\qquad X_{\alpha\beta}=\tfrac{1}{2}(u_{\alpha,\beta}+u_{\beta,\alpha}+u_{i,\alpha}u_{i,\beta})\,.$$

We see that the strain tensor has a *nonlinear* expression in terms of the components of the displacement vector $\mathbf{u}$.

v) *Linearized strain tensor.* If the displacement vector $\mathbf{u}=\mathbf{u}(a,t)$ varies slowly with a, the partial derivatives $u_{i,\alpha}$ are small and the strain tensor itself is small. We say then that we deal with small deformations. If we assume that the quantities $u_{i\alpha}$ are of order ε (ε a parameter tending towards 0), the quantities are of order ε^2 and therefore

$$(1.41)\qquad X_{\alpha\beta}=\tfrac{1}{2}(u_{\alpha,\beta}+u_{\beta,\alpha})+\text{terms of order }\varepsilon^2\,.$$

We say that the tensor with the components $\varepsilon_{\alpha\beta}(u)$

$$\varepsilon_{\alpha\beta}(u)=\tfrac{1}{2}(u_{\alpha,\beta}+u_{\beta,\alpha}) \tag{1.42}$$

is the *linearized strain tensor.* ☐

Remark 1.6. In (1.23), we introduced the tensor $\mathbb{D}$ of strain velocities. Now we can justify this terminology. We form the time derivative of the scalar product $\mathrm{d}\mathbf{M}.\delta\mathbf{M}$, keeping the point M_0 and the infinitesimal vectors $\mathrm{d}\mathbf{M}_0$ and $\delta\mathbf{M}_0$ fixed. It follows that

$$\frac{\mathrm{d}}{\mathrm{d}t}(\mathrm{d}\mathbf{M}.\delta\mathbf{M})=\frac{\mathrm{d}}{\mathrm{d}t}(\mathrm{d}\mathbf{M}).\delta\mathbf{M}+\mathrm{d}\mathbf{M}.\frac{\mathrm{d}}{\mathrm{d}t}(\delta M). \tag{1.43}$$

But we have

$$\left(\frac{\mathrm{d}}{\mathrm{d}t}\mathrm{d}\mathbf{M}\right)_i=\frac{\mathrm{d}}{\mathrm{d}t}\left(\frac{\partial x_i}{\partial a_\alpha}\mathrm{d}a_\alpha\right)=\frac{\partial^2 x_i}{\partial t\,\partial a_\alpha}\mathrm{d}a_\alpha=\frac{\partial^2 x_i}{\partial a_\alpha\,\partial t}\mathrm{d}a_\alpha,$$

and, since

$$\partial x_i(a,t)/\partial t=v_i,$$

it follows that

$$\frac{\mathrm{d}}{\mathrm{d}t}\mathrm{d}\mathbf{M}=v_{i,\alpha}\mathrm{d}a_\alpha=v_{i,j}x_{j,\alpha}\mathrm{d}a_\alpha=v_{i,j}\mathrm{d}x_j. \tag{1.44}$$

We do the same for $\left(\frac{\mathrm{d}}{\mathrm{d}t}\delta\mathbf{M}\right)$ and, substituting in (1.43), we obtain

$$\frac{\mathrm{d}}{\mathrm{d}t}(\mathrm{d}\mathbf{M}.\delta\mathbf{M})=2D_{ij}\mathrm{d}x_i\delta x_j, \tag{1.45}$$

which explains why the tensor with components D_{ij} is called the tensor of strain velocities. ☐

1.4. Constituent Laws

Constituent laws do not have the universal character of conservation laws which were discussed in Section 1.2. Rather, they characterize the behavior of each kind of continuous medium. Their origin is often experimental, though they have to obey certain rules of invariance. (W. Noll and C. Truesdell [1], C. Truesdell and R. Toupin [1]). Generally speaking, these are relations between the stress tensor, the strain tensor, the strain velocity tensor, temperature, and the flux of heat vector. ☐

We are not going to list here "all" types of continuous media and their respective constituent laws. In each chapter, we will introduce the constituent law to be taken into account, and we will write the system of equations and *inequalities* which govern the phenomena in question by making use of the general principles which we have just reviewed. This we proceed to do now for the phenomena treated in this first chapter.

2. Problems of Semi-Permeable Membranes and of Temperature Control

2.1. Formulation of Equations

2.1.1. Equations of Thermics**

Thermics is the study of the field of temperatures in a continuous medium, assuming that the phenomena of deformation are uncoupled[2] and the velocities negligible, which has the effect of linearizing equations (1.25)–(1.27). They become

$$\partial\rho/\partial t=0\,, \tag{2.1}$$

$$\sigma_{ij,j}+f_i=0\,, \tag{2.2}$$

$$\rho\,\partial e/\partial t=\rho w-q_{i,i}\,. \tag{2.3}$$

The retained constituent laws are:

(2.4) σ_{ij} are independent of temperature,

$$e=C\theta\,, \tag{2.5}$$

$$q=-\mathbb{K}\,\mathrm{grad}\,\theta\,. \tag{2.6}$$

The notations used are the following.

i) The coefficient C is a specific heat, a strictly positive scalar. In reality, C depends on the temperature, more or less according to the material. Here, we assume that C may depend on x only.

ii) Law (2.6) is known as *Fourier's law*. It is linear when the tensor of second order $\tilde{\mathbb{K}}$ does not depend on x. This does not perfectly correspond to reality; in certain materials, one can observe significant variations of $\tilde{\mathbb{K}}$ with temperature.

Here, we will retain the linear law only, either because it is satisfied by the materials under consideration, or because the variations of temperature are small enough to justify a linearization around the mean temperature. Furthermore, we will assume that

$$\tilde{K}_{ij}x_ix_j\geqslant k_0x_ix_i\,,\qquad k_0=\text{constant}>0\,, \tag{2.7}$$

** Fr. "la thermique" (translator).

[2] That is to say that a deformation of the medium does not cause a variation of temperature.

which is well satisfied physically. If the material is isotropic, then the tensor $\tilde{\mathbb{K}}$ is spherical, that is to say that

$$\tilde{K}_{ij} = \tilde{k}\,\delta_{ij},$$

and the scalar $\tilde{k}$ is then the *coefficient of thermal conductivity* of the material. □

Equation (2.1) shows that the specific mass ρ does not depend on time; it might possibly depend on x, but then it is a given quantity for the thermic problem. □

Equation (2.4) implies that Equations (2.2) and (2.3) are independent and, consequently, that the temperature has to satisfy the single equation

$$\rho C\,\partial\theta/\partial t - (\tilde{K}_{ij}\theta_{,j})_{,i} = \rho\omega\,. \tag{2.8}$$

If we make the assumption—which is not indispensable, but we will maintain it in what follows in order to simplify the proofs—that the medium is homogeneous and isotropic, we can divide (2.8) by ρC, hereafter constant, and thus obtain

$$\partial\theta/\partial t - k\,\Delta C = g\,, \tag{2.9}$$

where we put

$$k = \tilde{k}/\rho C\,, \qquad g = \omega/C\,,$$

where k is a positive constant and $\Delta = \partial^2/\partial x_i \partial x_i$. □

2.1.2. Equations of Mechanics of Fluids in Porous Media

Equation (2.9) also governs the phenomena of the flow of viscous fluids in porous media (A. Houpert [1], Muskat [1, 2]). For this reason, it is also called the *diffusion equation.*

In this situation, it is obtained starting from the law of conservation of mass which, for porous media, is expressed by

$$\varphi\,\partial\rho/\partial t + \operatorname{div}(\rho v) = g\,,$$

where the scalar φ represents the *porosity* of the medium.

For porous media, the field of velocities v is linked to the field of pressure $u(x,t)$ by a relation known as *Darcy's law*

$$v = K \operatorname{grad} u\,,$$

where K represents the permeability of the medium. Here, two cases are to be considered:

α) *The fluid is only slightly compressible (liquid).*

The relation between density and pressure is given by

$$\rho=\rho_0[1+c(u-u_0)],$$

where c is the compressibility coefficient, a scalar, small compared to unity; in the equation for conservation of mass, ρv differs little from $\rho_0 v$ and, therefore, the pressure u essentially satisfies the equation

$$c\varphi\,\partial u/\partial t-\operatorname{div}(K\operatorname{grad}u)=g,$$

that is identical with Equation (2.9) when the porous medium is homogeneous and isotropic.

β) *The fluid is a perfect gas in isothermal flow.*
The relation between pressure and density is then (law of Mariotte)

$$u=C\rho$$

where C is a constant. The law of conservation of mass then yields

$$\varphi\,\partial\rho/\partial t-\operatorname{div}(C\rho K\operatorname{grad}\rho)=g,$$

or else, in terms of the pressure u,

$$\varphi\,\partial u/\partial t-\operatorname{div}(Ku\operatorname{grad}u)=Cg,$$

which is *non-linear.* □

2.1.3. Equations of Electricity

The general equations of electromagnetism will be reviewed in Chapter VII. Here, we simply point out that, in a medium that conducts electricity, the equation of conservation of charge, in the absence of free electric charges, can be written

$$\operatorname{div}J=g,$$

where J represents the *electric current* vector.
Moreover,

$$J=\sigma E,$$

where E is the *electric field* vector and σ the conductivity of the medium. In the stationary case, the electric field is derived from an electric potential u, i.e.,

$$E=-\operatorname{grad}u,$$

so that the conservation of charge can finally be written

$$-\operatorname{div}(\rho \operatorname{grad} u) = g\,,$$

which, for a *stationary phenomenon*, is just an equation of the type (2.9). □

2.2. Semi-Permeable Walls

Remark 2.1 *(preliminary)*. In the expositions that follow, Equation (2.9) will be written with $k=1$, which amounts to normalizing the equation by a change of time scale. Furthermore, u will always denote the unknown function which, according to the case considered, represents either a temperature, a pressure, or an electric potential, and therefore satisfies equation

$$\partial u/\partial t - \Delta u = g\,. \tag{2.10}$$

In the classical problems with Equation (2.10), the boundary data are either u or $\partial u/\partial n$ (the normal derivative of u on the boundary) which represents a flux of heat, of matter (fluid) or of electricity. □

We will introduce new types of problems which are just as natural as the preceding ones, but which involve *inequalities* in the conditions (at the boundaries or otherwise) and, consequently, lead to what we will call "Variational Inequalities".

The language employed for comments of a physical character is taken from mechanics of fluids in porous media, since that appeared to us closest to intuition. Of course, it can be translated from one field to another with the help of the table of correspondences below.

Thermics	*Fluid mechanics*	*Electricity*
temperature	pressure	potential
heat	fluid	electricity
heat flux	outflow of fluid	flux of electricity
conducting medium	porous medium	conducting medium
semi-permeable	semi-permeable	semi-conducting

We observe that the correspondence in the field of electricity applies only to *stationary* solutions or phenomena.

By Ω, we will always denote the open set in $\mathbb{R}^n$ ($n \leqslant 3$) under consideration; its boundary $\partial\Omega = \Gamma$ is regular and has the exterior unit normal n. The region Ω is occupied by a porous medium and is the seat of a field of pressure of a viscous fluid which is only slightly compressible (see 2.1.2, α)).

2.2.1. Wall of Negligible Thickness

We assume that the boundary Γ consists of a semi-permeable membrane of negligible thickness, i.e., it allows the fluid which enters Ω to pass freely but, on the other hand, prevents all outflow of fluid.

We now apply a given fluid pressure $h(x)$, $(x \in \Gamma)$ to the boundary Γ on the outside of Ω. Two situations are then possible for the points x of the boundary:

i) $h(x) < u(x,t)$:

the outside pressure $h(x)$ is smaller than the inside pressure $u(x,t)$ at the same point x of the boundary. The fluid then tries to leave Ω, but the semi-permeable wall prevents it and, consequently, the loss of fluid through the wall is zero at this point, i.e., $q.n=0$, from which we derive, since $q=-k\,\mathrm{grad}\,u$,

$$\partial u/\partial n=0\,.$$

ii) $h(x) \geqslant u(x,t)$:

the outside pressure $h(x)$ is greater or equal to the pressure $u(x,t)$ at the same point x of the boundary, the fluid then tries to enter into Ω, which the wall freely permits, so that

$$q.n \leqslant 0\,.$$

But $q.n=-\tilde{k}\,\partial u/\partial n$ must be finite which implies that $u(x)$ is continuous on the normal n in the neighborhood of x and, therefore, since the wall has negligible thickness, that $h(x)=u(x,t)$. Thus, the properties of the wall *make it impossible* that the pressure $u(x,t)$ at a point of the boundary is actually smaller than the outside pressure $h(x)$.

Summing up, the stated problem: "*Semi-permeable wall*" is the following:

Find a $u(x,t)$ *that satisfies* (2.10) *in* Ω, *with the following conditions on* (Γ),

$$\begin{aligned} &h(x)<u(x,t) \Rightarrow \partial u/\partial n=0\\ &h(x)=u(x,t) \Rightarrow \partial u/\partial n \geqslant 0\,, \end{aligned} \tag{2.11}$$

and with the initial condition

$$u(x,0)=u_0(x)\,. \quad \square$$

Remark 2.2. We might ask ourselves whether such walls exist in reality:

i) it appears to us that a valve represents an aporoximate local realization of such a situation (see variant 1 below);

ii) one can imagine physical devices that are capable of approximately realizing the phenomenon described above. This leads us to the formulation of variants of the preceding problem.

2.2.2. Semi-Permeable Wall of Finite Thickness

As in the preceding case, the wall that allows the fluid to enter only is subjected to an exterior fluid pressure $h(x)$. Two situations are possible:

i) $h(x) < u(x,t)$:
the fluid has the tendency to leave Ω, but the semi-permeable wall prevents this, so that the outflow is zero in such a point of Γ, i.e.,

$$\partial u/\partial n = 0 .$$

ii) $h(x) \geqslant u(x,t)$:
the fluid tries to enter Ω through the wall with finite thickness. It is reasonable to assume that the outflow through this wall is proportional to the difference in pressure, i.e.,

$$-\partial u/\partial n = k(u-h) ,$$

where k, a positive scalar, is a measure of the conductivity of the wall.

Summing up, *the pressure u in such a reservoir satisfies* (2.10), *the initial condition and the following boundary conditions*

$$\begin{aligned} &u > h \Rightarrow \partial u/\partial n = 0 \\ &u \leqslant h \Rightarrow \partial u/\partial n = -k(u-h) . \quad \square \end{aligned} \tag{2.12}$$

Remark 2.3. If k, the *conductivity of the wall*, is zero, the fluid can neither enter nor leave and the pressure u is the solution of the classical Neuman problem, stationary or instantaneous as the case may be. □

Remark 2.4. If the conductivity k tends toward $+\infty$, conditions (2.12), in the limit, become conditions (2.11). Actually, we will show[3] that if u_k is the solution corresponding to the conductivity k, then u_k tends, in a certain sense, toward u as k tends toward $+\infty$; here, u denotes the solution of the problem: "thin semi-permeable wall". □

Let us point out two further natural variants of the preceding problems.

Variant 1. The semi-permeable wall does not constitute the total boundary Γ but only a non empty part Γ_2. A classical condition applies to $\Gamma - \Gamma_1$: Either the pressure is given or the outflow is given. □

Variant 2. The semi-permeable wall is "inverted", that is to say that it only permits the fluid to flow out. The conditions at the points of such a wall become then:

i) for a thin wall:

$$\begin{aligned} &u < h \Rightarrow \partial u/\partial n = 0 \\ &u = h \Rightarrow \partial u/\partial n \leqslant 0 ; \end{aligned} \tag{2.11a}$$

[3] See Sec. 5.2.

ii) for a wall of finite thickness:

(2.12a)
$$u<h \Rightarrow \partial u/\partial n=0$$
$$u \geqslant h \Rightarrow \partial u/\partial n=k(u-h). \quad \square$$

2.2.3. Semi-Permeable Partition in the Interior of Ω

The reservoir Ω possesses a boundary Γ with a prescribed pressure h (classical condition). Furthermore, there exists in the interior of Ω a semi-permeable partition, a two-dimensional manifold Σ that divides Ω, at least in the neighborhood of Σ, into two regions denoted by 1 and 2 (Fig. 1). The unit normal to Σ pointing towards region 2 is denoted by n. If we call u_1 and u_2 the pressures on Σ from the sides of the regions 1 and 2, respectively, and if we assume that the fluid can only pass through the partition in the direction from 1 to 2, the conditions on Σ are:

i) if the partition is thin:

(2.11b)
$$u_1<u_2, \quad \partial u/\partial n=0$$
$$u_1=u_2, \quad \partial u/\partial n \leqslant 0;$$

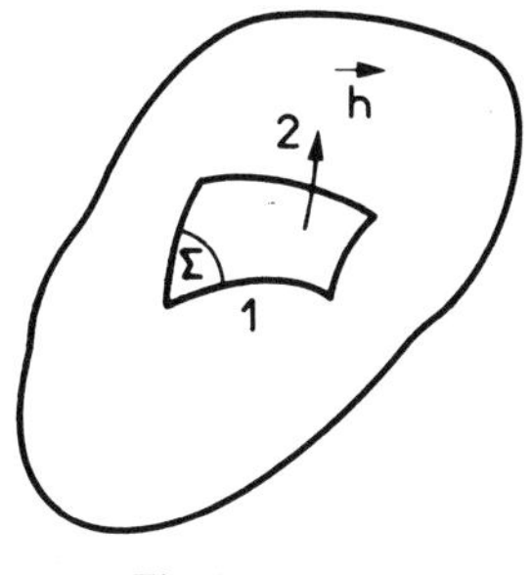

Fig. 1

ii) if the partition has finite thickness:

(2.12b)
$$u_1<u_2 \Rightarrow \partial u/\partial n=0$$
$$u_1 \geqslant u_2 \Rightarrow -\partial u/\partial n=k(u_1-u_2). \quad \square$$

2.2.4. Volume Injection Through a Semi-Permeable Wall

At its boundary, the reservoir Ω is subjected to classical conditions, for example: given pressure. Moreover, it is augmented by a prescribed volume influx of fluid $g(x,t)$ and by an additional volume influx of fluid $\tilde{g}(x,t)$ (through a semi-permeable partition or, more generally, regulated by a servo-mechanism) that obeys one of the following laws, where $h(x)$ is a field of prescribed pressures in Ω,

(2.13)
$$u>h \Rightarrow \tilde{g}=0$$
$$u=h \Rightarrow \tilde{g} \geqslant 0$$

or possibly

$$(2.14)\qquad \begin{aligned} u>h &\Rightarrow \tilde{g}=0 \\ u\leqslant h &\Rightarrow \tilde{g}=k(h-u). \end{aligned}$$

Naturally, in Ω the pressure u satisfies the equation

$$(2.15)\qquad \partial u/\partial t-\Delta u=g+\tilde{g}. \quad \square$$

Remark 2.5. The problems mendioned in 2.2 are essentially non-stationary. Nevertheless, if the initial condition is given up, they may, in certain cases, possess stationary solutions. The study of the latter is interesting from the point of view of physics, because they correspond to global thermal equilibrium positions.

On the basis of a physical argument, we observe that these stationary solutions do not always exist and, if they exist, that they are not necessarily unique. Indeed, let us consider the stationary problem: To find $u(x)$ such that

$$-\Delta u=g \quad \text{in} \quad \Omega,$$

$$\left.\begin{aligned} u>h &\Rightarrow \partial u/\partial n=0 \\ u=h &\Rightarrow \partial u/\partial n\geqslant 0 \end{aligned}\right| \quad \text{on } \Gamma.$$

Here, the semi-permeable partition only permits heat to enter; consequently:

i) if $\int_\Omega g\,dx>0$, that is to say, if globally there is a volume gain of heat, then the temperature must strictly increase in certain points, therefore, there is no solution;

ii) if $\int_\Omega g\,dx=0$: a solution u may possibly exist; this solution will also satisfy $\partial u/\partial n=0$ almost everywhere on Γ, therefore $u+C$ will also be a solution $\forall C>0$;

iii) if $\int_\Omega g\,dx<0$, a solution u can exist. It would correspond to a thermal balance zero, i.e.,

$$\int_\Omega g\,dx+\int_\Gamma(\partial u/\partial n)\,d\Gamma=0.$$

2.3. Temperature Control

The terminology used here is taken from thermal phenomena. The continuous medium under consideration occupies an open region Ω of $\mathbb{R}^n$ ($n\leqslant 3$) with a boundary Γ.

We distinguish two types of temperature control:

—Temperature control through the boundary, regulated by the temperature at the boundary (2.3.1);

—Temperature control through the interior, regulated by the temperature in the interior (2.3.3).

2.3.1. Temperature Control Through the Boundary, Regulated by the Temperature at the Boundary

We give two reference temperatures $h_1(x)$ and $h_2(x)$ with $x \in \Gamma$, $h_1(x) \leqslant h_2(x)$, and we require that the temperature $u(x)$ at the boundary deviates as little as possible from the interval $(h_1(x), h_2(x))$. For that purpose, we place "thermostatic controls", i.e., devices which are capable of injecting an appropriate heat flux (in the algebraic sense) through the boundary. Since the effectiveness of these devices is limited, the flux of injected heat, which can be measured by $-\partial u/\partial n$, is confined to the closed interval $[g_1, g_2]$ with $0 \in [g_1, g_2]$.

We regulate these controls in such a way that:

i) if $u(x,t) \in [h_1(x), h_2(x)]$, that is to say if the temperature is in the desired bracket, no correction needs to be made and, consequently, we postulate

$$\partial u/\partial n = 0 \; ; \tag{2.16}$$

ii) if $u(x,t) \notin [h_2(x), h_2(x)]$, we inject a quantity of heat proportional to the distance between $u(x,t)$ and the interval (h_1, h_2), if that is possible, in the form

$$\begin{aligned} u(x,t) > h_2(x) \Rightarrow & -\partial u/\partial n = +k_2(u-h_2) \\ & \quad \text{if} \quad +k_2(u-h_2) \leqslant g_2 \\ \text{and} \quad & -\partial u/\partial n = g_2 \quad \text{if} \quad +k_2(u-h_2) > g_2 \\ u(x,t) < h_1(x) \Rightarrow & -\partial u/\partial n = +k_1(u-h_1) \\ & \quad \text{if} \quad +k_1(u-h_1) \geqslant g_1 \\ \text{and} \quad & -\partial u/\partial n = g_1 \quad \text{if} \quad k_1(u-h_1) < g_1 , \end{aligned} \tag{2.17}$$

(k_1 and k_2 are positive scalars).

Conditions (2.16), (2.17) can be expressed in a simpler form if we introduce the function Φ as follows:

$$\begin{aligned} \Phi(\lambda) &= g_1 && \text{if} \quad \lambda \leqslant h_1 + g_1/k_1 \\ &= k_1(\lambda - h_1) && \text{if} \quad h_1 + g_1/h_1 < \lambda \leqslant h_1 \\ &= 0 && \text{if} \quad h_1 \leqslant \lambda \leqslant h_2 \\ &= k_2(\lambda - h_2) && \text{if} \quad h_2 < \lambda \leqslant h_2 + g_2/k_2 \\ &= g_2 && \text{if} \quad \lambda \geqslant h_2 + g_2/k_2 \end{aligned} \tag{2.18}$$

represented graphically by Fig. 2.

Conditions (2.16), (2.17) are then equivalent to

$$-\partial u/\partial n = \Phi(u) . \tag{2.19}$$

Summarizing, we have: *The solution u of the problem of temperature control of type 1 must satisfy* (2.10) *in* Ω, (2.19) *on* Γ *and the initial condition*

(2.20) $$u(x,0)=u_0(x).$$

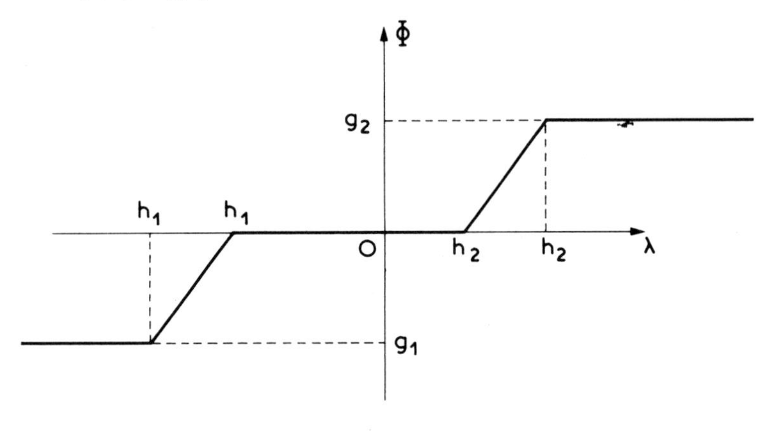

Fig. 2

This problem has several interesting *special cases* and *limiting cases*:

α) $h_1=h_2$. The temperature control is effected with a single reference temperature.

β) $h_1=h_2$, $k_1=k_2$, $g_2=+\infty$, $g_1=0$. We have returned to the conditions of a problem with a semi-permeable wall of finite thickness.

The graph of the corresponding function Φ is indicated in Fig. 3.

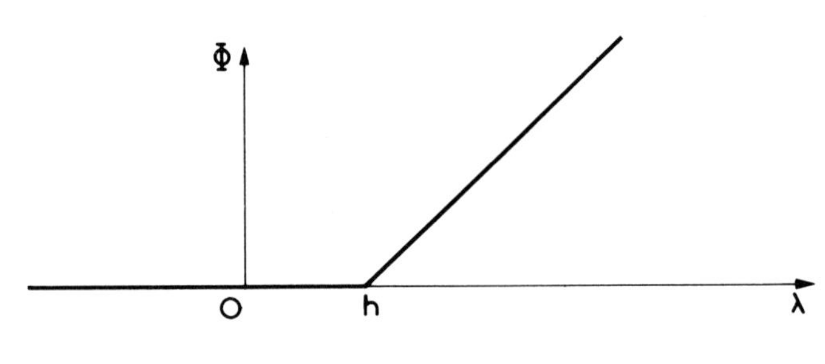

Fig. 3

γ) $h_1=h_2$, $k_1=k_2=+\infty$, $g_1=0$, $g_2=+\infty$. This brings us back to the conditions for the problem of a semi-permeable thin wall. The graph of the corresponding function Φ is indicated in Fig. 4. We observe that in this case $\Phi(u)$ is a **multiple-valued function.**

δ) $h_1<h_2$, $k_1=k_2=+\infty$, $g_2=-g_1=+\infty$. Heat control is achieved perfectly, because in this case u stays in the interval $[h_1,h_2]$.

The graph of the function Φ is indicated in Fig. 5.

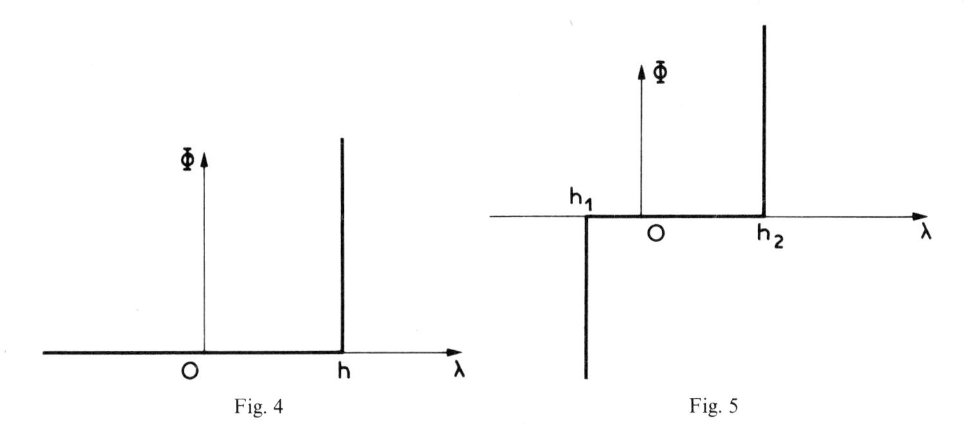

Fig. 4

Fig. 5

Here, the function Φ is again multiple-valued.

ε) $h_1 = h_2$, $k_1 = k_2 = +\infty$, $g_2 = -g_1 = +\infty$. The temperature u is prescribed on the boundary. This is the Dirichlet problem. □

Remark 2.6. *Generalization.* The preceding problem can be generalized by replacing (2.18) by (2.18a)

$$-\partial u/\partial n = \Phi(u)\,, \tag{2.18a}$$

where the function Φ is a continuous and increasing function of u or, still more generally, where Φ is a multiple-valued, increasing function with maximal graph (that is to say that the graph is a continuous curve in $\mathbb{R}^2$). □

2.3.2. Temperature Control Through the Interior, Regulated by the Temperature in the Interior

We prescribe two temperatures $h_1(x)$ and $h_2(x)$ for $x \in \Omega$, $h_1(x) \leqslant h_2(x)$, and we ask that the temperature $u(x)$, $x \in \Omega$, deviates as little as possible from the interval (h_1, h_2). For this purpose, we set up volume sources of heat (in the algebraic sense). These devices have limited power which, consequently, limits the heat flux $-\tilde{g}$; we assume therefore that the latter remains in the closed intervals $[g_1, g_2]$ with $0 \in [g_1, g_2]$.

The heat injection is regulated in the following manner:

i) if, at the time t, $u(x,t) \in [h_1(x), h_2(x)]$, that is to say that the temperature is contained in the chosen interval, there is no need to carry out a correction, so that

$$\tilde{g} = 0\,; \tag{2.21}$$

ii) if $u(x,t) \notin [h_1(x), h_2(x)]$, we inject a quantity of heat proportional to the distance between $u(x)$ and the interval $[h_1(x), h_2(x)]$, if that is possible. Hence

$$(2.22)\quad \begin{aligned} u(x,t) > h_2(x) \Rightarrow & \left| \begin{array}{l} -\tilde{g} = k_2(u-h_2) \\ \qquad \text{if} \quad k_2(u-h_2) \leqslant g_2, \\ -\tilde{g} = g_2 \quad \text{if} \quad k_2(u-h_2) > g_2, \end{array} \right. \\ u(x,t) < h_1(x) \Rightarrow & \left| \begin{array}{l} -\tilde{g} = k_1(u-h_1) \\ \qquad \text{if} \quad k_1(u-h_1) \geqslant g_1, \\ -\tilde{g} = g_1 \quad \text{if} \quad k_1(u-h_1) < g_1. \end{array} \right. \end{aligned}$$

We can combine (2.21) and (2.22) in

$$(2.23)\qquad -\tilde{g} = \Phi(u),$$

where $\Phi(u)$ is the function previously defined in (2.18).

The function u (temperature) thus satisfies

$$(2.24)\qquad \partial u/\partial t - \Delta u = g - \Phi(u) \quad \text{in} \quad \Omega \times]0, T[,$$

$$(2.25)\qquad u(x,0) = u_0(x) \qquad \text{in} \quad \Omega$$

and a classical boundary condition on the boundary Γ, as for example

$$(2.26)\qquad u(x,t) = \theta(x,t) \quad \text{for} \quad x \in \Gamma,$$

where θ is a given temperature on Γ. □

Remark 2.7. *Generalization.* As in the preceding paragraph, one can consider special cases and limit cases where the function $\Phi(u)$ is multiple-valued with an increasing and maximal graph. □

Remark 2.8. Other types of temperature control can be envisioned that involve mean temperatures and "*cost of production*" functions. They lead to problems of *optimal control.*

Example. Given the temperature interval $[\theta_1(t), \theta_2(t)]$ with $\forall(t)$, $\theta_1(t) < \theta_2(t)$; we want to maintain the average temperature $\bar{u}(t)$ in the open domain Ω in the interval $[\theta_1, \theta_2]$ at every instant. To achieve this, we set up adjustable heat fluxes through $\partial\Omega = \Gamma$.

We investigate how these (algebraic) fluxes should be prescribed in order that the desired result is obtained at the least expense. We can assume that the cost is proportional to the integral of an increasing function of the imposed flux.

The equations and conditions are:

$$(2.27)\qquad \partial u/\partial t - \Delta u = g,$$

(g is a given function of x and t).

We put

(2.28) $\quad \bar{u}(t) = (\text{measure } \Omega)^{-1} \int_\Omega u(x,t)\,dx$,

and we require

(2.29) $\quad \bar{u}(t) \in [\theta_1(t), \theta_2(t)]$.

We are then looking for a $\varphi(x,t)$ on Γ which permits (2.29) to hold, if we know that

(2.30) $\quad -\partial u/\partial n = \varphi$ a.e. on Γ, $\quad \forall t$,

and which, at every moment, minimizes a given functional $I(t,\varphi(t))$ of the form

(2.31) $\quad I(t,\varphi(t)) = \int_\Gamma F(t,\varphi(x,t))\,d\Gamma$,

where $\lambda \to F(t,\lambda)$, $\lambda \in \mathbb{R}$, is a given function. Possible choices of functions F are:

i) $\quad F(t,\lambda) = |\lambda|$,
ii) $\quad F(t,\lambda) = \lambda^2$,
iii) $\quad F(t,\lambda) = \lambda^+$,
iv) $\quad F(t,\lambda) = \lambda^-$, etc. □

3. Variational Formulations of Problems of Temperature Control and of Semi-Permeable Walls

We now reformulate the problems encountered in Section 2 in the form of "*variational inequalities*". The tools from Functional Analysis which we need to precisely formulate the problems that we will meet will be given in Section 4 below, while the solutions of the problems will be found in the 5th and following sections.

3.1. Notation

If u and v are two real-valued functions, defined in Ω, we put, whenever this has a meaning[4],

(3.1) $\quad a(u,v) = \int_\Omega u_{,i} v_{,i}\,dx$

(3.2) $\quad (u,v) = \int_\Omega uv\,dx$,

and for the functions φ, ψ, defined on Γ,

(3.3) $\quad (\varphi,\psi) = \int_\Gamma \varphi\psi\,d\Gamma$.

[4] This will be precisely formulated in Section 4.

We will deal with functions of $x \in \Omega$ and of $t \in]0, T[$. We introduce:

$$Q = \Omega \times]0, T[, \quad \Sigma = \Gamma \times]0, T[. \tag{3.4}$$

If u is a (real-valued) function defined in Q, we put:

$$\begin{aligned} &u(t) = u(0, t) = \text{function } x \to u(x, t), \\ &u'(t) = \partial u(t)/\partial t = \text{function } x \to \partial u(x, t)/\partial t, \quad \text{etc.} \quad \square \end{aligned} \tag{3.5}$$

The functions ψ

We will use functions $\lambda \to \psi(\lambda)$, $\lambda \in \mathbb{R}$, with the following properties:

(3.6) the function $\lambda \to \psi(\lambda)$ is convex, lower semi-continuous, with values in $]-\infty, +\infty]$[5], ψ not being identically $+\infty$.

We will distinguish three types of functions with the properties (3.6):

(3.7) *Function* ψ *of type* 1:
$\lambda \to \psi(\lambda)$ is once continuously differentiable.

Example 3.1 (Fig. 6)

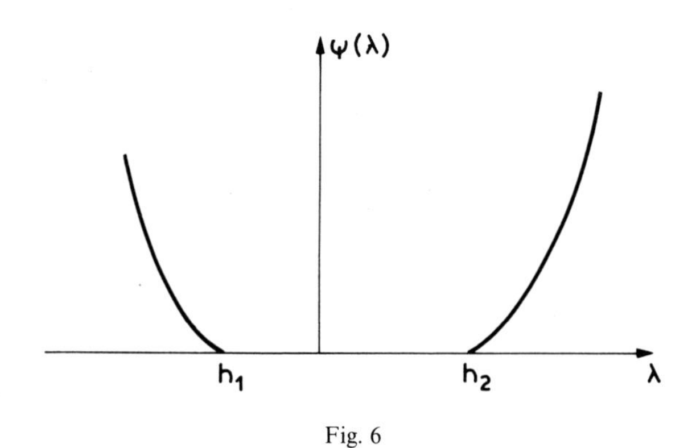

Fig. 6

$$\psi(\lambda) = \begin{vmatrix} a_1(\lambda - h_1)^2, & \lambda < h_1 \\ 0, & h_1 \leqslant \lambda \leqslant h_2 \\ a_2(\lambda - h_2)^2, & \lambda \geqslant h_2 \end{vmatrix}$$

$$0 < a_1 < a_2.$$

[5] Therefore, ψ can take the value "$+\infty$".

(3.8) | *Function* ψ *of type* 2:
$\lambda \to \psi(\lambda)$ is not continuously differentiable, but of finite value everywhere.

Example 3.2 (Fig. 7)

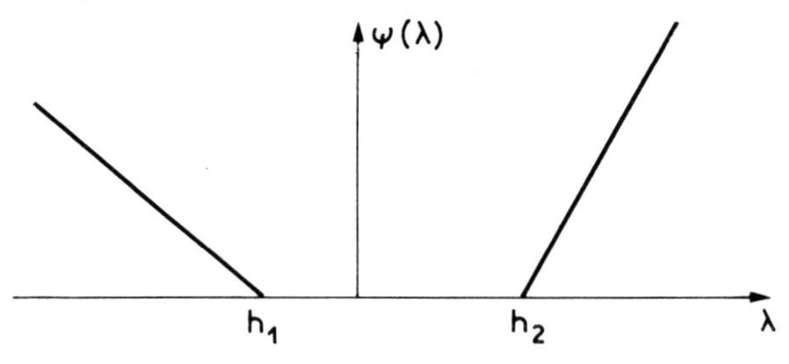

Fig. 7

$$\psi(\lambda) = \begin{cases} g_1(\lambda - h_1), & \lambda \leqslant h_1, \\ 0, & h_1 \leqslant \lambda \leqslant h_2, \\ g_2(\lambda - h_2), & \lambda \geqslant h_2, \end{cases}$$
$$g_1 < 0 < g_2. \quad \square$$

(3.9) *Function* ψ *of type* 3:
$\lambda \to \psi(\lambda)$ takes the value $+\infty$.

Example 3.3 (Fig. 8)

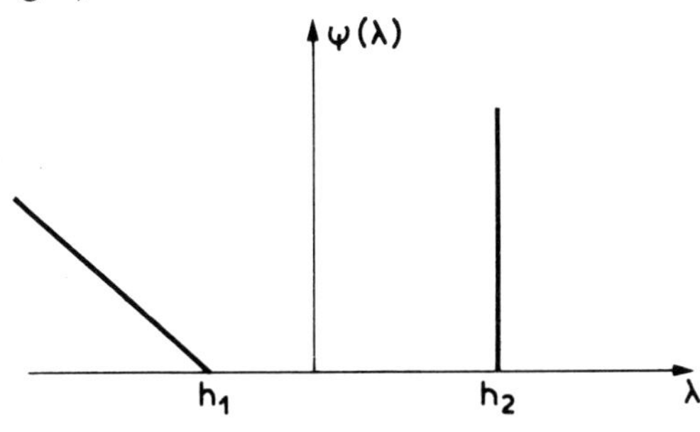

Fig. 8

$$\psi(\lambda) = \begin{cases} g_1(\lambda - h_1), & \lambda \leqslant h_1, \\ 0, & h_1 \leqslant \lambda \leqslant h_2, \\ +\infty, & \lambda > h_2, \end{cases}$$
$$g_1 < 0 < g_2. \quad \square$$

With the function ψ, we associate the *functional* Ψ. We set:

(3.10) $$\Psi(v) = \int_\Gamma \psi(v(x))\, d\Gamma,$$

where

(3.11) $\Psi(v)=\int_\Omega \psi(v(x))\,dx$,

whenever this has a meaning[6].

Remark 3.1. We observe at once that in the case when ψ is of type 3, Ψ is not defined "everywhere"; for example, in the case of Example 3.3, the functional (3.10) is only defined when $v(x)\leqslant h_2$ almost everywhere. □

Remark 3.2. One can also consider a family "$\lambda\to\psi(x;\lambda)$" of functions ψ depending, as the case may be, on $x\in\Gamma$ or $x\in\Omega$; one then introduces

(3.12) $\Psi(v)=\int_\Gamma \psi(x;v(x))\,d\Gamma$,

where

(3.13) $\Psi(v)=\int_\Omega \psi(x;v(x))\,dx$. □

3.2. Variational Inequalities

We consider, independently, problems of *variational inequalities*, both, problems of evolution and stationary problems. Subsequently, we will show in which way these problems are "equivalent" to the problems encountered in Section 2.

Variational inequalities of evolution[7]

We are seeking a function $t\to u(t)=u(.,t)$ such that

(3.14)
$$(u'(t),v-u(t))+a(u(t),v-u(t))+\Psi(v)-\Psi(u(t))\geqslant(f(t),v-u(t))\quad\forall v,$$
Ψ *given by* (3.10),

where the function $t\to f(t)=f(.,t)$ is prescribed, with

(3.15) $(u(0)=u_0$ given, (i.e., $u(x,0)=u_0(x)$, $x\in\Omega)$.

Remark 3.3. *If* Ψ *is given by* (3.11), we add boundary conditions for $u(t)$ and v, *for example*

(3.16) $u(t)=0$ on Γ, $v=0$ on Γ. □

Stationary variational inequalities (of "elliptic" nature)

It is sufficient to omit the dependence on t in (3.14). We seek a function u, defined in Ω, such that

(3.17)
$$a(u,v-u)+\Psi(v)-\Psi(u)\geqslant(f,v-u)\quad\forall v$$
Ψ *given by* (3.10),

[6] We will state this more precisely in Sections 4 and 5.

[7] We deal here with "*parabolic*" inequalities; we will encounter analogous problems of a "*hyperbolic*" nature in the following chapters.

with additional boundary conditions on u and v (as in Remark 3.3) in case Ψ is given by (3.11). □

3.3. Examples. Equivalence with the Problems of Section 2

3.3.1. Functions ψ of Type 1 (and Ψ given by (3.10))

In this case, we introduce

$$\Phi(\lambda)=\mathrm{d}\psi(\lambda)/\mathrm{d}\lambda\,. \tag{3.18}$$

Because of the *convexity* of ψ, we have:

$$\psi(\mu)-\psi(\lambda)-\Phi(\lambda)(\mu-\lambda)\geqslant 0 \qquad \forall \mu\,. \tag{3.19}$$

We will verify that, *with these conditions*, (3.14) *is equivalent to the* (variational) *equation*

$$(u'(t),v)+a(u(t),v)+\int_\Gamma \Phi(u(t))\,v\,\mathrm{d}\Gamma=(f(t),v) \qquad \forall v\,. \tag{3.20}$$

Indeed, if in (3.14) we first take

$$v=u(t)+\lambda w\,, \quad w \text{ "arbitrary"}, \quad \lambda>0\,,$$

we obtain, after division by λ:

$$(u'(t),w)+a(u(t),w)+\lambda^{-1}\left[\Psi(u(t)+\lambda w)-\Psi(u(t))\right]\geqslant(f(t),w)$$

and, letting λ tend toward 0:

$$(u'(t),w)+a(u(t),w)+\int_\Gamma \Phi(u(t))\,w\,\mathrm{d}\Gamma\geqslant(f(t),w)\,.$$

Changing w to $-w$, we obtain equality and therefore (3.20).

Conversely, if we have (3.20), then

$$\begin{aligned}(u'(t),v-u(t))&+a(u(t),v-u(t))+\Psi(v)-\Psi(u(t))\\ &=(f(t),v-u(t))+\Psi(v)-\Psi(u(t))-\int_\Gamma \Phi(u(t))(v-u(t))\,\mathrm{d}\Gamma\,.\end{aligned} \tag{3.21}$$

But, according to (3.19), we have:

$$\Psi(v)-\Psi(u(t))-\int_\Gamma \Phi(u(t))(v-u(t))\,\mathrm{d}\Gamma\geqslant 0\,, \tag{3.22}$$

so that (3.21) implies (3.14). □

Remark 3.4. If Ψ is given by (3.11), then (3.14) is equivalent to

$$(u'(t),v)+a(u(t),v)+\int_\Omega \Phi(u(t))\,v\,\mathrm{d}x=(f(t),v) \tag{3.23}$$

with an additional boundary condition, for example (3.16). □

It remains to "*interpret*" (3.20), which is an exercise in integration by parts. We take in (3.20)

$$v = \varphi = \text{regular function with compact support in } \Omega. \tag{3.24}$$

Then (3.20) reduces to

$$(u'(t), \varphi) + a(u(t), \varphi) = (f(t), \varphi)$$

where

$$\int_\Omega (\partial u/\partial t - \Delta u)\varphi\, dx = \int_\Omega f(x,t)\varphi(x)\, dx$$

and therefore—"*in the sense of distributions on* Ω"[8]—

$$\partial u/\partial t - \Delta u = f \quad \text{in} \quad Q = \Omega \times]0, T[\,. \tag{3.25}$$

Now, let v be an "arbitrary" function; we multiply both sides of (3.25) with $v = v(x)$ and integrate by parts over Ω; let $\partial/\partial n$ be the normal derivative on Γ *directed toward the exterior of* Ω. Then:

$$(u'(t), v) - \int_\Gamma \frac{\partial u(t)}{\partial n} v\, d\Gamma + a(u(t), v) = (f(t), v) \tag{3.26}$$

which, comparing with (3.20), implies

$$\int_\Gamma (\partial u(t)/\partial n + \Phi(u(t)))v\, d\Gamma = 0 \qquad \forall v\,, \tag{3.27}$$

i.e.,

$$\partial u(t)/\partial n + \Phi(u(t)) = 0 \quad \text{on } \Gamma\,. \tag{3.28}$$

Summarizing:

(3.29) *for a function* ψ *of type* 1 (and Ψ given by (3.10)), *the problem* (3.14), (3.15) *is equivalent*[9] *to the determination of a solution* $u = u(x,t)$ *of* (3.25), *with the boundary condition* (3.28) ($\Phi = d\psi/d\lambda$) *and the initial condition* (3.15). ☐

Remark 3.5. In the case (3.11), (3.16), the problem is equivalent to the determination of a solution u of

$$\partial u/\partial t - \Delta u + \Phi(u) = f \quad \text{in } Q, \tag{3.30}$$

$$u = 0 \quad \text{on } \Sigma \tag{3.31}$$

and the initial condition (3.15). ☐

---

[8] L. Schwartz [1,2].

[9] The fact that (3.28), (3.25) imply (3.14), (3.15) is immediately seen.

Remark 3.6. Except for the case where

$$\Phi(\lambda)=k\lambda\,, \qquad k>0\,,$$

the problem is *non-linear*. ▯

Example 3.4. Let Φ be *given by* (2.18); let us define ψ by

(3.32) $$\psi(\lambda)=\int_0^\lambda \Phi(\mu)\,\mathrm{d}\mu \quad \text{(see graph in Fig. 9)};$$

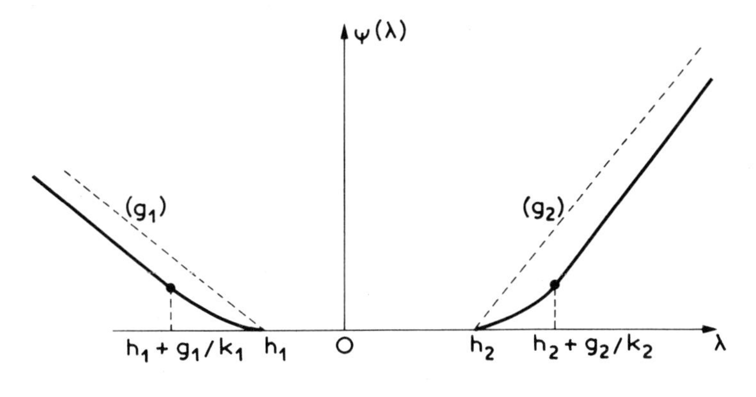

Fig. 9

i.e.,

(3.33) $$\psi(\lambda)=\begin{cases} g_1(\lambda-h_1)-g_1^2/2k_1\,, & \lambda\leqslant h_1+g_1/k_1\,,\\ \frac{1}{2}k_1(\lambda-h_1)^2\,, & h_1+g_1/k_1\leqslant\lambda\leqslant h_1\,,\\ 0\,, & h_1\leqslant\lambda\leqslant h_2\,,\\ \frac{1}{2}k_2(\lambda-h_2)^2\,, & h_2\leqslant\lambda\leqslant h_2+g_2/k_2\,. \end{cases}$$

The function ψ *thus defined satisfies* (3.6), *and is of type 1*. Therefore, we can apply (3.29) and see that *problem* (3.14), (3.15) *is equivalent to the problem of temperature control through the boundary*, Section 2.3.1.

Remark 3.7. In the case (3.11), (3.16), the problem is equivalent to the problem of temperature control through the interior, Section 2.3.2. ▯

Example 3.5. Let ψ be given by (Fig. 10)

(3.34) $$\psi(\lambda)=\begin{cases} \frac{1}{2}k(\lambda-h)^2\,, & \lambda\leqslant h\,,\\ 0\,, & \lambda\geqslant h\,. \end{cases}$$

The function ψ satisfies (3.6) and is of type 1; obviously, we have:

(3.35) $$\Phi(\lambda)=\begin{cases} k(\lambda-h)\,, & \lambda\leqslant h\,,\\ 0\,, & \lambda\geqslant h\,. \end{cases}$$

If we take Ψ as in (3.10), we obtain (3.28), i.e.,

$$(3.36) \qquad \begin{aligned} &\partial u(t)/\partial n = 0 \quad \text{if} \quad u(t) > h\,, \\ &\partial u(t)/\partial n + k(u(t) - h) = 0 \quad \text{if} \quad u(t) \leqslant h\,; \end{aligned}$$

this is the problem of the semi-permeable thick wall; see Section 2.2.2. ☐

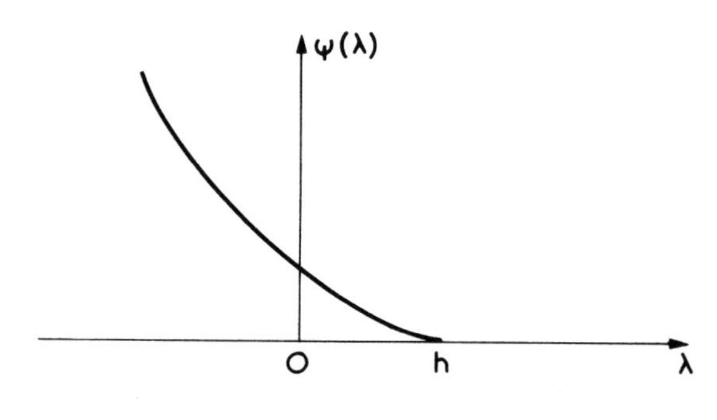

Fig. 10

3.3.2. Functions ψ of Type 2

We again denote by $\Phi(\lambda)$ the "derivative" of $\psi(\lambda)$, but this "derivative" may be multiple-valued. Generally, this multiple-valued function is called "*sub-differential*"; in the points λ where ψ *is not* differentiable, we denote by $\Phi(\lambda)$ the *contingent* of ψ at the point λ, i.e.,

$$(3.37) \qquad \Phi(\lambda) = \textbf{set} \text{ of limits } \frac{\psi(\mu) - \psi(\lambda)}{\mu - \lambda}, \qquad \mu \to \lambda\,.$$

We have (analogous to (3.19)):

$$(3.38) \qquad \chi \in \Phi(\lambda) \Leftrightarrow \psi(\mu) - \psi(\lambda) - \chi(\mu - \lambda) \geqslant 0\,, \qquad \forall \mu\,.$$

The reasoning that led to (3.28) is still valid, provided that we replace the *number* $\Phi(u(x,t))$ by the *set* $\Phi(u(x,t))$, (if $u(x,t) = a$ point where ψ is not differentiable); then we have to formulate

$$(3.39) \qquad -\partial u(t)/\partial n \in \Phi(u(t))$$

(naturally, (3.39) is equivalent to (3.28) in the points where ψ is differentiable). Therefore:

(3.40) *for a function* ψ *of type* 2 (and Ψ defined by (3.10)), the problem (3.14), (3.15) *is equivalent to the determination of a solution* $u = u(x,t)$ *of* (3.25) *with the "boundary condition"* (3.39) (Φ = *sub-differential or contingent of* ψ) *and the initial condition* (3.15). ☐

Remark 3.8. In case (3.11), (3.16), the problem is equivalent to the determination of a solution u of

$$-(\partial u(x,t)/\partial t) - \Delta u(x,t) - f(x,t)) \in \Phi(u(x,t)) \tag{3.41}$$

with (3.31) and (3.15). □

Example 3.6. Let us take ψ as given by Example 3.2 (Fig. 7). Then

$$\Phi(\lambda) = \left| \begin{array}{ll} g_1 & \text{if } \lambda < h_1\,, \\ [g_1, 0] & \text{if } \lambda = h_1\,, \\ 0 & \text{if } h_1 < \lambda < h_2\,, \\ [0, g_2] & \text{if } \lambda = h_2\,, \\ g_2 & \text{if } \lambda > h_2\,. \end{array} \right. \tag{3.42}$$

The *boundary condition* (3.39) on Σ can then be written:

$$\begin{array}{ll} -\partial u(t)/\partial n = g_1 & \text{if } u < h_1\,, \\ g_1 \leqslant -\partial u(t)/\partial n \leqslant 0 & \text{if } u = h_1\,, \\ \partial u(t)/\partial n = 0 & \text{if } h_1 < u < h_2\,, \\ 0 \leqslant -\partial u(t)/\partial n \leqslant g_2 & \text{if } u = h_2\,, \\ -\partial u(t)/\partial n = g_2 & \text{if } u > h_2\,. \quad \square \end{array} \tag{3.43}$$

3.3.3. Functions ψ of Type 3

We already observed (Remark 3.1) that when ψ is of type 3, the function Ψ is not defined everywhere—or rather, takes the value $+\infty$.

Accordingly, we formulate the following definition[10]:

(3.44) *K denotes the set of v for which* $\Psi(v) \neq +\infty$.

We will see later that K is a *convex closed set* in a suitable function space. *Then* (3.14) *is equivalent to*

$$(u'(t), v-u(t)) + a(u(t), v-u(t)) + \Psi(v) - \Psi(u(t)) \geqslant (f(t), v-u(t)) \qquad \forall v \in K\,, \tag{3.45}$$

$$u(t) \in K\,. \tag{3.46}$$

Let w be arbitrary in K; in (3.45), take

$$v = u(t) + \theta(w - u(t))\,, \qquad 0 < \theta < 1\,;$$

[10] Still somewhat formal. We will make this precise after choosing the classes of functionals to which v is restricted.

it follows that:

$$(u'(t), w-u(t))+a(u(t), w-u(t))+\theta^{-1}(\Psi(u(t)+\theta(w-u(t)))-\Psi(u(t))) \geqslant (f(t), w-u(t))$$

which implies, letting $\theta \to 0$ (and replacing w by v),

$$(u'(t), v-u(t))+a(u(t), v-u(t))+(\chi, v-u(t)) \geqslant (f(t), v-u(t)) \quad \forall v \in K \tag{3.47}$$

where

$$\chi \in \Phi(u(t)), \tag{3.48}$$

i.e.,

$$\chi = \Phi(u(t)) \quad \text{if } \Psi \text{ is differentiable at the "point" } u(t),$$

or $\chi \in$ contingent of Ψ at the point $u(t)$ if Ψ is not differentiable there; we can also say that χ is the set of elements such that

$$\Psi(v)-\Psi(u(t))-(\chi, v-u(t)) \geqslant 0 \qquad \forall v \in K. \tag{3.49}$$

Conversely, if u satisfies (3.46), (3.47), (3.48), then we have (3.45) (immediate consequence of (3.49)).

Hence:

(3.50) *For a function* ψ *of type* 3, *the problem is equivalent to the determination of a* $u(t) \in K$ (K defined in (3.44)) *satisfying* (3.47), (3.48), *with the initial condition* (3.15). □

When Ψ is given by (3.10), we can go further. First, we observe that if φ is given as in (3.24), then

$$v = u(t) \pm \varphi \in K \tag{3.51}$$

so that the choice (3.51) is permissible in (3.45) and yields

$$(u'(t), \varphi)+a(u(t), \varphi)=(f(t), \varphi)$$

which again leads to Equation (3.25).

Now, making use of (3.26), (with $v-u(t)$ instead of v) and comparing with (3.45), we conclude

$$\begin{aligned} &\int_\Gamma (\partial u(t)/\partial n+\chi)(v-u(t))\,d\Gamma \geqslant 0 \qquad \forall v \in K, \\ &\chi \in \Phi(u(t)). \end{aligned} \tag{3.52}$$

Summarizing:

(3.53) *For a function ψ of type 3 and Ψ defined by* (3.10), *the problem is equivalent to the determination of a solution u of* (3.25), *satisfying*

$$u(t)\in K \quad \forall t,$$

the conditions (3.52) *and the initial condition* (3.15). □

Example 3.7. Let us first take ψ as given by (Fig. 11):

$$\psi(\lambda)=\begin{cases} 0 & \text{on } [h_1,h_2], \\ +\infty & \text{elsewhere.}\end{cases} \tag{3.54}$$

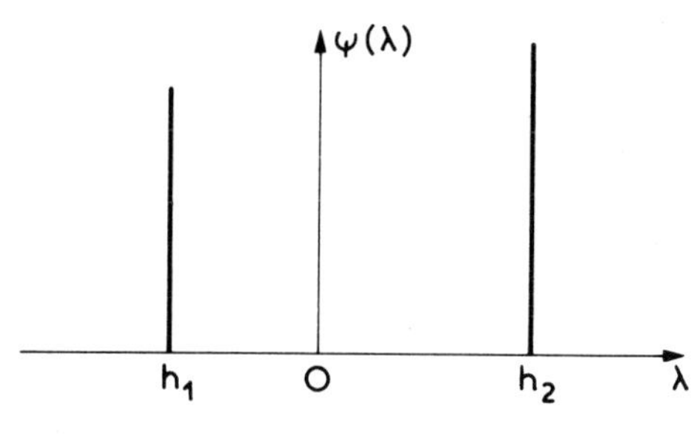

Fig. 11

Then

$$K=\{v \mid h_1\leqslant v\leqslant h_2 \text{ on } \Gamma\} \tag{3.55}$$

and therefore, the "boundary conditions" on Σ are:

$$\begin{aligned} & h_1\leqslant u(x,t)\leqslant h_2, \\ & \int_\Gamma \frac{\partial u(t)}{\partial n}(v-u(t))\geqslant 0 \quad \forall v\in K, \end{aligned} \tag{3.56}$$

(since $\Phi(u(t))=0$ (or $\{0\}$)), or again[11]

$$\begin{aligned} u=h_1 \quad & \Rightarrow \partial u(t)/\partial n\geqslant 0, \\ h_1<u<h_2 & \Rightarrow \partial u(t)/\partial n=0, \\ u=h_2 \quad & \Rightarrow \partial u(t)/\partial n\leqslant 0. \end{aligned} \tag{3.57}$$

[11] We observe that the inequality in (3.56) is equivalent to the pointwise inequality

$$\frac{\partial u(t)}{\partial n}(v-u(t))\geqslant 0 \text{ on } \Gamma.$$

Special case
We take[12]

(3.58) $$h_1 = h, \quad h_2 = +\infty \quad \text{(which is permitted);}$$

this is the *problem of semi-permeable thin walls* (see Sec. 2.2.1). □

Example 3.8. Let us take ψ as in Example 3.3. Then we obtain:

(3.59) $$K = \{v \mid v \leqslant h_2 \text{ on } \Gamma\}$$

and

(3.60) $$\begin{aligned} &(\partial u(t)/\partial n + \chi)(v - u(t)) \geqslant 0 \qquad \forall v \in K, \\ &\chi \in \Phi(u(t)), \end{aligned}$$

which implies

(3.61) $$\begin{aligned} &u(t) = h_2 &&\Rightarrow \partial u(t)/\partial n \leqslant 0, \\ &h_1 < u(t) < h_2 &&\Rightarrow \partial u(t)/\partial n = 0, \\ &u(t) = h_1 &&\Rightarrow \partial u(t)/\partial n + \chi = 0, \quad \chi \in [g_1, 0], \\ &u(t) < h_1 &&\Rightarrow \partial u(t)/\partial n + g_1 = 0. \end{aligned}$$ □

3.4. Some Extensions

If we take $a(u,v)$ as given by (instead of (3.1))

(3.62) $$a(u,v) = \int_\Omega a_{ij} u_{,j}(x) v_{,i}(x) \, dx,$$

where

(3.63) $$\begin{aligned} &a_{ij} \in L^\infty(\Omega) \quad \text{(measurable and bounded functions in } \Omega), \\ &a_{ij}(x)\xi_i\xi_j \geqslant \alpha \xi_i \xi_i, \quad \alpha > 0, \quad x \in \Omega \quad \forall \xi \in \mathbb{R}^n \end{aligned}$$

then we can consider—and solve, by methods described below,—problems analogous to the preceding ones. *The interpretation is analogous, with the following modifications*: the operator $-\Delta$ is replaced by the operator A given by

(3.64) $$Av = -(a_{ij}(x) v_{,j})_{,i}$$

and the normal derivative $\partial/\partial n$ (directed toward the exterior of Ω) is replaced by $\partial/\partial v_A$, given by

---

[12] In fact, $h_1 = h(x)$; see Remark 3.2.

(3.65)
$$\frac{\partial v}{\partial v_A} = a_{ij}(x)\frac{\partial v}{\partial x_j}\cos(n, x_i),$$

$\cos(n, x_i) = i$-th direction cosine of the normal n to Γ, directed toward the exterior of Ω. ▯

Remark 3.9. One can—but we are not going to do this here—generalize the case (3.62) further by taking:

i) *time-dependent* forms $a(t; u, v)$ (i.e., the functions a_{ij} in (3.62) are dependent on x and on t);

ii) certain *non-linear* forms in u. ▯

3.5. Stationary Cases

By making use of the notions introduced in Section 3.3 above, we can interpret (3.17) immediately and obtain the following results.

3.5.1. The Function Ψ is of Type 1

If Ψ is given by (3.10), then (3.17) *is equivalent to the determination of a solution u of*

(3.66) $$-\Delta u = f \quad \text{in } \Omega$$

with the boundary condition

(3.67) $$\partial u/\partial n + \Phi(u) = 0 \quad \text{on } \Gamma.$$

If Ψ is given by (3.11) with (3.16), then

(3.68)
$$-\Delta u + \Phi(u) = f \quad \text{in } \Omega,$$
$$u = 0 \quad \text{on } \Gamma. \quad ▯$$

3.5.2. The Function Ψ is of Type 2

If Ψ is given by (3.10), then (3.17) *is equivalent to* (3.66) *with the boundary condition*

(3.69) $$-\partial u/\partial n \in \Phi(u).$$

In the case where Ψ is given by (3.11) with (3.16), we have

(3.70) $$-(-\Delta u - f) \in \Phi(u), \quad u = 0 \quad \text{on } \Gamma. \quad ▯$$

3.5.3. The Function Ψ is of Type 3

K is again defined by (3.44).

If Ψ is given by (3.10), then (3.17) *is equivalent to the determination of u satisfying*

$$u \in K, \quad -\Delta u = f \ \text{ in } \Omega, \quad \int_\Gamma (\partial u/\partial n + \chi)(v-u)\,d\Gamma \geqslant 0 \quad \forall v \in K, \ \text{ where } \chi \in \Phi(u). \tag{3.71}$$

If Ψ is given by (3.11), then (3.17) is equivalent to the determination of u satisfying

$$u \in K, \quad a(u, v-u) + (\chi, v-u) \geqslant (f, v-u) \quad \forall v \in K, \ \chi \in \Phi(u). \tag{3.72}$$

3.5.4. Stationary Case and Problems of the Calculus of Variations

The following theorem holds:

Assuming that $a(u,v)=a(v,u)\ \forall u,v$ [13], *the determination of the solution* u of (3.17) *is equivalent to the determination of the u that furnishes the minimum (if it exists) of the functional*

$$J(v) = \tfrac{1}{2}a(v,v) + \Psi(v) - (f,v), \tag{3.73}$$

where the lower bound is taken over "all" functions v such that

$$\Psi(v) < +\infty$$

and to which one adds, for example, "$v=0$ on Γ" when Ψ is given by (3.11).

The verification of this result follows from the following general result (Lions [2], Chap. 1):

Let K be a convex closed set of a Hilbert space $\mathscr{H}$, and let J_1 and J_2 be two continuous and convex functions of $K \to \mathbb{R}$, where the function J_1 is differentiable, the function J_2 is or *is not* differentiable. *Then, the following two conditions are equivalent:*

$$u \in K, \quad J_1(u) + J_2(u) = \inf_{v \in K} [J_1(v) + J_2(v)], \tag{3.74}$$

$$u \in K, \quad (J_1'(u), v-u) + J_2(v) - J_2(u) \geqslant 0 \quad \forall v \in K. \tag{3.75}$$

---

[13] This is the case when $a(u,v)$ is given by (3.1), or by (3.62) if—and only if—$a_{ij} = a_{ji}\ \forall i,j$. Notice that the statement is wrong if

$$a(u,v) = a(v,u) \quad \forall u,v.$$

does not hold.

Before verifying this equivalence, we immediately observe that it implies the desired result, if we take

$$J_1(v)=\tfrac{1}{2}a(v,v)-(f,v)\,, \qquad J_2(v)=\Psi(v)\,, \tag{3.76}$$
$$K=\{v\,|\,\Psi(v)<\infty\}\,^{14}.$$

Equivalence of (3.74) *and* (3.75). Let u satisfy (3.74); we then have, for arbitrary v in K and $\theta\in]0,1[$,

$$J_1(u)+J_2(u)\leqslant J_1(u+\theta(v-u))+J_2(u+\theta(v-u))$$
$$\leqslant J_1(u+\theta(v-u))+(1-\theta)J_2(u)+\theta J_2(v)$$

which implies

$$\theta^{-1}[J_1(u+\theta(v-u))-J_1(u)]+J_2(v)-J_2(u)\geqslant 0$$

which leads to (3.75) if we let $\theta\to 0$.

Conversely, if u satisfies (3.75), then

$$J_1(v)+J_2(v)-J_1(u)-J_2(u) \tag{3.77}$$
$$=[(J_1'(u),v-u)+J_2(v)-J_2(u)+[J_1(v)-J_1(u)-(J_1'(u),v-u)]\,;$$

the first part of the right hand side of (3.77) is $\geqslant 0$ according to (3.75), and the second is $\geqslant 0$ because J_1 is convex; from this, (3.74) follows. □

4. Some Tools from Functional Analysis

Just as we did not intend to redevelop the complete theory of continuous media in Section 1, so our aim here is not at all to present the theory of Sobolev spaces in all its generality. We will simply introduce the *basic notions*, which are sufficient for the understanding of the solution of the problems of Section 3 (given in the following sections), while referring to the bibliography for proofs of these basic notions (for example, for the trace theorems, see below).

The ideas which we introuce in this section will be used in all subsequent chapters; at the appropriate time, we will recall the more elaborate results on Sobolev spaces that we need.

---

[14] Later on, we will more precisely define the class from which to take Ψ.

4.1. Sobolev Spaces

On the open set Ω of $\mathbb{R}^n$, we introduce the following spaces[15]:

1) $L^p(\Omega)$ = space of (classes of) measurable functions f such that (p being given with $1 \leqslant p \leqslant \infty$):

$$\|f\|_{L^p(\Omega)} = \left(\int_\Omega |f(x)|^p \, \mathrm{d}x\right)^{1/p} < \infty \qquad (p \neq \infty), \tag{4.1}$$

$$\|f\|_{L^\infty(\Omega)} = \operatorname*{ess\,sup}_{X \in \Omega} |f(x)| < \infty; \tag{4.2}$$

taken with the norm (4.1) or (4.2), $L^p(\Omega)$ is a Banach space; if $p=2$, $L^2(\Omega)$ is a Hilbert space, where the scalar product corresponding to the norm (4.1) (where $p=2$) is given by

$$(f,g) = \int_\Omega f(x) g(x) \, \mathrm{d}x. \tag{4.3}$$

2) $\mathscr{D}(\Omega)$ = space of functions C^∞ with compact support in Ω; for a given sequence[16] $\varphi_j \in \mathscr{D}(\Omega)$, we will say that $\varphi_j \to 0$ in $\mathscr{D}(\Omega)$ if:

i) the φ_j have their support in a fixed compact subset E of Ω;

ii) we have, $\forall \alpha$, $D^\alpha \varphi_j \to 0$ uniformly on Ω, where we put:

$$D^\alpha \varphi = \frac{\partial^{\alpha_1 + \cdots + \alpha_n}}{\partial x_1^{\alpha_1} \cdots \partial x_n^{\alpha_n}} \varphi, \qquad \alpha = \{\alpha_1, \ldots, \alpha_n\}; \tag{4.4}$$

3) $\mathscr{D}(\Omega)$ = space of functions C^∞ in $\bar{\Omega}$, with the family of norms

$$\sup_{x \in \Omega} |D^\alpha \varphi(x)|$$

which make it a Fréchet space;

4) $\mathscr{D}'(\Omega)$ = space *of distributions* on Ω = space of forms $p \to (f, \varphi)$, which are linear and continuous on $\mathscr{D}(\Omega)$ (i.e., $(f, \varphi_\alpha) \to 0$ when $\varphi_\alpha \to 0$ in $\mathscr{D}(\Omega)$). We will say that $f_j \to f$ in $\mathscr{D}'(\Omega)$, if $(f_j, \varphi) \to (f, \varphi)$ $\forall \varphi \in \mathscr{D}(\Omega)$. □

Examples of distributions. j) If $a \in \Omega$, then $\varphi \to \varphi(a)$ is continuous on $\mathscr{D}(\Omega)$; as one observes, this is the mass of Dirac concentrated at the point a:

$$\varphi(a) = (\delta_{(a)}, \varphi) = \int_\Omega \delta(x-a) \varphi(x) \, \mathrm{d}x$$

writing distributions like functions (which is what we shall do).

jj) If $f \in L^p(\Omega)$, then $\varphi \to \int_\Omega f(x) \, \mathrm{d}x$ is continuous on $\mathscr{D}(\Omega)$, hence defines a distribution $\tilde{f}$ by

$$(\tilde{f}, \varphi) = \int_\Omega f(x) \varphi(x) \, \mathrm{d}x.$$

---

[15] All functions considered here are real-valued.

[16] See L. Schwartz [1] for the topology of $\mathscr{D}(\Omega)$ defined by a fundamental system of neighborhoods of 0.

Since the mapping $f \to \tilde{f}$ is *injective*, we identify $\tilde{f}$ with f. Consequently, we have:

$$(4.5) \qquad L_p(\Omega) \subset \mathscr{D}'(\Omega)$$

where, moreover, the *injection is continuous*[17]; indeed, if $f_j \to 0$ in $L^p(\Omega)$, then $f_j \to 0$ in $\mathscr{D}'(\Omega)$.

One can *characterize* the subspace $L^p(\Omega)$ of $\mathscr{D}'(\Omega)$ for $1 < p \leqslant \infty$ by:

$$(4.6) \qquad f \in L^p(\Omega) \Leftrightarrow \left| \begin{array}{l} f \in \mathscr{D}'(\Omega) \text{ and there exists a } c \text{ such that} \\ |(f, \varphi)| \leqslant c \|\varphi\|_{L^{p'}(\Omega)} \qquad \forall \varphi \in \mathscr{D}(\Omega) \end{array} \right.$$

where p'—the exponent conjugate to p—is defined by

$$(4.7) \qquad \frac{1}{p} + \frac{1}{p'} = 1 .$$

The implication "$\Rightarrow$" results from the inequality

$$|(f, \varphi)| \leqslant \|f\|_{L^p(\Omega)} \|\varphi\|_{L^{p'}(\Omega)}$$

and is valid for $1 \leqslant p \leqslant \infty$.

For the implication "$\Leftarrow$", we note that when $p = 1$, then $p' \neq \infty$ and $\mathscr{D}(\Omega)$ is *dense* in $L^p(\Omega)$, so that there exists a *unique* f_* in $L^p(\Omega)$ (*that can be identified with the dual of* $L^p(\Omega)$) with $(f, \varphi) = (f_*, \varphi)$ $\forall \varphi \in \mathscr{D}(\Omega)$; identifying f_* with f, we have the desired result.

jjj) If $f \in \mathscr{D}'(\Omega)$, *we define* $D^\alpha f \in \mathscr{D}'(\Omega)$—this $\forall \alpha$—by:

$$(4.8) \qquad (D^\alpha f, \varphi) = (-1)^{|\alpha|} (f, D^\alpha \varphi), \qquad |\alpha| = \alpha_1 + \cdots + \alpha_n .$$

Furthermore, the linear mapping $f \to \mathscr{D}^\alpha f$ of $\mathscr{D}'(\Omega) \to \mathscr{D}'(\Omega)$ is *continuous*. ▯

Sobolev spaces $W^{m,p}(\Omega)$. One calls "Sobolev space of order m on $L^p(\Omega)$"—denoted by $W^{m,p}(\Omega)$—the space defined by

$$(4.9) \qquad W^{m,p}(\Omega) = \{ v \mid v \in L^p(\Omega),\, D^\alpha v \in L^p(\Omega),\, |\alpha| \leqslant m \} .$$

With the norm

$$(4.10) \qquad \|v\|_{W^{m,p}(\Omega)} = \left(\textstyle\sum_{|\alpha| \leqslant m} \|D^\alpha v\|^p_{L^p(\Omega)} \right)^{1/p} ,$$

it is *a Banach space*. ▯

The case "$p = 2$" is fundamental. To simplify the writing, we will put:

$$(4.11) \qquad W^{m,2}(\Omega) = H^m(\Omega);$$

[17] Henceforth, if X and Y denote two topological vector spaces, "$X \subset Y$" will mean: algebraic inclusion with continuous injection.

with the scalar product

(4.12) $$(u,v)_{H^m(\Omega)} = \sum_{|\alpha| \leqslant m} (D^\alpha u, D^\alpha v),$$

it is a Hilbert space. ▯

First trace theorem. One has to be aware that, except in one dimension, a function v of $H^1(\Omega)$ *is not necessarily continuous in* Ω, nor a fortiori in $\bar{\Omega}$; one can, however,—and *this is absolutely essential for applications*—define *the values of* v *on the boundary* Γ of Ω[18].

We will make the assumption:

(4.13) Ω is an open bounded set[19] with the boundary Γ which is a manifold of dimension $(n-1)$, once continuously differentiable, Ω lying locally on only one side of Γ.

Then one proves (see for example Lions-Magenes [1], Chap. 1) that if $\mathscr{C}^1(\Omega)$ denotes the space of once continuously differentiable functions in $\bar{\Omega}$,

(4.14) $\mathscr{C}^1(\bar{\Omega})$ is dense in $H^1(\Omega)$.

For $v \in \mathscr{C}^1(\bar{\Omega})$, we will put:

(4.15) $\gamma_0 v = \gamma v =$ "trace of v on Γ" $=$ value of v on Γ.

We put

(4.16) $L^2(\Gamma) = \{f \mid, f$ measurable and square integrable on Γ for the measure $\mathrm{d}\Gamma\}$,

(4.17) $$(f,g)_\Gamma = \int_\Gamma fg \, \mathrm{d}\Gamma.$$

One proves then (see for example Lions-Magenes, loc. cit.):

Theorem 4.1. *Under assumption* (4.13), *one can uniquely define the trace* $\gamma_0 v = \gamma v$ *of* $v \in H^1(\Omega)$ *on* Γ *in such a way that* γv *coincides with its usual definition* (4.15), *when* $v \in \mathscr{C}^1(\Omega)$; $\gamma_0 v \in L^2(\Gamma)$, *and the mapping* $v \to \gamma_0$ *is linear continuous from*

$$H^1(\Omega) \to L^2(\Gamma).$$ ▯

Remark 4.1. The mapping

$$v \to D^\alpha v$$

is linear continuous from $H^{|\alpha|+1}(\Omega) \to H^1(\Omega)$; then we can define

(4.18) $$\{\gamma_0(D^\alpha v) \mid |\alpha| \leqslant m-1\} \quad \text{if} \quad v \in H^m(\Omega).$$

[18] More generally, on a regular manifold of *dimension* $n-1$, contained in $\bar{\Omega}$.

[19] Theorem 4.1 below is valid without change if Ω is not bounded but has a bounded boundary Γ; if Γ is not bounded, then $\gamma_0 v$ is defined as in Theorem 4.1, and is *locally* square integrable on Γ.

In fact, since the knowledge of v on Γ implies the knowledge of its *tangential* derivatives on Γ, it is preferable to replace the set of derivatives which appear in (4.18) by the normal derivatives of order $\leqslant m-1$:

$$\gamma v = \{v, \partial v/\partial n, \ldots, \partial^{m-1} v/\partial n^{m-1}\} \in (L^2(\Gamma))^m \text{ if } v \in H^m(\Omega). \quad \Box \tag{4.19}$$

Remark 4.2. The kernel of γ_0 plays an essential role in what follows; we put:

$$H_0^1(\Omega) = \text{kernel of } \gamma_0 = \{v \mid \quad v \in H^1(\Omega), \quad \gamma_0 v = 0\} \tag{4.20}$$

and, more generally,

$$\begin{aligned} H_0^m(\Omega) &= \text{kernel of } \gamma \text{ defined by (4.19)} \\ &= \{v \mid \quad v \in H^m(\Omega) \quad \gamma v = 0\}. \end{aligned} \tag{4.21}$$

We have:

$$H_0^m(\Omega) \text{ is a } \textit{closed} \text{ sub-space of } H^m(\Omega). \tag{4.22}$$

Therefore, $H_0^m(\Omega)$ is a Hilbert space for the structure induced by that of $H^m(\Omega)$. $\Box$

Remark 4.3. One shows (Lions-Magenes, loc. cit.) that

$$\mathscr{D}(\Omega) \text{ is dense in } H_0^m(\Omega). \tag{4.23}$$

Then, every linear continuous form on $H_0^m(\Omega)$ can be identified with a *distribution* on Ω. By $H^{-m}(\Omega)$ we denote the space dual to $H_0^m(\Omega)$ in this identification; then

$$H_0^m(\Omega) \subset L^2(\Omega)(=H^0(\Omega)) \subset H^{-m}(\Omega). \quad \Box \tag{4.24}$$

Remark 4.4. Theorem 4.1 *is not the best possible one*, in the sense that the mapping $v \to \gamma_0 v$ *is not surjective* from $H^1(\Omega) \to L^2(\Gamma)$. In order to characterize the image space of $H^1(\Omega)$ under γ_0, we have to introduce some supplementary notions. $\Box$

The spaces $H^s(\mathbb{R}^n)$, s not necessarily an integer

In the special case when $\Omega = \mathbb{R}^n$, one can define $H^m(\mathbb{R}^n)$ by *Fourier transformation.* If v is a continuous function with compact support, its Fourier transform $\hat{v}$ is defined by

$$\hat{v}(\xi) = \int_{\mathbb{R}^n} \exp(-2\pi i x \bullet \xi) v(x) \, dx, \tag{4.25}$$

where $\quad \xi = \{\xi_1, \ldots, \xi_n\}, \quad x \bullet \xi = x_i \xi_i .$

One proves (Plancherel's theorem) that $\hat{v} \in L^2(\mathbb{R}_\xi^n)$ and that

$$\|\hat{v}\|_{L^2(\mathbb{R}_\xi^n)} = \|v\|_{L^2(\mathbb{R}_x^n)} . \tag{4.26}$$

By *continuous extension*, one can therefore define $\hat{v}$ for all $v \in L^2(\mathbb{R}^n)$. The mapping $v \to \hat{v}$ is *invertible;* when $\hat{v} = w$, then $v =$ limit in $L^2(\mathbb{R}^n_x)$ of v_j, $j \to +\infty$, where

$$v_j(x) = \int_{|\xi| \leqslant j} \exp(2\pi i x \cdot \xi) \hat{v}(\xi) d\xi \;;$$

We will also write, symbolically:

$$v(x) = \int_{\mathbb{R}^n_\xi} \exp(2\pi i x \cdot \xi) \hat{v}(\xi) d\xi \,.$$

This being understood, one verifies that

$$v \in H^m(\mathbb{R}^n) \Leftrightarrow (1 + |\xi|^2)^{m/2} \hat{v} \in L^2(\mathbb{R}^n_\xi) \,. \tag{4.27}$$

But then it is natural *to make the following definition:*

$$H^s(\mathbb{R}^n) = \{v | \quad (1 + |\xi|^2)^{s/2} \hat{v} \in L^2(\mathbb{R}^n_\xi)\} \,, \tag{4.28}$$

which is a *Hilbert space* for the norm

$$\|v\|_{H^s(\mathbb{R}^n)} = \|(1 + |\xi|^2)^{s/2} \hat{v}\|_{L^2(\mathbb{R}^n_\xi)} \,, \tag{4.29}$$

a norm which is *equivalent* to the norm corresponding to (4.12), when $s = m$. ☐

Remark 4.5. Definition (4.28) *is valid for* $s < 0$. When $H^0(\mathbb{R}^n)$ is identified with its dual, one has[20]:

$$H^s(\mathbb{R}^n) = H^{-s}(\mathbb{R}^n) \,. \tag{4.30}$$

For $s = -m$ and $Q = \mathbb{R}^n$, one is led back to the space introduced in Remark 4.3. ☐

Remark 4.6. The spaces $H^s(\mathbb{R}^n)$ have the following "localization" property:

(4.31) $\forall \varphi \in \mathscr{D}(\mathbb{R}^n)$, $\varphi v \in H^s(\mathbb{R}^n)$ $\forall v \in H^s(\mathbb{R}^n)$ and the mapping $v \to \varphi v$ is linear and continuous from $H^s(\mathbb{R}^n) \to H^s(\mathbb{R}^n)$. ☐

The properties (4.31) *permit the definition of* $H^s(\Gamma)$, $\forall s$.
We define $H^s(\Gamma)$ for $s \geqslant 0$ and then take

$$H^\sigma(\Gamma) = (H^{-\sigma}(\Gamma))' \quad \text{when} \quad \sigma < 0 \,. \tag{4.32}$$

For a function $v \in L^2(\Gamma)$, we define, with the help of a family of local charts and of a subordinated partition of unity $\{\theta_i\}$:

$$v = \sum \theta_i v \quad (\textit{finite}\ \text{sum});$$

---

[20] Generally, X' denotes the dual of the space X.

then one defines the images of the θ_i in $\mathbb{R}^{n-1}$; *we then say that* $v \in H^s(\Gamma)$, *if all the images of the* $\theta_i v$ are contained in $H^s(\mathbb{R}^{n-1})$; if w_i is the image of $\theta_i v$, we take as *(Hilbert)* norm on $H^s(\Gamma)$:

(4.33) $$\|v\|_{H^s(\Gamma)} = \left(\textstyle\sum_i \|w_i\|^2_{H^s(\mathbb{R}^{n-1})}\right)^{1/2}.$$

Due to (4.31), the space $H^s(\Gamma)$ thus defined *does not depend* on the choice of the family of local charts nor on that of the partition of unity; the norm (4.33) *depends on these*, but only within an equivalence of norms. The space $H^s(\Gamma)$ is thus defined *intrinsically* as is its *topology* but *not its norm.* □

Second trace theorem. Now we can complete Theorem 4.1 with

Theorem 4.2. *Under the assumption of Theorem* 4.1, *the mapping* $v \to \gamma_0 v$ *is linear, continuous and surjective from* $H^1(\Omega) \to H^{1/2}(\Gamma)$.

For the proof, see for example Lions-Magenes, loc. cit. □

4.2. Applications: The Convex Sets K

Let us go back to the notions of Section 3.

First, we note—we provided the justification—that the bilinear forms $a(u,v)$, defined by (3.1) or (3.62), are *continuous on* $H^1(\Omega)$. Moreover, condition (3.16) can be written

$$u(t) \in H_0^1(\Omega), \qquad v \in H_0^1(\Omega).$$

To make the meaning of the *stationary* inequalities (Sec. 3.5) precise, it remains then only to define $\Psi(v)$ for $v \in H^1(\Omega)$.

First, we consider the case when Ψ *is defined (formally) by* (3.10). We note that according to Theorem 4.1, $v(x)$ is defined almost everywhere on Γ, and therefore also $\psi(v(x))$. Let

(4.34) $$E = \{\lambda \mid \quad \psi(\lambda) \text{ finite}\};$$

the set E is closed and convex (bounded or not).

Then we put:

(4.35) $$K = \{v \mid \quad v \in H^1(\Omega),\ v(x) \in E \quad \text{a.e. for } x \in \Gamma.$$

Because of Theorem 4.1, we have:

(4.36) *the set K is closed convex in* $H^1(\Omega)$.

We make the assumption (not indispensable, see Remark 4.7 below):

(4.37) the function ψ is at most increasing quadratically at infinity on E (this is obviously meaningless when E is bounded, i.e., $\psi(\lambda) \leqslant c_1\lambda^2 + c_2$, where the c_i are suitable constants, $\lambda \in E$.

Then, for $v \in K$, $\psi(v(x)) \in L^1(\Gamma)$ and we can put

$$\Psi(v) = \int_\Gamma \psi(v(x))\,d\Gamma\,. \quad \square \tag{4.38}$$

Remark 4.7. If (4.37) does not hold, it is sufficient to *define* K by (4.35) *and* the condition "$\psi(v) \in L^1(\Gamma)$". We limit ourselves to the situation (4.37), because it simplifies the exposition to some extent and is sufficient for the applications. □

The function $v \to \Psi(v)$ is *continuous* from $K \to \mathbb{R}$; indeed, according to Krasnosel'skii [1] (Theorem 2.1, p. 22), the mapping $v \to \psi(v)$ is *continuous* from $L^2(\Gamma) \to L^1(\Gamma)$.

Moreover, because of the convexity of ψ, the function $v \to \Psi(v)$ is *convex*. □

In the case where Ψ is (formally) defined by (3.11), we introduce

$$K = \{v \mid \quad v \in H_0^1(\Omega), \quad v(x) \in E \quad \text{a.e. in } \Omega\}\,.$$

With assumption (4.37), we obtain the same properties as before. □

4.3. Spaces of Vector-Valued Functions

For a precise formulation of the problems *of evolution* posed in Section 3, we need additional tools which we will now introduce.

Let X be a Banach space with its norm denoted by $\| \quad \|_X$; by $L^p(0,T;X)$, we denote the space of (classes of) functions $t \to f(t)$ measurable from $[0,T] \to X$ (for the measure dt) such that

$$\left(\int_0^T \|f(t)\|_X^p\,dt\right)^{1/p} = \|f\|_{L^p(0,T;X)} < \infty \qquad (p \neq \infty)\,, \tag{4.39}$$

$$\|f\|_{L^\infty(0,T;X)} = \operatorname*{ess\,sup}_{t \in (0,T)} \|f(t)\|_X\,. \tag{4.40}$$

This is a *Banach space.*

By $\mathscr{D}'(]0,T[;X)$, we denote the space of distributions on $]0,T[$ with values in X defined by

$$\mathscr{D}'(]0,T[;X) = \mathscr{L}(\mathscr{D}(]0,T[;X) \tag{4.41}$$

where, generally, $\mathscr{L}(Y;X)$ denotes the space of linear continuous mappings of $Y \to X$.

To a function $f \in L^2(0,T;X)$ there corresponds a distribution $\tilde{f}$ on $]0,T[$ with values in X, defined by

$$\tilde{f}(\varphi) = \int_0^T f(t)\varphi(t)\,dt\,, \qquad ([0,T])\,, \tag{4.42}$$

(a well defined mapping $\varphi \to \tilde{f}(\varphi)$, linear continuous from $\mathscr{D}(]0,T[) \to X$). The mapping $f \to \tilde{f}$ is an *injection*; we will identify $\tilde{f}$ and f. Furthermore, if

$f \to 0$ in $L^p(0,T;X)$, then $\tilde{f} \to 0$ in $\mathscr{D}'(]0,T[;X)$, i.e., $\tilde{f}(\varphi) \to 0$ $\forall \varphi \in \mathscr{D}(]0,T[)$. Then we have

$$L^p(0,T;X) \subset \mathscr{D}'(]0,T[;X). \quad \square \tag{4.43}$$

For $f \in \mathscr{D}'(]0,T[;X)$, we define $\mathrm{d}^k f/\mathrm{d}t^k = f^{(k)}$ by

$$f^{(k)}(\varphi) = (-1)^k f(\varphi^{(k)}) \qquad \forall \varphi \in \mathscr{D}(]0,T[), \tag{4.44}$$

which defines $f^{(k)} \in \mathscr{D}'(]0,T[;X)$. Furthermore, the mapping $f \to f^{(k)}$ is linear continuous from $\mathscr{D}'(]0,T[;X)$ into itself. $\square$

In the following, we will constantly encounter situations of the *following type:* let X and Y be two Banach spaces with

$$X \subset Y. \tag{4.45}$$

Let v be given such that

$$v \in L^p(0,T;X); \tag{4.46}$$

then its derivative $\mathrm{d}v/\mathrm{d}t$ is defined as element of $\mathscr{D}'(]0,T[;X)$, hence in particular (since $\mathscr{D}'(]0,T[;X) \subset \mathscr{D}'(]0,T[;Y)$):

$$\mathrm{d}v/\mathrm{d}t \in \mathscr{D}'(]0,T[;Y). \tag{4.47}$$

Let us now *assume* that

$$\mathrm{d}v/\mathrm{d}t \in L^q(0,T;Y) \qquad (1 \leqslant q \leqslant \infty). \tag{4.48}$$

Then the function v is continuous from $[0,T] \to Y$, possibly after a modification on a set of measure zero.

This is not the best possible result[21]. We will state the result more precisely for a particular pair X, Y—a pair we will use very frequently in the remainder of the book.

Let V and H be two Hilbert spaces over $\mathbb{R}$, with norms $\| \ \|$, $| \ |$, respectively, their scalar product in H being written $(\ , \ )$; we assume that

$$V \subset H^{22}, \qquad V \text{ dense in } H. \tag{4.49}$$

Identifying H with its dual, H is then identified with a subspace of the dual V' of V, whence

$$V \subset H \subset V'. \tag{4.50}$$

[21] See J. L. Lions and J. Peetre [1].

[22] Therefore, there exists a constant c such that $|v| \leqslant c\|v\|$, $\forall v \in V$.

Example 4.1.:

$$V = H_0^m(\Omega), \qquad H = L^2(\Omega). \qquad V' = H^{-m}(\Omega).$$

Example 4.2.:

$$V = H^m(\Omega), \qquad H = L^2(\Omega);$$

then V' is not a space of distributions on Ω (since $\mathscr{D}(\Omega)$ is not dense in $H^m(\Omega)$). ☐

Let then v be given by

(4.51) $$v \in L^2(0,T;V), \qquad dv/dt \in L^2(0,T;V').$$

We show then (see Lions-Magenes, Chap. 1, loc. cit.) that

(4.52) *after possible modification on a set of measure zero, the function* $t \to v(t)$ *is continuous from* $[0,T] \to H$. ☐

5. Solution of the Variational Inequalities of Evolution of Section 3

5.1. Definitive Formulation of the Problems

For a neater presentation of the theory, we further generalize the formulation of the problems of Section 3.

5.1.1. Data V, H, V' and $a(u,v)$

Let a triplet of Hilbert spaces be given as in (4.49), (4.50) and a bilinear form $u, v \to (u,v)$, *continuous on* V and *coercive* in the sense that:

(5.1) there exist a c and $d > 0$ such that[23]

$$a(v,v) + c\,|v|^2 \geqslant \alpha\,\|v\|^2 \qquad \forall v \in V.$$

Example 5.1. $V = H^1(\Omega)$, where a is given by (3.1). Then (5.1) holds, if for example $c = 1$, $\alpha = 1$, ☐

Example 5.2. $V = H_0^1(\Omega)$, where a is given by (3.1). Then (5.1) holds with $c = 0$. ☐

5.1.2. The Functional Ψ

We prescribe a function $v \to \Psi(v)$ from $V \to \mathbb{R}$ with the following properties:

(5.2) the function $v \to \Psi(v)$ is convex, lower semi-continuous for the weak topology of V, with values in $]-\infty, +\infty]$;

[23] We recall that $|v|$ (respectively $\|v\|$) denotes the norm of v in H (resp. V).

there exists a family of functions Ψ_j that are *differentiable* on V such that:

(5.3)
$$\forall v \in L^2(0,T;V) \quad \text{one has:}$$
$$\int_0^T \Psi_j(v(t))\,dt \to \int_0^T \Psi(v(t))\,dt\,, \qquad j\to\infty$$

(5.4)
there exists a sequence φ_j *bounded in* V such that
$$\Psi_j'(\varphi_j)=0 \qquad \forall j\,,$$

(5.5)
if $v_j \to v$, $v_j' \to v'$ weakly in $L^2(0,T;V)$ and $\int_0^T \Psi_j(v_j)\,dt \leqslant$ constant, then $\liminf \int_0^T \Psi_j(v_j)\,dt \geqslant \int_0^T \Psi(v)\,dt$. ▯

Remark 5.1. These last assumptions are not indispensable, see H. Brézis [1, 2], but they serve for our purpose (see Sec. 5.3 below). ▯

5.1.3. Formulation of the Problem

With the above notions, we see that *all the problems of evolution encountered so far*[24] *are of the following type:* find a function u such that

(5.6)
$$u \in L^2(0,T);V), \qquad u' \in L^2(0,T;V')\,,$$

(5.7)
for almost all t and $\forall v \in V$, we have:
$$(u'(t), v-u(t)) + a(u(t), v-u(t)) + \Psi(v) - \Psi(u(t)) \geqslant (f(t), v-u(t)) \qquad \forall v \in V,$$

(5.8)
$$u(0)=u_0\,,$$

where u is given in V (with $\Psi(u_0)<\infty$ and f is given in $L^2(0,T;V')$.

5.2. Statement of the Principal Results

In the following, we are going to obtain "*strong*" solutions (expecially in improving (5.6)). (We do not present the theory of "weak" solutions here, but refer to Brézis [1,2] and Lions [1]). This explains the rather strong assumptions made for the data f and u:

(5.9)
$$f, f' \in L^2(0,T;V')\,,$$

(5.10)
there exist $u_{0j} \in V$ such that $u_{0j} \to u_0$ in V, and for every u_{0j} there exist $k_j \in H$ such that
$$a(u_{0j}, v) + (\Psi_j'(u_{0j}), v) = (k_j, v) \quad \forall v \in V$$
and
$$|k_j| \leqslant \text{constant when } j\to\infty.$$

[24] With the reservation that the hypotheses (5.2)–(5.5) have to be verified. This will be done in Section 5.3.

Theorem 5.1. *We assume that* (5.1), (5.5) *and* (5.9), (5.10) *hold. Then there exists one and only one function u satisfying*

$$(5.11)\qquad u\in L^2(0,T;V)\,,\quad u'\in L^2(0,T;V)\cap L^\infty(0,T;H)$$[25]

and (5.7), (5.8).

Theorem 5.2 (Approximation Theorem). *The assumptions are the same as in Theorem* 5.1. *Let u be the solution of*

$$(5.11)_j\qquad u_j\in L^2(0,T;V)\,,\quad u_j'\in L^2(0,T;V)\cap L^\infty(0,T;H)\,,$$

$$(5.7)_j\qquad (u_j'(t),v-u_j(t))+a(u_j(t),v-u_j(t))+\Psi_j(v)-\Psi_j(u_j(t)) \geqslant (f(t),v-u_j(t))\qquad \forall v\in V,$$

$$(5.8)_j\qquad u_j(0)=u_{0j}\,.$$

Then, as $j\to+\infty$, *we have, if u denotes the solution obtained from Theorem* 5.1,

$$(5.12)\qquad u_j\to u\quad \text{weakly in}\quad L^2(0,T;V),$$

$$(5.13)\qquad u_j'\to u'\quad \text{weakly in}\quad L^2(0,T;V)\quad \text{and weakly star in}\quad L^\infty(0,T;H).$$[26]

Remark 5.2. Since the function Ψ_j is differentiable, (5.7) is equivalent (see Sect. 3.3.1) to

$$(5.14)\qquad (u_j'(t),v)+a(u_j(t),v)+(\Phi_j(u_j(t)),v)=(f(t),v)\qquad \forall v\in V,$$

where we put

$$(5.15)\qquad \Phi_j=\Psi_j'.\ \square$$

Remark 5.3. Theorem 5.2 justifies the formal passages to the limit in Section 2. □

Before proving Theorems 5.1 and 5.2 (in Secs. 5.5 and 5.6 below), we verify assumptions (5.3), (5.4) and (5.5) for the examples of Section 3.

5.3. Verification of the Assumptions

We are going to carry out the verification for the following example, but the method is *absolutely general.*

We assume that ψ is given by Example 3.3 (Fig. 8) and that

$$(5.16)\qquad \Psi(v)=\int_\Gamma \psi(v)\,d\Gamma\,.$$

[25] Therefore u is *continuous* from $[0,T]\to V$.

[26] We say that $f_j\to f$ weakly star in $L^\infty(0,T;H)$ if

$$\int_0^T (f_j(t),\varphi(t))\,dt\to\int_0^T (f(t),\varphi(t))\,dt\qquad \forall\varphi\in L^1(0,T;H).$$

We introduce ψ_j as follows (Fig. 12):

$$(5.17)\qquad \psi_j(\lambda)=\begin{cases} g_1(\lambda-h_1)-g_1^2/2j\,, & \lambda\leqslant h_1+g_1/j\,,\\ \frac{1}{2}j(\lambda-h_1)^2\,, & h_1+g_1/j\leqslant\lambda\leqslant h_1\,,\\ 0\,, & h_1\leqslant\lambda\leqslant h_2\,,\\ \frac{1}{2}j(\lambda-h_2)^2\,, & h_2\leqslant\lambda\leqslant h_2+1\,,\\ j(\lambda-h_2)-j/2\,, & h_2+1\leqslant\lambda\,. \end{cases}$$

Then we define Ψ_j by

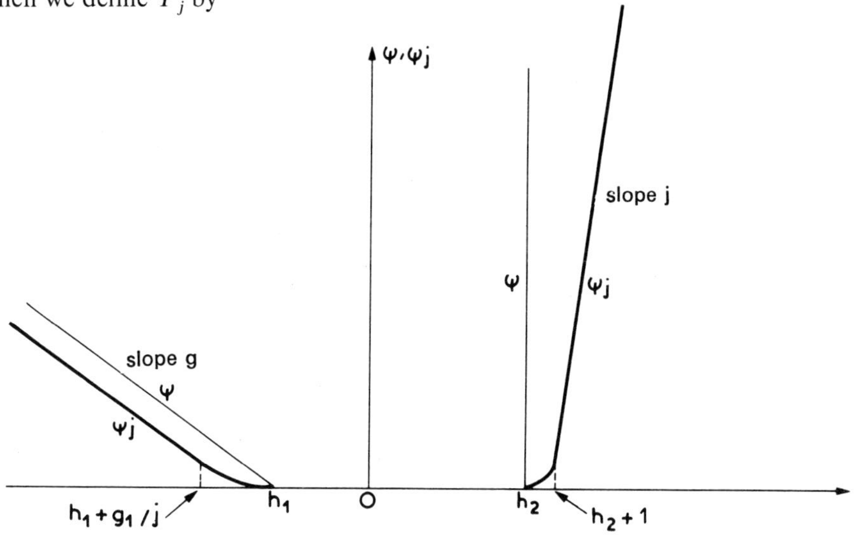

Fig. 12

$$(5.18)\qquad \Psi_j(v)=\int_\Gamma \psi_j(v)\,d\Gamma\,,\qquad v\in H^1(\Omega)\,.$$

Now (5.3) follows immediately and (5.4) holds, for example with $\varphi_j=0\ \forall j$. It remains to verify (5.5).

From *compactness theorems* (see for example Lions-Magenes [1], Chap. 1), it follows that when $v_j\to v$, $\partial v_j/\partial t\to\partial v/\partial t$ in $L^2(0,T;H^1(\Omega))$ weakly, then

$$(5.19)\qquad v_j\to v\quad\text{in}\quad L^2(\Sigma)\ \text{strongly.}$$

Furthermore, if we introduce M by

$$(5.20)\qquad M(\lambda)=\begin{cases} 0 & \text{if}\quad \lambda\leqslant h_2\,,\\ \frac{1}{2}(\lambda-h_2)^2 & \text{if}\quad h_2\leqslant\lambda\leqslant h_2+1\,,\\ \lambda-h_2-\frac{1}{2} & \text{if}\quad h_2+1\leqslant\lambda \end{cases}$$

then

$$(5.21)\qquad \int_0^T \Psi_j(v_j)\,dt=\int_\Sigma \psi_j(v_j)\,d\Sigma\geqslant j\int_\Sigma M(v_j)\,d\Sigma\,.$$

Accordingly, since

$$\int_0^T \Psi_j(v_j)\,\mathrm{d}t \leqslant c\,,$$

we have:

$$\int_\Sigma M(v_j)\,\mathrm{d}\Sigma \to 0$$

and from (5.19)

$$\int_\Sigma M(v_j)\,\mathrm{d}\Sigma \to \int_\Sigma M(v)\,\mathrm{d}\Sigma$$

from which

$$\int_\Sigma M(v)\,\mathrm{d}\Sigma = 0 \tag{5.22}$$

and consequently

$$v \leqslant h_2 \quad \text{a.e.} \quad \text{on } \Sigma\,. \tag{5.23}$$

If we now set

$$\theta(\lambda) = \begin{vmatrix} g_1(\lambda - h_1) & \text{if} \quad \lambda \leqslant h_1\,, \\ 0 & \text{if} \quad \lambda \geqslant h_1\,, \end{vmatrix}$$

$$\theta_j(\lambda) = \begin{vmatrix} g_1(\lambda - h_1) - g_1^2/2j & \text{if} \quad \lambda \leqslant h_1 + g_1/j\,, \\ \frac{1}{2}j(\lambda - h_1)^2 & \text{if} \quad h_1 + g_1/j \leqslant \lambda \leqslant h_1\,, \\ 0 & \text{if} \quad h_1 \leqslant \lambda\,. \end{vmatrix}$$

we have:

$$\int_\Sigma \psi_j(v_j)\,\mathrm{d}\Sigma \geqslant \int_\Sigma \theta_j(v_j)\,\mathrm{d}\Sigma = \int_\Sigma \theta(v_j)\,\mathrm{d}\Sigma + \int_\Sigma (\theta_j(v_j)) - (\theta(v_j))\,\mathrm{d}\Sigma\,; \tag{5.24}$$

but

$$\left|\int_\Sigma (\theta_j(v_j) - \theta(v_j))\,\mathrm{d}\Sigma\right| \leqslant c/j$$

and

$$\int_\Sigma \theta(v_j)\,\mathrm{d}\Sigma \to \int_\Sigma \theta(v)\,\mathrm{d}\Sigma = \text{(according to (5.23))} \int_0^T \Psi(v)\,\mathrm{d}t\,.$$

Therefore, (5.24) yields

$$\liminf \int_\Sigma \psi_j(v_j)\,\mathrm{d}\Sigma \geqslant \int_0^T \Psi(v)\,\mathrm{d}t\,. \quad \square$$

5.4. Other Methods of Approximation

In applications, Theorem 5.2 reduces to approximating (in a suitable sense) *functions* ψ *of type* 3 *by functions* ψ_j *of type* 1.

The approximation is effected **through regularization** (in the neighborhood of the points where ψ is not differentiable) and **through penalization** (where $\psi=+\infty$).

In Theorem 5.2, we can also assume that the functionals Ψ_j are everywhere finite, but not necessarily differentiable.

This amounts to approximating ψ **by penalization only.**

If we again consider Example 3.3 which we examined in Section 5.3 above, we now "approximate" ψ by ψ_j defined by (see Fig. 13)

(5.25) $$\psi_j(\lambda)=\begin{vmatrix}\psi(\lambda) & \text{if } \lambda\leqslant h_2\\ j(\lambda-h_2) & \text{if } \lambda\geqslant h_2.\end{vmatrix}$$

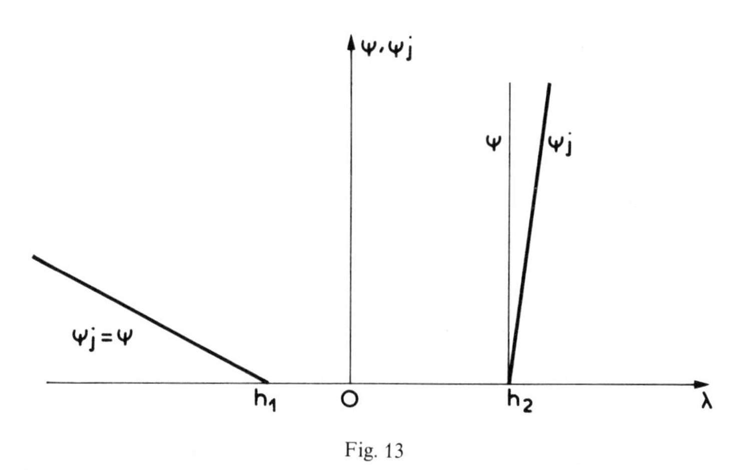

Fig. 13

5.5. Uniqueness Proof in Theorem 5.1 (and 5.2)

Let u and u_* be two possible solutions. If we take $v=u_*(t)$, which is permissible, (resp. $v=u(t)$) in the inequality (5.7) (resp. the analogous inequality with respect to u_*), and add the two inequalities, it follows, setting $w=u-u_*$

$$-(w'(t),w(t))-a(w(t),w(t)\geqslant 0$$

or again (from (5.1))

$$\frac{1}{2}\frac{\mathrm{d}}{\mathrm{d}t}|w(t)|^2+\alpha\|w(t)\|^2\leqslant c|w(t)|^2$$

therefore, in particular:

$$\frac{\mathrm{d}}{\mathrm{d}t}|w(t)|^2\leqslant 2c|w(t)|^2$$

which, together with $w(0)=0$, shows that $w(t)=0$. □

5.6. Proof of Theorems 5.1 and 5.2

Following is the outline of the proof:
1) Solution of (5.14), with $u_j(0)=u_{0j}$.
2) A priori estimates of the u_j, derivation of (5.12) and (5.13).
3) Proof of the fact that u, thus obtained, is a solution of (5.7). □

5.6.1. Solution of (5.14)

Let $w_1, \ldots, w_m$ be a "base" of V in the following sense: $w_1, \ldots, w_m \ldots$ are linearly independent $\forall m$ and the finite combinations $\Sigma \xi_j w_j$ are dense in V. Such "bases" always exist when V is *separable*, which we assume to be the case (actually, this assumption is in no way essential). We choose w_1, w_2 in such a way that

$$\{u_{0j} \text{ and } \varphi_j \text{ belong to the space } [w_1, w_2] \text{ spanned by } w_1 \text{ and } w_2\}. \tag{5.26}$$

We define $u_{jm}(t)$ as the solution of the following system of (ordinary) non linear differential equations by:

$$u_{jm}(t) \in [w_1, \ldots, w_m] = \text{space spanned by } w_1, \ldots, w_m, \tag{5.27}$$

$$(u'_{jm}(t), w_k) + a(u_{jm}(t) + (\Phi_j(u_{jm}(t), w_k) = (f(t), w_k), \quad 1 \leqslant k \leqslant m, \tag{5.28}$$

$$u_{jm}(0) = u_{0j} \quad \text{(this is permissible, since } m \geqslant 2 \text{ because of (5.26))}. \tag{5.29}$$

If

$$u_{jm}(t) = \sum_{k=1} g_{jk}(t) w_k,$$

we then have a differential system in the g_{jk}—which defines u_{jm} in an interval $[0, t_m], t_m > 0$.

The a priori estimates below show that $t_m = T$. □

A priori estimates (I). According to (5.26), we have:

$$(u'_{jm}(t), \varphi_j) + a(u_{jm}(t), \varphi_j) + (\Phi_j(u_{jm}(t)), \varphi_j) = (f(t), \varphi_j). \tag{5.30}$$

Furthermore, if we multiply (5.28) with $g_{jk}(t)$ and take the sum over k and then substract, we obtain (observing that $\Phi_j(\varphi_j) = 0$):

$$\begin{aligned}(u'_{jm}(t), u_{jm}(t) - \varphi_j) + a(u_{jm}(t), u_{jm}(t) - \varphi_j)& \\ + (\Phi_j(u_{jm}(t)) - \Phi_j(\varphi_j), u_{jm}(t) - \varphi_j) &= (f(t), u_{jm}(t) - \varphi_j).\end{aligned} \tag{5.31}$$

But Φ_j is *monotone*, i.e.,

$$(\Phi_j(u) - \Phi_j(v), u - v) \geqslant 0^{27}. \tag{5.32}$$

---

[27] The derivative of a convex function Ψ_j is monotone; in fact
$\Psi_j(v) - \Psi_j(u) - (\Psi'_j(u), v - u) \geqslant 0$.
Interchange u and v and add.

Therefore, it follows from (5.31) that

$$(u'_{jm}(t), u_{jm}(t)-\varphi_j)+a(u_{jm}(t), u_{jm}(t)-\varphi_j)\leqslant(f(t), u_{jm}(t)-\varphi_j)\,, \tag{5.33}$$

whence

$$\frac{1}{2}\frac{\mathrm{d}}{\mathrm{d}t}|u_{jm}(t)-\varphi_j|^2+a(u_{jm}(t), v_{jm}(t))\leqslant a(u_{jm}(t),\varphi_j)+(f(t),u_{jm}(t)-\varphi_j)\,.$$

Therefore (c designating various constants):

$$\frac{1}{2}\frac{\mathrm{d}}{\mathrm{d}t}|u_{jm}(t)-\varphi_j|^2+\alpha\,\|u_{jm}(t)\|^2$$
$$\leqslant c\,|u_{jm}(t)|^2+c\,\|u_{jm}(t)\|\,\|\varphi_j\|+c\,\|f(t)\|_*\,\|u_{jm}(t)-\varphi_j\|$$

(where $\|f\|_*=$ norm in V').

Since according to (5.4) $\|\varphi_j\|\leqslant c$, we have

$$\frac{\mathrm{d}}{\mathrm{d}t}|u_{jm}(t)-\varphi_j|^2+2\alpha\,\|u_{jm}(t)\|^2$$
$$\leqslant c\,|u_{jm}(t)-\varphi_j|^2+\alpha\,\|u_{jm}(t)\|^2+c(1+\|f(t)\|_*^2)$$

from which

$$|u_{jm}(t)-\varphi_j|^2+\alpha\textstyle\int_0^t\|u_{jm}(\sigma)\|^2\,\mathrm{d}\sigma$$
$$\leqslant c\textstyle\int_0^t|u_{jm}(\sigma)-\varphi_j|^2\,\mathrm{d}\sigma+c(t+\int_0^t\|f(\sigma)\|_*^2\,\mathrm{d}\sigma)+|u_{0j}-\varphi_j|^2\,. \tag{5.34}$$

Therefore, in particular, if we put $|u_{jm}(t)-\varphi_j(t)|^2=\eta(t)$:

$$\begin{aligned}&\eta(t)\leqslant c\textstyle\int_0^t\eta(\sigma)\,\mathrm{d}\sigma+d\,,\\ &d\leqslant c(T+\textstyle\int_0^T\|f(\sigma)\|_*^2\,\mathrm{d}\sigma)+|u_{0j}-\varphi_j|^2\,.\end{aligned} \tag{5.35}$$

But because of *Gronwall's inequality*, (5.35) *implies*:

$$\eta(t)\leqslant d\exp(ct) \tag{5.36}$$

from which we infer:

(5.37) $u_{jm}\in$ bounded subset of $L^\infty(0,T;H)$ *(bounded independent of j and m)*.

But taking (5.34) with $t=T$, we see that

(5.38) $u_{jm}\in$ bounded subset of $L^2(0,T;V)$ *(bounded independent of j and m)*. ▯

A priori estimates (II). We will now derive estimates for u'_{jm}, analogous to (5.37) and (5.38). First, we deduce from (5.28), (5.29) that

$$(u'_{jm}(0), w_k) = (f(0), w_k) - [a(u_{0j}, w_k) + (\Phi_j(u_{0j}), w_k)]$$
$$= (\text{according to } (5.10))(f(0) - k_j, w_k)$$

from which

$$|u'_{jm}(0)|^2 = (f(0) - k_j, u'_{jm}(0))$$

and, consequently:

$$|u'_{jm}(0)| \leqslant |f(0) - k_j| . \tag{5.39}$$

We now differentiate (5.28) with respect to t[28]; it follows that

$$(u''_{jm}(t), w_k) + a(u'_{jm}(t), w_k) + ((\Phi_j(u_{jm}(t)))', w_k) = (f'(t), w_k) . \tag{5.40}$$

We observe that

$$(\Phi_j(u_{jm}(t))', u'_{jm}(t)) \geqslant 0 \tag{5.41}$$

for, *because of the monotonicity of* Φ_j,

$$(\Phi_j(u_{jm}(t+h)) - \Phi_j(u_{jm}(t)), u_{jm}(t+h) - u_{jm}(t)) \geqslant 0 .$$

Therefore, we infer from (5.40), using (5.41), that

$$(u''_{jm}(t), u'_{jm}(t)) + a(u'_{jm}(t), u'_{jm}(t)) \leqslant (f'(t), u'_{jm}(t)) \tag{5.42}$$

from which

$$\frac{1}{2} \frac{\mathrm{d}}{\mathrm{d}t} |u_{jm}(t)|^2 + \alpha \|u'_{jm}(t)\|^2 \leqslant c |u'_{jm}(t)|^2 + \|f'(t)\|_* \|u'_{jm}(t)\| . \tag{5.43}$$

But from (5.39), (5.43), we deduce, as before for u_{jm}, that

(5.44) u'_{jm} remains in a bounded set of $L^2(0, T; V) \cap L^\infty(0, T; H)$ independent of j and m. ☐

Passage to the limit with m. According to (5.44) and since $u_{jm}(0) = u_{0j}$, $\|u_{0j}\| \leqslant C$, we have:

$$u_{jm} \in \text{bounded set of } L^\infty(0, T; V) \tag{5.45}$$

and therefore

$$\Phi_j(u_{jm}) \in \text{bounded set of } L^\infty(0, T; V') . \tag{5.46}$$

[28] What follows is correct if Φ_j is Lipschitz continuous. If it is not, we have to replace the derivative by a differential quotient.

We can then extract a sub-sequence $u_{j\mu}$ such that

$$\begin{aligned} &u_{j\mu}\to u_j \quad \text{weakly star in} \quad L^\infty(0,T;V),\\ (5.47)\qquad &u'_{j\mu}\to u'_j \quad \text{weakly in} \quad L^2(0,T;V) \quad \text{and weakly star in} \quad L^\infty(0,T;H),\\ &\Phi_j(u_{j\mu})\to \chi_j \quad \text{weakly star in} \quad L^\infty(0,T;V'),\end{aligned}$$

and

$$(5.48)\qquad \|u_j\|_{L^\infty(0,T;V)}+\|u'_j\|_{L^2(0,T;V)}+\|u'_j\|_{L^\infty(0,T;H)}\leqslant C.$$

According to (5.47), $u_{j\mu}(0)\to u_j(0)$ weakly in V, therefore

$$u_j(0)=u_{0j}.$$

Let us take (5.28) for $m=\mu$; then, passing to the limit in (5.28) with fixed $k(<\mu)$, we obtain

$$(u'_j,w_k)+a(u_j,w_k)+(\chi_j,w_k)=(f,w_k)$$

and this $\forall k$; hence, since the finite combinations $\sum \xi_k w_k$ are dense in V, we conclude that

$$(5.49)\qquad (u'_j,v)+a(u_j,v)+(\chi_j,v)=(f,v) \qquad \forall v\in V.$$

Therefore, we will have solved (5.14) if we show that

$$(5.50)\qquad \chi_j=\Phi_j(u_j).$$

For this purpose, we make use of a "*monotonicity argument*". Because of the monotonicity of Φ_j, we have, $\forall \varphi\in L^2(0,T;V)$:

$$(5.51)\qquad X_m=\int_0^T(\Phi_j(u_{jm})-\Phi_j(\varphi),u_{jm}-\varphi)\,dt\geqslant 0.$$

But according to (5.28)

$$\begin{aligned} X_m &= -\int_0^T[(u'_{jm},u_{jm})+a(u_{jm},u_{jm})-(f,u_{jm})]\,dt\\ &\quad -\int_0^T(\Phi_j(u_{jm}),\varphi)\,dt-\int_0^T(\Phi_j(\varphi),u_{jm}-\varphi)\,dt\\ &= -\tfrac{1}{2}|u_{jm}(T)|^2+\tfrac{1}{2}|u_{0j}|^2-\int_0^T a(u_{jm},u_{jm})\,dt\\ &\quad +\int_0^T(f,u_{jm})\,dt-\int_0^T(\Phi_j(u_{jm}),\varphi)\,dt-\int_0^T(\Phi_j(\varphi),u_{jm}-\varphi)\,dt.\end{aligned}$$

Since $u_{jm}(T)\to u_j(T)$ in V weakly, we have:

$$\limsup\,(-|u_{jm}(T)|^2)\leqslant -|u_j(T)|^2$$

and therefore

(5.52)
$$0 \leqslant \limsup X_m \leqslant -\tfrac{1}{2}|u_j(T)|^2 + \tfrac{1}{2}|u_{0j}|^2 - \int_0^T a(u_j,u_j)\,dt + \int_0^T (f,u_j)\,dt - \int_0^T (\chi_j,\varphi)\,dt - \int_0^T (\Phi_j(\varphi),u_j-\varphi)\,dt\,.$$

But we conclude from (5.49) that

$$-\tfrac{1}{2}|u_j(T)|^2 + \tfrac{1}{2}|u_{0j}|^2 - \int_0^T a(u_j,u_j)\,dt + \int_0^T (f,u_j)\,dt = \int_0^T (\chi_j,u_j)\,dt$$

and therefore (5.52) implies:

(5.53)
$$\int_0^T (\chi_j - \Phi_j(\varphi), u_j - \varphi)\,dt \geqslant 0\,.$$

We take: $\varphi = u_j - \lambda\theta, \lambda > 0, \theta \in L^2(0,T;V)$; then, after division by λ

$$\int_0^T (X_j - \Phi_j(u_j - \lambda\theta),\theta)\,dt \geqslant 0$$

and letting $\lambda \to 0$:

$$\int_0^T (X_j - \Phi_j(u_j),\theta)\,dt \geqslant 0 \qquad \forall \theta \in L^2(0,T;V)$$

from which (5.50) follows. ▯

5.6.2. Estimates for u_j and u'_j

We already have (5.48). Moreover, $(5.7)_j$ follows from (5.14). Let

$$v_0 \in L^2(0,T;V)$$

such that

$$\int_0^T \Psi(v_0)\,dt < \infty\,.$$

Taking $v = v(t) = v_0(t)$ in $(5.7)_j$, we see that

(5.54)
$$\int_0^T \Psi_j(u_j(t))\,dt \leqslant \int_0^T \{u'_j, v_0 - u_j) + a(u_j, v_0 - u_j) + \Psi_j(v_0) - (f, v_0 - u_j)\}\,dt\,.$$

According to (5.48) and (5.3), we conclude from (5.54) that

(5.55)
$$\int_0^T \Psi_j(u_j(t))\,dt \leqslant C\,.$$

Hence, we can select a sub-sequence, again denoted by u_j, such that

(5.56)
$$\begin{aligned} &u_j \to u && \text{weakly in } L^2(0,T;V)\\ &u'_j \to u' && \text{weakly in } L^2(0,T;V) \text{ and weakly star in } L^\infty(0,T;H)\,, \end{aligned}$$

and according to (5.5)

$$\liminf \int_0^T \Psi_j(u_j)\,dt \geqslant \int_0^T \Psi(u)\,dt\,. \tag{5.57}$$

Since $u_{0j} \to u$ in V and $u_j(0) \to u(0)$ weakly in V and $u_j(0) = u_{0j}$, we have (5.8).

5.6.3. Proof of (5.7)

In $(5.7)_j$ we take $v = v(t)$ where $t \to v(t)$ is arbitrary in $L^2(0,T;V)$. From this we conclude:

$$\int_0^T [(u_j',v) + a(u_j,v) + \Psi_j(v) - (f,v-u_j)]\,dt \geqslant \int_0^T [(u_j',u_j) + a(u_j,u_j) + \Psi_j(u_j)]\,dt\,. \tag{5.58}$$

The right hand side of (5.58) equals

$$Y_j = \tfrac{1}{2}|u_j(T)|^2 - \tfrac{1}{2}|u_{0j}|^2 + \int_0^T a(u_j,u_j)\,dt + \int_0^T \Psi_j(u_j)\,dt$$

and, using in particular (5.57), we have:

$$\begin{aligned}\liminf Y_j &\geqslant \tfrac{1}{2}|u(T)|^2 - \tfrac{1}{2}|u_0|^2 + \int_0^T a(u,u)\,dt + \int_0^T \Psi(u)\,dt \\ &= \int_0^T [(u',u) + a(u,u) + \Psi(u)]\,dt\,.\end{aligned}$$

Therefore, we deduce from (5.58) that

$$\int_0^T [(u',v-u) + a(u,v-u) + \Psi(v) - \Psi(u) - (f,v-u)]\,dt \geqslant 0 \qquad \forall v \in L^2(0,T;V)\,, \quad \square \tag{5.59}$$

Let $s \in]0,T[$ be fixed arbitrarily (for the moment) and let $w \in V$ be arbitrary. We take the family $\mathcal{O}_k$ of neighborhoods of s:

$$\mathcal{O}_k =]s - 1/k,\, s + 1/k[$$

and let v be defined by

$$v(t) = \left|\begin{array}{ll} u(t) & \text{if } t \notin \mathcal{O}_k \\ w & \text{if } t \in \mathcal{O}_k \end{array}\right.$$

Then (5.59) yields

$$\begin{aligned}&\int_{\mathcal{O}_k} [(u',w) + a(u,w) + \Psi(w) - (f,w)]\,dt \\ &\qquad - \int_{\mathcal{O}_k} [(u',u) + a(u,u) + \Psi(u) - (f,u)]\,dt \geqslant 0\,,\end{aligned} \tag{5.60}$$

from which again follows—denoting the measure of $\mathcal{O}_k$ by $|\mathcal{O}_k|$

$$\begin{aligned}&(|\mathcal{O}_k|^{-1} \textstyle\int_{\mathcal{O}_k} u'(t)\,dt, w) + a(|\mathcal{O}_k|^{-1} \int_{\mathcal{O}_k} u(t)\,dt, w) + \Psi(w) \\ &- (|\mathcal{O}_k|^{-1} \textstyle\int_{\mathcal{O}_k} f(t)\,dt, w) - |\mathcal{O}_k|^{-1} \int_{\mathcal{O}_k} [(u',u) + a(u,u) + \Psi(u) - (f,u)]\,dt \geqslant 0\,.\end{aligned} \tag{5.61}$$

But, generally, for a scalar-valued or vector-valued measurable function g, we have *(Theorem of Lebesgue)*

$$|\mathcal{O}_k|^{-1} \int_{\mathcal{O}_k} g(t)\,dt \to g(s)$$

for almost all s.

Therefore, we conclude from (5.61) that, except possibly for s in an exceptional set of measure 0, we have:

$$(u'(s), w-u(s)) + a(u(s), w-u(s)) + \Psi(w) - \Psi(u(s)) \geqslant (f(s), w-u(s))$$

from which (5.7) follows. Since the limit u is *unique*, the extraction of a sub-sequence of u_j is unnecessary, and Theorems 5.1 and 5.2 are proved. □

6. Properties of Positivity and of Comparison of Solutions

6.1. Positivity of Solutions

Generally, for any real-valued function v, one sets

(6.1) $$v^+ = \sup(v,0), \qquad v^- = \sup(-v,0).$$

We are going to apply Theorem 5.1 where, in all cases,

(6.2) $$V = H^1(\Omega)$$[29]

and $a(u,v)$ given by (3.62), i.e.,

(6.3) $$a(u,v) = \int_\Omega a_{ij}(x) u v \, dx$$

with (3.63).

We observe that $v \to v^+$ and $v \to v^-$ are continuous mappings (in fact contractions) of $H^1(\Omega)$ into itself. Having stated this, we shall prove

Theorem 6.1. *We make the same assumptions as for Theorem* 5.1, *with* (6.2) *and* (6.3). *Let the functional* Ψ *satisfy*

(6.4) $$\Psi(v) \geqslant \Psi(v^+) \qquad \forall v \in V.$$

If we assume in addition that

(6.5) $$f \geqslant 0 \text{ a.e. in } Q, \qquad u_0 \geqslant 0 \text{ a.e. in } \Omega,$$

[29] Easy modifications occur when $V = H_0^1(\Omega)$ or, for example, when V is the space of functions of $H^1(\Omega)$ which vanish *on a part* Γ_0 of Γ.

then the u, furnished by Theorem 5.1, *satisfies*

$$u \geqslant 0 \quad \text{a.e.} \quad \text{in } Q. \tag{6.6}$$

Proof. In the inequality

$$(u'(t), v-u(t)) + a(u(t), v-u(t)) + \Psi(v) - \Psi(u(t)) \geqslant (f(t), v-u(t)) \tag{6.7}$$

we take

$$v = u^+(t).$$

Then, since $u^+(t) - u(t) = u^-(t)$, it follows that

$$(u'(t), u^-(t)) + a(u(t), u^-(t)) + \Psi(u^+(t)) - \Psi(u(t)) \geqslant (f(t), u^-(t)). \tag{6.8}$$

But

$$\begin{aligned} &a(\varphi, \varphi^-) = -a(\varphi^-, \varphi^-) \quad (\text{since } a(\varphi^+, \varphi^-) = 0) \\ &(u'(t), u^-(t)) = -((u^-(t))', u^-(t)) \end{aligned} \tag{6.9}$$

so that (6.8) is equivalent to

$$((u^-(t))', u^-(t)) + a(u^-(t), u^-(t)) + \Psi(u(t)) - \Psi(u^+(t)) + (f(t), u^-(t)) \leqslant 0. \tag{6.10}$$

But, according to (6.4), $\Psi(u(t)) - \Psi(u^+(t)) \geqslant 0$, and since, according to (6.5), $f(t) \geqslant 0$ and since by definition $u^-(t) \geqslant 0$, we have:

$$(f(t), u^-(t)) \geqslant 0,$$

so that (6.10) implies

$$((u^-(t))', u^-(t)) + a(u^-(t), u^-(t)) \leqslant 0. \tag{6.11}$$

We conclude from (6.11) that

$$\frac{1}{2} \frac{\mathrm{d}}{\mathrm{d}t} |u^-(t)|^2 + \alpha \|u^-(t)\|^2 \leqslant c |u^-(t)|^2 \tag{6.12}$$

therefore, in particular, that

$$\frac{\mathrm{d}}{\mathrm{d}t} |u^-(t)|^2 \leqslant 2c |u^-(t)|^2. \tag{6.13}$$

But, according to (6.5), $u_0 \geqslant 0$, therefore $u^-(0) = 0$ which, together with (6.13), proves that $u^-(t) = 0$, therefore (6.6) follows. □

Example 6.1. We assume (see Sec. 3) that

$$\Psi(v) = \int_\Gamma \psi(v)\,d\Gamma \tag{6.14}$$

with

$$\psi(\lambda) \geqslant \psi(0) \quad \text{if} \quad \lambda \leqslant 0 . \tag{6.15}$$

Then we have $\psi(\lambda) \geqslant \psi(\lambda^+)\ \forall \lambda \in \mathbb{R}$ *and therefore* (6.4) *holds*. Property (6.15) can be verified in all examples occurring in Section 3. Thus:

Example 6.1.1. Let ψ be given by (3.33), Fig. 9. We see that the solution u *does not depend* on the values of ψ for $\lambda \leqslant 0$; in other words, if $f \geqslant 0$ and $u_0 \geqslant 0$, *the temperature stays* $\geqslant 0$ (and the mechanism for temperature control for temperatures <0 never has to intervene; in this case, the temperature control always consists of an addition of cold, never of heat).

Example 6.1.2. Let us assume that $\psi(\lambda)=0$ if $\lambda \geqslant 0$ and $\psi(\lambda) \geqslant 0$ if $\lambda \leqslant 0$[30], since $u \geqslant 0$, we have: $\Psi(u)=0$, $\Psi(v) \geqslant 0$ and the *solution* u *coincides with the solution of the problem for* $\Psi=0$, i. e.,

$$\begin{aligned} &(u',v)+a(u,v)=(f,v) \qquad \forall v \in H^1(\Omega),\\ &u(0)=u_0 . \end{aligned} \tag{6.16}$$

Therefore

$$\begin{aligned} &\partial u/\partial t - (a_{ij}u_{,j})_{,i} = f,\\ &\partial u/\partial \nu_A = 0 \qquad \text{on } \Sigma,\\ &u(x,0)=u_0(x) \qquad \text{in } \Omega . \quad \square \end{aligned}$$

6.2. Comparison of Solutions (I)

We will now compare solutions u and $\hat{u}$ corresponding to two different functionals Ψ and $\hat{\Psi}$.

Theorem 6.2. *We make the same assumptions as in Theorem* 5.1 *with* (6.2) *and* (6.3). *Let* Ψ *and* $\hat{\Psi}$ *be two functionals* (which have the properties of the functional Ψ in Theorem 5.1) *such that* $\forall v$ and $\hat{v} \in H^1(\Omega)$:

$$\Psi(\sup(v,\hat{v})) + \hat{\Psi}(\inf(v,\hat{v})) \leqslant \Psi(v) + \hat{\Psi}(\hat{v}) . \tag{6.17}$$

Let u (resp. $\hat{u}$) *be the solution given by Theorem* 5.1 *corresponding to* Ψ (resp. $\hat{\Psi}$), *all other circumstances being the same. Then we have*

$$u \geqslant \hat{u} \quad \text{a.e.} \quad \text{in } Q . \tag{6.18}$$

[30] This is the case of a semi-permeable partition that only lets heat enter and that is in contact with an exterior medium at temperature 0.

Proof. We introduce

$$w=\sup(u,\hat{u})=u+(\hat{u}-u)^{+}, \qquad \hat{w}=\inf(u,\hat{u})=\hat{u}-(\hat{u}-u)^{+}. \tag{6.19}$$

Observe that $w+\hat{w}=u+\hat{u}$.

In (6.7) we take $v=w$ and, in the analogous inequality corresponding to $\hat{u}$, we take $v=\hat{w}$. Setting

$$\theta=\hat{u}-u \tag{6.20}$$

it follows that

$$-(\theta',\theta^{+})-a(\theta,\theta^{+})+\Psi(w)+\hat{\Psi}(\hat{w})-\Psi(u)-\hat{\Psi}(\hat{u})\geqslant 0$$

from which, taking (6.17) into account:

$$(\theta',\theta^{+})+a(\theta,\theta^{+})\leqslant 0. \tag{6.21}$$

Therefore

$$\frac{1}{2}\frac{\mathrm{d}}{\mathrm{d}t}|\theta^{+}(t)|^{2}+a(\theta^{+},\theta^{+})\leqslant 0$$

from which

$$\frac{1}{2}\frac{\mathrm{d}}{\mathrm{d}t}|\theta^{+}(t)|^{2}+\alpha\|\theta^{+}(t)\|^{2}\leqslant c|\theta^{+}(t)|^{2}.$$

Since $u(0)=\hat{u}(0)=u_0$, we have $\theta^{+}(0)=0$ and therefore $\theta^{+}=0$ from which (6.18) follows. □

Example 6.2. We assume that Ψ is given by (6.14) and $\hat{\Psi}$ by

$$\hat{\Psi}(v)=\int_{\Gamma}\hat{\psi}(v)\,\mathrm{d}\Gamma. \tag{6.22}$$

We verify without difficulty *that* (6.17) *is satisfied if*

$$\psi(\mu)-\psi(\lambda)\leqslant\hat{\psi}(\mu)-\hat{\psi}(\lambda) \qquad \forall\lambda,\mu, \quad \mu\geqslant\lambda. \tag{6.23}$$

Example 6.2.1. Property (6.13) occurs if, *for example,*

$$\psi(\lambda)=\hat{\psi}(\lambda-\lambda_0), \qquad \lambda_0>0, \tag{6.24}$$

where the derivative of $\hat{\psi}$ is increasing.

This remark, applied to the problem of temperature control, gives the following—physically obvious—result; *if for a same right hand side and same initial temperatures, we regulate in the same way around "brackets"* (h_1, h_2) *and* $(\hat{h}_1+\lambda_0, \hat{h}_2+\lambda_0)$, where $\lambda_0>0$, *then the temperature corresponding to* $(\hat{h}_1+\lambda_0, \hat{h}_2+\lambda_0)$ *is higher than that corresponding to* $(\hat{h}_1, \hat{h}_2)$. □

6.3. Comparison of Solutions (II)

We now compare the solutions corresponding to *a same* functional Ψ but corresponding to different data $\{f, u_0\}$. We have

Theorem 6.3. *With the same assumptions as in Theorem* 5.1 *with* (6.2) *and* (6.3), *let* u (resp. $\hat{u}$) *be the solution given by Theorem* 5.1 *corresponding to* $\{f, u_0\}$ (resp. $\{\hat{f}, \hat{u}_0\}$). *We assume that*

(6.25) $$f \geqslant \hat{f} \quad \text{a.e.} \quad \text{in } Q, \qquad u_0 \geqslant \hat{u}_0 \quad \text{a.e.} \quad \text{in } \Omega$$

and that the functional Ψ *satisfies,* $\forall v, \hat{v} \in H^1(\Omega)$:

(6.26) $$\Psi(v) + \Psi(\hat{v}) - \Psi(\sup(v, \hat{v})) - \Psi(\inf(v, \hat{v})) \geqslant 0 .$$

Then

(6.27) $$u \geqslant \hat{u} \quad \text{a.e.} \quad \textit{in } Q .$$

Proof. The principle is the same as in the proof of Theorem 6.2. Using the same notation, we obtain:

$$-(\theta', \theta^+) - a(\theta, \theta^+) + \Psi(w) + \Psi(\hat{w}) - \Psi(u) - \Psi(\hat{u}) \geqslant (f - \hat{f}, \theta^+),$$

from which we conclude, taking (6.26) into account:

(6.28) $$(\theta', \theta^+) + a(\theta, \theta^+) + (f - \hat{f}, \theta^+) \leqslant 0 .$$

But according to (6.25), $(f - \hat{f}, \theta^+) \geqslant 0$, therefore (6.28) leads to

(6.29) $$(\theta', \theta^+) + a(\theta, \theta^+) \leqslant 0 .$$

But $\theta^+(0) = (\hat{u}_0 - u_0)^+ = 0$ (according to (6.25)) and therefore $\theta^+(t) = 0$ from which (6.27) follows. □

Example 6.3. If we assume Ψ to be given by (6.14), *then* (6.26) *holds* and, more precisely:

(6.30) $$\Psi(v) + \Psi(\hat{v}) - \Psi(\sup(v, \hat{v})) - \Psi(\inf(v, \hat{v})) = 0 .$$

Example 6.3.1. Applied to the problem of temperature control, this result shows—which, here again, is physically obvious—that if *one regulates in the same manner and if* u (resp. $\hat{u}$) *are the temperatures corresponding to a right hand side* f (resp. $\hat{f}$) *and an initial temperature* u_0 (resp. $\hat{u}_0$), then, for $f \geqslant \hat{f}$ and $u_0 \geqslant \hat{u}_0$, *we have:* $u \geqslant \hat{u}$ *almost everywhere.*

In particular, if ψ is given by Fig. 9 and if $u_0(x) \geqslant u_1(x)$ with $0 \leqslant u_1(x) \leqslant h_2(x)$, then $u(x,t) \geqslant u_1(x)$. □

7. Stationary Problems

We now return to the situation of Section 3.5, with

(7.1) $$a(u,v) = \int_\Omega a_{ij}(x) u_{,j} v_{,i} \, \mathrm{d}x,$$

where the $a_{ij} \in L^\infty(\Omega)$ satisfy (3.63).

We start with the case where $a(u,v)$ is replaced by $a(u,v) + c(u,v), c > 0$.

7.1. The Strictly Coercive Case

In this entire section, we take

(7.2) $$V = H^1(\Omega), \quad \Omega \text{ an open bounded set of } \mathbb{R}^n.$$

Then, under assumption (3.63), we have:

(7.3) $$a(v,v) \geqslant \int_\Omega v_{,i} v_{,i} \, \mathrm{d}x$$

but *we do not have* $a(v,v) \geqslant \alpha \|v\|^2_{H^1(\Omega)}$; moreover, $a(1,1) = 0$.

In contrast, we obviously have

(7.4) $$a(v,v) + c|v|^2 \geqslant \min(\alpha,c) \|v\|^2_{H^1(\Omega)} = \min(\alpha,c) \|v\|^2$$

where we set $|v|^2 = \int_\Omega v^2 \, \mathrm{d}x$ and $\|v\| = \|v\|_{H^1(\Omega)}$.

In this section, we take Ψ as in (3.10)[31]:

(7.5) $$\Psi(v) = \int_\Gamma \psi(v(x)) \, \mathrm{d}\Gamma$$

or, more generally,

(7.6) $$\Psi(v) = \int_\Gamma \psi(x; v(x)) \, \mathrm{d}\Gamma \quad \text{(cf. Remark 3.2).}$$

---

[31] Without restriction of generality. We have analogous results to the ones which follow, when Ψ is given by (3.11).

We make assumptions analogous to (5.3), (5.4), (5.5): we assume that there exists a family of functions Ψ_j, *differentiable* on V, such that

(7.7) $$\forall v \in V, \quad \Psi_j(v) \to \Psi(v),$$

(7.8) (identical with (5.4)) there exists a sequence φ_j bounded in V such that $\Psi'_j(\varphi_j) = 0 \ \forall j$,

(7.9) if $v_j \to v$ in V weakly, $\Psi_j(v_j) \leqslant$ constant, then $\liminf \Psi_j(v_j) \geqslant \Psi_j(v)$.

Theorem 7.1. *We make the assumptions* (3.63), (7.7), (7.8), (7.9). *Let* $c > 0$. *Let* $v \to (f, v)$ *be a linear form, continuous on* V. *There exists a unique* $u \in V$ *such that*

(7.10) $$a(u, v-u) + c(u, v-u) + \Psi(v) - \Psi(u) \geqslant (f, v-u) \quad \forall v \in V.$$

Remark 7.1. In the case where $a(u,v) = a(v,u)\ \forall u, v$, problem (7.10) is equivalent (see (3.73)) to finding u which minimizes the functional

(7.11) $$J(v) = \tfrac{1}{2}\left[a(v,v) + c\,|v|^2\right] + \Psi(v) - (f, v);$$

in this case, the existence proof for a solution follows immediately (without introducing Ψ_j). Indeed, the problem is equivalent to minimizing the functional J on the closed convex set (possibly identical with V) of those v for which $\Psi(v) < \infty$; now, the function $v \to J(v)$ is *convex, lower semicontinuous for the weak topology of* V and, finally, (due to the introduction of the term $c|v|^2$), $J(v) \to +\infty$ as $\|v\| \to \infty$. From this follows the existence of a solution. Furthermore, since the function $v \to J(v)$ is *strictly* convex, we also have uniqueness, from which the theorem follows for the symmetric case. □

Uniqueness proof. The uniqueness proof is essentially analogous to the one of Section 5.5. Let u and u_* be two possible solutions; if we take $v = u_*$ (resp. $v = u$) in (7.10) (resp. in the analogous inequality with respect to u_*) and add, it follows, setting $w = u - u_*$, that

$$-a(w, w) - c\,|w|^2 \geqslant 0$$

from which $w = 0$ because of (7.4). □

Sketch of the existence proof[32]. 1) First, we replace Ψ by Ψ_j in (7.10); we are then looking for a solution u_j of

(7.12) $$a(u_j, v-u_j) + c(u_j, v-u_j) + \Psi_j(v) - \Psi_j(u_j) \geqslant (f, v-u_j) \quad \forall v \in V.$$

[32] For the understanding of the remainder of the book, this proof is not important. For this reason, we only give the essential steps of the proof.

Since Ψ_j is *differentiable*, the *inequality* (7.12) is equivalent to the *equation*:

$$a(u_j, v) + c(u_j, v) + (\Phi_j(u_j), v) = (f, v) \qquad \forall v \in V \tag{7.13}$$

where we set

$$\Phi_j = \Psi'_j . \tag{7.14}$$

To solve (7.13), we use the Galerkin method.

Taking a "base" $w_1, \ldots, w_m, \ldots$ of V as in Section 5.6.1 (with $\varphi_j \in$ space spanned by w_1), we seek u_{jm} with

$$u_{jm} \in [w_1, \ldots, w_m] , \tag{7.15}$$

$$a(u_{jm}, w_k) + c(u_{jm}, w_k) + (\Phi_j(u_{jm}), w_k) = (f, w_k) , \qquad 1 \leqslant k \leqslant m . \tag{7.16}$$

The existence of a solution u_{jm} of (7.15), (7.16) follows from Brouwer's fixed point Theorem (see for example Lions [1], Lemma 4.3, p. 53).

A priori estimates for u_{jm}: we conclude from (7.16), taking into account that $\varphi_j = \xi w_1$ for suitable real ξ, that

$$\begin{aligned} &a(u_{jm}, u_{jm} - \varphi_j) + c(u_{jm}, u_{jm} - \varphi_j) + (\Phi_j(u_{jm}) - \Phi_j(\varphi_j), u_{jm} - \varphi_j) \\ &\qquad = (f, u_{jm} - \varphi_j) \quad (\text{since } \Phi_j(\varphi_j) = 0) . \end{aligned} \tag{7.17}$$

Since Φ_j is monotone, we conclude from (7.17) that

$$a(u_{jm}, u_{jm} - \varphi_j) + c(u_{jm}, u_{jm} - \varphi_j) \leqslant (f, u_{jm} - \varphi_j) \tag{7.18}$$

from which $(\alpha_1 = \min(\alpha, c))$

$$\alpha_1 \|u_{jm}\|^2 \leqslant c_1 \|u_{jm}\| + c_2$$

and therefore

$$u_{jm} \leqslant c_3 = \text{constant independent of } j \text{ and } m . \tag{7.19}$$

Due to (7.19), we can select a sequence u_{jm} from the $u_{j\mu}$ such that

$$u_{j\mu} \to u_j \quad \text{in } V \text{ weakly} . \tag{7.20}$$

Using *the monotonicity of* Φ_j, we show that (see Minty [1], Browder [1], Lions [1]):

$$\Phi_j(u_{j\mu}) \to \Phi_j(u_j) \quad \text{in } V' \text{ weakly} \tag{7.21}$$

and that u_j satisfies (7.13). Furthermore, we conclude from (7.19) that

$$\|u_j\| \leqslant c_3 . \tag{7.22}$$

According to (7.22), we can select from the u_j a sequence, again denoted by u_j, such that

$$u_j \to u \quad \text{in } V \text{ weakly}. \tag{7.23}$$

According to (7.12), where we fix $v = v_0$ such that $\Psi(v_0) < \infty$ (and therefore, according to (7.7), $\Psi_j(v_0) \leqslant$ constant, we have

$$\Psi_j(u_j) \leqslant \text{constant}. \tag{7.24}$$

Therefore, we have according to (7.9)

$$\liminf_j \Psi_j(u_j) \geqslant \Psi(u). \tag{7.25}$$

We write (7.12) in the form

$$a(u_j, v) + c(u_j, v) + \Psi_j(v) - (f, v - u_j) \geqslant a(u_j, u_j) + c|u_j|^2 + \Psi_j(u_j); \tag{7.26}$$

but $\liminf_j [a(u_j, u_j) + c|u_j|^2] \geqslant a(u, u) + c|u|^2$; thus we conclude from (7.26):

$$a(u, v) + c(u, v) + \Psi(v) - (f, v - u) \geqslant a(u, u) + c|u|^2 + \Psi(u)$$

from which (7.10) follows. □

Remark 7.2. The preceding proof also yields an *approximation result*: according to (7.23), *we approximate u by solutions u_j of the* (regularized and penalized) *equations* (7.13). We also note that remarks analogous to those of Section 5.4 apply here. □

7.2. Approximation of the Stationary Condition by the Solution of Problems of Evolution when $t \to +\infty$

We note that assumptions (5.3), (5.5) *imply* (7.7), (7.9) ((5.4) and (7.8) are identical); indeed, it is sufficient to take functions *independent of* t in (5.3), (5.5). We will prove

Theorem 7.2. *We make the same assumptions as in Theorem* 5.1 *with*

$$f(t) = f \quad \textit{independent of } t. \tag{7.27}$$

Let $u(t) = u$ be the solution of

$$\begin{aligned}(u'(t), v - u(t)) + a(u(t), v - u(t)) + c(u(t), v - u(t)) + \Psi(v) - \Psi(u(t))& \\ \geqslant (f, v - u(t)) \qquad \forall v \in V&,\end{aligned} \tag{7.28}$$

$$u(0) = u_0 \tag{7.29}$$

and let w be the solution given by Theorem 7.1, *i.e.*,

(7.30) $$a(w, v-w)+c(w, v-w)+\Psi(v)-\Psi(w) \geqslant (f, v-w) \qquad \forall v \in V.$$

Then

(7.31) $$u(t) \to w \quad \text{in} \quad H = L^2(\Omega) \quad \textit{when} \quad t \to +\infty$$

and more precisely

(7.32) $$|u(t)-w| \leqslant c_1 \exp(-c_2 t), \qquad c_i > 0.$$

Proof. Taking $v = w$ (resp. $u(t)$) in (7.28) (resp. (7.30)) and setting

$$m(t) = u(t) - w,$$

we find

$$-(m'(t), m(t)) - [a(m(t), m(t)) + c|m(t)|^2] \geqslant 0$$

from which

(7.33) $$\frac{\mathrm{d}}{\mathrm{d}t}|m(t)|^2 + 2\alpha_1 \|m(t)\|^2 \leqslant 0.$$

But $\|v\| \geqslant \mathrm{d}|v|$, therefore, we conclude from (7.33) that

$$\frac{\mathrm{d}}{\mathrm{d}t}(\exp(2\alpha_1 \mathrm{d}t)|m(t)|^2) \leqslant 0$$

from which

(7.34) $$|m(t)|^2 \leqslant |u_0 - w|^2 \exp(-2\alpha_1 \mathrm{d}t)$$

which proves (7.32) (if we choose suitable constants). □

Corollary 7.1. *For the solution of stationary problems, we have properties of positivity and of comparison of solutions analogous to those of Theorems* 6.1, 6.2, 6.3.

Proof. Indeed, we pass to the limit in t with the properties in Theorems 6.1, 6.2, 6.3. □

Remark 7.3. Naturally, we can also *directly* prove the properties given in the preceding corollary with the methods given in Theorems 6.1, 6.2, 6.3. □

7.3. The not Strictly Coercive Case

We are now going to examine the case where "$c = 0$". We limit ourselves to *the symmetric case*:

(7.35) $$a(u, v) = a(v, u) \qquad \forall u, v \in V$$

so that we are in the situation of Remark 7.1, with $c=0$, i.e., *the two following problems are equivalent*:

$$\text{(7.36)}\qquad \begin{aligned}&\text{find } u\in V \text{ such that}\\ &a(u,v-u)+\Psi(v)-\Psi(u)\geqslant (f,f-u),\end{aligned}$$

$$\text{(7.37)}\qquad \begin{aligned}&\text{find } u\in V \text{ minimizing}\\ &J(v)=\tfrac{1}{2}a(v,v)+\Psi(v)-(f,v).\end{aligned}$$

Since $a(1,1)=0$, it may *not be correct* that $J(v)\to+\infty$ when $\|v\|\to\infty$; *therefore, in general, we do not have existence* (moreover, see Remark 2.5 above). □

Let us first give a technical lemma which is essential for what follows:

Lemma 7.1. *There exist constants $\beta_i>0$ such that $\forall v\in H^1(\Omega)$ we have*:

$$\text{(7.38)}\qquad a(v,v)+\int_\Gamma |v|^2\,\mathrm{d}\Gamma\geqslant\beta_1\|v\|^2 \qquad (\|v\|=\|v\|_{H^1(\Omega)}),$$

$$\text{(7.39)}\qquad a(v,v)+\left(\int_\Gamma v\,\mathrm{d}\Gamma\right)^2\geqslant\beta_2\|v\|^2.$$

Proof. Let us prove for example (7.39); the proof of (7.38) is analogous in all points. The quantity

$$|||v|||=\left(a(v,v)+\left(\int_\Gamma v\,\mathrm{d}\Gamma\right)^2\right)^{1/2}$$

is a *norm* on V; indeed, it is obviously a semi-norm and if $|||v|||=0$, then $a(v,v)=0$, therefore, according to (7.3), $v=$constant and $\int_\Gamma v\,\mathrm{d}\Gamma=0$ and therefore $v=0$.

The space $H^1(\Omega)=V$ is *complete* with respect to the norm $|||\ \ |||$. Indeed, let v_m be a Cauchy sequence for this norm. Then, $\forall i$, $v_{m,i}$ is a Cauchy sequence in $L^2(\Omega)$ and

$$\text{(7.40)}\qquad \int_\Gamma v_m\,\mathrm{d}\Gamma\to\xi\in\mathbb{R}.$$

Since the $v_{m,i}$ converge in $L^2(\Omega)$, it follows from Deny-Lions [1] that we can find constants k_m such that

$$\text{(7.41)}\qquad \begin{aligned}&v_m+k_m\to v &&\text{in } L^2(\Omega)\\ &v_{m,i}\to v_{,i} &&\text{in } L^2(\Omega).\end{aligned}$$

It follows from (7.41) that

$$\int_\Gamma (v_m+k_m)\,\mathrm{d}\Gamma\to\int_\Gamma v\,\mathrm{d}\Gamma$$

which, together with (7.40), proves that $k_m\to k$ and therefore $v_m\to v$ in $H^1(\Omega)$.

Let us now consider the mapping I (=identity) of V taken with the norm $|||\ \ |||$ into V taken with the norm $\|\ \ \|$. This linear mapping has a *closed graph*, for the topology defined by $\|\ \ \|$ is finer than the one defined by $|||\ \ |||$. Since V

is a Banach space (in fact, a Hilbert space) for each of the norms, it follows from the *closed graph theorem* (see for example Bourbaki [1]) that the mapping is continuous, therefore there exists a constant c such that

$$\|v\| \leqslant c |||v|||$$

from which (7.39) follows. □

We immediately conclude from (7.39) that when $\psi(\lambda)$ *increases faster than linearly as* $\lambda \to \pm\infty$ (i.e., $\psi(\lambda)/\lambda \to \pm\infty$ when $\lambda \to \pm\infty$) (therefore, a fortiori, when $\psi(\lambda) = +\infty$ for sufficiently large $|\lambda|$), then $J(v) \to +\infty$ for $\|v\| \to \infty$, from which *the existence* of a solution follows immediately.

The interesting case—and moreover, the case which is useful for applications—thus is the case where $\psi(\lambda)$ is of "*linear growth*" for $\lambda \to +\infty$ or (and) $\lambda \to -\infty$. We introduce

$$\psi_{\pm} = \lim_{\lambda \to \pm\infty} \psi(\lambda)/\lambda \tag{7.42}$$

assuming that *at least one* of the numbers ψ_+ or ψ_- is *finite*[33]; in all cases

$$\psi_- < 0 < \psi_+ . \tag{7.43}$$

Remark 7.4. Similarly, if $\psi(\lambda) = \psi(x;\lambda)$:

$$\psi_{\pm}(x) = \lim_{\lambda \to \pm\infty} \psi(x;\lambda)/\lambda \tag{7.44}$$

where we assume these limits to be *uniform* for $x \in \Gamma$. □

7.3.1. Necessary Conditions for the Existence of Solutions

We will prove

Theorem 7.3. *We assume* (7.35), (7.42). *In order that a solution of problem* (7.36) *or* (7.37) *exists, it is necessary that*

$$\int_\Gamma \psi_- \, d\Gamma \leqslant (f,1) \leqslant \int_\Gamma \psi_+ \, d\Gamma . \tag{7.45}$$

Remark 7.5. Naturally, (7.45) is equivalent to

$$|\Gamma|\psi_- \leqslant (f,1) \leqslant |\Gamma|\psi_+ , \qquad |\Gamma| = \text{measure of } \Gamma .$$

Condition (7.45) can be extended to the case (7.44) in the form

$$\int_\Gamma \psi_-(x) \, d\Gamma \leqslant (f,1) \leqslant \int_\Gamma \psi_+(x) \, d\Gamma . \quad \square \tag{7.45bis}$$

---

[33] *In the following, the condition concerning* ψ_+ (resp. ψ_-) *becomes irrelevant when* $\psi_+ = +\infty$ (resp. $\psi_- = -\infty$).

Remark 7.6. We can take $f \in V'$ given by

(7.46) $$(f, v) = \int_\Omega f_0 v \, dx + \int_\Gamma f_1 v \, d\Gamma$$

where

$$f_0 \in L^2(\Omega), \quad f_1 \in L^2(\Gamma) \quad (\text{and even } \in H^{-1/2}(\Gamma)).$$

Then

$$(f, 1) = \int_\Omega f_0 \, dx + \int_\Gamma f_1 \, d\Gamma. \quad \square$$

Remark 7.7. In the situation of Remark 2.5, we have

$$\psi(\lambda) = \begin{cases} +\infty & \text{when } \lambda < h \\ 0 & \text{when } \lambda \geq h \end{cases}$$

therefore $\psi_+ = 0$, $\psi_- = -\infty$ and (7.45) reduces to

(7.47) $$(f, 1) \leq 0.$$

If we take (7.46) with $f_1 = 0$, we obtain

$$\int_\Omega f_0 \, dx \leq 0. \quad \square$$

Proof of Theorem 7.3. Let us take $v = \lambda \in \mathbb{R}$; then

$$J(\lambda) = \Psi(\lambda) - \lambda(f, 1) = \lambda[\lambda^{-1} \int_\Gamma \psi(\lambda) d\Gamma - (f, 1)]$$

and $J(\lambda)$ is bounded below for $\lambda \to \pm\infty$ only if (7.45) is satisfied.

7.3.2. Sufficient Conditions for the Existence of a Solution

Theorem 7.4. *We assume* (7.35), (7.42) *and also that* (*strict* conditions corresponding to (7.45)):

(7.48) $$\int_\Gamma \psi_- d\Gamma < (f, 1) < \int_\Gamma \psi_+ d\Gamma,$$

and that

(7.49) *ψ is a Lipschitz continuous function on the set where $\psi(\lambda) \neq \infty$. Then there exists a solution u of problem* (7.36) *or* (7.37).

Proof. For $v \in H^1(\Omega)$, we put

(7.50) $$\bar{v} = |\Gamma|^{-1} \int v \, d\Gamma$$

and

(7.51) $$\tilde{v} = v - \bar{v}.$$

Then $\int_\Gamma \tilde{v}\,d\Gamma = 0$ and (7.39) applied to $\tilde{v}$ yields

(7.52) $$a(\tilde{v},\tilde{v}) \geqslant \beta_2 \|\tilde{v}\|^2 .$$

Making use of (7.51), we have:

(7.53) $$J(v) = \tfrac{1}{2} a(\tilde{v},\tilde{v}) - (f,\tilde{v}) + \Psi(\bar{v}+\tilde{v}) - \bar{v}(f,1) .$$

We will show (which is sufficient for proving the Theorem) that

(7.54) $$J(v) \to +\infty \quad \text{when} \quad \|v\| \to \infty .$$

We carry out the proof for the case where

(7.55) $$\psi(\lambda) = \infty \quad \text{when} \quad \lambda < h , \qquad \psi_+ > 0 \quad \text{finite};$$

the proof is analogous when ψ_- and ψ_+ are finite.

In accordance with (7.55), we limit ourselves to functions $v \in K$, where K is defined by

(7.56) $$K = \{v \mid v(x) \geqslant h \text{ almost everywhere on } \Gamma\} .$$

We note that then

(7.57) $$\bar{v} \geqslant h .$$

Since ψ is Lipschitz continuous, there exists a constant c such that

$$\psi(\bar{v} + \tilde{v}(x)) - \psi(\bar{v}) \geqslant -c|\tilde{v}(x)|$$

and consequently—the c's denoting various constants—

(7.58) $$\Psi(\bar{v}+\tilde{v}) - \Psi(\bar{v}) \geqslant -c\|\tilde{v}\| .$$

We conclude from (7.52) and (7.58) that

(7.59) $$J(v) \geqslant \tfrac{1}{2}\beta_2\|\tilde{v}\|^2 - c\|\tilde{v}\| + \bar{v}\left[\int_\Gamma (\psi(\bar{v})/\bar{v})\,d\Gamma - (f,1)\right] .$$

But, according to (7.48), we have:

$$\int_\Gamma (\psi(\bar{v})/\bar{v})\,d\Gamma - (f,1) > 0$$

for $\bar{v}$ sufficiently large, from which (7.54) follows. □

7.3.3. The Problem of Uniqueness under Assumption (7.48)

We begin with a *counter-example* showing *that the solution is not necessarily unique.* Indeed, let us take ψ as in Example 3.2 (Fig. 7). Then:

(7.60) $$\psi_+ = g_2 , \qquad \psi_- = g_1 .$$

Let f be given with

(7.61) $$(f,1)=0;$$

then (7.48) holds.

Then there exists a solution u of the Neumann problem

(7.62) $$a(u,v)=(f,v) \quad \forall v \in H_1(\Omega)$$

and all the solutions of (7.62) are given by $u+c$. Therefore

$$Au=f, \quad \partial u/\partial v_A=0 \quad \text{on } \Gamma.$$

Let us assume that u is *bounded* on Γ and let us assume that the constants h_1 and h_2 which occur in ψ, satisfy

(7.63) $$h_1 \leqslant u(x) \leqslant h_2, \quad x \in \Gamma.$$

Then, *for every constant c, such that* $w=u+c$ *satisfies*

$$h_1 \leqslant w(x) \leqslant h_2, \quad x \in \Gamma,$$

we have

$$a(w,v-w)+\Psi(v)-\Psi(w) \geqslant (f,v-w).$$

Therefore, Problem (7.36) *can have an infinity of solutions.* □

Thus, the problem of possible uniqueness can only be solved by examining each particular case. We give an example for uniqueness.

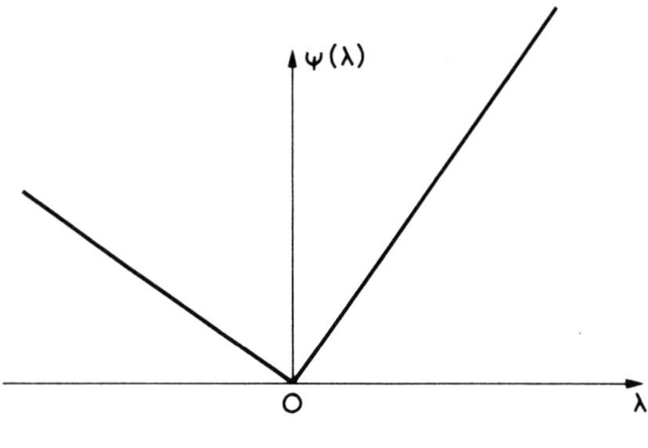

Fig. 14

We will assume that ψ is given by (Fig. 14)

(7.64) $$\psi(\lambda)=\begin{cases} g_1\lambda & \text{when } \lambda \leqslant 0 \\ g_2\lambda & \text{when } \lambda \geqslant 0. \end{cases}$$

Again we have (7.60). We will prove

Theorem 7.5. *We assume that* (7.35) *holds and that* ψ *is given by* (7.64). *We assume that* (7.48) *holds*[34] *and that*

$$(f,v)=\int_\Omega f_0 v\,\mathrm{d}x\,, \qquad f_0\in L^2(\Omega)\,[35]\,.$$

Then we have uniqueness.

Proof[36]. If u and u_* are two possible solutions of (7.36), then, setting $v=u_*$ (resp. $v=u$) in (7.36), (resp. the analogous inequality for u_*), we obtain by addition

$$a(u-u_*,u-u_*)\leqslant 0\,,$$

therefore

$$u-u_*=c\,[37]\,.$$

Let then u and $u+c$ be two possible solutions. We have to show that $c=0$. The boundary conditions (see (3.69)) on Γ are:

$$\text{(7.65)}\qquad \begin{aligned} &u>0\Rightarrow \partial u/\partial \nu_A+g_2=0\,,\\ &u=0\Rightarrow -g_2<\partial u/\partial \nu_A<-g_1\,,\\ &u<0\Rightarrow \partial u/\partial \nu_A+g_1=0\,, \end{aligned}$$

and

$$\text{(7.66)}\qquad \begin{aligned} &u+c>0\Rightarrow \partial u/\partial \nu_A+g_2=0\,,\\ &u+c=0\Rightarrow -g_2<\partial u/\partial \nu_A<-g_1\,,\\ &u+c<0\Rightarrow \partial u/\partial \nu_A+g_1=0\,. \end{aligned}$$

We first verify that the following alternatives are either impossible or imply $c=0$:

$$\text{(7.67)}\qquad \partial u/\partial \nu_A+g_1=0 \quad \text{a.e. on } \Gamma \;\; (\text{or } \partial u/\partial \nu_A+g_2=0 \text{ a.e. on } \Gamma)\,,$$

$$\text{(7.68)}\qquad -g_2<\partial u/\partial \nu_A<-g_1 \quad \text{on} \quad E\subset\Gamma\,, \quad \text{measure}(E)>0.$$

Indeed, from $Au=f_0$ and (7.67), we conclude, by means of Green's formula, that

$$(f_0,1)=\int_\Gamma g_1\,\mathrm{d}\Gamma\,,$$

---

[34] I.e. $g_1|\Gamma|<(f,1)<g_2|\Gamma|$.
[35] Solely for the purpose of simplifying the exposition.
[36] This proof can be passed over.
[37] This is true *in general*, i.e., independent of the fact that ψ is given by (7.64).

which is impossible according to (7.48); if (7.68) holds, then, according to (7.65), (7.66), we have:

$$u = u + c = 0 \quad \text{on } E\text{, therefore} \quad c = 0\,.$$

The only case that remains to be examined is then[38] the one in which

$$(7.69)\quad \begin{cases} \partial u/\partial \nu_A = -g_1 & \text{on } \Gamma_1, \quad \text{measure}(\Gamma_1) > 0 \\ \partial u/\partial \nu_A = -g_2 & \text{on } \Gamma_2, \quad \text{measure}(\Gamma_2) > 0 \\ \Gamma = \Gamma_1 \cup \Gamma_2 & \text{(except for a set of measure zero)} \end{cases}$$

and we have to show that (7.69) implies $c = 0$.

According to (7.65), (7.66), we have:

$$u < 0\,, \quad u + c < 0 \quad \text{on} \quad \Gamma_1\,,$$

$$u > 0\,, \quad u + c > 0 \quad \text{on} \quad \Gamma_2\,.$$

Let us assume $c > 0$ (the argument is the same for $c < 0$); then

$$(7.70)\qquad u < -c \quad \text{on } \Gamma_1\,, \qquad u > 0 \quad \text{on } \Gamma_2\,.$$

But, according to (7.69), $\partial u/\partial \nu_A \in L^2(\Gamma)$ which, according to the theory of non-homogeneous boundary value problems, implies (see Lions-Magenes [1][39])

$$(7.71)\qquad u \in H^1(\Gamma)\,.$$

We are now going to show[40] that (7.70), (7.71) imply $c = 0$.

Let P be the projection operator from $\mathbb{R} \to [-c, 0]$; if $\varphi \in L^2(\Gamma)$, we denote by P_φ the function $x \to P(\varphi(x))$. The mapping P transforms $H^1(\Gamma)$ into itself (and is a contraction). Now, according to (7.70), we have:

$$(7.72)\qquad Pu = \{-c \text{ on } \Gamma_1,\ 0 \text{ on } \Gamma_2\}\,.$$

But (7.62) implies that $c = 0$, since $\Gamma_1 \cup \Gamma_2 = \Gamma$ (except for a set of measure zero); indeed, the first order derivatives of Pu coincide when taken either in the sense of distributions on Γ or in the ordinary sense (see for example Deny-Lions [1]); according to (7.72), the ordinary first derivatives of Pu are zero almost everywhere, therefore $Pu = \text{const.}$ on Γ, therefore $c = 0$. □

---

[38] This is the analogue to the Bang Bang of optimal control theory.

[39] We assume that Γ and the coefficients a_{ij} are sufficiently regular.

[40] The argument which follows, pointed out to us by J. Deny, has been taken from the theory of "Dirichlet spaces"; see A. Beurling and J. Deny [1].

7.3.4. The Limiting Cases in (7.48)[41]

In the context of Theorem 7.5, we now examine the case in which equality holds in (7.48). We have:

Theorem 7.6. *We assume that* (7.35) *holds, that* ψ *is given by* (7.64) *and that*

$$(f,v)=\int_\Omega f_0 v\,\mathrm{d}x\,,\qquad f_0\in L^2(\Omega)\,.$$

We also assume that

$$(f,1)=\int_\Gamma \psi_-\,\mathrm{d}\Gamma\qquad (=g_1|\Gamma|) \tag{7.73}$$

(resp. *that*

$$(f,1)=\int_\Gamma \psi_+\,\mathrm{d}\Gamma\qquad (=g_2|\Gamma|))\,. \tag{7.74}$$

Let w_1 (resp. w_2) *be the solution*[42] *of:*

$$A w_1=f_0\,,\qquad \partial w_1/\partial \nu_A+g_1=0\,,\qquad \int_\Gamma w_1\,\mathrm{d}\Gamma=0\,^{43} \tag{7.75}$$

(resp. of

$$A w_2=f_0\,,\qquad \partial w_2/\partial \nu_A+g_2=0\,,\qquad \int_\Gamma w_2\,\mathrm{d}\Gamma=0)\,. \tag{7.76}$$

Problem (7.36) (*or* (7.37)) *has a solution if and only if*

$$w_1 \text{ (resp. } w_2\text{) is bounded above (resp. below) on } \Gamma\,. \tag{7.77}$$

If (7.77) *holds, all solutions are given by*

$$u=w_i+c\,, \tag{7.78}$$

where c is any constant such that

$$w_1+c\leqslant 0\quad \text{(resp. } w_2+c\geqslant 0\text{) on } \Gamma\,. \tag{7.79}$$

Remark 7.8. Since $f_0\in L^2(\Omega)$ and $g_i\in H^{1/2}(\Gamma)$, we have: $w_i\in H^2(\Omega)$, therefore $w_i|_\Gamma\in H^{3/2}(\Gamma)$ and one can prove (see J. Peetre [1]) that this implies

$$w_i\in L^\infty(\Gamma)\quad \text{—and even } w_i\in\mathscr{C}^0(\Gamma)\text{—}, \tag{7.80}$$

if $\dfrac{1}{2}-\dfrac{3}{2(n-1)}<0$, i.e., $n<4$.

---

[41] This section may be passed over.

[42] w_1 (resp. w_2) exists due to (7.73) (resp. (7.74)).

[43] A condition which determines w_1 uniquely and which can be replaced by any other linear condition determining w_1 uniquely.

Therefore, *assumption* (7.77) *is always satisfied if* $n \leqslant 3$—i.e., in all cases arising in practice. ☐

Proof of Theorem 7.6. We prove the Theorem for the case (7.73). If u is a possible solution, we have:

$$Au = f_0 \tag{7.81}$$

and (see (7.65)) $-g_2 \leqslant \partial u/\partial \nu_A \leqslant -g_1$.

But (7.81) and Green's formula imply

$$(f,1) + \int_\Gamma (\partial u/\partial \nu_A)\, d\Gamma = 0$$

which, together with (7.73), has the consequence that necessarily

$$\partial u/\partial \nu_A + g_1 = 0 \quad \text{almost everywhere on } \Gamma .$$

Comparing with (7.75), it follows that necessarily

$$u = w_1 + c$$

that, actually, is a solution only if (see (7.65)) $u \leqslant 0$ on Γ, i.e., $w_1 \leqslant -c$, where c is a suitable constant. From this the result follows. ☐

8. Commentary

The bibliography for the recapitulation of Mechanics in Section 1 is given in the text.

The problems of heat control and of semi-permeable walls formulated in Section 2 were introduced by the authors: Duvaut-Lions [1,2].

The problems of variational inequalities of evolution in the *parabolic* case were introduced in Lions-Stampacchia [1], in a less general form than the one given in Section 3 (which is indispensable for the solution of the problems posed in Section 2). We have attempted to give as elementary a presentation as possible, both of the theory and, in Section 5, of the solution of the problems. For other aspects of the theory, see H. Brézis [1,2] and J. L. Lions [1].

The summaries in Section 4 provide the indispensable minimum for the understanding of this book. We might add to the bibliography indicated in the text the work of S. L. Sobolev [1] and the book by J. Nečas [1].

The methods of Section 6 are modifications of those of Y. Haugazeau [1]. Other results can be found in M. Schatzman [1].

The results of Sections 7.2 and 7.3 are given in detail here for the first time. For additional properties concerning regularity of solutions, see H. Brézis [2].

Chapter II

Problems of Heat Control

Orientation. In this chapter, it is assumed that the reader is familiar with Sections 1–5 of Chapter I.

The reader who is not especially interested in the problems of the Theory of Heat can pass over this (short) chapter.

1. Heat Control

We examine the field of temperatures in the interior of an open region Ω with the boundary Γ, a region which is occupied by a continuous medium. In all problems considered in this section, the temperature at certain points is "controlled" to change in a certain way as t increases.

We will distinguish two classes of controls depending on the control being *instantaneous* or *delayed.*

1.1. Instantaneous Control

1.1.1. Temperature Control at the Boundary

If we "control" the temperature $u(x,t)$ solely at the boundary, the function u satsfies the heat equation in the interior:

(1.1) $$\partial u/\partial t-\Delta u=f, \quad x\in\Omega, \quad t\in]0,T]$$

or, more generally,

(1.2) $$\partial u/\partial t+Au=f,$$

with

(1.3) $$A\varphi=-(a_{ij}(x)\varphi_{,j})_{,i},$$

where the functions a_{ij} satisfy:

(1.4)
$$a_{ij} \in L^\infty(\Omega), \quad a_{ij}=a_{ji} \quad \forall i,j$$
$$a_{ij}(x)\xi_i\xi_j \geqslant \alpha\xi_i\xi_i \quad \forall \xi_i \in R.$$

Furthermore, the *initial temperature* is given:

(1.5)
$$u(x,0)=u_0(x).$$

The *control* then appears in the *boundary conditions.*

We give two examples:

Example 1.1 *(Thin walls).* The temperature on Γ is made to rise by injections of heat through the wall which imposes the following boundary conditions:

(1.6)
$$\partial u(x,t)/\partial t>0 \Rightarrow \partial u(x,t)/\partial v_A=0, \text{ [1]}$$
$$\partial u(x,t)/\partial t=0 \Rightarrow \partial u(x,t)/\partial v_A \geqslant 0. \quad \square$$

Example 1.2 *(Thick walls).* When the wall is thick, the heat flow can no longer be instantaneous and the control is effected by the following conditions:

(1.7)
$$\partial u(x,t)/\partial t>0 \Rightarrow \partial u(x,t)/\partial v_A=0,$$
$$\partial u(x,t)/\partial t \leqslant 0 \Rightarrow \partial u(x,t)/\partial v_A=-k\,\partial u(x,t)/\partial t$$

where k is a positive scalar which depends on the wall and on the control system. □

Remark 1.1. *One* of the examples of heat control (Sec. I.2.3) led to the boundary conditions:

(1.8)
$$u>0 \Rightarrow \partial u/\partial v_A=0$$
$$u=0 \Rightarrow \partial u/\partial v_A \geqslant 0.$$

Comparing with (1.6), we see that *these are boundary conditions in the form of inequalities that cannot be reduced to each other.* □

Remark 1.2. If, in Example 1.2, $k \to +\infty$, conditions (1.7) in the limit (formally) yield conditions (1.6). Indeed, we shall see (in Sec. 3) that this is justified and that the solution u_k for Example 1.2 converges (in a suitable topology) toward the solution u of Example 1.1, when $k \to +\infty$. □

Remark 1.3. Quite obviously, the temperature can be controlled so that it will decrease!

[1] $\partial\varphi/\partial v_A = a_{ij}(x)\varphi_{,j}\cos(n,x_i)$, where the normal n to Γ points *toward the exterior of* Ω.

Remark 1.4. We could also impose a control on only a part of Γ, say Γ_1. If $\Gamma=\Gamma_1\cup\Gamma_2$, we may, for example, arrive at the following conditions:

$$u=0 \quad \text{if} \quad x\in\Gamma_2, \quad t>0, \tag{1.9}$$

and, on Γ_1, at conditions analogous to (1.6) or (1.7) depending on whether the wall Γ_1 is thin or thick. □

1.1.2. Temperature Control in the Interior

We can equally well imagine a control system which, by injection of a volume density of heat g *in* Ω, causes the temperature $u=u(x,t)$ *in* Ω to rise.

Therefore, for example:

$$\begin{aligned} &\partial u(x,t)/\partial t>0 \Rightarrow g(x,t)=0,\\ &\partial u(x,t)/\partial t\leqslant 0 \Rightarrow g(x,t)=-k\,\partial u(x,t)/\partial t, \qquad k>0. \end{aligned} \tag{1.10}$$

The field of temperatures u is then a solution of

$$\partial u/\partial t+Au=f+g \quad \text{in} \quad Q=\Omega\times]0,T[, \tag{1.11}$$

still retaining (1.5), with the boundary condition, *for example:*

$$u(x,t)=\theta(x,t)=\text{temperature imposed on } \Sigma=\Gamma\times]0,T[. \tag{1.12}$$

1.1.3. Properties of the Solutions

It is physically "evident" that in the case (1.6), if $f\geqslant 0$ and $u_0\geqslant 0$, the temperature u is positive in Q.

Indeed, we will prove this property (and others of this type) in Section 4.

1.1.4. Other Controls

We can imagine many other controls. For example, we can control the temperature at the boundary so that it will rise, but "not too fast"; this leads, for example, to the conditions:

$$\begin{aligned} 0<\partial u(x,t)/\partial t<1 &\Rightarrow \partial u(x,t)/\partial \nu_A=0,\\ \partial u(x,t)/\partial t=0 &\Rightarrow \partial u(x,t)/\partial \nu_A\geqslant 0,\\ \partial u(x,t)/\partial t=1 &\Rightarrow \partial u(x,t)/\partial \nu_A\leqslant 0, \end{aligned} \tag{1.13}$$

In Section 2, we will give a general formulation comprising all these cases (and others). □

1.2. Delayed Control

Let us now assume that the control mechanism controls the boundary temperatures only in discrete fashion, i. e., at the instants $n\tau$, where n is an integer and

τ is a finite interval of time. Then, if one makes adjustments in the controls to increase the temperatures, one is led to the boundary conditions:

$$\begin{aligned} &u(x,t)-u(x,t-\tau)>0 \Rightarrow \partial u(x,t)/\partial v_A=0, \\ &u(x,t)-u(x,t-\tau)\leqslant 0 \Rightarrow \partial u(x,t)/\partial v_A=-k[u(x,t)-(x,t-\tau)]. \quad \square \end{aligned} \tag{1.14}$$

Remark 1.5. The "initial condition" must now be replaced by an "extended initial condition", for example:

$$u(x,t)=u_0(x) \quad (\text{or } u_0(x,t)) \quad \text{in} \quad -\tau\leqslant t\leqslant 0. \quad \square \tag{1.15}$$

Remark 1.6. Obviously, analogous conditions can be devised *for the interior of* Ω. $\square$

2. Variational Formulation of Control Problems

2.1. Notation

As in Chapter I, we set:

$$a(u,v)=\int_\Omega a_{ij}u_{,j}v_{,i}\,\mathrm{d}x. \tag{2.1}$$

We observe that

$$a(u,v)=a(v,u). \tag{2.2}$$

As in Section I.3.1, we introduce the functions $\psi=\psi(\lambda)$ and the functionals

$$\Psi(v)=\int_\Gamma \psi(v)\,\mathrm{d}\Gamma \tag{2.3}$$

or

$$\Psi(v)=\int_\Omega \psi(v)\,\mathrm{d}x. \tag{2.4}$$

2.2. Variational Inequalities

2.2.1. Instantaneous Control

We seek a function $u=u(x,t)$ such that[2] $u'(t)\in H^1(\Omega)$ and

$$(u'(t),v-u'(t))+a(u(t),v-u'(t))+\Psi(v)-\Psi(u'(t))\geqslant(f(t),v-u'(t)) \quad \forall v\in H^1(\Omega) \tag{2.5}$$

[2] $u(t)=u(\cdot,t)$, $u'(t)=\partial u(\cdot,t)\partial t$.

with

(2.6) $u(0)=u_0$.

Remark 2.1. In Section 3, we will precisely state the properties imposed on the function $t \to u(t)$. □

Remark 2.2. In the case where Ψ is given by (2.4), we could replace $H^1(\Omega)$ by $H_0^1(\Omega)$ (cf. Chap. I). We also can (to solve the problem considered in Remark 1.4) replace $H^1(\Omega)$ be the space

(2.7) $V=\{v|\ v \in H^1(\Omega),\ v=0 \text{ on } \Gamma_2\}$. □

Remark 2.3. In we compare (2.5) with (3.14) of Chap. I, we see that $v-u(t)$, which appears in (3.14), is here replaced by $v-u'(t)$, and that the expression $\Psi(v)-\Psi(u(t))$, which appears in (3.14), is here replaced by $\Psi(v)-\Psi(u'(t))$.

Examples are given in Section 2.3 below. □

2.2.2. Delayed Control

We set:

(2.8) $Mu(t)=[u(t)-u(t-\tau)]/\tau$.

We seek a function u such that $u'(t) \in H^1(\Omega)$ and

(2.9)
$$(u'(t), v-Mu(t))+a(u(t), v-Mu(t))+\Psi(v)-\Psi(Mu(t)) \geqslant (f(t), v-Mu(t)) \qquad \forall v \in H^1(\Omega),$$

with

(2.10) $u(t)=u_0(t), \qquad -\tau \leqslant t \leqslant 0$. □

There exist variations analogous to those indicated in Remark 2.2.

2.3. Examples

2.3.1. The Function ψ of Type 1[3]

We introduce:

(2.11) $\Phi(\lambda)=\mathrm{d}\psi(\lambda)/\mathrm{d}\lambda$

and verify, exactly as in Section I.3.3.1, that (in the case (2.3)) (2.5) *is equivalent to:*

(2.12) $(u'(t), v)+a(u(t), v)+\int_\Gamma \Phi(u(t))\, v\, \mathrm{d}\Gamma=(f(t), v) \qquad \forall v \in H^1(\Omega)$

[3] Cf. definitions in Sec. I.3.

or, in the case (2.4), to

$$(2.13)\qquad (u'(t),v)+a(u(t),v)+\int_\Omega \Phi(u(t))\,v\,\mathrm{d}x=(f(t),v)\quad \forall v\in H_0^1(\Omega)\ \text{(for example)}.\ \square$$

Interpretation of (2.12). We proceed exactly as in Section I.3.3.1, and find that (2.5), (2.6) are equivalent to:

$$(2.14)\qquad \begin{aligned} &\partial u/\partial t+Au=f && \text{in}\quad Q=\Omega\times]0,T[,\\ &\partial u/\partial v_A+\Phi(u')=0 && \text{on}\quad \Sigma=\Gamma\times]0,T[,\\ &u(x,0)=u_0(x), && x\in\Omega. \end{aligned}$$

Interpretation of (2.13). We find:

$$(2.15)\qquad \begin{aligned} &\partial u/\partial t+Au+\Phi(\partial u/\partial t)=f \quad \text{in}\quad Q,\\ &u=0\quad \text{on}\quad \Sigma,\\ &u(x,0)=u_0(x),\qquad x\in\Omega.\ \square \end{aligned}$$

Example 2.1. Let us take for ψ:

$$(2.16)\qquad \psi(\lambda)=\left|\begin{array}{ll} \frac{1}{2}k\lambda^2 & \text{if}\quad \lambda\leqslant 0,\\ 0 & \text{if}\quad \lambda\geqslant 0. \end{array}\right.$$

Then the boundary conditions of (2.14) are equivalent to:

$$\begin{aligned} &\partial u/\partial t\geqslant 0\Rightarrow \partial u/\partial v_A=0,\\ &\partial u/\partial t\leqslant 0\Rightarrow \partial u/\partial v_A+k\,\partial u/\partial t=0, \end{aligned}$$

i. e., conditions (1.7). $\square$

The case of delayed control. We verify, always by the method of Section I.3, that (2.9) is equivalent:

in the case (2.3), *to*

$$(2.17)\qquad (u'(t),v)+a(u(t),v)+\int_\Gamma \Phi(Mu(t))\,v\,\mathrm{d}\Gamma=(f(t),v)\qquad \forall v\in H^1(\Omega),$$

in the case (2.4), *to*

$$(2.18)\qquad (u'(t),v)+a(u(t),v)+\int_\Omega \Phi(Mu(t))\,v\,\mathrm{d}x=(f(t),v)\qquad \forall v\in H_0^1(\Omega);$$

for example, the interpretation of (2.17), together with (2.10), is

$$(2.19)\qquad \begin{aligned} &\partial u/\partial t+Au=f\quad \text{in}\ Q,\\ &\frac{\partial u}{\partial v_A}(t)+\Phi\left(\frac{u(t)-u(t-\tau)}{\tau}\right)=0\quad \text{on}\quad \Sigma,\\ &u(x,t)=u_0(x,t),\qquad -\tau\leqslant t\leqslant 0.\ \square \end{aligned}$$

2.3.2. The Function ψ of Type 2

Again we introduce (cf. Sec. I.3.3.2) the derivative $\Phi(\lambda)$ of $\psi(\lambda)$, which may be *multiple-valued.*

We are then led to the following interpretations:

in the case (2.3), the problem is equivalent to finding a solution u of

$$\begin{aligned} &\partial u/\partial t + Au = f \quad \text{in} \quad Q, \\ &-\partial u(t)/\partial v_A \in \Phi(\partial u(t)/\partial t) \quad \text{on} \quad \Sigma, \\ &u(x,0) = u_0(x), \qquad x \in \Omega, \end{aligned} \tag{2.20}$$

and, *in the case* (2.4), the problem is equivalent to:

$$\begin{aligned} &-\partial u(t)/\partial t + Au(t) - f(t) \in \Phi(\partial u(t)/\partial t), \\ &u = 0 \quad \text{on} \quad \Sigma, \\ &u(x,0) = u_0(x), \qquad x \in \Omega. \quad \square \end{aligned} \tag{2.21}$$

Example 2.2. Let us take ψ as defined in Example 3.2, Chapter I. Then $\Phi(\lambda)$ is given by (3.42), Chapter I, and the boundary conditions of (2.20) become:

$$\begin{aligned} &-\partial u(t)/\partial v_A = g_1 && \text{if} \quad \partial u(t)/\partial t \leqslant h_1, \\ &g_1 \leqslant -\partial u(t)/\partial v_A \leqslant 0 && \text{if} \quad \partial u(t)/\partial t = h_1, \\ &\partial u(t)/\partial v_A = 0 && \text{if} \quad h_1 < \partial u(t)/\partial t < h_2, \\ &0 \leqslant -\partial u(t)/\partial v_A \leqslant g_2 && \text{if} \quad \partial u(t)/\partial t = h_2, \\ &-\partial u(t)/\partial v_A = g_2 && \text{if} \quad \partial u(t)/\partial t > h_2. \quad \square \end{aligned} \tag{2.22}$$

2.3.3. The Function ψ of Type 3

We now introduce (as in (3.44), Chap. I):

$$K = \{v \mid v \in H^1(\Omega),\ \Psi(v) < \infty\}. \tag{2.23}$$

Then inequality (2.5) is equivalent to:

$$\begin{aligned} &u'(t) \in K \\ &(u'(t), v - u'(t)) + a(u(t), v - u'(t)) + \Psi(v) - \Psi(u'(t)) \geqslant (f(t), v - u'(t)) \quad \forall v \in K, \end{aligned} \tag{2.24}$$

or again (cf. Sec. I.3.3.3):

$$\begin{aligned} &u'(t) \in K, \\ &(u'(t), v - u'(t)) + a(u(t), v - u'(t)) + (\chi, v - u'(t)) \geqslant (f(t), v - u'(t)) \qquad \forall v \in K, \\ &\text{and where } \chi \text{ is in the set of elements such that} \\ &\Psi(v) - \Psi(u'(t)) - (\chi, v - u'(t)) \geqslant 0 \qquad \forall v \in K. \end{aligned} \tag{2.25}$$

When Ψ is defined by (2.3), the problem is equivalent to:

$$(2.26)\qquad \begin{aligned} &\partial u/\partial t + Au = f \quad \text{in} \quad Q,\\ &\partial u(t)/\partial t \in K, \quad \int_\Gamma (\partial u(t)/\partial t + \chi)(v - \partial u(t)/\partial t)\,\mathrm{d}\Gamma \geqslant 0 \qquad \forall v \in K,\\ &u(0) = u_0. \end{aligned}$$

Example 2.3. Let us take ψ as defined by

$$(2.27)\qquad \psi(\lambda) = \begin{vmatrix} 0 & \text{on} \quad [h_1, h_2], \\ +\infty & \text{elsewhere.} \end{vmatrix}$$

Then $K = \{v| \quad h_1 \leqslant v \leqslant h_2 \text{ on } \Gamma\}$ and we can state (2.26) explicitly in the following manner (cf. Chap. I, (3.57)):

$$(2.28)\qquad \begin{aligned} &\partial u/\partial t + Au = f \quad \text{in} \quad Q,\\ &h_1 \leqslant \partial u(t)/\partial t \leqslant h_2 \quad \text{on} \quad \Sigma,\\ &\partial u(t)/\partial t = h_1 \Rightarrow \partial u(t)/\partial v_A \geqslant 0,\\ &h_1 < \partial u(t)/\partial t < h_2 \Rightarrow \partial u(t)/\partial v_A = 0,\\ &\partial u(t)/\partial t = h_2 \Rightarrow \partial u(t)/\partial v_A \leqslant 0,\\ &u(x,0) = u_0(x), \qquad x \in \Omega. \end{aligned}$$

Special case 1

$$h_1 = 0, \qquad h_2 = +\infty;$$

we obtain conditions (1.6).

Special case 2

$$h_1 = 0, \qquad h_2 = 1;$$

we obtain conditions (1.13). □

The case of delayed control. In the case (2.3), we are led to the problem:

$$(2.29)\qquad \left| \begin{aligned} &\partial u/\partial t + Au = f \quad \text{in} \quad Q,\\ &\frac{u(t) - u(t-\tau)}{\tau} \in K,\\ &\int_\Gamma \left(\frac{u(t) - u(t-\tau)}{\tau} + \chi\right)\left(v - \frac{(u(t) - u(t-\tau))}{\tau}\right)\mathrm{d}\Gamma \geqslant 0 \qquad \forall v \in K,\\ &\chi \text{ satisfying } \Psi(v) - \Psi\left(\frac{u(t) - u(t-\tau)}{\tau}\right) - \left(\chi, v - \left(\frac{u(t) - u(t\tau)}{\tau}\right)\right) \geqslant 0,\\ &u(t) = u_0(t), \qquad -\tau \leqslant t \leqslant 0. \quad \square \end{aligned} \right.$$

2.4. Orientation

We will now solve the problems of instantaneous control (Sec. 3) and then give some properties of the solution (in Sec. 4). The case of delayed control will be discussed in Section 5.

3. Solution of the Problems of Instantaneous Control

3.1. Statement of the Principal Results

To be more definite, we take

(3.1) $V = H^1(\Omega)$,

where $a(u,v)$ is defined by (2.1) such that

(3.2) $a(u,v) = a(v,u) \quad \forall u, v \in V$,

(3.3) $a(v,v) + c|v|^2 \geqslant \alpha \|v\|^2, \quad \alpha > 0,\ c > 0.$ [4]

We take a functional $v \to \Psi(v) = \int_\Gamma \psi(v)\,d\Gamma$ with the following properties[5]:

(3.4) $v \to \Psi(v)$ is convex from $V \to R$, lower semi-continuous for the weak topology of V, with values in $]-\infty, +\infty]$,

and there exists a family of functionals Ψ_j, *which are differentiable on* V, such that:

(3.5) $\forall v \in L^2(0,T;V)$ we have:
$\int_0^T \Psi_j(v)\,dt \to \int_0^T \Psi(v)\,dt, \quad j \to \infty$,

(3.6) there exists a sequence φ_j, *which is bounded in* V, such that:
$\Psi_j'(\varphi_j) = 0 \quad \forall j, \quad$ (or $\Phi_j(\varphi_j) = 0$ if $\Phi_j = \Psi_j'$),

(3.7) if $v_j \to v$ weakly star in $L^\infty(0,T;V)$ and if $v_j' > v'$ weakly in $L^2(0,T;H)$ $(H = L^2(\Omega))$, with $\int_0^T \Psi(v)\,dt$ constant, then
$\liminf \int_0^T \Psi_j(v)\,dt \geqslant \int_0^T \Psi(v)\,dt.$ ☐

Remark 3.1. We saw in Section I.5.3, how the preceding assumptions are satisfied in all applications. ☐

---

[4] $|v|$ = norm of v in $L^2(\Omega)$, $\|v\|$ = norm of v in $H^1(\Omega)$.

[5] These are analogous to (5.2), (5.3), (5.4), (5.5), Chapter I, with only a small technical difference between (5.5), Chapter I and (3.7), a difference with no significance for applications.

We will derive existence theorems for "strong" solutions by means of rather restrictive assumptions on the data f and u_0.

We assume:

(3.8) $\quad f, f' \in L^2(0,T; H)$,

(3.9) $\quad u_0$ is given in V with $Au_0 \in L^2(\Omega)$; we can find a sequence $u_{0j} \in V$ with $Au_{0j} \in L^2(\Omega)$ and $u_{0j} \to u_0$ in V, $Au_{0j} \to Au_0$ in $L^2(\Omega)$, where each u_{0j} satisfies $f(0)+Au_{0j} \in H^1(\Omega)$ and $\partial u_{0j}/\partial v_A + \Phi_j(f(0)-Au_{0j})=0$ on Γ and $f(0)+Au_{0j} \to f(0)+Au_0$ in $H^1(\Omega)$.

Then we have the following results:

Theorem 3.1. *We assume* (3.1)–(3.9) *to hold. Then there exists one and only one function u such that*

(3.10) $\quad u \in L^\infty(0,T; V), \qquad u' \in L^\infty(0,T; V), \qquad u'' \in L^2(0,T; H)$,

(3.11) $$(u'(t), v-u'(t)) + a(u(t)), v-u'(t)) + \Psi(v) - \Psi(u'(t)) \geq (f(t), v-u'(t)) \quad \forall v \in V \text{ (almost everywhere in } t),$$

(3.12) $\quad u(0)=u_0$.

Theorem 3.2 (Approximation theorem). *The assumptions here are the same as for Theorem* 3.1. *Let u_j be the solution of*

$(3.10)_j$ $\quad u_j \in L^\infty(0,T; V), \qquad u_j' \in L^\infty(0,T; V), \qquad u_j'' \in L^2(0,T; H)$,

$(3.11)_j$ $\quad (u_j'(t), v-u_j'(t)) + a(u_j(t), v-u_j'(t)) + \Psi_j(v) - \Psi_j(u_j'(t)) \geq (f(t), v-v_j'(t)) \quad \forall v \in V$,

$(3.12)_j$ $\quad u_j(0)=u_{0j}$.

Then, as $j \to \infty$, u designating the solution furnished by Theorem 3.1, *we have*

(3.13) $\quad u_j \to u, \qquad u_j' \to u'$ *weakly star in* $L^\infty(0,T; V)$,

(3.14) $\quad u_j'' \to u''$ *weakly in* $L^2(0,T; H)$.

Remark 3.2. Since Ψ_j is differentiable, (3.11) *is equivalent* to the equation

(3.15) $\quad (u_j'(t), v) + a(u_j(t), v) + (\Phi_j(u_j'(t)), v) = (f(t), v)$. □

Remark 3.3. Theorem 3.2 justifies Remark 1.2 (and expresses it more precisely). □

Remark 3.4. The observations which we made in Section I.5.4, apply to the present situation. □

3.2. Uniqueness Proof for Theorem 3.1 (and 3.2)

The uniqueness proof is a modification of the proof carried out in Section I.5.5.

Let u and u_* be two possible solutions of (3.11); if we take $v=u'_*(t)$ in (3.11) (resp. $v=u'(t)$ in the inequality analogous to (3.11) with respect to u_*), add the two inequalities, and set $w=u-u_*$, it follows that

$$-|w'(t)|^2-a(w(t),w'(t))\geqslant 0,$$

or again, *due to the fact that* $a(u,v)=a(v,u)$, $\forall u, v\in V$,

$$|w'(t)|^2+\frac{1}{2}\frac{\mathrm{d}}{\mathrm{d}t}a(w(t),w(t))\leqslant 0. \tag{3.16}$$

In general, we will write

$$a(\varphi,\varphi)=a(\varphi). \tag{3.17}$$

Since $w(0)=0$, it follows from (3.16) that

$$a(w(t))+2\int_0^t|w'(\sigma)|^2\,\mathrm{d}\sigma\leqslant 0. \tag{3.18}$$

Taking account of (3.3)[6], it follows from (3.18) that

$$\alpha\|w(t)\|^2+2\int_0^t|w'(\sigma)|^2\,\mathrm{d}\sigma\leqslant c|w(t)|^2=c\left|\int_0^t w'(\sigma)\,\mathrm{d}\sigma\right|^2\leqslant ct\int_0^t|w'(\sigma)|^2\,\mathrm{d}\sigma \tag{3.19}$$

whence $w(t)=0$ in $[0,t_0]$, $t_0=2/c$; then one integrates (3.16) from 0 to t and obtains $w(t)=0$ in $[t_0,2t_0]$, and so on, successively. □

3.3. Proof of Theorems 3.1 and 3.2

Following is the *outline of the proof*:

1) solution of the "Galerkin approximation" of (3.15);
2) solution of (3.15) with $u_j(0)=u_{0j}$;
3) a priori estimates for the u_j;
4) proof of the statements of the theorems.

3.3.1. Solutions of the Galerkin Approximation of (3.15)

We introduce a "base" $w_1,\dots,w_m,\dots$ for V such that

$$u_{0j},\ f(0)+Au_{0j},\ \varphi_j\in\text{space }[w_1,w_2,w_3]\text{ spanned by }w_1,\ w_2\text{ and }w_3. \tag{3.20}$$

---

[6] In fact, in the applications which we have seen, $a(v,v)\geqslant 0$ also holds, and then the proof follows immediately.

The "Galerkin approximation" for (3.15), $u_m(t)$, is then defined as the solution of

(3.21) $$(u_m'(t), w_k) + a(u_m(t), w_k) + (\Phi_j(u_m'(t)), w_k) = (f(t), w_k), \qquad 1 \leqslant k \leqslant m,$$

(3.22) $$u_m(t) \in [w_1, \ldots, w_m] = \text{space spanned by } w_1, \ldots, w_m,$$

(3.23) $$u_m(0) = u_{0j} \text{ (this is possible due to (3.20), as long as } m \geqslant 3).$$

But, because of the presence of the term $(\Phi_j(u_m'(t)), w_k)$, the system of (nonlinear) differential Equations (3.21) could, on the face of it, be singular. Therefore, we must first determine that this is not the case and that (3.21), (3.22), (3.23) define u_m uniquely in an interval $[0, t_m]$ with actually $t_m = T$—due to the a priori estimates.

There are several possible methods; one of them consists in "approximating" (3.21) by a system of equations *of second order* in t; for $\varepsilon > 0$, we seek $\psi_{\varepsilon m}(t) = \psi_\varepsilon(t)$ satisfying:

(3.24) $$\psi_\varepsilon(t) \in [w_1, \ldots, w_m],$$

(3.25) $$\varepsilon(\psi_\varepsilon''(t), w_k) + (\psi_\varepsilon'(t), w_k) + a(\psi_\varepsilon(t), w_k) + (\Phi_j(\psi_\varepsilon'(t)), w_k) = (f(t), w_k), \quad 1 \leqslant k \leqslant m,$$

(3.26) $$\begin{aligned} &\psi_\varepsilon(0) = u_{0j}, \\ &\psi_\varepsilon'(0) = u_{ij} = f(0) - A u_{0j}; \end{aligned}$$

due to (3.20), conditions (3.26) are possible as long as $m \geqslant 3$; we note that (3.21), taken for $t = 0$, yields:

(3.27) $$\begin{aligned} (u_m'(0), w_k) + (\Phi_j(u_m'(0)), w_k) &= (f(0), w_k) - a(u_{0j}, w_k) \\ &= (f(0) - A u_{0j}, w_k) - \int_\Gamma \frac{\partial u_{0j}}{\partial \nu_A} w_k \, d\Gamma \\ &= \text{(according to (3.9))} \quad (f(0) - A u_{0j}, w_k) + \int_\Gamma (\Phi_j(f(0) - A u_{0j}) w_k \, d\Gamma \end{aligned}$$

from which it follows that $u_m'(0) = f(0) - A u_{0j}$ is *a* solution; but if we consider (3.27) as an *equation in* $u'(0)$, uniqueness results; indeed, if φ and ψ are two possible solutions, it follows by subtraction:

$$(\varphi - \psi, w_k) + (\Phi_j(\varphi) - \Phi_j(\psi), w_k) = 0, \qquad 1 \leqslant k \leqslant m$$

from which, since φ and $\psi \in [w_1, \ldots, w_m]$,

(3.28) $$|\varphi - \psi|^2 + (\Phi_j(\varphi) - \Phi_j(\psi), \varphi - \psi) = 0.$$

Since Φ_j is monotone, it follows from (3.28) that $\varphi = \psi$ and consequently:

(3.29) $$u_m'(0) = f(0) - A u_{0j},$$

which justifies the second condition (3.26).

The system of differential Equations (3.24), (3.25), (3.26) is non-singular and therefore defines ψ_ε uniquely in an interval $[0, t_m]$. ☐

A priori estimate for ψ_ε (1st part). Due to (3.20), we deduce from (3.25) (observing also that $\Phi_j(\varphi_j)=0$) that

$$\varepsilon(\psi_\varepsilon''(t), \psi_\varepsilon'(t)-\varphi_j)+(\psi_\varepsilon'(t), \psi_\varepsilon'(t)-\varphi_j)+a(\psi_\varepsilon'(t), \psi_\varepsilon'(t)-\varphi_j) + (\Phi_j(\psi_\varepsilon'(t))-\Phi_j(\varphi_j), \psi_\varepsilon'(t)-\varphi_j)=(f(t), \psi_\varepsilon'(t)-\varphi_j) \tag{3.30}$$

whence, since Φ_j is monotone:

$$\frac{1}{2}\varepsilon\frac{d}{dt}|\psi_\varepsilon'(t)|^2+|\psi_\varepsilon'(t)|^2+\frac{1}{2}\frac{d}{dt}a(\psi_\varepsilon(t)) \leqslant \varepsilon(\psi_\varepsilon''(t), \varphi_j) + (\psi_\varepsilon'(t), \varphi_j)+a(\psi_\varepsilon(t), \varphi_j)+(f(t), \psi_\varepsilon'(t)-\varphi_j)$$

whence, by integration with respect to t:

$$\begin{aligned}\tfrac{1}{2}\varepsilon|\psi_\varepsilon'(t)|^2+\tfrac{1}{2}a(\psi_\varepsilon(t))+\textstyle\int_0^t|\psi_\varepsilon'(\sigma)|^2\,d\sigma &\leqslant \varepsilon(\psi_\varepsilon'(t), \varphi_j)-\varepsilon(u_{ij}, \varphi_j)\\ &\quad+\textstyle\int_0^t(\psi_\varepsilon'(\sigma), \varphi_j)\,d\sigma+\int_0^t a(\psi_\varepsilon(\sigma), \varphi_j)\,d\sigma+\int_0^t(f(\sigma), \psi_\varepsilon'(\sigma)-\varphi_j)\,d\sigma\\ &\quad+\tfrac{1}{2}\varepsilon|u_{ij}|^2+\tfrac{1}{2}a(u_{0j})\\ &\leqslant \tfrac{1}{4}\varepsilon|\psi_\varepsilon'(t)|^2+\tfrac{1}{2}\textstyle\int_0^t|\psi_\varepsilon'(\sigma)|^2\,d\sigma+\int_0^t a(\psi_\varepsilon(\sigma), \varphi_j)\,d\sigma+c(1+\int_0^t|f(\sigma)|^2\,d\sigma)\end{aligned}$$ [7]

whence

$$\tfrac{1}{2}\varepsilon|\psi_\varepsilon'(t)|^2+a(\psi_\varepsilon(t))+\textstyle\int_0^t|\psi_\varepsilon'(\sigma)|^2\,d\sigma \leqslant c(1+\int_0^t|f(\sigma)|^2\,d\sigma)+\int_0^t a(\psi_\varepsilon(\sigma), \varphi_j)\,d\sigma. \tag{3.31}$$

In order to simplify the course of the proof, let us assume (but this is not essential) that:

$$a(v, v)=a(v)\geqslant 0. \tag{3.32}$$

Then

$$a(\psi_\varepsilon(\sigma), \varphi_j)\leqslant \tfrac{1}{2}a(\psi_\varepsilon(\sigma))+\tfrac{1}{2}a(\varphi_j)$$

and (3.31) yields

$$\tfrac{1}{2}\varepsilon|\psi_\varepsilon'(t)|^2+a(\psi_\varepsilon(t))+\textstyle\int_0^t|\psi_\varepsilon'(\sigma)|^2\,d\sigma \leqslant c(1+\int_0^t|f(\sigma)|^2\,d\sigma)+\tfrac{1}{2}\int_0^t a(\psi_\varepsilon(\sigma))\,d\sigma. \tag{3.33}$$

In particular, it follows from (3.33) that:

$$a(\psi_\varepsilon(t))\leqslant c+\tfrac{1}{2}\textstyle\int_0^t a(\psi_\varepsilon(\sigma))\,d\sigma$$

whence, due to Gronwall's inequality

$$a(\psi_\varepsilon(t))\leqslant c. \tag{3.34}$$

---

[7] c denoting various constants.

But, applying (3.34) in (3.33), it follows that:

(3.35) $\varepsilon \psi'_\varepsilon(t)|^2 \leqslant c,$

(3.36) $\int_0^T |\psi'_\varepsilon(\sigma)|^2 \mathrm{d}\sigma \leqslant c.$

But then $\psi_\varepsilon(t) = u_{0j} + \int_0^T \psi'_\varepsilon(\sigma)\mathrm{d}\sigma$ satisfies, due to (3.36):

$$|\psi_\varepsilon(t)| \leqslant c$$

which, together with (3.34), gives:

(3.37) $\|\psi_\varepsilon(t)\| \leqslant c.$ ☐

Remark 3.5 *(fundamental). The above introduced constants c do not depend on either ε, m or j.* ☐

A priori estimates for ψ_ε (2nd part). We note that u_{1j} was chosen so that (3.25) would give

(3.38) $\psi''_\varepsilon(0) = 0.$

We differentiate (3.25) with respect to t—this differentiation is only formal, unless we assume the function Φ_j to be Lipschitz continuous. The calculation that follows can be justified if we take difference quotients instead of the derivative. We obtain

(3.39) $\varepsilon(\psi'''_\varepsilon, w_k) + (\psi''_\varepsilon, w_k) + a(\psi'_\varepsilon, w_k) + ((\Phi_j(\psi'_\varepsilon))', w_k) = (f', w_k).$

From this it follows that:

(3.40) $\varepsilon(\psi'''_\varepsilon, \psi''_\varepsilon) + |\psi''_\varepsilon|^2 + a(\psi'_\varepsilon, \psi''_\varepsilon) + ((\Phi_j(\psi'_\varepsilon))', \psi''_\varepsilon) = (f', \psi''_\varepsilon).$

But since Φ_j is monotone (cf. (5.41), Chap. I):

$$(\Phi_j(\psi'_\varepsilon(t))', \psi''_\varepsilon(t)) \geqslant 0$$

and therefore, (3.40) leads to:

(3.41) $\frac{\varepsilon}{2} \frac{\mathrm{d}}{\mathrm{d}t} |\psi''_\varepsilon(t)|^2 + |\psi''_\varepsilon(t)|^2 + \frac{1}{2} \frac{\mathrm{d}}{\mathrm{d}t} a(\psi'_\varepsilon(t)) \leqslant (f'(t), \psi''_\varepsilon(t))$

whence, taking account of (3.38):

(3.42) $\frac{1}{2}\varepsilon |\psi''_\varepsilon(t)|^2 + \int_0^t |\psi''_\varepsilon(\sigma)|^2 \mathrm{d}\sigma + \frac{1}{2} a(\psi'_\varepsilon(t)) \leqslant \frac{1}{2} a(u_{1j}) + \int_0^t |f'(\sigma)|\, |\psi''_\varepsilon(\sigma)| \mathrm{d}\sigma.$

According to (3.9), $a(u_{1j}) \leqslant c$, and therefore (3.42) leads to:

$$\varepsilon |\psi_\varepsilon''(t)|^2 + \int_0^t |\psi_\varepsilon''(\sigma)|^2 \, d\sigma + a(\psi_\varepsilon'(t)) \leqslant c + \int_0^t |f'(\sigma)|^2 \, d\sigma. \tag{3.43}$$

Consequently,—the c's here are again independent of ε, m, j—

$$\varepsilon |\psi_\varepsilon''(t)|^2 \leqslant c, \tag{3.44}$$

$$\int_0^t |\psi_\varepsilon''(\sigma)|^2 \, d\sigma \leqslant c, \tag{3.45}$$

$$\|\psi_\varepsilon'(t)\| \leqslant c. \quad \square \tag{3.46}$$

Passage to the limit in ε. According to the a priori estimates obtained, and the fact that (3.46) has the consequence:

$$\|\Phi_j(\psi_\varepsilon'(t))\|_* \leqslant c, \tag{3.47}$$

we can extract a sequence, again denoted by ψ_ε, such that, when $\varepsilon \to 0$, we have:

$$\begin{aligned} &\psi_\varepsilon \to u_m, \quad \psi_\varepsilon' \to u_m' \quad \text{weakly star in } L^\infty(0,T;V) \\ &\psi_\varepsilon'' \to u_m'' \quad \text{weakly in } L^2(0,T;H), \end{aligned} \tag{3.48}$$

$$\Phi_j(\psi_\varepsilon') \to \chi \quad \text{weakly star in } L^\infty(0,T;V'). \tag{3.49}$$

Due to (3.44), we can then pass to the limit with ε in (3.25), for fixed k,

$$\varepsilon(\psi_k'', w_k) \to 0 \quad \text{in } L^\infty(0,T) \quad \text{and we obtain}$$

$$(u_m', w_k) + a(u_m, w_k) + (\chi, w_k) = (f, w_k), \qquad 1 \leqslant k \leqslant m. \tag{3.50}$$

Since $\psi_\varepsilon(0) = u_{0j} \to u_m(0)$ weakly in V, we have:

$$u_m(0) = u_{0j};$$

thus we will have solved (3.21), (3.22), (3.23) if we can show that:

$$\chi = \Phi_j(u_m'). \tag{3.51}$$

For this purpose, we make use of a "monotonicity argument". Let φ be a function $\in L^2(0,T;V)$, with values in $[w_1, \ldots, w_m]$. We set:

$$X_\varepsilon = \int_0^T (\Phi_j(\psi_\varepsilon') - \Phi_j(\varphi), \psi_\varepsilon' - \varphi) \, dt. \tag{3.52}$$

Due to the monotonicity of the Φ_j, we have: $X_\varepsilon \geqslant 0$. But, making use of (3.25), we have:

$$\begin{aligned} X_\varepsilon = &-\varepsilon \int_0^T (\psi_\varepsilon'', \psi_\varepsilon') \, dt - \int_0^T |\psi_\varepsilon'|^2 \, dt - \int_0^T a(\psi_\varepsilon, \psi_\varepsilon') \, dt + \int_0^T (f, \psi_\varepsilon') \, dt \\ &- \int_0^T (\Phi_j(\psi_\varepsilon'), \varphi) \, dt - \int_0^T (\Phi_j(\varphi), \psi_\varepsilon' - \varphi) \, dt. \end{aligned} \tag{3.53}$$

But $\varepsilon \int_0^T (\psi_\varepsilon'', \psi_\varepsilon')\,dt \to 0$,

$$\limsup \left(-\int_0^T |\psi_\varepsilon'|^2\,dt\right) \leqslant -\int_0^T |u_m'|^2\,dt, \ \limsup\left[-\int_0^T a(\psi_\varepsilon, \psi_\varepsilon')\,dt\right]$$
$$-\limsup\left[-\tfrac{1}{2}a(\psi_\varepsilon(T)) + \tfrac{1}{2}a(u_{0j})\right]$$
$$\leqslant -\tfrac{1}{2}a(u_m(T)) + \tfrac{1}{2}a(u_{0j}) = \int_0^T a(u_m, u_m')\,dt,$$

whence

$$\begin{aligned} 0 \leqslant \limsup X_\varepsilon \leqslant &-\int_0^T \left[|u_m'(t)|^2 + a(u_m, u_m') - (f, u_m')\right] dt \\ &-\int_0^T (\chi, \varphi)\,dt - \int_0^T (\Phi_j(\varphi), u_m' - \varphi)\,dt. \end{aligned} \tag{3.54}$$

But, according to (3.50):

$$-\int_0^T \left[|u_m'(t)|^2 + a(u_m, u_m') - (f, u_m')\right] dt = \int_0^T (\chi, u_m')\,dt$$

and (3.54) gives

$$\int_0^T (\chi - \Phi_j(\varphi), u_m' - \varphi)\,dt \geqslant 0. \tag{3.55}$$

In (3.55), let us take $\varphi = u_m' - \lambda\psi$, $\lambda > 0$, $\psi \in L^2(0,T;V)$ with values in $[w_1, \dots, w_m]$; then, after dividing by λ:,

$$\int_0^T (\chi - \Phi_j(u_m' - \lambda\psi), \psi)\,dt \geqslant 0$$

from which, letting $\lambda \to 0$, it follows that:

$$\int_0^T (\chi - \Phi_j(u_m'), \psi)\,dt \geqslant 0;$$

hence (3.51). □

We note that from estimates (3.37), (3.45), (3.46), we can deduce that:

$$\begin{aligned} &\|u_m(t)\| + \|u_m'(t)\| \leqslant c, \\ &\int_0^T |u_m''(t)|^2\,dt \leqslant c, \end{aligned} \tag{3.56}$$

where the c's are constants *independent of m and j.* □

3.3.2. Solution of (3.15) and a Priori Estimates for u_j

According to (3.56), we can extract from $u_m = u_{jm}$ a sequence u_μ such that

$$\begin{aligned} &u_\mu \to u_j, \quad u_\mu' \to u_j' \quad \text{weakly star in} \quad L^\infty(0,T;V) \\ &u_\mu'' \to u_j'' \qquad\qquad\quad \text{weakly in} \quad L^2(0,T;H) \end{aligned} \tag{3.57}$$

and

$$\Phi_j(u_\mu') = \chi_j \quad \text{weakly star in} \quad L^\infty(0,T;V'). \tag{3.58}$$

Furthermore, according to (3.56):

$$\begin{aligned}&\|u_j(t)\|+\|u_j'(t)\|\leqslant c,\\&\textstyle\int_0^T |u_j''(t)|^2\,dt\leqslant c.\end{aligned}\tag{3.59}$$

Passing to the limit in (3.21) with $m=\mu$ (for fixed k), we find that

$$(u_j',w_k)+a(u_j,w_k)+(\chi_j,w_k)=(f,w_k)$$

and this $\forall k$; therefore:

$$(u_j',v)+a(u_j,v)+(\chi_j,v)=(f,v)\qquad \forall v\in V.\tag{3.60}$$

But we can prove—exactly as we just did for (3.51)—that

$$\chi_j=\Phi_j(u_j').\tag{3.61}$$

(We introduce $y_\mu=\int_0^T(\Phi_j(u_\mu')-\Phi_j(\varphi),u_\mu'-\varphi)\,dt\geqslant 0$, $\varphi\in L^2(0,T;V)$.)

Thus we have constructed a solution u_j of (3.15) *with* $(3.12)_j$ *and have obtained the estimates* (3.59).

Furthermore, (a consequence of the second condition (3.26)):

$$u_j'(0)=f(0)-Au_{0j}.\quad\Box\tag{3.62}$$

3.3.3. Proof of the Statements of the Theorems

According to (3.59), we can extract a sequence, which we again denote by u_j, such that (3.13), (3.14) hold.

Since (3.15) is equivalent to $(3.11)_j$, it only remains to show that u is a solution of (3.11), (then, because of the uniqueness of the solution, the extraction of a sub-sequence of u_j is unnecessary).

In (3.11), we choose $v=v_0$ such that $\Psi(v_0)<\infty$ and obtain (since $\Psi_j(v_0)\to\Psi(v_0)$) $\Psi_j(v_0)\leqslant c$. From $(3.11)_j$, it follows that:

$$\textstyle\int_0^T \Psi_j(u_j'(t))\,dt\leqslant C.\tag{3.63}$$

Thus we can apply (3.7) with $v_j=u_j'$, $v=u'$. Then

$$\liminf \textstyle\int_0^T \Psi_j(u_j')\,dt\geqslant \int_0^T \Psi(u')\,dt.\tag{3.64}$$

In (3.11), let us take $v=v(t)$, where $v\in L^2(0,T;V)$. Then

$$\begin{aligned}\textstyle\int_0^T[(u_j',v)+a(u_j,v)+\Psi_j(v)-(f,v-u_j')]\,dt&\geqslant\textstyle\int_0^T[|u_j'|^2+a(u_j,u_j')+\Psi_j(u_j')]\,dt\\&=\textstyle\int_0^T|u'|^2\,dt+\frac12 a(u_j(T))-\frac12 a(u_{0j})+\int_0^T\Psi_j(u_j')\,dt\end{aligned}$$

from which it follows, using notably (3.63), that

$$\int_0^T [(u',v)+a(u,v)+\Psi(v)-(f,v-u')]\,dt \geq \int_0^T |u'|^2\,dt+\tfrac{1}{2}a(u(T)) -\tfrac{1}{2}a(u_0)+\int_0^T \Psi(u')\,dt$$

whence

$$\int_0^T [(u',v-u')+a(u,v-u')+\Psi(v)-\Psi(u')-(f,v-u')]\,dt \geq 0 \quad \forall v\in L^2(0,T;V). \quad \square \tag{3.65}$$

But, by exactly the same argument that follows (5.59) in Chapter I, we verify that (3.64) implies (and is equivalent to) (3.11). □

Remark 3.6. From (3.62) it follows that the solution u furnished by Theorem 3.1 satisfies

$$u'(0)=f(0)-Au_0. \quad \square \tag{3.66}$$

4. A Property of the Solution of Instantaneous Control at a Thin Wall

We consider the solution u of the problem in Example 1.1, therefore (cf. Example 2.3, special Case 1) furnished by Theorem 3.1 with

$$\psi(\lambda)=\begin{cases} +\infty & \text{if } \lambda<0 \\ 0 & \text{if } \lambda\geq 0, \end{cases} \tag{4.1}$$

$$\Psi(v)=\int_\Gamma \psi(v)\,d\Gamma. \tag{4.2}$$

We shall prove

Theorem 4.1. *We make the same assumptions as for Theorem* 3.1 *with* (4.1), (4.2). *We assume that the data f and u_0 satisfy*

$$\partial f/\partial t \geq 0 \quad \text{a. e. in } Q, \tag{4.3}$$

$$f(0)-Au_0\geq 0 \quad \text{a. e. in } \Omega, \tag{4.4}$$

$$\partial u_0/\partial \nu_A=0. \tag{4.5}$$

Then the solution u of the corresponding problem satisfies

$$\partial u/\partial t\geq 0 \quad \text{a. e. in } Q. \tag{4.6}$$

Proof. 1) We take

$$\psi_j(\lambda) = \begin{vmatrix} \frac{1}{2} j\lambda^2 & \text{if} \quad \lambda \leqslant 0 \\ 0 & \text{if} \quad \lambda \geqslant 0, \end{vmatrix} \tag{4.7}$$

$$\Psi_j(v) = \int_\Gamma \psi_j(v)\,\mathrm{d}\Gamma, \tag{4.8}$$

in such a way that

$$(\Phi_j(u), v) = \int_\Gamma \Phi_j(u(x))\,v(x)\,\mathrm{d}\Gamma, \tag{4.9}$$

where

$$\Phi_j(\lambda) = \begin{vmatrix} j\lambda & \text{if} \quad \lambda \leqslant 0 \\ 0 & \text{if} \quad \lambda \geqslant 0. \end{vmatrix} \tag{4.10}$$

Then we take in (3.9)

$$u_{0j} = u_0. \tag{4.11}$$

The conditions for Theorem 3.2 are satisfied because, in particular:

$$\begin{aligned} \partial u_0/\partial v_A + \Phi_j(f(0) - Au_0) &= \partial u_0/\partial v_A \quad (\text{since } f(0) - Au_0 \geqslant 0) \\ &= 0 \qquad\qquad\quad (\text{according to (4.5)}). \end{aligned}$$

Now we can apply Theorem 3.2. We will have proved (4.6), if we can show—which, moreover, is of intrinsic interest—that

$$\partial u_j/\partial t \geqslant 0 \quad \text{a. e. in} \quad Q. \tag{4.12}$$

2) But the function Φ_j is Lipschitz continuous, and we can differentiate Equation (3.15) with respect to t (with the above choice (4.9)). Then:

$$(u_j'', v) + a(u_j', v) + ((\Phi_j(u_j'))', v) = (f', v). \tag{4.13}$$

We take $v = (u_j')^-$ in (4.13); observing that

$$(u_j'', (u_j')^-) = -\int_\Omega \frac{\partial}{\partial t}(u_j')^-\,(u_j')^-\,\mathrm{d}x = -\frac{1}{0}\frac{\mathrm{d}}{\mathrm{d}t}|u_j'^-(t)|^2,$$

$$a(u_j', u_j'^-) = -a(u_j'^-, u_j'^-),$$

we then deduce from (4.13) that

$$\frac{1}{2}\frac{\mathrm{d}}{\mathrm{d}t}|u_j'^-(t)|^2 + a(u_j'^-) - ((\Phi_j(u_j'))', u_j'^-) + (f', (u_j')^-) = 0. \tag{4.14}$$

But according to (4.9), (4.10), we have:

$$-((\Phi_j(u'_j))' u_j'^-) = -\int_\Gamma \frac{\partial}{\partial t}(ju'_j)\cdot(u'_j)^- \,d\Gamma$$

(because we integrate only where $u'_j \leqslant 0$)

$$= j\int_\Gamma \left(\frac{\partial}{\partial t}(u'_j)^-\right)(u'_j)^- \,d\Gamma$$

$$= \frac{j}{2}\frac{d}{dt}\int_\Gamma ((u'_j)^-)^2 \,d\Gamma$$

and therefore (4.14) can be written:

(4.15) $$\frac{1}{2}\frac{d}{dt}|(u_j'^-(t)|^2 + a(u_j'^-(t)) + \frac{j}{2}\frac{d}{dt}(\int_\Gamma (u_j'^-)^2 \,d\Gamma) + (f'(t), u'_j(t)^-) = 0.$$

According to (4.3), $(f'(t), u'_j(t)^-) \geqslant 0$ and therefore, (since, according to (3.62),

$$u'_j(0) = f(0) - Au_{0j} = (\text{by } (4.11)) \quad f(0) - Au_0 \geqslant 0,$$

we have: $u'_j(0)^- = 0$), (4.15) gives:

(4.16) $$\tfrac{1}{2}|u_j'^-(t)|^2 + \int_0^t a(u_j'^-(\sigma))\,d\sigma + \tfrac{1}{2}j\int_\Gamma u_j'^-(t)^2 \,d\Gamma \leqslant 0 \quad \text{whence} \quad u_j'^- = 0$$

whence (4.12). □

5. Partial Results for Delayed Control[8]

5.1. Statement of a Result

The general problem of delayed control is far from having been solved completely. Here we will give a partial result, established under very restrictive assumptions on the functional Ψ, defined by

(5.1) $$\Psi(v) = \int_\Gamma \psi(v(x))\,d\Gamma;$$

we assume that:

(5.2) the function ψ is differentiable[9] and its derivative $d\psi/d\lambda = \Phi$ is Lipschitz continuous.

[8] This section may be passed over.
[9] Therefore of type 1 in the terminology of Sec. I.3.3.

Then the "inequality" (2.9) is, indeed, equivalent to the equation:

$$(u',v)+a(u,v)+(\Phi(Mu),v)=(f,v) \quad \forall v\in H^1(\Omega)=V \tag{5.3}$$

where

$$(\Phi(w),v)=\int_\Gamma \Phi(w(x))\,v(x)\,d\Gamma. \tag{5.4}$$

Therefore, this is a problem for a parabolic non linear partial differential *equation* with delay[10].

We shall prove

Theorem 5.1. *We assume that the form* $a(u,v)$, *not necessarily symmetric*[11] *is coercive in the sense*

$$a(v,v)+c_3|v|^2\geqslant\alpha\|v\|^2, \quad \alpha>0. \tag{5.5}$$

We assume that (5.2) *holds. We give* f *and* u_0 *with*

$$f\in L^2(0,T;V'), \tag{5.6}$$

$$u_0\in H^1(\Omega). \tag{5.7}$$

Then there exists one and only one function u *such that*

$$\begin{aligned} &u\in L^2(0,T;V),\\ &\partial u/\partial t\in L^2(0,T;V'), \end{aligned} \tag{5.8}$$

$$u(t)=u_0 \quad \text{for} \quad -\tau\leqslant t\leqslant 0, \tag{5.9}$$

and such that (5.3) *is satisfied* almost everywhere *for* $t\in(0,T)$.

5.2. Proof of Existence in Theorem 5.1

We again take a "base" $w_1,\dots,w_m,\dots$ of V, with

$$w_1=u_0 \quad (\text{if } u\neq 0). \tag{5.10}$$

---

[10] Cf. M. Artola [1].

[11] Here, there is an essential difference from the results of Section 3. It is likely (but has not been proved) that the problems of Section 3 are generally improperly posed, if $a(u,v)$ (or, at least, its "principal part") is not symmetric. Therefore, it is probably impossible to obtain the results of Section 3 starting from Theorem 5.1 by letting $\tau\to 0$. Moreover, we have not been able to effect this passage to the limit in τ, even when assuming that $a(u,v)$ is symmetric.

We define $u_m(t)$ by

$$u_m(t) \in [w_1, \ldots, w_m] \quad \text{(the space spanned by } w_1, \ldots, w_m), \tag{5.11}$$

$$(u_m'(t), w_k) + a(u_m(t), w_k) + (\Phi(M u_m(t)), w_k) = (f(t), w_k), \qquad 1 \leqslant k \leqslant m, \tag{5.12}$$

$$u_m(t) = u_0, \qquad -\tau \leqslant t \leqslant 0. \tag{5.13}$$

Here, we deal with a system of differential equations with delay, defining u_m (cf. Bellman-Cooke [1], Halanay [1]) in an interval $[0, t_m]$; the following a priori estimates show that we can take $t_m = T$. □

A priori estimates (I). It follows from (5.12) that

$$(u_m'(t), u_m(t)) + a(u_m(t), u_m(t)) + (\Phi(M u_m(t)), u_m(t)) = (f(t), u_m(t)) \tag{5.14}$$

whence

$$\frac{1}{2}\frac{\mathrm{d}}{\mathrm{d}t}|u_m(t)|^2 + \alpha\|u_m(t)\|^2 - c_3|u_m(t)|^2 \leqslant \|f(t)\|_* \|u_m(t)\| + |(\Phi(M u_m(t)), u_m(t))|. \tag{5.15}$$

But according to (5.2), we have $|\Phi(\lambda)| \leqslant c_1|\lambda| + c_2$ whence

$$\begin{aligned} |(\Phi(M u_m(t)), u_m(t))| &\leqslant \tau^{-1} \int_\Gamma (c_1|u_m(t)| + |u_m(t-\tau)| + c_2) \times |u_m(t)| \,\mathrm{d}\Gamma \\ &\leqslant c_4(\tau) \int_\Gamma |u_m(t)|^2 \,\mathrm{d}\Gamma + c_5(\tau) \int_\Gamma |u_m(t-\tau)|^2 \,\mathrm{d}\Gamma \end{aligned} \tag{5.16}$$

where the constants $c_4(\tau)$, $c_5(\tau)$ *depend on* τ.

We particularly make use of the following result (cf., for example, Deny-Lions [1]): for every $\varepsilon > 0$, there exists a c_ε such that

$$\int_\Gamma |v|^2 \,\mathrm{d}\Gamma \leqslant \varepsilon\|v\|^2 + c_\varepsilon|v|^2 \qquad \forall v \in H^1(\Omega). \tag{5.17}$$

Applying this inequality to (5.16), we obtain

$$\begin{aligned} |(\Phi(M u_m(t)), u_m(t))| &\leqslant \tfrac{1}{4}\alpha(\|u_m(t)\|^2 + \|u_m(t-\tau)\|^2) \\ &\quad + c_6(\tau)(|u_m(t)|^2 + |u_m(t-\tau)|^2). \end{aligned} \tag{5.18}$$

Making use of the inequality

$$\|f(t)\|_* \|u_m(t)\| \leqslant \tfrac{1}{4}\alpha\|u_m(t)\|^2 + 4\alpha^{-1}\|f(t)\|_*^2$$

we can conclude from (5.15) that

$$\begin{aligned} \frac{1}{2}\frac{\mathrm{d}}{\mathrm{d}t}|u_m(t)|^2 + \frac{\alpha}{2}\|u_m(t)\|^2 - c_3|u_m(t)|^2 &\leqslant \frac{4}{\alpha}\|f(t)\|_*^2 \\ &\quad + \tfrac{1}{4}\alpha\|u_m(t-\tau)\|^2 + c_6(T)(|u_m(t)|^2 + |u_m(t-\tau)|^2). \end{aligned} \tag{5.19}$$

We integrate (5.19) from 0 to t; it follows that:

$$
\begin{aligned}
\tfrac{1}{2}|u_m(t)|^2 + \tfrac{1}{2}\alpha \int_0^t \|u_m(\sigma)\|^2 \mathrm{d}\sigma \leqslant c_7(\tau) \int_0^t |u_m(\sigma)|^2 \mathrm{d}\sigma & \\
+ c_6(\tau) \int_0^t |u_m(t-\tau)|^2 \mathrm{d}\tau & \\
+ \tfrac{1}{4}\alpha \int_0^t \|u_m(\sigma-\tau)\|^2 \mathrm{d}\sigma + 4\alpha^{-1} \int_0^t \|f(\sigma)\|_*^2 \mathrm{d}\sigma. &
\end{aligned} \tag{5.20}
$$

But

$$
\begin{aligned}
\int_0^t \|u_m(\sigma-\tau)\|^2 \mathrm{d}\sigma &= \tau \|u_0\|^2 + \int_0^{t-\tau} \|u_m(\sigma)\|^2 \mathrm{d}\sigma \\
&\leqslant \tau \|u_0\|^2 + \int_0^t \|u_m(\sigma)\|^2 \mathrm{d}\sigma
\end{aligned}
$$

(and the analogous inequality with the norm in H). Then (5.20) gives:

$$
\tfrac{1}{2}|u_m(t)|^2 + \tfrac{1}{4}\alpha \int_0^t \|u_m(\sigma)\|^2 \mathrm{d}\sigma \leqslant c_8 + 4\alpha^{-1} \int_0^t \|f(\sigma)\|_*^2 \mathrm{d}\sigma + c_9(\tau) \int_0^t |u_m(\sigma)|^2 \mathrm{d}\sigma\,. \tag{5.21}
$$

Thus, in particular:

$$
\tfrac{1}{2}|u_m(t)|^2 \leqslant c_8 + 4\alpha^{-1} \int_0^t \|f(\sigma)\|_*^2 \mathrm{d}\sigma + c_9(\tau) \int_0^t |u_m(\sigma)|^2 \mathrm{d}\sigma\,,
$$

which has the consequence, according to Gronwall's inequality, that

$$
u_m(t) \leqslant c(\tau) \quad \text{(a constant independent of } m\text{, but dependent on } \tau), \tag{5.22}
$$

and this, together with (5.21), implies

$$
\int_0^T \|u_m(t)\|^2 \mathrm{d}t \leqslant c(\tau)\,. \quad \square \tag{5.23}
$$

A priori estimates (II). We now obtain estimates for the *fractional derivatives* of u_m in t by Fourier transformation (cf. Lions [1], Sec. I.6.5).

It follows from (5.12) that

$$
(u'_m(t), w_k) = (\xi_m(t), w_k) \tag{5.24}
$$

and, according to (5.23), we verify that

$$
\int_0^T \|\xi_m(t)\|_*^2 \mathrm{d}t \leqslant c(\tau)\,. \tag{5.25}
$$

Let us extend u_m, ξ_m to $\tilde{u}_m, \tilde{\xi}_m$ as 0 outside $]0, T[$; then we have:

$$
\frac{\mathrm{d}}{\mathrm{d}t}(\tilde{u}_m(t), w_k) = (\tilde{\xi}_m(t), w_k) + (u_0, w_k)\,\delta(t) - (u_m(T), w_k)\,\delta(t-T) \tag{5.26}
$$

whence by Fourier transformation in t (we set: $\hat{\varphi}(s) = \int_{-\infty}^{+\infty} \exp(-2\pi i t s)\,\hat{\varphi}(t)\,\mathrm{d}t$):

$$
2\pi i s(\hat{u}_m(s), w_k) = (\hat{\xi}_m(s), w_k) + (u_0, w_k) - (u_m(T), w_k)\exp(-2\pi i s T) \tag{5.27}
$$

from which we conclude:

$$2\pi is|\hat{u}_m(s)|^2 = (\hat{\xi}_m(s), \hat{u}_m(s)) + (u_0, \hat{u}_m(s)) - (u_m(\tau), \hat{u}_m(s))\exp(-2\pi isT) \tag{5.28}$$

whence, formally, ($\beta > 0$ to be chosen):

$$\begin{aligned} \int_{-\infty}^{+\infty} \frac{|s|}{1+|s|^\beta} |\hat{u}_m(s)|^2 \, ds &\leq c \int_{-\infty}^{+\infty} \frac{1}{1+|s|^\beta} \|\hat{\xi}_m(s)\|_* \|\hat{u}_m(s)\| \, ds \\ &+ c \int_{-\infty}^{+\infty} \frac{1}{1+|s|^\beta} |\hat{u}_m(s)| \, ds \end{aligned} \tag{5.29}$$

(where the c's *depend on* τ).

But, according to (5.23) and (5.25), we have:

$$\int_{-\infty}^{+\infty} \|\hat{\xi}_m(s)\|_* \|\hat{u}_m(s)\| \, ds \leq c$$

such that (5.29) leads to

$$\begin{aligned} \int_{-\infty}^{+\infty} \frac{|s|}{1+|s|^\beta} |\hat{u}_m(s)|^2 \, ds &\leq c + c\left(\int_{-\infty}^{+\infty} \frac{ds}{(1+|s|^\beta)^2}\right)^{1/2} \\ &\times \left(\int_{-\infty}^{+\infty} |\hat{u}_m(s)|^2 \, ds\right)^{1/2} \leq c \quad \text{if} \quad \beta > \tfrac{1}{2}. \end{aligned}$$

Consequently:

$$\int_{-\infty}^{+\infty} |s|^\gamma |\hat{u}_m(s)|^2 \, ds \leq c \quad \text{if} \quad 0 < \gamma < \tfrac{1}{2}. \tag{5.30}$$

Passage to the limit. Using estimates (5.23), (5.30), it follows from Lions, loc. cit., Theorem 5.2, Chap. I, that we can select a sequence u_μ such that

$$\begin{aligned} &u_\mu \to u \quad \text{weakly star in } L^\infty(0,T; L^2(\Omega)) \\ &u_\mu \to u \quad \text{weakly in } L^2(0,T; H^1(\Omega)), \end{aligned} \tag{5.31}$$

$$u_\mu \to u \quad \text{strongly in } L^2(0,T; H^\rho(\Omega)),\ \rho \text{ fixed with } \tfrac{1}{2} < \rho < 1. \tag{5.32}$$

Thus, since the mapping "trace" $u \to u|_\Gamma = \gamma_0 u$ is linear and continuous from

$$H^\rho(\Omega) \to L^2(\Gamma)$$

(cf. Lions-Magenes [1], Chap. 1), we have:

$$u_\mu \to u \quad \textit{strongly} \text{ in } L^2(\Sigma) \tag{5.33}$$

and consequently

$$\Phi(Mu_\mu) \to \Phi(Mu) \quad \textit{strongly} \text{ in } L^2(\Sigma). \tag{5.34}$$

We can now pass to the limit in (5.12), taken for $m=\mu$ with fixed k ($\leqslant \mu$). We obtain

$$\frac{\mathrm{d}}{\mathrm{d}t}(u(t), w_k) + a(u(t), w_k) + (\Phi(Mu(t)), w_k) = (f(t), w_k)$$

and this $\forall k$, from which it follows that u satisfies (5.3). Since, obviously, (5.8) and (5.9) hold, we have proved the existence of a solution.

5.3. Proof of Uniqueness in Theorem 5.1

Let u and u_* be two possible solutions. If we set $w=u-u_*$, we have:

$$(w', v) + a(w, v) + (\Phi(Mu) - \Phi(Mu_*), v) = 0 \qquad \forall v$$

from which

$$\frac{1}{2}\frac{\mathrm{d}}{\mathrm{d}t}|w(t)|^2 + a(w(t), w(t)) + (\Phi(Mu) - \Phi(Mu_*), w(t)) = 0\,. \tag{5.35}$$

Since Φ is Lipschitz continuous, we have:

$$\begin{aligned}|(\Phi(Mu) - \Phi(Mu_*), w(t))| &\leqslant c \textstyle\int_\Gamma |w(t) - w(t-\tau)|\,|w(t)|\,\mathrm{d}\Gamma \\ &\leqslant \text{(according to (5.17))}\ \tfrac{1}{4}\alpha \|w(t)\|^2 \\ &\quad + \tfrac{1}{4}\alpha \|w(t-\tau)\|^2 + c|w(t)|^2 + c|w(t-\tau)|^2\,.\end{aligned}$$

Thus we deduce from (5.35) that

$$\frac{1}{2}\frac{\mathrm{d}}{\mathrm{d}t}|w(t)|^2 + \frac{3\alpha}{4}\|w(t)\|^2 \leqslant c|w(t)|^2 + \frac{\alpha}{4}\|w(t-\tau)\|^2 + c|w(t-\tau)|^2\,. \tag{5.36}$$

We integrate (5.36) from 0 to t; observing that

$$\textstyle\int_0^t \|w(\sigma-\tau)\|^2\,\mathrm{d}\sigma \leqslant \int_0^t \|w(\sigma)\|^2\,\mathrm{d}\sigma$$

(and the analogous inequality with the norm $|\ \ |$), it follows that

$$\tfrac{1}{2}|w(t)|^2 + \tfrac{1}{2}\alpha \textstyle\int_0^t \|w(\sigma)\|^2\,\mathrm{d}\sigma \leqslant c\int_0^t |w(\sigma)|^2\,\mathrm{d}\sigma$$

whence $w=0$.

6. Commentary

The problems which we examined in this chapter were introduced by the authors in Duvaut-Lions [2]. Details of the proofs are given here for the first time. Other methods for the solution of these problems are given in H. Brézis [2].

As indicated in the text, the results of Section 5 are very incomplete. For further results and related models, cf. D. Viaud [1].

Chapter III

Classical Problems and Problems with Friction in Elasticity and Visco-Elasticity

The present chapter assumes familarity with only Sections 1 and 4 of Chapter I.

1. Introduction

In this chapter, we take up problems of elasticity and visco-elasticity. We recall (of course, this will be stated more precisely in what follows) that visco-elasticity differs from elasticity by the fact that the state of stress at present depends, in the visco-elastic case, on all the deformations undergone in the past (and does not depend on them in the elastic case).

The principal object of this chapter is the investigation of problems with *conditions of friction at the boundary*, which leads to *new problems involving inequalities*: this represents the subject of Section 5 et seq., while the first Sections (2 to 4) treat classical problems fairly completely (notably with a proof of *Korn's inequality*, which is the basic tool).

2. Classical Linear Elasticity

2.1. The Constituent Law

In the linear theory which we consider to begin with, the constituent law expresses a linear relation between the stress tensor σ_{ij} (Sec. I.1.1) and the linearized strain tensors $\varepsilon_{ij}(u)$ (Sec. I.1.3), i.e.,

$$\sigma_{ij} = a_{ijkh} \varepsilon_{kh}(u) ; \tag{2.1}$$

the a_{ijkh} in (2.1) are the *coefficients of elasticity*, independent of the strain tensor. The coefficients of elasticity have properties of *symmetry*

$$a_{ijkh} = a_{jihk} = a_{khij} \tag{2.2}$$

and of *ellipticity*

(2.3) $$a_{ijkh}\varepsilon_{ij}\varepsilon_{kh} \geqslant \alpha_1 \varepsilon_{ij}\varepsilon_{ij}\,, \quad \alpha_1 \text{ a constant } > 0\,, \quad \forall \varepsilon_{ij}\,.$$

Generally, the law (2.1) corresponds to an unisotropic material. In the case where the material is *inhomogeneous*, the coefficients of elasticity depend on x; but assuming that they are measurable and bounded in x, all that follows can be adapted without difficulty (one would have to make *differentiability* assumptions in x, if one wanted to study the *regularity* of the solutions).

With the above assumptions, we can "invert" (2.1):

(2.4) $$\varepsilon_{ij}(u) = A_{ijkh}\sigma_{kh}$$

where

(2.5) $$A_{ijkh} = A_{khij} = A_{jikh}$$

and

(2.6) $$A_{ijkh}\sigma_{ij}\sigma_{kh} \geqslant \alpha_2 \sigma_{ij}\sigma_{ij}\,, \quad \alpha_2 > 0\,, \quad \forall \sigma_{ij}\,.$$

Setting

(2.7) $$\alpha = \min(\alpha_1, \alpha_2)$$

we shall replace the relations (2.3) and (2.6) by

(2.8) $$\begin{aligned} &a_{ijkh}\varepsilon_{ij}\varepsilon_{kh} \geqslant \alpha \varepsilon_{ij}\varepsilon_{ih}\,, \quad \alpha > 0\,, \\ &A_{ijkh}\sigma_{ij}{}_{kh} \geqslant \alpha \sigma_{ij}\sigma_{ij}\,. \quad \square \end{aligned}$$

Isotropic case (cf. P. Germain [1], W. Prager [1]). In the isotropic case, the coefficients a_{ijkh} are given by

(2.9) $$a_{ijkh} = \lambda \delta_{ij}\delta_{kh} + \mu(\delta_{ik}\delta_{jh} + \delta_{ih}\delta_{jk})\,,$$

where the scalars λ and μ are the *Lamé coefficients.* Then (2.1) becomes

(2.10) $$\sigma_{ij} = \lambda \varepsilon_{kh}\delta_{ij} + 2\mu\varepsilon_{ij} \quad (\varepsilon_{ij} = \varepsilon_{ij}(u))\,.$$

From this follows

(2.11) $$\sigma_{kh} = (3\lambda + 2\mu)\,\varepsilon_{kh}$$

such that the relations inverse to (2.10) become

(2.12) $$\varepsilon_{ij} = \frac{1}{2\mu}\sigma_{ij} - \frac{\lambda}{3\lambda + 2\mu}\sigma_{kh}\delta_{ij}\,.$$

Frequently, one sets:

$$(2.13) \qquad 3K = 3\lambda + 2\mu, \qquad \frac{1}{E} = \frac{\lambda+\mu}{\mu(3\lambda+2\mu)}, \qquad \nu = \frac{\lambda}{2(\lambda+\mu)}$$

where K is the *modulus of rigidity under compression*, E *Young's modulus* and ν *Poisson's ratio.*

The relations (2.12) can be expressed in terms of ν and E by

$$(2.14) \qquad \varepsilon_{ij} = \frac{1+\nu}{E}\sigma_{ij} - \frac{\nu}{E}\sigma_{kh}\delta_{ij}. \quad \square$$

Remark 2.1. From their physical interpretation, K and μ (shear modulus) satisfy

$$K \geqslant 0 \qquad \mu \geqslant 0$$

which implies

$$\sigma_{ij}\varepsilon_{ij} \geqslant 0$$

since σ_{ij} and ε_{ij} are linked by the constituent law. In the nonisotropic case, this implies:

$$\sigma_{ij}\varepsilon_{ij} = A_{ijkh}\sigma_{ij}\sigma_{kh} = a_{ijkh}\varepsilon_{ij}\varepsilon_{kh} \geqslant 0.$$

In practice, the "*stronger*" inequalities (2.8) are usually satisfied, but they do not have a "basic" physical character. □

Remark 2.2 (Non linear elasticity). One can develop *non linear*[1] theories of elasticity in the following two different ways:

i) The constituent law is a relation between the stress tensor and the nonlinear strain tensor (cf. Sec. I.1.3); depending on the case, one then obtains materials of "*harmonic type*" (cf. F. John [1, 2]) or the Mooney material [1].

ii) The constituent law is a *nonlinear* relation between the stress tensor and the linearized strain tensor (cf. Dinca [1]).

There is no doubt that some of the problems solved in Section 5 et seq. can be handled in these situations, but this remains to be further clarified.

2.2. Classical Problems of Linear Elasticity

2.2.1. Linearization of the Equation of Conservation of Mass and of the Equation of Motion

Let Ω be an open set in $\mathbb{R}^3$ occupied by the elastic body in its non deformed state. We assume that Ω is bounded by a regular boundary Γ. Let us denote by

[1] It should be especially noted that the problems investigated at the beginning of Section 5 are *non linear though applying to a material with a linear constituent law.*

$\{x_i\}$, an element of $\bar{\Omega}$, the coordinates of a material particle in the non deformed state and let $\{X_i\}$ be the coordinates of the same particle at the time t. We have

(2.15) $$X_i = x_i + u_i(x,t),$$

where $\{u_i\}$ is the displacement vector of the particle $x = \{x_i\}$. *Assuming that the displacements are "small"*, we linearize with respect to the u_i. Then, for every function $\xi, t \to f(\xi, t)$, regular on $\mathbb{R}^3$ and in $t \geq 0$, we have:

(2.16) $$f(X_i,t) = f(x_i,t) + \frac{\partial f}{\partial X_i}(X_i,t)_{X=x} u_i + \cdots,$$

and

(2.17) $$\partial f(X_i,t)/\partial X_i = \partial f(x_i,t)/\partial x_i + \cdots$$

where the ... denote terms in the u_i of higher order than the last term written out.

The equation of conservation of mass which has the form (Sec. I.1.2)

(2.18) $$\mathrm{d}\rho/\mathrm{d}t + \rho\, \partial v_i(X,t)/\partial X_i = 0$$

then becomes, *after linearization* (in the u_i)

(2.19) $$\partial \rho(x,t)/\partial t + \rho(x,t)\, \partial v_i(x,t)/\partial x_i = 0.$$

For constant x, we then have

$$\mathrm{d}\rho/\rho + v_{i,i}\, \mathrm{d}t = 0,$$

or, after integration over $[0,t]$,

$$\log(\rho(x,t)/\rho(x,0)) + \int_0^t v_{i,i}(x,\tau)\, \mathrm{d}\tau = 0.$$

Since $u_i(x,t)$ is small, it suggests itself to assume that $v_{i,i} = \partial^2 u_i/\partial x\, \partial t$ is also small and, consequently, the preceding relation yields, to first order,

(2.20) $$\rho(x,t) = \rho_0(x)\left[1 - \int_0^t v_{i,i}(x,\tau)\, \mathrm{d}\tau\right],$$

where we set $\rho_0(x) = \rho(x,0)$.

The term $\rho\gamma_i$ in the equations of motion then linearize to $\rho_0\, \partial^2 u_i/\partial t^2$ and the equations become

(2.21) $$\rho_0(x)(\partial^2 u_i/\partial t^2) = \sigma_{ij,j} + f_i.$$

In what follows, we will make the assumption:

$$\rho_0 \equiv 0,$$

thus assuming that, in its non deformed state, the material has constant density. All that follows can be extended without difficulty, other than technicalities, to the case where $\rho_0(x)$ is a measurable and bounded function such that

$$\rho_0(x) \geqslant \rho_0 > 0$$

(replacing in the following the scalar product $\int_\Omega fg\,dx$ by the *equivalent* scalar product $\int_\Omega \rho_0(x) fg\,dx$).

To sum up, the equations of linear elasticity are:

$$\partial^2 u_i/\partial t^2 = \sigma_{ij,j} + f_i \quad \text{in} \quad \Omega\,, \tag{2.22}$$

$$\sigma_{ij} = a_{ijkh}\varepsilon_{kh}(u) \tag{2.23}$$

where the density after deformation is given by (2.20) (with $\rho_0(x)=1$); the vector $f=\{f_i\}$ represents a volume density of prescribed forces.

2.2.2. Boundary Conditions[2]

We assume that the *displacements* are given on a part Γ_U of Γ and that the *surface forces* are given on the remainder Γ_F of the boundary. **We assume that Γ_U and Γ_F do not depend on time**[3]. Then

$$\Gamma = \Gamma_U \cup \Gamma_F\,, \qquad \Gamma_U \cap \Gamma_F = \emptyset\,, \tag{2.24}$$

(2.25) $u_i = U_i$ on Γ_U, $\{U_i\}$ = vector field prescribed on Γ_U, possibly dependent on time.

(2.26) $\sigma_{ij}n_j = F_i$ on Γ_F, $\{F_i\}$ = surface density of forces prescribed on Γ_F, possibly dependent on time. □

Remark 2.3. One of the sets Γ_U or Γ_F may be empty.

Remark 2.4. The boundary data (2.25) and (2.26) can in fact be generalized without any new fundamental difficulty: we can assume that in each point of Γ the component of displacement (resp. of traction) in one direction and the components of traction (resp. of displacement) in the plane perpendicular to that direction are prescribed.

2.2.3. Summary

Summing up, we seek u and σ_{ij}, linked by Equations (2.22), (2.23), with *boundary conditions* (2.25), (2.26), to which we naturally add the *initial conditions*

$$\begin{aligned} &u_i(x,0) = u_{0i}(x)\,, \\ &\partial u_i(x,0)/\partial t = u_{1i}(x)\,. \end{aligned} \tag{2.27}$$

---

[2] These boundary conditions will have to be modified when we introduce friction (in Sec. 5).

[3] Giving up this hypothesis appears to lead to very interesting open questions.

In the *stationary* case (discussed in Sec. 3), we put $\partial^2 u_i/\partial t^2 = 0$ in (2.22). Then, obviously, conditions (2.27) become irrelevant.

We will now proceed to give the *variational formulation* of the problem of evolution.

2.3. Variational Formulation of the Problem of Evolution

2.3.1. Green's Formula

We set

$$(2.28) \qquad (Au)_i = -\partial/\partial x_j \; (a_{ijkh}\varepsilon_{kh}(u)),$$

the differential system A being the system of elasticity.

Generally, we set:

$$(f,g) = \int_\Omega f_i g_i \, dx \quad \text{for} \quad f,g \in (L^2(\Omega))^3$$

and, for two vectors u and v, we set:

$$(2.29) \qquad a(u,v) = \int_\Omega a_{ijkh}\varepsilon_{kh}(u)\varepsilon_{ij}(v)\, dx .$$

Then, with σ_{ij} and u connected by (2.1), we have

$$\begin{aligned} (Au)_i &= -\sigma_{ij,j}, \\ a(u,v) &= \int_\Omega \sigma_{ij}\varepsilon_{ij}(v)\, dx = \int_\Omega \sigma_{ij}(\partial v_i/\partial x_j)\, dx \\ &= -\int_\Omega \sigma_{ij,j} v_i \, dx + \int_\Gamma (\sigma_{ij} n_j) v_i \, d\Gamma . \end{aligned}$$

Thus, we obtain *Green's formula*:

$$(2.30) \qquad \begin{aligned} &(Au,v) = a(u,v) - \int_\Gamma (\sigma_{ij}n_j) v_i \, d\Gamma \\ &\sigma_{ij} \text{ linked to } u \text{ be (2.1).} \quad \square \end{aligned}$$

Remark 2.5. Further on, we will use the following (classical) notations. We introduce:

$$(2.31) \qquad \begin{aligned} \sigma_N &= \sigma_{ij} n_i n_j, \\ \sigma_T &= \{\sigma_{iT}\}, \\ \sigma_{iT} &= \sigma_{ij} n_j - \sigma_N n_i \end{aligned}$$

and

$$(2.32) \qquad v_N = v_i n_i, \qquad v_T = v - n v_N \qquad (n = \{n_i\}).$$

Then

$$(\sigma_{ij}n_j) v_i = \sigma_T v + \sigma_N v_N = \sigma_T v_T + \sigma_N v_N$$

from which, substituting in (2.30),

(2.33) $$(Au,v)=a(u,v)-\int_\Gamma(\sigma_T v_T+\sigma_N v_N)\,d\Gamma\,. \quad \square$$

Remark 2.6. From (2.2), (2.8), it follows that

(2.34) $$a(u,v)=a(v,u) \qquad \forall u,v\,,$$

(2.35) $$a(v,v)\geqslant\alpha\int_\Omega \varepsilon_{ij}(v)\,\varepsilon_{ij}(v)\,dx\,.$$

2.3.2. Variational Formulation

With the notation of the preceding Section 2.3.1, (2.22) becomes:

(2.36) $$\partial^2u/\partial t^2+Au=f \quad \text{in} \quad Q=\Omega\times]0,T[$$ [4]

with the boundary conditions (2.25), (2.26) and the initial conditions which we write, in the notation already used in the preceding chapters,

$$(u(t)=\text{function } x\to u(x,t),\ u'(t)=\partial u(t)/\partial t):$$

(2.37) $$u(0)=u_0\,,$$

(2.38) $$u'(0)=u_1\,.$$

Taking the scalar product of (2.36) with $v-u(t)$, where v *is a test function such that*

(2.39) $$v_i=U_i \quad \text{on} \quad \Gamma_U\,,$$ [5]

we obtain (a condition *equivalent* to (2.36)):

(2.40) $$(u''(t),v-u(t))+(Au(t),v-u(t))=(f(t),v-u(t))\;;$$

of course, as always in the first stage of weak formulations, we assume that all functions are regular, so that integration by parts is ligitimate.

Making use of (2.30), we have, since $v=u$ on Γ_U:

$$(Au(t),v-u(t))=a(u(t),v-u(t))-\int_{\Gamma_F}(\sigma_{ij}n_j)(v_i-u_i(t))\,d\Gamma;$$

from this it follows, if we take (2.36) into consideration, that

(2.41) $$(Au(t),v-u(t))=a(u(t),v-u(t))-\int_{\Gamma_F}F(t)(v-u(t))\,d\Gamma\,,$$

[4] T arbitrary but fixed. We could let $T\to+\infty$.

[5] We will see later on why it is of interest to take the scalar product with $v-u$. Observe that v varies over an affine space.

and thus (2.40) leads to

$$(u''(t), v - u(t)) + a(u(t), v - u(t)) = (f(t), v - u(t)) + \int_{\Gamma_F} F(t)(v - u(t))\,d\Gamma \quad \forall v \ \text{ with (2.39)}. \tag{2.42}$$

Conversely, if $u = u(t)$ is a (regular) function *which satisfies* (2.25) ($u_i = U_i$ on Γ_U), (2.42) and (2.37), (2.38), *then* u *is the looked for solution of the problem* (it is sufficient to reverse the steps).

Orientation. In the following (in fact in Sec. 4), we will solve this problem, in a suitable sense. Before that, in Section 3, we will investigate the *stationary case*, which is important by itself and which necessitates the introduction of *basic tools* for the solution of the problem of evolution.

3. Static Problems

3.1. Classical Formulation

In the static case, in the notation of Section 2, the problem consists in finding a function u which is a solution of

$$Au = f \quad \text{in } \Omega, \tag{3.1}$$

$$u_i = U_i \quad \text{on } \Gamma_U, \tag{3.2}$$

$$\sigma_{ij} n_j = F_i \quad \text{on } \Gamma_F. \tag{3.3}$$

3.2. Variational Formulation

Formally (cf. Sec. 2.3.2), the problem is *equivalent* to the determination of a u *satisfying* (3.2) *and such that*

$$a(u, v - u) = (f, v - u) + \int_{\Gamma_F} F(v - u)\,d\Gamma \tag{3.4}$$

$\forall v$ such that

$$v_i = U_i \quad \text{on } \Gamma_U. \tag{3.5}$$

Interpretation. A field of vectors $v = \{v_i\}$ which satisfies (3.5) is called kinematically admissible (abbreviated, we say v is a k.a.f.). Then u is a k.a.f. satisfying (3.4) for all k.a.f.'s, *or also: u is a* k.a.f. *which, among all the* k.a.f.'s, *minimizes the potential energy of v, defined by*

$$I(v) = \tfrac{1}{2} a(v, v) - (f, v) - \int_{\Gamma_F} Fv\,d\Gamma. \tag{3.6}$$

Naturally, we have to express this a little more precisely. We will put[6]

(3.7) $$V=\{v \mid v=\{v_i\},\ v_i \in H^1(\Omega)\}=(H^1(\Omega))^3 .$$

which is a Hilbert space for the scalar product

(3.8) $$(u,v)=(u_i,v_i)_{H^1(\Omega)}=\int_\Omega (u_i v_i+u_{i,j}v_{i,j})\,\mathrm{d}x .$$

For $v\in V$, we can define the trace of v_i on Γ; here $v_i\in H^{1/2}(\Gamma)$. Then, we define the K.a.f. by $v\in U_{\mathrm{ad}}$ where

(3.9) $$\mathscr{U}_{\mathrm{ad}}=\{v \mid v\in V,\ v_i=U_i \text{ on } \Gamma_U\}$$

which makes sense (i.e., it defines a non empty set) provided that

(3.10) $$U_i\in H^{1/2}(\Gamma) \quad \forall i .$$

Then $\mathscr{U}_{\mathrm{ad}}$ is *a closed linear affine manifold in* V.

Furthermore, for $v\in\mathscr{U}_{\mathrm{ad}}$, we can define v_i on Γ_F, and then v_i belongs in particular to $H^{1/2}(\Gamma_F)$, but does not range over all of $H^{1/2}(\Gamma_F)$; the v_i must satisfy a "compatibility condition" at the interface between Γ_F and Γ_U, because $v_i=u_i=U_i$ on Γ_U; these compatibility conditions are technically complicated (for analogous questions, cf. Lions-Magenes [1], Vol. 1, Chap. 1 and 2); to avoid these difficulties (of a technical nature), we will assume that

(3.11) $$F\in(L^2(\Gamma_F))^3 ;$$

then $v\to\int_{\Gamma_F} Fv\,\mathrm{d}\Gamma$ is continuous on V and we can state the static problem in the following (definitive) way:

Problem 3.1. *To minimize on* $\mathscr{U}_{\mathrm{ad}}$ *the "potential energy" functional defined by* $I(v)$, or, what amounts to the same thing: *to find* $u\in\mathscr{U}_{\mathrm{ad}}$ such that (3.4) holds $\forall v\in\mathscr{U}_{\mathrm{ad}}$.

The *basic* problem is now the *coerciveness of* $a(v,v)$; this question rests on *Korn's inequality* which is the subject of the following Section 3.3.

3.3. Korn's Inequality and its Consequences

Theorem 3.1. *Let Ω be a bounded open set with regular boundary*[7]. *There exists a constant $c>0$ (dependent on Ω) such that*

(3.12) $$\int_\Omega \varepsilon_{ij}(v)\,\varepsilon_{ij}(v)\,\mathrm{d}x+\int_\Omega v_i v_i\,\mathrm{d}x \geqslant c\|v\|_V^2 \qquad \forall v\in V .$$

[6] We are now making use of Sec. I.4.

[7] More generally, the result is valid when Ω is an open set, bounded or *not*, with a boundary that can be described by a *finite* number of "local" charts that are functions once continuously differentiable and *bounded* in both directions. This is the case for an open set Ω with regular boundary that is either bounded or "goes regularly to infinity".

This result is not trivial: indeed, the left hand side of (3.12) only involves *certain combinations* of first derivatives, namely $v_{i,j}+v_{j,i}$, while the right hand side of (3.12) involves *all* first derivatives; clearly, the "inverse" inequality is obvious, so that (3.12) is equivalent to saying that

(3.13) $\quad (\int_\Omega \varepsilon_{ij}(v)\,\varepsilon_{ij}(v)\,\mathrm{d}x + \int_\Omega v_i v_i\,\mathrm{d}x)^{1/2}$ is a norm on V equivalent to $\|v\|_V$.

Theorem 3.1 is a simple consequence (as we shall see) of

Theorem 3.2. *Let Ω be an open bounded set with regular boundary*[8]. *Let v be a distribution on Ω such that*

(3.14) $\quad v\in H^{-1}(\Omega)\,, \quad v_{,i}\in H^{-1}(\Omega) \quad \forall i\,.$

Then

(3.15) $\quad v\in L^2(\Omega)\,.$

Proof of Theorem 3.1 *using Theorem* 3.2. Let E be the space of $v\in(L^2(\Omega))^3$ such that

$$\varepsilon_{ij}(v)\in L^2(\Omega) \qquad \forall i,j\,;$$

E is a Hilbert space for the norm

$$(\int_\Omega \varepsilon_{ij}(v)\,\varepsilon_{ij}(v)\,\mathrm{d}x + \int_\Omega v_i v_i\,\mathrm{d}x)^{1/2}\,.$$

We have

(3.16) $$\frac{\partial^2 v_i}{\partial x_j \partial x_k} = \frac{\partial}{\partial x_j}\varepsilon_{ik}(v) + \frac{\partial}{\partial x_k}\varepsilon_{ij}(v) - \frac{\partial}{\partial x_i}\varepsilon_{jk}(v)\,.$$

If $v\in E$, then $\varepsilon_{ij}(v)\in L^2(\Omega)$, therefore $\partial\varepsilon_{ij}(v)/\partial x_k\in H^{-1}(\Omega)$ and therefore (3.16) yields

(3.17) $\quad \partial^2 v_i/\partial x_j \partial x_k \in H^{-1}(\Omega) \qquad \forall i,j,k\,.$

Applying Theorem 3.2 to $v_{i,k}$, we see that (3.17) implies

$$v_{i,k}\in L^2(\Omega) \qquad \forall i,k$$

therefore $v\in(H^1(\Omega))^3$. Thus, we have the *algebraic equality* $E=(H^1(\Omega))^3$. Since the injection of $(H^1(\Omega))^3\to E$ is *continuous* and, as we shall see, *surjective*, this is (according to the closed graph theorem) an *isomorphism*, whence (3.12). □

[8] Cf. footnote [7] of Theorem 3.1.

Remark 3.1. A variant of the preceding proof shows that Theorem 3.2 implies the existence of a constant c_1 such that

(3.18) $$\|v\|_{L^2(\Omega)} \leqslant c_1(\|v\|_{H^{-1}(\Omega)} + \sum_{i=1}^n \|v_{,i}\|_{H^{-1}(\Omega)}),$$

from which we can again deduce (3.12). ☐

Proof of Theorem 3.2. For the proof, we introduce the space

(3.19) $$X(\Omega) = \{v \mid v \in H^{-1}(\Omega),\ v_{,i} \in H^{-1}(\Omega)\ \forall i\},$$

(which is a Hilbert space for the norm

$$(\|v\|_{H^{-1}(\Omega)}^2 + \sum_{i=1}^n \|v_{,i}\|_{H^{-1}(\Omega)}^2)^{1/2}.$$

We have to show that

(3.20) $$X(\Omega) = L^2(\Omega).$$

The proof proceeds in several steps.

i) Relation (3.20) is true for $\Omega = R^3$.[9]

Indeed, by a Fourier transformation (cf. Sec. I.4), assumption (3.14) is then equivalent to

$$(1+|\xi|^2)^{-1/2}\hat{v} \in L^2(R_\xi^3), \qquad (1+|\xi|^2)^{-1/2}\xi_i\hat{v} \in L^2(R_\xi^3)$$

therefore

$$\int_{R^3}(1+|\xi|^2)^{-1}(1+\sum_{i=1}^n \xi_i^2)|\hat{v}|^2\,d\xi < \infty$$

i.e., $$v \in L^2(R^3) = L^2(\Omega).$$

ii) It is sufficient to prove (3.20) for a half-space

(3.21) $$\Omega = \{x \mid x_3 > 0\}.$$

Indeed, let $\alpha_0, \alpha_1, \ldots, \alpha_N$ be such that

(3.22) $$\alpha_0 \in \mathscr{D}(\Omega), \quad \alpha_i \in \mathscr{D}(\bar{\Omega}), \quad i = 1, \ldots, N, \quad \sum_{i=0}^N \alpha_i = 1,$$
α_i with support in a local chart defining Γ[10].

In general, if $\varphi \in \mathscr{D}(\bar{\Omega})$, $v \to \varphi v$ maps $X(\Omega) \to X(\Omega)$; then

(3.23) $$v = \sum_{i=0}^N \alpha_i v;$$

---

[9] Naturally, the fact that the dimension equals 3 is not significant here.

[10] If Γ is not bounded, certain α_i do not have compact support but are, by assumption, in a chart where we can introduce the image in the half-space (3.21).

but we can consider $\alpha_0 v$ as an element of $X(R^3)$ (extended as 0 outside of Ω), thus, according to i), $\alpha_0 v \in L^2(\Omega)$. According to (3.23), we will have the result if we prove that $\alpha_i v \in L^2(\Omega)$. But $\alpha_i v$ has support in a local chart; therefore, we can consider its *image* in the half-space (3.21), *if we assume* Γ *to be a once continuously differentiable manifold of dimension* $(n-1)$. The image of $\alpha_i v$ is in the space $X(\Omega)$ with Ω given by (3.21). From this, assertion ii) follows.

Henceforth, we can thus assume Ω *to be defined by* (3.21).

iii) We introduce

$$H_0^1(0,\infty; L^2(R^2)) = \{\varphi| \ \varphi, \ d\varphi/dx_3 \in L^2(0,\infty; L^2(R^2)), \ \varphi(x',0)=0\}$$

(where $x' = \{x_1, x_2\}$),

$H^{-1}(0,\infty; L^2(R^2))$ = dual of $H_0^1(0,\infty; L^2(R^2))$ (when $L^2(R^3)$ is identified with its dual),

$$Y(\Omega) = \{v| \quad v, \quad dv/dx_3 \in H^{-1}(0,\infty; L^2(R^2))\}. \tag{3.24}$$

We have: $Y(\Omega) \subset X(\Omega)$ and, more precisely,

$$Y(\Omega) \text{ is dense in } X(\Omega). \tag{3.25}$$

Indeed, let ρ_m be a regularizing sequence for $\mathcal{D}(R^2_{x'})$ [11]; for $v \in X(\Omega)$, we define

$$v_m = v \underset{(x')}{*} \rho_m \tag{3.26}$$

(for every distribution f on $]0,\infty[$ with values in $L^2(R^2_{x'})$ and every linear continuous operator π of $L^2(R^2_{x'})$ into itself, we define πf as distribution on $]0,\infty[$ with values in $L^2(R^2_{x'})$); formally

$$v_m(x) = \int_{R^2} v(x'-y', x_3)\rho_m(y')\,dy', \qquad y' = \{y_1, y_2\}.$$

Then, as $m \to \infty$, we have: $v_m \to v$ in $X(\Omega)$. In particular, v_m is contained in $Y(\Omega)$; (indeed, v_m and dv_m/dx_3 are contained in $H^{-1}(0,\infty; H^k(R^2))$ $\forall k$).

iv) $\mathcal{D}(\Omega)$ is dense in $X(\Omega)$.
By virtue of (3.25), it is sufficient to prove that

$$\mathcal{D}(\bar{\Omega}) \text{ is dense in } Y(\Omega). \tag{3.27}$$

For this purpose, we make use of the Hahn-Banach Theorem. Let $v \to M(v)$ be a linear form, continuous on $Y(\Omega)$ and therefore of the form

$$\begin{aligned} &M(v) = \int_0^\infty [(f,v) + (g, dv/dx_3)]\,dx_3, \\ &f, g \in H_0^1(0,\infty; L^2(R^2_{x'})). \end{aligned} \tag{3.28}$$

[11] Then ρ_m is a sequence of functions $\geqslant 0$ of $\mathcal{D}(R^2)$, $\int \rho_m(x')\,dx' = 1$, with support of ρ_m contained in a ball having the origin as center and the radius ε_m, $\varepsilon_m \to 0$, when $m \to \infty$.

We assume that $M(v)=0\ \forall v\in\mathscr{D}(\bar{\Omega})$. We have to prove that $M=0$. But if $\tilde{f},\tilde{g}$ denote the continuations of f and g as 0 for $x_3<0$, the condition

$$\text{“}M(v)=0 \quad \forall v\in\mathscr{D}(\bar{\Omega})\text{”}$$

is equivalent to

$$\tilde{f}-\mathrm{d}\tilde{g}/\mathrm{d}x_3=0\,, \tag{3.29}$$

therefore

$$\mathrm{d}\tilde{g}/\mathrm{d}x_3\in H^1(-\infty,+\infty;L^2(R^2_{x'}))$$

and consequently

$$g\in H^2_0(0,\infty;L^2(R^2_{x'}))\,. \tag{3.30}$$

But then

$$\int_0^\infty (g,\mathrm{d}v/\mathrm{d}x_3)\,\mathrm{d}x_3=-\int_0^\infty (\mathrm{d}g/\mathrm{d}x_3,v)\,\mathrm{d}x_3 \qquad \forall v\in Y(\Omega)$$

and therefore $M(v)=0\ \forall v\in Y(\Omega)$.

v) For $v\in\mathscr{D}(\bar{\Omega})$, we now set

$$Pv(x)=\left|\begin{array}{l} v(x)\quad \text{if}\quad x_3>0\\ a_1 v(x',-x_3)+a_2(x',-2x_3)\end{array}\right. \tag{3.31}$$

where

$$a_1+a_2=1,\quad a_1+a_2/2=-1,\quad (\text{i.e. } a_1=-3,\ a_2=4)\,. \tag{3.32}$$

We verify that

(3.33) $v\to Pv$ is continuous in $\mathscr{D}(\bar{\Omega})$, provided with the topology induced by $X(\Omega)\to X(R^3)$.

Conceding this point for a moment, we see that, (according to iv) we can extend $v\to Pv$ to a linearly continuous mapping of $X(\Omega)\to X(R^3)$ and such that

(3.34) Pv, restricted to Ω, equals v.

Then, for $v\in X(\Omega)$, $Pv\in X(R^3)$, therefore, as a consequence of i), $Pv\in L^2(R^3)$, therefore, by (3.34),

$$v\in L^2(\Omega)\,.$$

It remains to prove (3.33).

Since $a_1+a_2=1$, we have:

$$\frac{\partial}{\partial x_3} Pv = \begin{cases} \partial u/\partial x_3, & x_3>0 \\ -a_1 \dfrac{\partial v}{\partial x_3}(x',-x_3)-2a_2\dfrac{\partial v}{\partial x_3}(x',-2x_3), & x_3<0. \end{cases}$$

Setting $\partial v/\partial x_3=w$, we introduce

(3.35) $$Qw = \begin{cases} w \text{ if } x_3>0 \\ -a_1 w(x',-x_3)-2a_2 w(x',-2x_3) \text{ if } x_3<0. \end{cases}$$

Thus, we have to prove that P (resp. Q) is continuous on $\mathscr{D}(\bar{\Omega})$, with the topology induced by $H^{-1}(\Omega)\to H^{-1}(R^3)$, thus, *by transposition*, that ${}^tP$ (resp. ${}^tQ$) is linearly continuous from $H^1(R^3)\to H_0^1(\Omega)$.

Now, for $\varphi\in H^1(R^3)$, we have:

$$\begin{aligned} {}^tP\varphi(x) &= \varphi(x)+a_1\varphi(x',-x_3)+\tfrac{1}{2}a_2\varphi(x',-x_3/2), \\ {}^tQ\varphi(x) &= \varphi(x)-a_1\varphi(x',-x_3)-a_2\varphi(x',-x_3/2). \end{aligned}$$

Then ${}^tP\varphi(x',0)={}^tQ\varphi(x',0)=0$ because of (3.32), from which the result follows. □

Remark 3.2. In Gobert [1], a different proof can be found that is valid for more general conditions on Ω (it is sufficient that Ω has the "cone property"), but that makes use of the theory of singular integrals. □

From the preceding results, we will now draw consequences that are fundamental for the remainder of the chapter.

Theorem 3.3. *We make the same assumptions as for Theorem* 3.1. *Let* $\Gamma_U\subset\Gamma$ *and* Γ_U *have positive measure. Let*

(3.36) $$V_0=\{v|\quad v\in(H^1(\Omega))^3,\quad v=0 \text{ on } \Gamma_U\}$$

Then there exists an $\alpha_0>0$ *such that*

(3.37) $$a(v,v)\geqslant\alpha_0\|v\|_V^2 \qquad \forall v\in V_0.$$

Proof. 1) In general

(3.38) $$a(v,v)=0 \Leftrightarrow v\in\mathscr{R}, \quad v\in V$$

where

(3.39) $$\mathscr{R}=\{v|\quad v(x)=a+b\wedge x,\quad a,b\in R^3\}\ ^{12}.$$

[12] $\mathscr{R}$ = the set of rigid displacements.

Because Γ_U has positive measure,

$$v \in \mathscr{R} \cap V_0 \Rightarrow v = 0$$

and consequently

(3.40) $$a(v,v)=0, \quad v \in V_0 \Leftrightarrow v=0.$$

2) Thus we see that $a(v,v)$ is a norm on V_0 and we have to show that this norm is equivalent to $\|v\|_V$. To simplify the notation, we set

(3.41) $$\varepsilon(v) = \int_\Omega \varepsilon_{ij}(v)\,\varepsilon_{ij}(v)\,dx.$$

According to (2.35) and Theorem 3.1, it all comes back to proving the existence of a $c_0 > 0$ such that

(3.42) $$\varepsilon(v) \geqslant c_0 |v|^2, \quad |v|^2 = \int_\Omega v_i v_i \, dx, \quad \forall v \in V_0.$$

If we replace v by $v|v|^{-1}$, we can assume that $|v|=1$; then we have to prove the existence of a $c_0>0$ with $\varepsilon(v) \geqslant c_0$. We argue by contradiction. If the result were false, there would exist a sequence $v_n \in V_0$ with:

(3.43) $$|v_n| = 1, \quad \varepsilon(v_n) \to 0.$$

According to Theorem 3.1, we then have: $\|v_n\|_V \leqslant$ constant; then we can select a subsequence, again denoted by v_n such that

(3.44) $$v_n \to v \quad \text{weakly in } V.$$

But then:

$$\liminf \varepsilon(v_n) \geqslant \varepsilon(v),$$

therefore $\varepsilon(v)=0$, therefore, according to 1), $v=0$.

Furthermore, since Ω is bounded by a regular boundary, the injection $V \to (L^2(\Omega))^3$ is *compact* and, consequently, $v_n \to 0$ strongly in $(L^2(\Omega))^3$ which contradicts the assumption $v_n = 1$, from which the result follows. □

Corollary 3.1. *We assume that Ω is bounded by a regular boundary and that Γ_U has positive measure. Then, with $I(v)$ given by (3.6), we have*

(3.45) $$I(v) \to +\infty \quad \text{when} \quad \|v\|_V \to +\infty, \quad v \in \mathscr{U}_{ad}.$$

Proof. Let $\Phi \in V$, with

$$\Phi = U \quad \text{on} \quad \Gamma_U.$$

Then, if $v \in \mathscr{U}_{ad}$, we have $v - \Phi \in V_0$. We set $v - \Phi = v_0$. Then

$$I(v) = \tfrac{1}{2}a(v_0, v_0) + a(v_0, \Phi) + \tfrac{1}{2}a(\Phi, \Phi) - (f, \Phi + v_0) - \int_{\Gamma_F} F(v_0 + \Phi)\,d\Gamma .$$

According to Theorem 3.3, we thus have

$$I(v) \geqslant \tfrac{1}{2}\alpha_0 \|v_0\|_V^2 - c_1 \|v_0\|_V - c_2 ,$$

from which

$$\text{(3.46)} \qquad I(v) \geqslant \alpha_1 \|v\|_V^2 - c_3 \|v\|_V - c_4 \qquad \forall v \in \mathscr{U}_{ad} ,$$

from which, in particular, (3.45) follows. □

Let us now examine the case when $\Gamma_U = \Phi$. Then $a(v,v)^{1/2}$ is no longer a norm on $V_0 = V$ but a semi-norm. In accordance with (3.38), we pass to the quotient with $\mathscr{R}$; we introduce thus:

$$\text{(3.47)} \qquad V^{\cdot} = V/\mathscr{R} .$$

For $u^{\cdot}, v^{\cdot} \in V^{\cdot}$, we define

$$\text{(3.48)} \qquad a(u^{\cdot}, v^{\cdot}) = a(u, v), \qquad u \in u^{\cdot}, \qquad v \in v^{\cdot} .$$

We have

Theorem 3.4. *With the assumptions of Theorem* 3.1, *we have*

$$\text{(3.49)} \qquad a(v^{\cdot}, v^{\cdot}) \geqslant \alpha \|v^{\cdot}\|_{V^{\cdot}}^2 , \qquad \alpha > 0 , \qquad \forall v^{\cdot} \in V^{\cdot} .$$

Proof. According to Theorem 3.1, we can take

$$(\varepsilon(v) + |v|^2)^{1/2}$$

as norm on V. Then (3.49) is equivalent to

$$\text{(3.50)} \qquad \varepsilon(v) \geqslant \alpha \left[\inf_{\rho \in \mathscr{R}} |v + \rho|^2 + \varepsilon(v) \right] .$$

We define:

(3.51) $P =$ the orthogonal projection operator in $(L^2(\Omega))^3$ (for the scalar product corresponding to $|v|$) of $(L^2(\Omega))^3 \to \mathscr{R}$.

Then

$$\inf_{\rho \in \mathscr{R}} |v + \rho|^2 = |v - Pv|^2$$

and the proof of (3.50) reduces to the proof of (compare with (3.42))

$$\varepsilon(v) \geqslant c_1 |v - Pv|^2 \qquad \forall v \in V. \tag{3.52}$$

We argue as for (3.42). Replacing v by $v|v-Pv|^{-1}$, we are reduced to the case

$$|v - Pv| = 1, \tag{3.53}$$

and we then have to prove the existence of a $c_1 > 0$ such that

$$\varepsilon(v) \geqslant c_1. \tag{3.54}$$

We argue by contradiction. If (3.54) were true, we can find a v_n of V such that

$$|v_n - Pv_n| = 1, \qquad \varepsilon(v_n) \to 0. \tag{3.55}$$

But then, if we set

$$w_n = v_n - Pv_n, \tag{3.56}$$

we have

$$|w_n| = 1, \qquad \varepsilon(w_n) \to 0,$$

from which we deduce (as for (3.43)) that we can select a subsequence, again denoted by w_n, such that $w_n \rightharpoonup w$ weakly in V, $\varepsilon(w) = 0$. Thus $w \in \mathscr{R}$. On the other hand, $w_n \in \mathscr{R}^\perp$ [13] therefore $w \in \mathscr{R}^\perp$ and therefore $w = 0$. But then $w_n \to 0$ strongly in $(L^2(\Omega))^3$ which is absurd since $|w_n| = 1$. □

3.4. Results

3.4.1. The Case "Γ_U has Positive Measure"

From Corollary 3.1 and the *strict convexity* of the (continuous) functional $v \to I(v)$ on $\mathscr{U}_{ad}$, we deduce at once:

Theorem 3.5. *Under the assumptions of Theorem* 3.1, *Problem* 3.1 *has a unique solution.*

Remark 3.3. We return to formulation (3.4), (3.5). We can reduce this formulation to the case of vector spaces; introducing a $\Phi \in V$ with $\Phi = U$ on Γ_U, we put:

$$u - \Phi = u_0, \quad v - \Phi = v_0, \quad u_0, v_0 \in V_0, \quad V_0 \text{ defined in (3.36)}. \tag{3.57}$$

Then (3.4), (3.5) are equivalent to the determination of a $u_0 \in V_0$ such that

$$a(u_0, v_0) = (f, v_0) + \int_{\Gamma_F} F v_0 \, d\Gamma - a(\Phi, v_0) \qquad \forall v_0 \in V_0. \quad \square \tag{3.58}$$

[13] $\mathscr{R}^\perp$ is the subspace of V, orthogonal to $\mathscr{R}$, in the Hilbertian structure induced by $(L^2(\Omega))^3$.

3.4.2. The Case "Γ_U is empty" [14]

We saw that, when Γ_U is empty, it is necessary to pass to the quotient with $\mathscr{R}$. In this case, $\mathscr{U}_{ad}=V=$ vector space so that (3.4), (3.5) are *equivalent* (under transformation (3.57)) *to the determination of a* $u\in V$ *such that*

$$a(u,v)=(f,v)+\int_\Gamma Fv\,\mathrm{d}\Gamma \qquad \forall v\in V, \qquad (\Gamma_F=\Gamma)\,. \tag{3.59}$$

The problem is only possible if the linear form on the right hand side of (3.59) is zero on $\mathscr{R}$, i.e., if

$$\int_\Omega f_i\rho_i\,\mathrm{d}x+\int_\Gamma F_i\rho_i\,\mathrm{d}\Gamma=0 \qquad \forall\rho\in\mathscr{R}\,. \tag{3.60}$$

From the point of view of mechanics, (3.60) *expresses the fact that the system of forces* $\{f_i\}$ and $\{F_i\}$ (the only given quantities) *is statically equivalent to zero.*

If we put

$$L(v)=(f,v)+\int_\Gamma Fv\,\mathrm{d}\Gamma\,, \tag{3.61}$$

we can "pass to the quotient with $\mathscr{R}$": we define $\dot{L}$, a linear continuous form on $\dot{V}=V/\mathscr{R}$, by

$$\dot{L}(\dot{v})=L(v)\,, \qquad v\in\dot{v}\,; \tag{3.62}$$

then (3.59) is equivalent to

$$a(\dot{u},\dot{v})=\dot{L}(\dot{v}) \qquad \forall\dot{v}\in\dot{V}\,. \tag{3.63}$$

According to Theorem 3.4, this problem has a unique solution; therefore

Theorem 3.6. *We assume* $\Gamma_U=\emptyset$. *The forces* f_i *and* F_i *satisfy* (3.60). *Then Problem* 3.1 *has a solution u, defined within an arbitrary rigid displacement.*

Remark 3.4. In Theorem 3.6, *the fields of strains and stresses are unique.* □

Remark 3.5. As we already indicated, there certainly are intermediate cases between "Γ_U with positive measure" and "$\Gamma_U=\emptyset$".

For example, if, instead of force densities, we consider *point forces*, then Γ_U reduces to a finite number of points. □

Remark 3.6. We have solely considered *weak solutions.* In order to study regularity of solutions, we refer to the general works on elliptic equations: Lions-Magenes [1], Nécas [1].

3.5. Dual Formulations

We will examine dual formulations from two slightly different points of view.

[14] Evidently, there are intermediate possibilities between the cases 3.4.1 and 3.4.2. Cf. Remark 3.5.

3.5.1. Statically Admissible Fields and Potential Energy

We reconsider Problem (3.1), (3.2), (3.3), *or again*

$$\begin{aligned} &\sigma_{ij,j}+f_i=0\,, \\ &\sigma_{ij}=a_{ijkh}\varepsilon_{kh}(u) \quad \text{in} \quad \Omega\,, \\ &\sigma_{ij}n=F_i \quad \text{on} \quad \Gamma_F\,, \\ &u_i=U_i \quad \text{on} \quad \Gamma_U\,. \end{aligned} \tag{3.64}$$

We introduce the set K of *statically admissible fields* (s.a.f.) as follows:

$$K=\{\tau \mid \quad \tau_{ij}\in L^2(\Omega) \quad \forall i,j, \quad \tau_{ij}=\tau_{ji}, \quad \tau_{ij,j}+f_i=0 \text{ in } \Omega, \quad \tau_{ij}n=F_i \text{ on } \Gamma_F\}\,. \tag{3.65}$$

Remark 3.7. Let us introduce the space

$$H=\{\tau \mid \tau_{ij}\in L^2(\Omega) \;\; \forall i,j\}\,, \tag{3.66}$$

provided with the *Hilbert* structure:

$$(\sigma,\tau)=\int_\Omega \sigma_{ij}\tau_{ij}\mathrm{d}x\,.$$

We verify that if $\tau\in H$ and satisfies

$$\tau_{ij,j}+f_i=0\,,$$

i.e., $\tau_{ij,j}\in L^2(\Omega)$, then we can uniquely define

$$\tau_{ij}n_j\in H^{-1/2}(\Gamma) \tag{3.67}$$

in such a way that definition (3.65) of K *has a meaning*. The set K *is an affine closed manifold of* H. ▯

For $\tau\in K$, we define its potential energy $J(\tau)$ by

$$J(\tau)=\tfrac{1}{2}\mathscr{A}(\tau,\tau)-\int_{\Gamma_U}\tau_{ij}n_jU_i\,\mathrm{d}\Gamma \tag{3.68}$$

where generally

$$\mathscr{A}(\sigma,\tau)=\int_\Omega A_{ijkh}\sigma_{kh}\tau_{ij}\mathrm{d}x\,. \tag{3.69}$$

A dual problem to the initial problem is then:

Problem 3.2. *To minimize* $J(\tau)$ *on* K. ▯

Remark 3.8. In Problem 3.2, we assume that K is not empty.

If Γ_U has positive measure, K is not empty since $U_i\in H^{1/2}(\Gamma_U)$ (more precisely: U_i is a restriction to Γ_U of elements of $H^{1/2}(\Gamma)$).

If $\Gamma_U=\emptyset$, K is not empty if and only if the system of forces f_i and F_i is statically equivalent to zero. □

According to (2.8), we have:

$$\mathscr{A}(\tau,\tau)\geqslant\alpha\int_\Omega \tau_{ij}\tau_{ij}\,\mathrm{d}x=\alpha\|\tau\|_H^2$$

from which at once follows

Theorem 3.7. *Assuming that K is not empty* (cf. Remark 3.8), *Problem* 3.2 *has one and only one solution.*

It remains to clarify the relation between the initial (or primal) problem and the "dual" problem (that up to now has been only a *definition!*); this is the object of

Theorem 3.8. *We assume that K is not empty. Let σ be the solution of Problem* 3.2. *Then equations*

$$\varepsilon_{ij}(u)=A_{ijkh}\sigma_{kh} \tag{3.70}$$

and the boundary conditions

$$u=U \quad on \quad \Gamma_U \tag{3.71}$$

(resp. *without boundary conditions if $\Gamma_U=\emptyset$) define u uniquely* (resp. *within a rigid displacement*), *where u is the* (resp. *a*) *solution of Problem* 3.1.

Proof. We only have to prove that if u is the (or a) solution of Problem 3.1, then σ, defined by $\sigma_{ij}=a_{ijkh}\varepsilon_{kh}(u)$, minimizes J on K, i.e., satisfies

$$\mathscr{A}(\sigma,\tau-\sigma)-\int_{\Gamma_U}U_i(\tau_{ij}n_j-\sigma_{ij}n_j)\,\mathrm{d}\Gamma=0\,.$$

Now, applying (3.70) to (3.69), we have

$$\begin{aligned}\mathscr{A}(\sigma,\tau-\sigma)&=\int_\Omega \varepsilon_{ij}(u)(\tau_{ij}-\sigma_{ij})\,\mathrm{d}x=\int_\Omega u_{i,j}(\tau_{ij}-\sigma_{ij})\,\mathrm{d}x\\&=\int_\Gamma u_i(\tau_{ij}n_j-\sigma_{ij}n_j)\,\mathrm{d}\Gamma-\int_\Omega u_i(\tau_{ij,j}-\sigma_{ij,j})\,\mathrm{d}x\\&=\int_{\Gamma_U}u_i(\tau_{ij}n_j-\sigma_{ij}n_j)\,\mathrm{d}\Gamma \quad \text{since} \quad \sigma,\tau\in K\end{aligned}$$

from which the result follows, since $u_i=U_i$ on Γ_U. □

3.5.2. Duality and Lagrange Multipliers

We will now derive the preceding results from a different vantage point, namely by making use of "Lagrange multipliers", in the following denoted by q_{ij} [15].

[15] We give a *direct* derivation, without recourse to the general theorems on duality for which we refer, in particular, to Rockafellar [1] and Teman [2].

We start with the functional $I(v)$ and consider the equations

$$a_{ijkh}\varepsilon_{kh}(v)=\tau_{ij} \tag{3.72}$$

as constraints for which we introduce the multipliers q_{ij}.

We note that, with the constraint (3.72), we have:

$$a(v,v)=\mathscr{A}(\tau,\tau)\;;$$

we then introduce the functional

$$I(\tau,v,q)=\tfrac{1}{2}\mathscr{A}(\tau,\tau)-(f,v)-\int_{\Gamma_F}Fv\,\mathrm{d}\Gamma-\int_\Omega q_{ij}(\tau_{ij}-a_{ijkh}\varepsilon_{kh}(v))\,\mathrm{d}x\,, \tag{3.73}$$

where τ,v,q are *independent variables*, $\tau,q\in H$, $v\in V$, with $v_i=U_i$ on Γ_U, i.e., $v\in\mathscr{U}_{\mathrm{ad}}$ [16].

We observe that

$$\inf_{\tau,v\in H} I(\tau,v,q)=\hat{I}(q)\leqslant\inf\; I(\tau,v,q)=\inf\; I(v)=I(u) \tag{3.74}$$

$[\tau,v$ satisfying (3.72)$]$,

and therefore

$$\sup\;\hat{I}(q)\leqslant I(u)\,. \tag{3.75}$$

We calculate $I(q)$ explicitly. We have:

$$I(\tau,v,q)=I_1(\tau,v,q)+I_2(\tau,v,q)\,,$$

where

$$I_1(\tau,v,q)=I_1(\tau,q)=\tfrac{1}{2}\mathscr{A}(\tau,\tau)-\int_\Omega q_{ij}\tau_{ij}\,\mathrm{d}x\,,$$

$$I_2(\tau,v,q)=I_2(v,q)=-(f,v)-\int_{\Gamma_F}Fv\,\mathrm{d}\Gamma+\int_\Omega q_{ij}a_{ijkh}\varepsilon_{kh}(v)\,\mathrm{d}x\,.$$

We easily verify that

$$\inf\; I_1(\tau,q)=-\tfrac{1}{2}\mathscr{A}(\tau,\tau),\quad\text{where}\quad A_{ijkh}\tau_{kh}=q_{ij}\,. \tag{3.76}$$

Then

$$I_2(v,q)=-(f,v)-\int_{\Gamma_F}Fv\,\mathrm{d}\Gamma+\int_\Omega \tau_{kh}v_{k,h}\,\mathrm{d}x$$

and we verify that

$$\inf_{v\in\mathscr{U}_{\mathrm{ad}}} I_2(v,q)=-\infty\quad\text{unless}\quad \tau\in K,\quad\text{in which case}$$

$$\inf\; I_2(v,q)=\int_{\Gamma_U}\tau_{kh}n_hU_k\,\mathrm{d}\Gamma\,.$$

---

[16] With adequate multipliers, we also could consider as constraints the conditions $v_i=U_i$ on Γ; *there are several dual formulations.*

Therefore:

$$\hat{I}(q) = \begin{vmatrix} -\frac{1}{2}\mathscr{A}(\tau,\tau) + \int_{\Gamma_U} \tau_{kh} n_h U_k \, d\Gamma & \text{if } \tau \in K, \text{ where } q_{ij} = A_{ijkh}\tau_{kh}, \\ -\infty & \text{otherwise}. \end{vmatrix} \tag{3.77}$$

But

$$I(u) = -\tfrac{1}{2} a(u,u) + \int_{\Gamma_U} \sigma_{ij} n_j U_i \, d\Gamma \tag{3.78}$$

such that

$$I(u) = -\tfrac{1}{2}\mathscr{A}(\sigma,\sigma) + \int_{\Gamma_U} \sigma_{ij} n_j U_i \, d\Gamma$$

and therefore

$$\begin{aligned} \sup \hat{I}(q) &= \sup_{\tau \in H} \left(-\tfrac{1}{2}\mathscr{A}(\tau,\tau) + \int_{\Gamma_U} \tau_{kh} n_h U_k \, d\Gamma\right) \\ &\geqslant -\tfrac{1}{2}\mathscr{A}(\sigma,\sigma) + \int_{\Gamma_U} \sigma_{kh} n_h U_k \, d\Gamma = I(u) \end{aligned}$$

which shows, comparing with (3.75), that

$$\sup \hat{I}(q) = I(u). \tag{3.79}$$

Replacing $\hat{I}(q)$ by its value (3.77), we then have:

$$\inf_{\tau \in K} \left[\tfrac{1}{2}\mathscr{A}(\tau,\tau) - \int_{\Gamma_U} \tau_{kh} n_h U_k \, d\Gamma\right] + I(u) = 0 \tag{3.80}$$

or again

$$\inf_{\tau \in K} \left[\tfrac{1}{2}\mathscr{A}(\tau,\tau) - \int_{\Gamma_U} \tau_{kh} n_h U_k \, d\Gamma\right] + \inf_{v \in \mathscr{U}_{ad}} I(v) = 0. \tag{3.81}$$

Moreover, if the first inf (resp. the second) *is attained at* σ (resp. u), σ *and* u *are linked to each other by the relation*

$$\sigma_{ij} = a_{ijkh} \varepsilon_{kh}(u). \quad \square$$

Remark 3.9. An exhaustive treatment of variational formulations of the problems of elasticity can be found in Tonti [1].

4. Dynamic Problems

4.1. Statement of the Principal Results

Applying the tools of Section 3, we are now able to solve the dynamic problems considered in Section 2.

We introduce a function $\Phi(t)$ such that[17]

$$\Phi(t) \in (H^1(\Omega))^3, \quad \Phi_i(t) = U_i(t) \quad \text{on} \quad \Gamma_U. \tag{4.1}$$

Then, *replacing* $u(t)$ by $u(t) - \Phi(t)$ and keeping the notation $u(t)$, problem (2.42) is equivalent to the following: we define

$$V_0 = \{v \mid \quad v \in (H^1(\Omega))^3, \quad v_i = 0 \text{ on } \Gamma_U\} \tag{4.2}$$

(then $V_0 = V = (H^1(\Omega))^3$ if $\Gamma_U = \emptyset$); we are looking for a function $t \to u(t)$ of $[0, T] \to V_0$ such that

$$(u''(t), v) + a(u(t), v) = (\Psi(t), v) \qquad \forall v \in V_0 \tag{4.3}$$

where

$$(\Psi(t), v) = (f(t), v) + \int_{\Gamma_F} F(t)\, v\, d\Gamma + (\Phi''(t), v) + a(\Phi(t), v), \tag{4.4}$$

with the initial conditions

$$u(0) = u_0, \quad u'(0) = u_1. \tag{4.5}$$

We introduce

$$H = (L^2(\Omega))^3; \tag{4.6}$$

we note that

$$V_0 \subset H, \quad V_0 \text{ is dense in } H. \tag{4.7}$$

We will denote the norm in V_0 (resp. H) by $\| \ \|$ (resp. $| \ |$) and the scalar product in H by $(\ , \ )$.

We identify H with its dual: then

$$H \subset V_0', \quad V_0' \text{ dual of } V_0. \tag{4.8}$$

By $(\ , \ )$, we denote the scalar product of V_0' and V_0, which is compatible with the scalar product in H.

By $\| \ \|_*$ we denote the norm in V_0', dual to $\| \ \|$, then:

$$\|f\|_* = \sup |(f, v)|, \quad v \in V_0, \quad \|v\| \leqslant 1.$$

---

[17] The mode of dependence on t will be precisely stated later on.

We will now prove

Theorem 4.1. *We assume that*

(4.9) $\Psi, \Psi' \in L^2(0,T; V_0')$ [18],

(4.10) $u_0 \in V_0, \quad u_1 \in H.$

There exists one and only one function u such that

(4.11) $u \in L^\infty(0,T; V_0),$

(4.12) $u' \in L^\infty(0,T; H),$

(4.13) $u'' \in L^\infty(0,T; V_0')$

and which satisfies (4.3) *and* (4.5).

Remark 4.1. Let us assume that

(4.14) $f_i, \ \partial f_i/\partial t \in L^2(Q), \quad Q = \Omega \times]0,T[,$

(4.15) $F_i, \ \partial F_i/\partial t \in L^2(\Sigma), \quad \Sigma = \Gamma \times]0,T[,$

(4.16) U_i is the restriction to $\Gamma_U \times]0,T[$ of $\tilde{U}_i$ with
$\tilde{U}_i, \ \partial \tilde{U}_i/\partial t, \ \partial^2 \tilde{U}_i/\partial t^2, \ \partial^3 \tilde{U}_i/\partial t^3 \in L^2(0,T; H^{1/2}(\Gamma)).$

Then we can choose Φ in such a way that Ψ, which we defined in (4.4), *satisfies* (4.9).

Indeed, we can choose Φ in such a way that

(4.17) $\Phi, \ldots, \Phi''' \in L^2(0,T; (H^1(\Omega))^3)$ [19].

We verify that then $\Psi' \in L^2(0,T; V_0')$. We have:

$$(\Psi'(t), v) = (f'(t), v) + \int_{\Gamma_F} F'(t) v \, d\Gamma + (\Phi'''(t), v) + a(\Phi'(t), v)$$

from which

$$|(\Psi'(t), v)| \leqslant c_1 [|f'(t)| \, |v| + \|F''(t)\|_{(L^2(\Gamma))^3} \|v\| + |\Phi'''(t)| \, |v| + \|\Phi'(t)\| \, \|v\|]$$

from which

$$\|\Psi'(t)\|_* \leqslant c_2 [|f'(t)| + \|F'(t)\|_{(L^2(\Gamma))^3} + \|\Phi'(t)\| + |\Phi'''(t)|]$$

from which the result follows. ▯

[18] Sufficient conditions for this to hold are indicated in Remark 4.1.

[19] We could generalize the assumptions for U_i; what we need for Φ is:

$$\Phi, \Phi' \in L^2(0,T; (H^1(\Omega))^3), \qquad \Phi'', \Phi''' \in L^2(0,T; H).$$

Remark 4.2. As seen in Section 3, we have:

(4.18) $a(u,v)=a(v,u) \quad \forall u,v\in V_0,$

(4.19) $a(v,v)\geqslant\alpha\|v\|^2 \quad \forall v\in V, \quad \alpha>0 \text{ if } \Gamma_U\neq\emptyset,$

(4.20) $\forall\lambda>0$, there exists an $\alpha>0$ such that $a(v,v)+\lambda|v|^2\geqslant\alpha\|v\|^2 \quad \forall v\in V,$ if $\Gamma_U=\emptyset \quad (V_0=V)$. □

In the following, we will prove Theorem 4.1 *in an abstract setting where only* V_0, H, V_0' *and* $a(u,v)$ *which satisfy* (4.20) *are involved.* □

Remark 4.3. It can happen in certain applications (for example in linearized thermo-elasticity, when we uncouple the temperature) that the coefficients a_{ijkh} are functions of x and t:

$$a_{ijkh}=a_{ijkh}(x,t).$$

In that case, we have to consider a bilinear form $a(t;u,v)$, continuous on V_0 and *dependent on* t; then Theorem 4.1 is valid (with an analogous proof), if

(4.21) $a(t;u,v)=a(t;v,u) \quad \forall u,v\in V_0,$

(4.22) there exists a $\lambda>0$ such that $a(t;v,v)+\lambda|v|^2\geqslant\alpha\|v\|^2, \quad \alpha>0, \quad \forall v\in V_0,$

(4.23) $\forall u,v\in V_0$, the function $t\to a(t;u,v)$ is once continuously differentiable in $[0,T]$. □

Remark 4.4. The case $\Gamma_U=\emptyset$.

The formulation of Theorem 4.1 *does not make any distinction, when* $\Gamma_U=\emptyset$, between the case where, $\forall t$, the forces $f_i(t)$ and $F_i(t)$ constitute a system statically equivalent to zero, i.e.,

(4.24) $(f(t),\rho)+\int_{\Gamma_F}F_i(t)\rho_i\,d\Gamma=0,$

and the case where (4.24) does not hold.

Let us examine the two cases more closely.

i) (4.24) *holds.*

Then we can more precisely describe *the structure of the solution* u in the following way[20]:

(4.25) $u(t)=u_0+tu_1+w(t),$

[20] We assume $u_1\in V(=V_0)$. Otherwise, we have to use a more complicated start than u_0+tu_1 in (4.25).

where w satisfies

$$(4.26)\qquad (w(t),\rho)=0 \quad \forall \rho\in\mathscr{R}.$$

Indeed, if we define $w(t)$ by (4.25) and substitute it in (4.3), it follows that:

$$(4.27)\qquad (w''(t),v)+a(u_0+tu_1+w(t),v)=(f(t),v)+\int_{\Gamma_F} F(t)v\,d\Gamma,$$

(since $\Gamma_U=\Phi$, hence $\Phi(t)=0$).

Taking $v=\rho\in\mathscr{R}$ in (4.27), we deduce, since $a(u(t),\rho)=0$ and since (4.24) holds:

$$(w''(t),\rho)=0,$$

which, together with $w(0)=0$, $w'(0)=0$, proves (4.26).

We now turn to the second case:

ii) *Relation* (4.24) *does not hold.*

In this case, the forces applied to the elastic body cause it to describe a solid rigid motion with superimposed elastic deformations.

Since then the motion involves large amplitudes, linearization is no longer legitimate. Therefore, Theorem 4.1 is always correct, but the model is no longer adequate. We must then, *to begin with*, determine the solid rigid motion. *Subsequently*, in a reference system tied to the thus determined motion, we treat the problem of elastic deformations *with, this time legitimate, linearization*, but *adding to the forces f_i and F_i the inertial forces of transportation and complementary forces*, say $\mathscr{F}$. The totality of forces $f(t)$, $F(t)$ and $\mathscr{F}(t)$ then forms, at each moment, a system statically equivalent to zero and we have returned to case i). □

4.2. Proof of Theorem 4.1

In order to prove *the existence* of a solution in Theorem 4.1, we use a method of approximation analogous to the one used in Section I.5.6.1.

The space V_0 is *separable*[21]. Then we can choose a sequence $w_1, \dots, w_m, \dots$, such that, $\forall m, w_1, \dots, w_m$ are linearly independent and such that the finite linear combinations of the w_j are dense in V_0. Furthermore, we assume that

$$w_1=u_0 \quad (\text{if } u_0\neq 0).$$

We define u_m, the "approximate solution of order m" by:

$$(4.28)\qquad \begin{aligned} &u_m(t)\in[w_1,\dots,w_m]=\text{space spanned by } w_1,\dots,w_m\\ &(u_m''(t),v)+a(u_m(t),v)=(\Psi(t),v) \qquad \forall v\in[w_1,\dots,w_m],\\ &u_m(0)=u_0,\\ &u_m'(0)=u_{1m},\quad u_{1m}\in[w_1,\dots,w_m],\quad u_{1m}\to u_1 \text{ in } H \text{ when } m\to\infty. \end{aligned}$$

---

[21] This is an assumption (incidentally, not an essential one) in the "abstract" case.

The system (4.28) is a system of m linear differential equations of second order, that is non singular since $w_1, \ldots, w_m$ are linearly independent; therefore, (4.28) defines the u_m uniquely.

A priori estimates for u_m. If we take $v = u'_m(t)$ in (4.28) which is permissible, it follows that

$$\frac{1}{2}\frac{\mathrm{d}}{\mathrm{d}t}|u'_m(t)|^2 + \frac{1}{2}\frac{\mathrm{d}}{\mathrm{d}t}a(u_m(t), u_m(t)) = (\Psi(t), u'_m(t))$$

from which

$$\begin{aligned}(4.29)\qquad |u'_m(t)|^2 + a(u_m(t), u_m(t)) &= |u_{1m}|^2 + a(u_0, u_0) + 2\textstyle\int_0^t (\Psi(\sigma), u'_m(\sigma))\,\mathrm{d}\sigma \\ &= |u_{1m}|^2 + a(u_0, u_0) + 2(\Psi(t), u_m(t)) - 2(\Psi(0), u_0) \\ &\quad - 2\textstyle\int_0^t (\Psi'(\sigma), u_m(\sigma))\,\mathrm{d}\sigma\,.\end{aligned}$$

But we have, with the c's denoting various constants >0,

$$\begin{aligned}&a(v,v) \geqslant \alpha\|v\|^2 - c|v|^2\,,\\ &|u_{1m}| \leqslant c|u_1|\,,\\ &2|(\Psi(t), u_m(t))| \leqslant \tfrac{1}{2}\alpha\|u_m(t)\|^2 + c\|\Psi(t)\|_*^2\,,\end{aligned}$$

such that (4.29) gives

$$\begin{aligned}(4.30)\qquad |u'_m(t)|^2 + \tfrac{1}{2}\alpha\|u_m(t)\|^2 &\leqslant c(|u_1|^2 + \|u_0\|^2 + \|\Psi(0)\|_*^2) \\ &\quad + c\|\Psi(t)\|_*^2 + c|u_m(t)|^2 + c\textstyle\int_0^t \|\Psi'(\sigma)\|_*\,\|u_m(\sigma)\|\,\mathrm{d}\sigma\,.\end{aligned}$$

But

$$\begin{aligned}&u_m(t) = u_0 + \textstyle\int_0^t u'_m(\sigma)\,\mathrm{d}\sigma \quad \text{gives}\\ &|u_m(t)|^2 \leqslant 2|u_0|^2 + c\textstyle\int_0^t |u'_m(\sigma)|^2\,\mathrm{d}\sigma\,,\end{aligned}$$

such that (4.30) gives

$$\begin{aligned}(4.31)\qquad |u'_m(t)|^2 + \|u_m(t)\|^2 &\leqslant c(|u_1|^2 + \|u_0\|^2 + \|\Psi(0)\|_*^2 + \|\Psi(t)\|_*^2 \\ &\quad + \textstyle\int_0^t \|\Psi'(\sigma)\|^2\,\mathrm{d}\sigma) + c\textstyle\int_0^t (|u'_m(\sigma)|^2 + \|u_m(\sigma)\|^2)\,\mathrm{d}\sigma\,.\end{aligned}$$

We set

$$(4.32)\qquad |||\Psi|||^2 = \textstyle\int_0^t (\|\Psi(t)\|_*^2 + \|\Psi'(t)\|_*^2)\,\mathrm{d}t$$

and

$$(4.33)\qquad \varphi_m(t) = |u'_m(t)|^2 + \|u_m(t)\|^2\,.$$

Then (4.31) gives

(4.34) $$\varphi_m(t) \leqslant c(|u_1|^2 + \|u_0\|^2 + |||\Psi|||^2) + c\int_0^t \varphi_m(\sigma)\,d\sigma$$

from which, according to Gronwall's inequality:

(4.35) $$\varphi_m(t) \leqslant c(|u_1|^2 + \|u_0\|^2 + |||\Psi|||^2)\exp(ct).$$

Conclusion

(4.36) u_m (resp. u'_m) remains in a bounded subset of $L^\infty(0, T; V_0)$ (resp. $L^\infty(0, T; H)$ when $m \to \infty$. □

Therefore, we conclude from (4.36) that we can select from the u_m a subsequence u_μ such that:

(4.37) u_μ (resp. u'_μ) $\to u$ (resp. u') weakly star in $L^\infty(0, T; V_0)$ (resp. $L^\infty(0, T; H)$)[22].

We will now verify that u is a solution of the problem. We introduce the space E of functions φ of the form

(4.38) $$\varphi(t) = \sum_{j=1}^{\mu_0} \varphi_j(t) w_j, \quad \varphi_j \in C^1([0, T]), \quad \varphi_j(T) = 0,$$
μ_0 arbitrary finite.

From (4.28), we conclude for $m = \mu > \mu_0$ that

$$(u''_\mu, \varphi) + a(u_\mu, \varphi) - (\Psi, \varphi) = 0 \quad \text{for } \varphi \text{ given by (4.38)},$$

from which

(4.39) $$\int_0^T [-(u'_\mu, \varphi') + a(u_\mu, \varphi) - (\Psi, \varphi)]\,dt = (u_{1\mu}, \varphi(0)).$$

We can pass to the limit with μ in (4.39), from which $\forall \varphi \in E$:

(4.40) $$\int_0^T [-(u', \varphi') + a(u, \varphi) - (\Psi, \varphi)]\,dt = (u_1, \varphi(0)).$$

Since the finite linear combinations of the w_j are dense in V_0, we obtain (4.40)

$$\forall \varphi \in C^1([0, T]; V), \quad \varphi(T) = 0.$$

From this, we deduce that, in the sense of distributions on $]0, T[$ with values in V_0,

(4.41) $$u'' + Au = \Psi$$

[22] I. e., for example
$\int_0^T ((u_\mu, \varphi))\,dt \to \int_0^T ((u, \varphi))\,dt \quad \forall \varphi \in L^1(0, T; V_0)$.

where $A \in \mathscr{L}(V_0; V_0')$ is *defined* by

$$a(u,v) = (Au, v), \qquad \forall u, v \in V_0. \tag{4.42}$$

Therefore $u'' = \Psi - Au \in L^\infty(0, T; V_0')$ [23]. Taking the scalar products of the two sides of (4.41) with $\varphi \in E$ (for example) and comparing with (4.40), we conclude that

$$(u_1, \varphi(0)) = (u'(0), \varphi(0)) \qquad \forall \varphi \in E$$

from which $u'(0) = u_1$ follows. Since (4.37) implies that $u_\mu(0) = u_0 \to u(0)$, we have: $u(0) = u_0$ and u satisfies the conditions of Theorem 4.1. □

We now prove the *uniqueness* of the solution[24]. Let u satisfy (4.11), (4.12), (4.13) with

$$u'' + Au = 0, \qquad u(0) = 0, \qquad u'(0) = 0. \tag{4.43}$$

For $\varphi \in C^1([0, T]; V_0')$ (for example), *there exists*, according to the first part of the proof[25], a function w such that

$$w \in L^\infty(0, T; V_0), \qquad w' \in L^\infty(0, T; H), \qquad w'' \in L^\infty(0, T; V_0'), \tag{4.44}$$

$$w'' + Aw = \varphi, \tag{4.45}$$

$$w(T) = 0, \qquad w'(T) = 0. \tag{4.46}$$

The formula for integration by parts *is valid*:

$$\int_0^T (u'', w)\,dt = \int_0^T (u, w'')\,dt\,; \tag{4.47}$$

thus, taking the scalar product of the two sides of (4.43) with w and making use of (4.47), we obtain (since $a(u,v) = a(v,u)$ $\forall u, v$);

$$\int_0^T (u, w'' + Aw)\,dt = 0$$

i. e.

$$\int_0^T (u, \varphi)\,dt = 0 \qquad \forall \varphi \in C^1([0, T]; V_0'), \quad \text{therefore} \quad u = 0. \quad \square$$

4.3. Other Boundary Conditions

We consider some variants of the boundary problems studied above.

[23] Since assumptions (4.9) imply that $\Psi \in C^0([0, T]; V_0') \subset L^\infty(0, T; V_0')$.

[24] This proof is due to L. Tartar.

[25] Changing t to $T - t$.

4.3.1. **Variant I** (for Example, a Body on a Rigid Support)

We look for a solution u of

(4.48) $$\partial^2 u/\partial t^2 + Au = f \quad \text{in } Q,$$

with the boundary conditions:

(4.49) $$\sigma_{ij} n_j = F_i \quad \text{for} \quad i=1,2 \quad \text{on} \quad \Gamma \times]0,T[= \Sigma,$$

(4.50) $$u_3 = 0 \quad \text{on } \Sigma$$

and, of course, the *initial conditions* (4.5).

The *mechanical interpretation* of the problem is the following:
on the boundary Γ, there are given the first two components F_1 and F_2 of the surface density of forces and the third component u_3 of the displacement. This type of boundary conditions occurs for example in the following problem: an elastic body, having a plane face (example: hemisphere bounded by an equatorial plane) rests without friction on a fixed horizontal plane. If we assume that, at the time of the deformation, there is no "detachment", then the boundary conditions on the plane face are (4.49), (4.50). On the remainder of the boundary, we have, for example, $F_i = 0$, $i = 1, 2, 3$. ☐

The corresponding *static problem* is:

(4.51) $$Au = f$$

with (4.49), (4.50). ☐

For the variational formulation, we introduce

(4.52) $$V_1 = \{v | \quad v \in V = H^1(\Omega))^3, \quad v_3 = 0 \text{ on } \Gamma\}.$$

Then we can verify without effort that the *dynamic problem* (resp. *static*) is equivalent to the determination of a solution $u = u(t)$ of

(4.53) $$(u''(t), v) + a(u(t), v) = (f(t), v) + \int_\Gamma [F_1(t) v_1 + F_2(t) v_2] \, d\Gamma \qquad \forall v \in V_1,$$

with

(4.54) $$u \in L^\infty(0, T; V_1), \quad u' \in L^\infty(0, T; H), \quad u'' \in L^\infty(0, T; V_1')$$

and (4.5) (resp. the determination of a solution $u \in V_1$ of

(4.55) $$a(u, v) = (f, v) + \int_\Gamma [F_1 v_1 + F_2 v_2] \, d\Gamma \qquad \forall v \in V_1).$$

For the solution of these problems, we note that $a(u,v)$ is not coercive on V_1, since

(4.56) $$a(v,v)=0\,, \qquad v\in V_1 \Leftrightarrow v\in\mathscr{R}_1\,,$$

where

(4.57) $$\mathscr{R}_1=\{\rho| \quad \rho\in R, \quad \rho_3=0 \text{ on } \Gamma\}\,.$$

Now, for the static problem we introduce

$$V_1^{\bullet}=V_1/\mathscr{R}_1\,;$$

$$a(u^{\bullet},v^{\bullet})=a(u,v)\,, \qquad u\in u^{\bullet}\,, \qquad v\in v^{\bullet}\,, \qquad u^{\bullet},v^{\bullet}\in V_1^{\bullet}\,;$$

problem (4.55) is possible only if

(4.58) $$(f,\rho)+\int_\Gamma[F_1\rho_1+F_2\rho_2]\,d\Gamma=0 \qquad \forall\rho\in\mathscr{R}_1\,;$$

the *mechanical interpretation of* (4.58) is: the system of volume forces $(f_1,f_2,0)$ and surface forces $(F_1,F_2,0)$ is statically equivalent to zero.

Then we can prove, as in Section 3.4.2, that if (4.58) holds, the *static problem* (4.55) *has a solution which is determined except for the addition of an element of* $\mathscr{R}_1$. □

For the dynamic problem, since $\forall\lambda>0$ there exists an $\alpha>0$ such that

$$a(v,v)+\lambda|v|^2\geqslant\alpha\|v\|^2 \qquad \forall v\in V_1\,,$$

we can apply Theorem 4.1 which proves the existence and the uniqueness of a solution of (4.53), (4.54), (4.55).

If, $\forall t\in[0,T]$, we have:

(4.59) $$(f(t),\rho)+\int_\Gamma[F_1(t)\rho_1+F_2(t)\rho_2]\,d\Gamma=0 \qquad \forall\rho\in\mathscr{R}_1$$

then, assuming that $u_1\in V_1$, we have:

(4.60) $$u(t)=u_0+tu_1+w(t)\,,$$

where

(4.61) $$(w(t),\rho)=0 \qquad \forall\rho\in\mathscr{R}_1\,.$$

The proof is the same as in Remark 4.4.

If (4.59) does not hold, then, as in Remark 4.4, the model is no longer adequate and we have to make modifications analogous to those indicated in ii) of Remark 4.4. □

4.3.2. **Variant II** (a Body Placed in an Elastic Envelope)

We are looking for a solution of (4.48) *with the boundary conditions*

(4.62) $\sigma_T = 0$ on Σ,[26]

(4.63) $\sigma_N + k u_N = 0 \quad (k > 0)$ on Σ,

with the initial conditions (4.5); the *static case* corresponds to (4.51) and (4.62), (4.63).

The *mechanical interpretation* of (4.62), (4.63) is as follows: the tangential displacements of the points of Γ are free while the normal forces are forces of elastic rebound whose absolute value is proportional to the normal displacement. Boundary conditions of this type can be encountered in the case of a body restrained by elastic ties or in the case of a body in the interior of an elastic enclosure.

Variational formulation. We can show without difficulty that the *static problem* is equivalent to the determination of

$$u \in V = (H^1(Q))^3$$

such that

(4.64) $a_1(u, v) = (f, v) \quad \forall v \in V$,

where

(4.65) $a_1(u, v) = a(u, v) + k \int_\Gamma u_N v_N \, d\Gamma$.

The *dynamic problem* is equivalent to the determination of $u = u(t)$ such that

(4.66) $(u''(t), v) + a_1(u(t), v) = (f(t), v) \quad \forall v \in V$

with

(4.67) $u \in L^\infty(0, T; V), \quad u' \in L^\infty(0, T; H), \quad u'' \in L^\infty(0, T; V')$,

and (4.5).

For the solution of these problems, we prove

Theorem 4.2. *Let Ω be an open bounded set of R^3. Then, with $a_1(u, v)$ defined by* (4.65), *we have:*

(4.68) $a_1(v, v) \geq \alpha_1 \|v\|^2 \quad \forall v \in V$.

---

[26] The notation is the same as in Remark 2.5.

Proof. The entire proof amounts to showing (with the notation already used:

$$\varepsilon(v) = \int_\Omega \varepsilon_{ij}(v)\,\varepsilon_{ij}(v)\,\mathrm{d}x)$$

that

(4.69) $$\varepsilon(v) + \int_\Gamma v_N^2\,\mathrm{d}\Gamma \geqslant c\,|v|^2\,.$$

Replacing v by $v|v|^{-1}$, this reduces to proving

(4.70) $$\varepsilon(v) + \int_\Gamma v_N^2\,\mathrm{d}\Gamma \geqslant c > 0\,, \quad |v| = 1\,, \quad v \in V\,.$$

Assuming that (4.70) is false, there would exist a sequence v_α such that

(4.71) $$v_\alpha \in V\,, \quad |v_\alpha| = 1\,, \quad \varepsilon(v_\alpha) + \int_\Gamma v_{\alpha N}^2\,\mathrm{d}\Gamma \to 0\,.$$

Then $\|v_\alpha\| \leqslant$ constant, and we can select a subsequence, again denoted by v_α, such that

(4.72) $$v_\alpha \to v \quad \text{weakly in } V$$

and since the injection $V \to (L^2(\Omega))^3$ is compact,

(4.73) $$v_\alpha \to v \quad \text{strongly in } (L^2(\Omega))^3\,.$$

But

$$\liminf\left[\varepsilon(v_\alpha) + \int_\Gamma v_{\alpha N}^2\,\mathrm{d}\Gamma\right] = 0 \geqslant \varepsilon(v) + \int_\Gamma v_N^2\,\mathrm{d}\Gamma\,,$$

therefore $\varepsilon(v) = 0$ and $v_N = 0$ on Γ. Then $v \in \mathcal{R}$, say $v = a + b \wedge x$, and $v_N = 0$ is equivalent to $an + (b \wedge x)n = 0$ on Γ, which is impossible (Γ cannot be a plane). Therefore $v = 0$ and thus (4.3) is equivalent to $|v_\alpha| \to 0$ which is absurd since

$$|v_\alpha| = 1 \quad \forall \alpha\,. \quad \square$$

From Theorem 4.2 and previous results, we immediately derive

Corollary 4.1. *The dynamic problem* (4.66), (4.67), (4.5) (resp. *the static problem* (4.64)) *has a unique solution.* $\square$

5. Linear Elasticity with Friction or Unilateral Constraints

5.1. First Laws of Friction. Dynamic Case

Outline. In this section, we consider deformations of a linear elastic body whose boundaries may possibly be subjected to classical boundary conditions of the type considered previously but also, at least on part of the boundary, to *conditions of friction.*

To begin with, we retain *Coulomb's law* (cf. 5.1.1 below); other laws will be investigated in Section 5.4.

5.1.1. Coulomb's Law

Let us consider two solid elastic bodies S_1 and S_2 in contact (cf. Fig. 15) at C, where the exterior unit normal to S_1 is $\mathbf{n}$. Let $\mathbf{F}$ be the force exerted by S_2 on S_1 at C (or the force density, if the contact extends over a whole region around C). We decompose $\mathbf{F}$:

$$\mathbf{F}=F_N\mathbf{n}+\mathbf{F}_T\,,\qquad F_N=\mathbf{F}.\mathbf{n}\,. \tag{5.1}$$

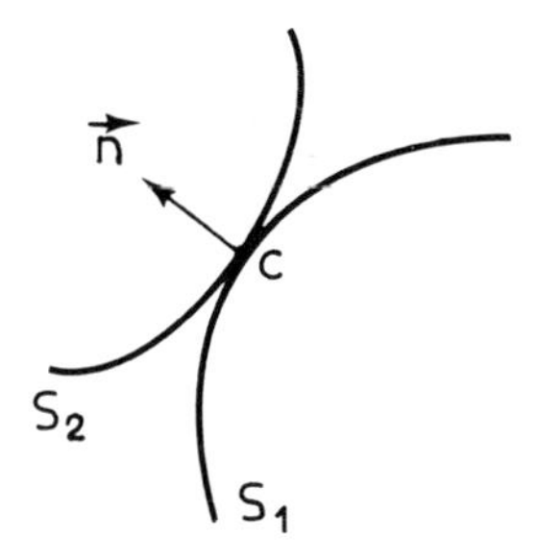

Fig. 15

Furthermore, let $\mathscr{F}=\mathscr{F}(C)$ be a positive scalar coefficient, called *coefficient of friction at the point* C (and that, of course, depends on the nature of the boundaries of S_1 and S_2 at C).

We have to distinguish *two types of contact* at C:

i) *bilateral contact,* i.e., the contact is maintained regardless of the direction of the forces;

ii) *unilateral contact,* i.e., the contact is maintained only if the forces press the solids against each other.

We now formulate *Coulomb's law* for each of these cases:

(5.2) (Coulomb's law for bilateral contact.)
At the instant t:

$$|\mathbf{F}_T(C)|<\mathscr{F}(C)|F_N(C)| \Rightarrow \partial\mathbf{u}_T(C)/\partial t=0\,,$$ [27]

$$|\mathbf{F}_T(C)|=\mathscr{F}(C)|F_N(C)| \Rightarrow \text{there exists } \lambda\geqslant 0 \text{ such that}$$

$$\partial\mathbf{u}_T(C)/\partial t=-\lambda\mathbf{F}_T(C)\,.$$

[27] We decomposed $\vec{u}(C)$ like $\vec{F}$: $\quad \vec{u}(C)=u_N(C)\vec{n}+\vec{u}_T(C)$.

(5.3)
(Coulomb's law for unilateral contact.)
It is necessary that $F_N(C) \leqslant 0$ and

$$|\mathbf{F}_T(C)| < -\mathscr{F}(C) F_N(C) \Rightarrow \partial u_T(C)/\partial t = 0,$$
$$|\mathbf{F}_T(C)| = -\mathscr{F}(C) F_N(C) \Rightarrow \text{there exists } \lambda \geqslant 0 \text{ such that}$$
$$\partial \mathbf{u}_T/\partial t = -\lambda \mathbf{F}_T. \quad \square$$

Remark 5.1. In the case of bilateral contact, we can also consider the situation where the friction coefficient $\mathscr{F}(C)$ is different (say $\mathscr{F}_1(C)$, $\mathscr{F}_2(C)$, distinct) according as $F_N(C) < 0$ or $F_N(C) > 0$, which leads to the law[28]:

$$(5.4)\quad \begin{cases} \left.\begin{aligned} &|\mathbf{F}_T(C)| < \mathscr{F}_1(C) F_N(C)^- \\ &\text{or just as well} \\ &|\mathbf{F}_T(C)| > \mathscr{F}_2(C) F_N(C)^+ \end{aligned}\right\} \Rightarrow \partial \mathbf{u}_T(C) \partial t = 0, \\ |\mathbf{F}_T(C)| = \mathscr{F}_1(C) F_N(C)^- \Rightarrow \text{there exists } \lambda_1 \geqslant 0 \text{ such that} \\ \qquad \partial \mathbf{u}_T(C)/\partial t = -\lambda_1 \mathbf{F}_T(C), \\ |\mathbf{F}_T(C)| = \mathscr{F}_2(C) F_N(C)^+ \Rightarrow \text{there exists } \lambda_2 \geqslant 0 \text{ such that} \\ \qquad \partial \mathbf{u}_T(C)/\partial t = -\lambda_2 \mathbf{F}_T(C). \end{cases}$$

This variant introduces only minor modifications in the considerations that follow: cf. Remark 5.2. □

5.1.2. Problems under Consideration

A part $\Gamma_{\mathscr{F}}$ of the boundary Γ of Ω (the elastic body under consideration) *shall be subjected to conditions of friction.*

The complementary part $\Gamma - \Gamma_{\mathscr{F}}$ shall be subjected to conditions of classical type, for example, to given forces or to given displacements.

In order to *simplify the discussion* somewhat, we will assume that, on $\Gamma - \Gamma_{\mathscr{F}}$, it is *the displacements that are given* (the case where one gives displacements on a part of $\Gamma - \Gamma_{\mathscr{F}}$ and forces on the remainder of $\Gamma - \Gamma_{\mathscr{F}}$ introduces no new difficulties except in writing). We set

$$(5.5)\qquad \Gamma - \Gamma_{\mathscr{F}} = \Gamma_U.$$

Then, if $u = u(x,t)$ denotes the field of displacements (which we are seeking), the *boundary conditions* on Γ_U are:

$$(5.6)\qquad u_i = U_i \quad \text{on} \quad \Sigma_U = \Gamma_U \times]0, T[.$$

On $\Sigma_{\mathscr{F}} = \Gamma_{\mathscr{F}} \times]0, T[$, the normal stress is given, say:

$$(5.7)\qquad \sigma_N = F_N \quad \text{on} \quad \Sigma_{\mathscr{F}}$$ [29]

[28] $x^+ = \sup(x, 0)$, $\quad x^- = \sup(-x, 0)$.

[29] We assume F_N to be measurable and bounded on $\Gamma_{\mathscr{F}}$. In order to simplify the discussion a little, we assume F_N to be independent of t; but we could just as well consider the case where F_N depends on t.

then we have, in the bilateral case (cf. (5.2))

$$
\begin{aligned}
&|\boldsymbol{\sigma}_T| < \mathscr{F}(x)|F_N| \Rightarrow \partial \mathbf{u}_T/\partial t = 0\,,\\
&|\boldsymbol{\sigma}_T| = \mathscr{F}(x)|F_N| \Rightarrow \text{there exists } \lambda \geqslant 0 \text{ such that } \partial \mathbf{u}_T/\partial t = -\lambda \boldsymbol{\sigma}_T\,,
\end{aligned}
\tag{5.8}
$$

where $x \in \Gamma_{\mathscr{F}}$, $\mathscr{F}(x) =$ friction coefficient at x; we assume that the function $x \to \mathscr{F}(x)$ is measurable and bounded on $\Gamma_{\mathscr{F}}$ and that

$$
\mathscr{F}(x) \geqslant \mathscr{F}_0 > 0\,, \qquad x \in \Gamma_{\mathscr{F}}\,.
\tag{5.9}
$$

Setting, for simplicity,

$$
\mathscr{F}\,|F_N| = g\,, \qquad g \in L^\infty(\Gamma_{\mathscr{F}})\,,
\tag{5.10}
$$

we arrive at the following problem:

Problem 5.1. (Dynamic with friction). The field of displacements $u = u(x,t)$ satisfies

$$
\begin{aligned}
&\partial^2 u_i/\partial t^2 = \sigma_{ij,j} + f_i \quad \text{in } Q\,,\\
&\sigma_{ij} = a_{ijkh}\varepsilon_{kh}(u)
\end{aligned}
\tag{5.11}
$$

with the *boundary conditions*

$$
u_i = U_i \quad \text{on} \quad \Gamma_U \times]0,T[= \Sigma_U\,,
\tag{5.12}
$$

and

$$
\begin{aligned}
&\sigma_N = F_N \quad \text{on} \quad \Sigma_{\mathscr{F}} = \Gamma_{\mathscr{F}} \times]0,T[\,,\\
&|\boldsymbol{\sigma}_T| < g \Rightarrow \partial u_T/\partial t = 0\\
&|\boldsymbol{\sigma}_T| = g \Rightarrow \text{there exists } \lambda \geqslant 0 \text{ such that } \partial u_T/\partial t = -\lambda \sigma_T
\end{aligned}
\tag{5.13}
$$

and the initial conditions

$$
u(x,0) = u_0(x)\,, \qquad \partial u(x,0)/\partial t = u_1(x)\,, \qquad x \in \Omega\,. \quad \square
\tag{5.14}
$$

Remark 5.2. In the bilateral case with different coefficients according to the direction of F_N (cf. Remark 5.1), we are led to the problem: find a u (and σ) with (5.11), (5.12) and (conditions (5.14) being unchanged), instead of (5.13):

$$
\sigma_N = F_N \quad \text{on } \Sigma_{\mathscr{F}}\,,
\tag{5.15}
$$

$$
\left|\begin{aligned}
&\left.\begin{aligned}&|\sigma_T| < \mathscr{F}_1 F_N^-\\ &\text{or}\\ &|\sigma_T| < \mathscr{F}_2 F_N^+\end{aligned}\right\} \Rightarrow \partial u_T/\partial t = 0\\
&\sigma_T = \mathscr{F}_1 F_N^- \Rightarrow \text{there exists } \lambda_1 \geqslant 0 \text{ such that } \partial u_T/\partial t = -\lambda_1 \sigma_T\,,\\
&\sigma_T = \mathscr{F}_2 F_N^+ \Rightarrow \text{there exists } \lambda_2 \geqslant 0 \text{ such that } \partial u_T/\partial t = -\lambda_2 \sigma_T\,. \quad \square
\end{aligned}\right.
\tag{5.16}
$$

Remark 5.3. In the unilateral case, if $F_N<0$ in every point $x\in\Gamma$, then (5.13) persists. If, on the contrary, $F_N\geqslant 0$ in certain points, then the contact which causes the friction may cease and we have a problem of the type of "Signorini with friction" (cf. Sec. 5.4.5). □

Outline. We shall now consider static *problems* associated with the laws of friction analogous to the preceding ones and then develop the tools necessary for the solution of the static case.

Subsequently, we will examine other laws of friction and then return to the dynamic case. □

5.2. Coulomb's Law. Static Case

5.2.1. Problems under Consideration

The *static* problem which corresponds to Problem 5.1 is the following:

Problem 5.2. Find a field of displacements $u=u(x)$ such that

$$\begin{aligned}&\sigma_{ij,j}+f_i=0\,,\\ &\sigma_{ij}=a_{ijkh}\varepsilon_{kh}(u)\quad\text{in }\Omega\,,\end{aligned}\tag{5.17}$$

$$u_i=U_i\quad\text{on }\Gamma_U\,,\tag{5.18}$$

$$\begin{aligned}&\sigma_N=F_N\quad\text{on }\Gamma_{\mathscr{F}}\,,\\ &|\sigma_T|<g\Rightarrow u_T=0\,,\\ &|\sigma_T|=g\Rightarrow\text{there exists }\lambda\geqslant 0\text{ such that }u_T=-\lambda\sigma_T\,.\quad\square\end{aligned}\tag{5.19}$$

Remark 5.4. The static analogue to the situation of Remark 5.2 is the following problem: find a u with (5.17), (5.18) and, instead of (5.19):

$$\sigma_N=F_N\quad\text{on }\Gamma_{\mathscr{F}}\,,\tag{5.20}$$

$$\left|\begin{aligned}&\left.\begin{aligned}&|\sigma_T|<\mathscr{F}_1F_N^-\\ &\text{or}\\ &|\sigma_T|<\mathscr{F}_2F_N^+\end{aligned}\right\}\Rightarrow u_T=0\,,\\ &|\sigma_T|=\mathscr{F}_1F_N^-\Rightarrow\exists\lambda_1>0\text{ such that }u_T=-\lambda_1\sigma_T\,,\\ &|\sigma_T|=\mathscr{F}_2F_N^+\Rightarrow\exists\lambda_2>0\text{ such that }u_T=-\lambda_2\sigma_T\,.\quad\square\end{aligned}\right.\tag{5.21}$$

Remark 5.5. The static analogue to the situation of Remark 5.3 is identical to (5.19), since $F_N<0$ at every point of Γ. □

5.2.2. Variational Formulation

We use the notation of Section 2.3.

Furthermore, we introduce the functional

$$j(v) = \int_{\Gamma_{\mathscr{F}}} g(x)|v_T(x)|\,d\Gamma\,, \tag{5.22}$$

which is a *continuous convex non-differentiable functional* on the space $V=(H^1(\Omega))^3$.

We will show that Problem 5.2 is "equivalent"[30] to the following problem:

Problem 5.3. Find the $u \in V$ such that

$$u = U \quad \text{on } \Gamma_U \tag{5.23}$$

$$\begin{cases} a(u, v-u) + j(v) - j(u) \geq (f, v-u) + \int_{\Gamma_{\mathscr{F}}} F_N(v_N - u_N)\,d\Gamma \\ \forall v \in V \quad \text{such that} \quad v = U \quad \text{on } \Gamma_U\,. \end{cases} \tag{5.24}$$

Indeed, if (5.19) holds, then

$$\sigma_T(v_T - u_T) + g(|v_T| - |u_T|) \geq 0 \quad \text{on } \Gamma_{\mathscr{F}}\,. \tag{5.25}$$

Taking the scalar product of (5.17) with $v-u$ and using Green's formula in the form (2.33), we obtain:

$$\begin{aligned} a(u, v-u) - \int_{\Gamma_{\mathscr{F}}} [\sigma_T(v_T - u_T) + \sigma_N(v_N - u_N)]\,d\Gamma \\ - \int_{\Gamma_U} \sigma(v-u)\,d\Gamma = (f, v-u)\,. \end{aligned} \tag{5.26}$$

But $\sigma_N = F_N$ on $\Gamma_{\mathscr{F}}$ and $v = u = U$ on Γ_U, therefore:

$$a(u, v-u) - (f, v-u) - \int_{\Gamma_{\mathscr{F}}} F_N(v_N - u_N)\,d\Gamma = \int_{\Gamma_{\mathscr{F}}} \sigma_T(v_T - u_T)\,d\Gamma$$

from which

$$\begin{aligned} a(u, v-u) + j(v) - j(u) - (f, v-u) - \int_{\Gamma_{\mathscr{F}}} F_N(v_N - u_N)\,d\Gamma \\ = \int_{\Gamma_{\mathscr{F}}} [\sigma_T(v_T - u_T) + g|v_T| - g|u_T|]\,d\Gamma \geq 0 \end{aligned}$$

according to (5.25), from which (5.24) follows.

Conversely, if u satisfies (5.23), (5.24), then, taking first $v = u \pm \varphi$, $\varphi \in (\mathscr{D}(\Omega))^3$, we find $Au = f$, from which (5.17) follows and from that (5.26) (formal utilization of Green's formula, which however can be justified as in Lions-Magenes [1], Volume 1, Chapter 2; besides, there is no logical difficulty to just carry out this calculation in a purely formal manner, since subsequently we will take the statement of Problem 5.3 as *definition*.)

But with (5.26), since $v = u = U$ on Γ_U, (5.24) leads to:

$$\begin{aligned} &\int_{\Gamma_{\mathscr{F}}} [\sigma_T(v_T - u_T) + g(|v_T| - |u_T|)]\,d\Gamma + \int_{\Gamma_{\mathscr{F}}} (\sigma_N - F_N)(v_N - u_N)\,d\Gamma \geq 0\,, \\ &\forall v \in (H^1(\Omega))^3 \quad \text{such that} \quad v = U \text{ on } \Gamma_U\,. \end{aligned} \tag{5.27}$$

[30] The equivalence is formal since in Problem 5.2 we did not state precisely in which class we seek the solution u.

We note that as v ranges over $(H^1(\Omega))^3$, v_N ranges over $H^{1/2}(\Gamma)$ and v_T ranges over the subspace of $(H^{1/2}(\Gamma))^3$ of functions satisfying $n v_T = 0$.

Thus, in (5.27), we can take $v_T = u_T$ and $v_N = any$ function of $H^{1/2}(\Gamma)$ with support in the interior of $\Gamma_{\mathscr{F}}$ [31] so that

$$\sigma_N = F_N .$$

Then (5.27) reduces to

(5.28) $$\int_{\Gamma_{\mathscr{F}}} [\sigma_T(v_T - u_T) + g(|v_T| - |u_T|)] \, d\Gamma \geqslant 0 .$$

Let

(5.29) Ψ = space of functions $\psi \in (H^{1/2}(\Gamma))^3$ with support in the interior of $\Gamma_{\mathscr{F}}$.

If $\psi \in \Psi$, we decompose ψ into:

$$\psi = \psi_N n + \psi_T , \qquad \psi_N = n\psi ,$$

and we can take in (5.28): $v_T = \psi_T$. But since $\sigma_T n = 0$, we have:

$$\sigma_T \psi_T = \sigma_T \psi ,$$

therefore (5.28) gives

(5.30) $$\int_{\Gamma_{\mathscr{F}}} [\sigma_T \psi + g|\psi_T|] \, d\Gamma - \int_{\Gamma_{\mathscr{F}}} [\sigma_T u_T + g|u_T|] \, d\Gamma \geqslant 0 .$$

Since $|\psi_T| \leqslant |\psi|$, we conclude from (5.30) that

(5.31) $$\int_{\Gamma_{\mathscr{F}}} [\sigma_T \psi + g|\psi|] \, d\Gamma - \int_{\Gamma_{\mathscr{F}}} [\sigma_T u_T + g|u_T|] \, d\Gamma \geqslant 0 , \qquad \forall \psi \in \Psi .$$

Replacing ψ in (5.31) by $\pm \lambda \psi$, $\lambda \geqslant 0$, we find that

$$\lambda \int_{\Gamma_{\mathscr{F}}} [\pm \sigma_T \psi + g|\psi|] \, d\Gamma - \int_{\Gamma_{\mathscr{F}}} [\sigma_T u_T + g|u_T|] \, d\Gamma \geqslant 0 \qquad \forall \lambda \geqslant 0 ,$$

from which

$$\int_{\Gamma_{\mathscr{F}}} [\pm \sigma_T \psi + g|\psi|] \, d\Gamma \geqslant 0$$

i.e.,

(5.32) $$\left| \int_{\Gamma_{\mathscr{F}}} \sigma_T \psi \, d\Gamma \right| \leqslant \int_{\Gamma_{\mathscr{F}}} g|\psi| \, d\Gamma \qquad \forall \psi \in \Psi$$

and

(5.33) $$\int_{\Gamma_{\mathscr{F}}} [\sigma_T u_T + g|u_T|] \, d\Gamma \leqslant 0 .$$

---

[31] We assume that the boundary of $\Gamma_{\mathscr{F}}$ in Γ is regular.

But (5.32) expresses that the form

$$\psi \to \int_{\Gamma_{\mathscr{F}}} \sigma_T \psi \, d\Gamma = \int_{\Gamma_{\mathscr{F}}} (g^{-1}\sigma_T) g\psi \, d\Gamma$$

is continuous on Ψ, having the topology induced by $(L^1(\Gamma_{\mathscr{F}}))^3$, and has norm $\leqslant 1$ for the norm

$$\int_{\Gamma_{\mathscr{F}}} g|\psi| \, d\Gamma \quad \text{on } (L^1(\Gamma))^3 .$$

Since Ψ is *dense* in $(L^1(\Gamma_{\mathscr{F}}))^3$, we have:

$$g^{-1}\sigma_T \in (L^\infty(\Gamma_{\mathscr{F}}))^3, \quad \text{with norm } \leqslant 1,$$

i.e.,

(5.34) $$|\sigma_T| \leqslant g \quad \text{a.e. on } \Gamma_{\mathscr{F}}$$

But then $\sigma_T u_T + g|u_T| \geqslant 0$ which, combined with (5.33), proves that

(5.35) $$\sigma_T u_T + g|u_T| = 0 \quad \text{a.e. on } \Gamma_{\mathscr{F}} .$$

But (5.34), (5.35) are equivalent to the last conditions (5.19) from which the claimed equivalence follows. □

From now on, we shall take the statement of Problem 5.3 as the *definition of the problem: this is the formulation in the form of a variational inequality of the static problem of friction*, with Coulomb's law and bilateral contact. □

Remark 5.6. Naturally, everything that has just been said, applies to $\Gamma_U = \emptyset$. □

Remark 5.7. If $g=0$ on $\Gamma_{\mathscr{F}}$, (5.24) reduces to

(5.36) $$a(u, v-u) = (f, v-u) + \int_{\Gamma_{\mathscr{F}}} F_N(v_N - u_N) \, d\Gamma ;$$

the corresponding problem is of the type of problems that were discussed in Section 3:

(5.37) $$Au = f,$$

(5.38) $$u = U \quad \text{on } \Gamma_U$$

(5.39) $$\sigma_N = F_N, \quad \sigma_T = 0 \quad \text{on } \Gamma_{\mathscr{F}} = \Gamma - \Gamma_U . \quad □$$

Remark 5.8. The variational formulation of the problem of Remark 5.4 is obtained by decomposing $\Gamma_{\mathscr{F}}$ into

$$\Gamma_{\mathscr{F}+} \quad \text{where } F_N \geqslant 0 \quad \text{and} \quad \Gamma_{\mathscr{F}-} \quad \text{where } F_N < 0 . \quad □$$

5.2.3. Results. The Case "Γ_U with Positive Measure"

We note that, since $a(u,v)$ is symmetric, Problem 5.3 is equivalent to the minimization on the set $\mathscr{U}_{\text{ad}}$

$$\mathscr{U}_{\text{ad}} = \{v| \quad v \in V = (H^1(\Omega))^3, \quad v = U \text{ on } \Gamma_U\} \tag{5.40}$$

of the functional

$$I(v) = \tfrac{1}{2}a(v,v) + j(v) - (f,v) - \int_{\Gamma_{\mathscr{F}}} F_N v_N \,\mathrm{d}\Gamma\,. \tag{5.41}$$

The expression $I(v)$ represents the *potential energy* of the kinematically admissible field v.

We assume:

$$F_N \in L^\infty(\Gamma_{\mathscr{F}}),\ \text{[32]} \tag{5.42}$$

$$f \in L^2(\Omega)\,. \tag{5.43}$$

Then, since $j(v) \geqslant 0$, Corollary 3.1 proves

Theorem 5.1. *Problem* 5.3 *has a unique solution.* □

Remark 5.9 (about the dependence on g). In order to indicate the dependence on g, we write $j_g(v)$ instead of $j(v)$. Let u_g be the corresponding solution of (5.23), (5.24). Then we have:

$$\|u_{g_1} - u_{g_2}\| \leqslant c\|g_1 - g_2\|_{L^2(\Gamma_{\mathscr{F}})}\,. \tag{5.44}$$

Indeed, taking $v = u_{g_2}$ (resp. $v = u_{g_1}$) in the inequality analogous to (5.24) with respect to u_{g_1} (resp. u_{g_2}) and adding, we have, setting $w = u_{g_1} - u_{g_2}$:

$$-a(w,w) - \int_{\Gamma_{\mathscr{F}}} (g_1 - g_2)(|u_{g_1}| - |u_{g_2}|)\,\mathrm{d}\Gamma \geqslant 0 \tag{5.45}$$

from which follows, since $w \in V_0$ (in the notation of (3.36)) and using (3.37):

$$\alpha_0 \|w\|^2 \leqslant \|g_1 - g_2\|_{L^2(\Gamma_{\mathscr{F}})} \|u_{g_1} - u_{g_2}\|_{L^2(\Gamma_{\mathscr{F}})}\,.$$

Since $\|\varphi\|_{L^2(\Gamma_{\mathscr{F}})} \leqslant c\|\varphi\|$, we conclude (5.44). The function $g \to u_g$ is thus *Lipschitz continuous* in a certain sense, precisely given in (5.44).

We will prove that when $g \to 0$ in $L^2(\Gamma_{\mathscr{F}})$, then $u_g \to u$ in $(H^1(\Omega))^3$, where u is the solution of the problem formulated in Remark 5.7. □

5.2.4. Results. The Case "$\Gamma_U = \emptyset$"

If $\Gamma_U = \emptyset$, the problem is: to find the $u \in V = (H^1(\Omega))^3$ such that

$$a(u, v-u) + j(v) - j(u) \geqslant (f, v-u) + \int_\Gamma F_N(v_N - u_N)\,\mathrm{d}\Gamma \qquad \forall v \in V\,. \tag{5.46}$$

[32] Which can be extended.

Setting $v=0$ and then $v=2u$ in (5.46), we see that

(5.47) $$a(u,u)+j(u)=(f,u)+\int_\Gamma F_N u_N \,\mathrm{d}\Gamma$$

such that (5.46) is equivalent to (5.47) *and* to

(5.48) $$a(u,v)+j(v)\geqslant (f,v)+\int_\Gamma F_N v_N \,\mathrm{d}\Gamma\,,\qquad \forall v\in V.$$

Changing v to $-v$, *we finally see that* (5.46) *is equivalent to*

(5.49) $$|a(u,v)-(f,v)-\int_\Gamma F_N v_N \,\mathrm{d}\Gamma|\leqslant j(v)\qquad \forall v\in V$$
with equality when $v=u$.

Now, taking $v=\rho\in\mathscr{R}$ in (5.49), we see that *the problem can only have a solution if*

(5.50) $$|(f,\rho)+\int_\Gamma F_N \rho_N \,\mathrm{d}\Gamma|\leqslant j(\rho)=\int_\Gamma g|\rho_T|\,\mathrm{d}\Gamma\qquad \forall\rho\in\mathscr{R}.$$

In the following, we will *solve the problem under the stronger assumption that the inequality in* (5.50) *is strict*, i.e.,

(5.51) $$\int_\Gamma g|\rho_T|\,\mathrm{d}\Gamma-|(f,\rho)+\int_\Gamma F_N \rho_N \,\mathrm{d}\Gamma|>0\qquad \forall\rho\in R,\quad \rho\neq 0$$

which is equivalent, since the space $\mathscr{R}$ has a finite dimension, to the existence of a $c>0$ such that

(5.32) $$\int_\Gamma g|\rho_T|\,\mathrm{d}\Gamma-|(f,\rho)+\int_\Gamma F_N \rho_N \,\mathrm{d}\Gamma|\geqslant c|\rho|$$ [33]

Thus we have

Theorem 5.2. *We assume that* $\Gamma_U=\emptyset$ *and that* (5.52) *holds. Then there exists a* $u\in V$ *which is a solution of* (5.46) *or, which amounts to the same thing, which on* V *minimizes the functional* $I(v)$ *given by* (5.41) (with $\Gamma_{\mathscr{F}}=\Gamma$).

Proof. The function $v\to I(v)$ is continuous and convex on V; therefore it is sufficient to prove that

(5.53) $$I(v)\to+\infty\quad\text{if}\quad \|v\|\to\infty\,.$$

We set

(5.54) $$w=v-Pv,\quad Pv=\text{orthogonal projection in } (L^2(\Omega))^3 \text{ of } V\to\mathscr{R},$$

and

(5.55) $$v=w+\rho\,,\qquad \rho\in\mathscr{R}\,.$$

---

[33] Where, as in the previous sections,
$|f|=\|f\|_{(L^2(\Omega))^3}=(\int_\Omega f_i f_i \,\mathrm{d}x)^{1/2}$.

Then, according to (3.52), $\varepsilon(w) \geqslant c_1 |w|^2$ such that

(5.56) $$a(w,w) \geqslant \alpha \|w\|^2, \quad \alpha > 0,$$

and furthermore

$$\|v\|^2 = \varepsilon(v) + |v|^2 = \varepsilon(w) + |w|^2 + |\rho|^2,$$

such that

(5.57) $\|v\|$ *is a norm equivalent to* $\|w\| + |\rho|$.

We have

$$\begin{aligned} I(v) = I(w+\rho) &= \tfrac{1}{2} a(w,w) + j(w+\rho) - (f,w) - \textstyle\int_\Gamma F_N w_N \, \mathrm{d}\Gamma \\ &\quad - [(f,\rho) + \textstyle\int_\Gamma F_N \rho_N \, \mathrm{d}\Gamma] \geqslant \text{(according to (5.52))} \\ &\geqslant \tfrac{1}{2} a(w,w) + j(w+\rho) - j(\rho) + c|\rho| - (f,w) - \textstyle\int_\Gamma F_N w_N \, \mathrm{d}\Gamma \\ &\geqslant c(\|w\|^2 + |\rho|) - c\|w\| \end{aligned}$$

from which the result follows, according to (5.57). □

The problem of uniqueness of the solution of (5.46) is open; in Section I.7, Theorem 7.5, we encountered a somewhat analogous situation where we were able to conclude uniqueness; we have not been able to extend the argument to the present situation. One obvious observation follows:

If u_1 and u_2 are two possible solutions of (5.46), then, taking $v = u_2$ (resp. $v = u_1$) in the inequality with respect to u_2 (resp. u_1) and adding, we conclude that:

$$a(u_1 - u_2, u_1 - u_2) = 0, \quad \text{therefore} \quad u_1 - u_2 \in \mathcal{R}, \quad \text{hence}$$

(5.58) *under the condition of Theorem 5.2, we have uniqueness of the field of stresses and of the field of strains.* □

5.3. Dual Variational Formulation

As in Section 3.5, we will examine duality from two slightly different points of view.

5.3.1. Statically Admissible Fields and Potential Energy

By K, we denote the *set of statically* admissible *fields* (of stresses), (s.a.f.), defined as follows:

(5.59) $$K = \{\tau \mid \quad \tau_{ij} = \tau_{ji} \in L^2(\Omega), \quad \tau_{ij,j} + f_i = 0 \text{ in } \Omega, \quad \tau_N = F_N \text{ on } \Gamma_{\mathscr{F}}, \quad |\tau_T| \leqslant g \text{ on } \Gamma_{\mathscr{F}}\}.$$ [34]

[34] Cf. Remark 5.12 below.

Remark 5.10. For τ such that $\tau_{ij,j}+f_i=0$, we can define (cf. Remark 3.8)

$$\tau_{ij}n_j \in H^{-1/2}(\Gamma) \qquad \forall i\,.$$

Then

(5.60) $\qquad \tau_N=\tau_{ij}n_jn_i$ is defined in $H^{-1/2}(\Gamma)$

and

(5.61) $$\tau_{iT}=\tau_{ij}n_j-\tau_N n_i \in H^{-1/2}(\Gamma),$$

such that

(5.62) $$\tau_T=\{\tau_{iT}\}\in (H^{-1/2}(\Gamma))^3\,.$$

The condition "$|\tau_T|\leqslant g$ on $\Gamma_{\mathscr{F}}$" signifies: the restriction of τ_T to $\Gamma_{\mathscr{F}}$ is in $(L^\infty(\Gamma_{\mathscr{F}}))^3$, with

$$|\tau_T(x)|\leqslant g(x) \quad \text{a.e. on } \Gamma_{\mathscr{F}}\,.$$

Then the set K defined by (5.59) is *a closed convex subset of the space* H defined by

(5.63) $$H=\{\tau | \quad \tau_{ij}=\tau_{ji}\in L^2(\Omega)\}$$

(with the scalar product $\int_\Omega \sigma_{ij}\tau_{ij}\mathrm{d}x$). ☐

For $\tau\in K$, we define *its potential energy* $J(\tau)$ by

(5.64) $$J(\tau)=\tfrac{1}{2}\mathscr{A}(\tau,\tau)-\int_{\Gamma_U}\tau_{ij}n_jU_i\mathrm{d}\Gamma$$

(as in (3.68), with the notation of (3.69)).

A dual problem to the original problem is then:

Problem 5.4. *Minimize* $J(\tau)$ *on* K. ☐

Remark 5.11. Comparing with Problem 3.2, we see that the presence of friction *only changes the convex set* K (significantly, we have to admit, because one passes from a convex *linear* affine manifold to a "non linear" convex set). ☐

Remark 5.12. Naturally, Problem 5.4 *assumes that* K *is not empty*. We note that if u is a solution of (5.23), (5.24) and measure $\Gamma_U>0$ (resp. of (5.46) if $\Gamma_U=\emptyset$), then

$$\sigma_{ij}=a_{ijkh}\varepsilon_{kh}(u)\in K$$

such that, according to the results of the previous sections, we have:

(5.65) $\qquad$ if measure $\Gamma_U>0$ (resp. if $\Gamma_U=\emptyset$ and if (5.62) holds), then $K\neq\emptyset$.

From this we conclude at once

Theorem 5.3. *If* measure $\Gamma_U > 0$ (resp. *if* $\Gamma_U = \emptyset$ *and if* (5.52) *holds*), *then there exists a unique* σ *in* K *such that*

(5.66) $$J(\sigma) \leqslant J(\tau) \qquad \forall \tau \in K$$

or again

(5.67) $$\mathscr{A}(\sigma, \tau - \sigma) \geqslant \int_{\Gamma_U} U_i(\tau_{ij} n_j - \sigma_{ij} n_j) \, \mathrm{d}\Gamma, \qquad \forall \tau \in K.$$

It remains to clarify the relation between the initial (or "primal") problem and the dual problem.

We will show that the primal problem and the dual problem are connected by a relation analogous to (3.81) (cf. (5.76) below) and that the solutions u and σ of these problems are connected by

(5.68) $$\sigma_{ij} = a_{ijkh} \varepsilon_{kh}(u) \quad \text{in } \Omega.$$

(5.69) $$\sigma_T u_T + g |u_T| = 0 \quad \text{on } \Gamma_{\mathscr{F}}.$$

5.3.2. Duality and Lagrange Multipliers

We now extend the discussion of Section 3.5.2 to the present situation "with friction".

With the notation of Section 3.5.2, we introduce (using (3.73)):

(5.70) $$I_*(\tau, v, q) = I(\tau, v, q) + j(v) \quad {}^{35}$$

(5.71) $$I_*(q) = \sup_{\tau, v} I_*(\tau, v, q), \qquad v = U, \qquad v = U \quad \text{on } \Gamma_U.$$

As in (3.75), we have:

(5.72) $$\sup_q I_*(q) \leqslant I(u).$$

With the notation of Section 3.5.2, we have

$$I_*(\tau, v, q) = I_1(\tau, q) + I_3(v, q)$$

$$I_3(v, q) = j(v) - (f, v) - \int_{\Gamma_{\mathscr{F}}} F_N v_N \, \mathrm{d}\Gamma + \int_\Omega q_{ij} a_{ijkh} \varepsilon_{kh}(v) \, \mathrm{d}x.$$

We have (cf. (3.76))

(5.73) $$\inf I_1(\tau, q) = -\tfrac{1}{2} \mathscr{A}(\tau, \tau), \quad \text{where} \quad A_{ijkh} \tau_{kh} = q_{ij},$$

then

$$I_3(v, q) = j(v) - (f, v) - \int_{\Gamma_{\mathscr{F}}} F_N v_N \, \mathrm{d}\Gamma + \int_\Omega q_{ij} a_{ijkh} (\partial v_k / \partial x_h) \, \mathrm{d}x,$$

[35] With $\int_{\Gamma_{\mathscr{F}}} F v \, \mathrm{d}\Gamma$ replaced by $\int_{\Gamma_{\mathscr{F}}} F_N v_N \, \mathrm{d}\Gamma$.

from which it follows that

$$\inf_v I_3(v,q) = -\infty \quad \text{if} \quad \tau \notin K, \quad v = U \quad \text{on } \Gamma_U,$$

$$\inf I_3(v,q) = \int_{\Gamma_U} \tau_{kh} n_h U_k \, d\Gamma \quad \text{if} \quad \tau \in K.$$

Therefore:

$$\hat{I}(q) = \begin{cases} -\tfrac{1}{2}\mathscr{A}(\tau,\tau) + \int_{\Gamma_U} \tau_{kh} n_h U_k \, d\Gamma & \text{if} \quad \tau \in K, \\ \text{where} \quad q_{ij} = A_{ijkh} \tau_{kk}, \\ -\infty \quad \text{otherwise.} \end{cases} \tag{5.74}$$

But

$$I(u) = -\tfrac{1}{2} a(u,u) + \int_{\Gamma_U} \sigma_{ij} n_j U_i \, d\Gamma. \tag{5.75}$$

Indeed, from $Au = f$, $u = U$ on Γ_U, $\sigma_N = F_N$ on $\Gamma_{\mathscr{F}}$,

$$|\sigma_T| \leqslant g, \quad \sigma_T U_T + g|u_T| = 0 \quad \text{on } \Gamma_{\mathscr{F}},$$

it follows that

$$a(u,u) = \int_{\Gamma_U} \sigma_{ij} n_j U_i \, d\Gamma - \int_{\Gamma_{\mathscr{F}}} F_N u_N \, d\Gamma - \int_{\Gamma_{\mathscr{F}}} \sigma_T u_T \, d\Gamma = (f,u)$$

and therefore, since

$$-\int_{\Gamma_{\mathscr{F}}} \sigma_T u_T \, d\Gamma = j(u),$$

(5.75) follows.

But then

$$\sup \hat{I}(q) \geqslant -\tfrac{1}{2}\mathscr{A}(\sigma,\sigma) + \int_{\Gamma_U} \sigma_{kh} n_h U_k \, d\Gamma = I(u)$$

from which follows the conclusion analogous to (3.81).

Theorem 5.4. *We have:*

$$\inf_{\tau \in K} \left[\tfrac{1}{2}\mathscr{A}(\tau,\tau) - \int_{\Gamma_U} \tau_{kh} n_h U_k \, d\Gamma\right] + \inf_{v = U \text{ on } \Gamma_U} I(v) = 0. \tag{5.76}$$

Furthermore, the solutions σ and u are connected by (5.68), (5.69). ▯

5.4. Other Boundary Conditions and Open Questions

Orientation. In this section, we will indicate other boundary conditions which might be encountered with or without the restriction of Coulomb's law of friction.

Thus we are led to problems some of which are simple variants of preceding problems while others appear to be open.

5.4.1. Normal Displacement with Friction

In the problems treated in Section 5.1.2, friction acted on the *tangential displacement*. We now consider the case where it is the *normal displacement* which takes place with friction[36].

Then the conditions of the problem are, keeping the preceding notation and assumptions: *find a solution u of* (5.17) *with* (5.18) and with the following conditions on $\Gamma_{\mathscr{F}}$: let g_1 and g_2 be given in $L^\infty(\Gamma_{\mathscr{F}})$ (or, more generally, in $H^{-1/2}(\Gamma)$ whose restrictions to $\Gamma_{\mathscr{F}}$ we consider); we assume that

$$g_1 \leqslant 0 \leqslant g_2\,; \tag{5.77}$$

then, for $x \in \Gamma_{\mathscr{F}}$, *we have*

$$\begin{aligned} g_1 < \sigma_N(x) < g_2 &\Rightarrow u_N(x) = 0\,,\\ \sigma_N(x) = g_1 &\Rightarrow u_N(x) \geqslant 0\,,\\ \sigma_N(x) = g_2 &\Rightarrow u_N(x) \leqslant 0 \end{aligned} \tag{5.78}$$

to which we add, for example, the classical condition

$$\tau_T = 0 \quad \text{on } \Gamma_{\mathscr{F}}\,. \tag{5.79}$$

Then we can prove without difficulty that the preceding problem is "equivalent"[37] to the minimization of the functional

$$I_1(v) = \tfrac{1}{2}a(v,v) + j_1(v) - (f,v) \tag{5.80}$$

where

$$j_1(v) = \int_{\Gamma_{\mathscr{F}}} (-g_1 v_N^+ + g_2 v_N^-)\,d\Gamma\,, \tag{5.81}$$

where the minimization is effected on the set $\mathscr{U}_{\text{ad}}$ of functions $v \in (H^1(\Omega))^3$ such that

$$v = U \quad \text{on } \Gamma_U\,.$$

Indeed, if u is a solution in $\mathscr{U}_{\text{ad}}$ minimizing $I_2(v)$, we have:

$$a(u, v-u) + j_1(v) - j_1(u) - (f, v-u) \geqslant 0 \qquad \forall v \in \mathscr{U}_{\text{ad}}\,. \tag{5.82}$$

---

[36] This situation may arise in certain mechanical arrangements.

[37] This is always a formal equivalence, as we already indicated on several occasions, since in Problem (5.17), (5.18), (5.78), (5.79) we did not make precise assumptions on the differentiability of u.

It follows that (taking $v=u\pm\varphi$, $\varphi\in(\mathscr{D}(\Omega))^3$:

$$Au=f,$$

from which, according to Green's formula (2.33) (since $v=u=U$ on Γ_U):

$$a(u,v-u)-\int_{\Gamma_{\mathscr{F}}}[\sigma_T(v_T-u_T)+\sigma_N(v_N-u_N)]\,d\Gamma=(f,v-u)$$
$$\leqslant a(u,v-u)+j_1(v)-j_1(u)$$

from which

$$j_1(v)-j_1(u)+\int_{\Gamma_{\mathscr{F}}}[\sigma_T(v_T-u_T)+\sigma_N(v_N-u_N)]\,d\Gamma\geqslant 0\,. \tag{5.83}$$

Taking $v_T-u_T=\psi$, $\psi\in(H^{1/2}(\Gamma))^3$, $n\psi=0$ on Γ, and taking $v_N-u_N=0$ (which is permissible), we conclude from (5.83) that (5.79) holds, and then (5.83) reduces to

$$j_1(v)-j_1(u)+\int_{\Gamma_{\mathscr{F}}}\sigma_N(v_N-u_N)\,d\Gamma\geqslant 0\,, \tag{5.84}$$

or again:

$$\begin{aligned}&\int_{\Gamma_{\mathscr{F}}}[(\sigma_N-g_1)v_N^+ +(g_2-\sigma_N)v_N^-]\,d\Gamma\\ &\qquad -\int_{\Gamma_{\mathscr{F}}}[(\sigma_N-g_1)u_N^+ +(g_2-\sigma_N)u_N^-\,d\Gamma\geqslant 0\,.\end{aligned} \tag{5.85}$$

Taking $v_N=\pm\lambda\varphi$, $\varphi\in H^{1/2}(\Gamma)$, $\varphi\geqslant 0$, then letting $\lambda\to+\infty$, we find that

$$\sigma_N-g_1\geqslant 0\,,\qquad g_2-\sigma_N\geqslant 0 \tag{5.86}$$

and

$$\int_{\Gamma_{\mathscr{F}}}[(\sigma_N-g_1)u_N^+ +(g_2-\sigma_N)u_N^-]\,d\Gamma\leqslant 0$$

which, together with (5.86) is equivalent to

$$(\sigma_N-g_1)u_N^+ +(g_2-\sigma_N)u_N^- =0\,. \tag{5.87}$$

But (5.86), (5.87) are equivalent to (5.78), from which the result follows. □

Consequences. With our previous methods, we see that, if measure $\Gamma_U>0$, *the problem has a unique solution.*

In the case, where $\Gamma_U=\emptyset$, it is *necessary*, in order that the problem has a solution, that

$$j_1(\rho)-(f,\rho)\geqslant 0\qquad\forall\rho\in\mathscr{R} \tag{5.88}$$

(from which follows (changing ρ to $-\rho$)

$$-\int_\Gamma(-g_1\rho_N^- +g_2\rho_N^+)\,d\Gamma\leqslant(f,\rho)\leqslant\int_\Gamma(-g_1\rho_N^+ +g_2\rho_N^-)\,d\Gamma\,,\qquad\forall\rho\in\mathscr{R}\,. \tag{5.89}$$

Then we prove—by methods analogous to those of Theorem 5.2—that, *if we have a strenghthened condition* (5.88), i.e., if

(5.90) $$j_1(\rho)-(f,\rho)>0 \quad \forall \rho\in\mathscr{R}, \quad \rho\neq 0$$ [38]

then the problem has at most one solution. The fields of strain and stress are unique. ☐

5.4.2. Signorini's Problem as Limit Case of Problems with Friction

First, we consider the *special case* of Section 5.4.3 where $g_1=0$ and $g_2=g>0$. We introduce

(5.91) $$\mathscr{J}(v)=\int_\Gamma v_N^- \,\mathrm{d}\Gamma .$$

The problem then corresponds to the minimization (on the space of v's such that $v=U$ on Γ_U) of

(5.92) $$\tfrac{1}{2}a(v,v)+g\mathscr{J}(v)-(f,v) .$$

We consider the case—the most delicate one—where

(5.93) $$\Gamma_U=\emptyset .$$

Then, if we assume that (5.90) holds, i.e. here:

(5.94) $$-(f,\rho)+g\mathscr{J}(\rho)>0 \quad \forall\rho\in\mathscr{R}, \quad \rho\neq 0,$$

there exists a u_g minimizing (5.92) on $V=(H^1(\Omega))^3$, where u_g is characterized by

(5.95) $$a(u_g,v-u_g)+g\mathscr{J}(v)-g\mathscr{J}(u_g)-(f,v-u_g)\geqslant 0 \quad \forall v\in V;$$

the field of strains and stresses is unique.

Now, we let g tend towards $+\infty$.

We introduce the set

(5.96) $$K=\{v| \quad v\in V,\ v_N\geqslant 0 \text{ on } \Gamma\}$$

which is convex and closed in V.

We assume

(5.97) $$-(f,\rho)+g_0\mathscr{J}(\rho)>0 \quad \forall\rho\in\mathscr{R}, \quad \rho\neq 0, \quad g_0>0 \text{ fixed.}$$

[38] Or again, $\mathscr{R}$ being of finite dimension,
$j_1(\rho)-(f,\rho)\geqslant\beta|\rho|, \quad \beta>0, \quad \rho\in\mathscr{R}$.

Then we note that (5.94) holds $\forall g \geqslant g_0$, therefore u_g exists for $g \geqslant g_0$. We will show that

(5.98) *we can finde a sequence $g \to +\infty$ such that u_g converges weakly in V toward a solution u in K of*

$$a(u, v-u) \geqslant (f, v-u) \qquad \forall v \in K.$$

Let us interpret the inequality in (5.98); we find:

$$(5.99) \qquad \begin{aligned} &Au = f, \\ &\sigma_T = 0 \quad \text{on } \Gamma, \\ &u_N \geqslant 0, \quad \sigma_N \geqslant 0, \quad \sigma_N u_N = 0 \quad \text{on } \Gamma; \end{aligned}$$

this is the *problem of Signorini* (cf. G. Fichera [1]).

We note that the proof of (5.98) *shows the existence* of a solution of Signorini's problem *under assumption* (5.97), *which implies*

$$(5.100) \qquad -(f, \rho) > 0 \quad \forall \rho \in \mathcal{R} \cap K, \quad \rho \neq 0.$$

In G. Fichera [1] (cf. also Lions-Stampacchia [1]), the existence of a solution of Signorini's problem[39] has been proved under the *sole* assumption (5.100). Here, we obtain that, under the stronger assumption (5.97), *Signorini's problem is a limit case of friction problems.* □

Proof of (5.98). From (5.97), it follows that

$$-(f, \rho) + g_0 \mathcal{J}(\rho) \geqslant c_0 \|\rho\|$$

where $\| \quad \|$ denotes the norm in V. Therefore, if we decompose $v \in V$ into

$$v = w + \rho$$

(as in Theorem 5.2), we have:

$$\begin{aligned} \tfrac{1}{2} a(v,v) + g\mathcal{J}(v) - (f,v) &= \tfrac{1}{2} a(w,w) - (f,w) + g\mathcal{J}(w+\rho) - (f,\rho) \\ &= \tfrac{1}{2} a(w,w) - (f,w) - (f,\rho) + g_0 \mathcal{J}(\rho) \\ &\quad + (g-g_0)\mathcal{J}(w+\rho) + g_0(\mathcal{J}(w+\rho) - \mathcal{J}(\rho)) \\ &\geqslant \text{(using (5.56) and assuming that } g \geqslant g_0) \\ &\geqslant \tfrac{1}{2}\alpha \|w\|^2 - c_1 \|w\| + c_0 \|\rho\| + g_0(\mathcal{J}(w+\rho) - \mathcal{J}(\rho)). \end{aligned}$$

But

$$\mathcal{J}(w+\rho) - \mathcal{J}(\rho) \geqslant -\mathcal{J}(w) \geqslant -c_2 \|w\|,$$

[39] In G. Fichera, *loc. cit.*, it is proved that condition (5.100) is *necessary*. For numerous extensions of all these questions, cf. M. Schatzman [1].

from which

$$\tfrac{1}{2}a(v,v)+g\mathscr{J}(v)-(f,v)\geqslant c_3(\|w\|^2+\|\rho\|)-c_4\|w\|$$

where c_3 and c_4 are constants, *independent* of g. Thus u_g—which is one of the elements minimizing $\tfrac{1}{2}a(v,v)+g\mathscr{J}(v)-(f,v)$—remains in a bounded set of V. But then it follows from (5.95) (for example, setting $v=0$), that

$$g\mathscr{J}(u_g)\leqslant c_5,$$

therefore

(5.101) $\quad \mathscr{J}(u_g)\to 0 \quad \text{when} \quad g\to+\infty;$

thus we can select a subsequence, again denoted by u_g, such that $u_g\to u$ weakly in V, and, according to (5.101), $\mathscr{J}(u)=0$, hence $u\in K$.

If, in (5.95), we take $v\in K$, we have (since then $\mathscr{J}(v)$ is zero):

(5.102) $\quad a(u_g,v)-(f,v-u_g)\geqslant a(u_g,u_g).$

Passing to the limit in (5.102), we obtain

$$a(u,v)-(f,v-u)\geqslant a(u,u) \qquad \forall v\in K$$

from which the result follows. ▯

5.4.3. Another Condition for Friction with Imposed Normal Displacement

Under certain boundary conditions with friction, it can happen that the normal displacement is imposed and that, *simultaneously*, one knows, at least approximately, the normal stress. This then leads to the following boundary conditions on $\Gamma_{\mathscr{F}}$ [40]:

(5.103) $\quad u_N=0 \quad \text{on } \Gamma_{\mathscr{F}},$

(5.104)
$$|\sigma_T|<g \Rightarrow u_T=0$$
$$|\sigma_T|=g \Rightarrow \exists\lambda\geqslant 0 \qquad u_T=-\lambda\sigma_T.$$

Condition (5.104) represents a law of friction which is different from Coulomb's law. The corresponding variational formulation is

(5.105)
$$a(u,v-u)+\int_{\Gamma_{\mathscr{F}}} g(|v_T|-|u_T|)\,d\Gamma\geqslant(f,v-u)$$
$$\forall v\in\mathscr{U}_{ad} \text{ and satisfying (5.103).}$$

The proof is carried out as in Section 5.2.

[40] The boundary conditions on Γ_U remain unchanged.

5.4.4. Coulomb Friction with Imposed Normal Displacement

We go back to Coulomb's law, which was discussed in Section 5.1.2, but this time we assume that it is the *normal displacement* that is given on $\Gamma_{\mathscr{F}}=\Gamma-\Gamma_U$ *instead of the normal stress.*

Thus, we are led to the following problem (cf. (5.7), (5.8)):

Find a solution u of (5.17), (5.18) with

$$u_N=0 \quad \text{on } \Gamma_{\mathscr{F}} \tag{5.106}$$

and, still on $\Gamma_{\mathscr{F}}$:

$$\begin{aligned} &|\sigma_T|\leqslant \mathscr{F}\,|\sigma_N|, \quad \text{with}\\ &|\sigma_T|<\mathscr{F}\,|\sigma_N| \Rightarrow u_T=0,\\ &|\sigma_T|=\mathscr{F}\,|\sigma_N| \Rightarrow \text{there exists } \lambda\geqslant 0 \text{ such that } \quad u_T=-\lambda\sigma_T. \end{aligned} \tag{5.107}$$

This problem appears to be *open.* We limit ourselves to the following remark: if we introduce

$$\mathscr{U}_{\text{ad}}=\{v| \quad v\in(H^1(\Omega))^3, \quad v=U \text{ on } \Gamma_U, \quad v_N=0 \text{ on } \Gamma_{\mathscr{F}}\}, \tag{5.108}$$

then u is a solution of (5.17), (5.18), (5.106), (5.107) if and only if u *is a solution in* $\mathscr{U}_{\text{ad}}$ *of the following inequality*:

$$a(u,v-u)+\int_{\Gamma_{\mathscr{F}}}|\sigma_N(u)|(|v_T|-|u_T|)\,d\Gamma\geqslant(f,v-u) \qquad \forall v\in\mathscr{U}_{\text{ad}}. \tag{5.109}$$

The formal equivalence is easy to verify. But Problem (5.109) presents several difficulties:

i) We have to give meaning to (5.109); for $u\in\mathscr{U}_{\text{ad}}$, $\sigma_N(u)$ is not defined on $\Gamma_{\mathscr{F}}$; however, if u is a "solution" of (5.109), then, necessarily, we have $Au=f$ and it is *possible* (but not proved) that, when $u\in(H^1(\Omega))^3$ and $Au\in(L^2(\Omega))^3$, we can define $|\sigma_N(u)|$ in $H^{-1/2}(\Gamma)$;

ii) Even if i) is overcome, the inequality (5.109)—*which is not an inequality of the usual type*—has not been entirely solved.

We can prove the existence of a solution u of the regularized *inequality*

$$\begin{aligned} &\varepsilon(Au,A(v-u))+a(u,v-u)+\int_{\Gamma_{\mathscr{F}}}\mathscr{F}\,|\sigma_N(u)|(|v_T|-|u_T|)\,d\Gamma\geqslant(f,v-u),\\ &u,v\in(H^2(\Omega))^3, \quad u=v=U \quad \text{on } \Gamma_U, \quad u_N=v_N=0 \quad \text{on } \Gamma_{\mathscr{F}}. \quad \square \end{aligned} \tag{5.110}$$

5.4.5. Signorini's Problem with Friction (G. Duvaut [5])

We deal here with the study of the deformation of an elastic body which rests on a rigid support. In the initial stage without stress, the body occupies a region Ω of $\mathbb{R}^3$ with the boundary Γ, one part of which, Γ_1, is in contact with a rigid support S. The body Ω is subjected to volume forces $\{f_i\}$ and to surface forces

$\{F_i\}$ on $\Gamma-\Gamma_1$ $(=\Gamma_2)$. These forces produce a deformation of the elastic body, and the corresponding displacements on Γ_1 are such that the normal component is negative or zero (as in the classic Signorini problem), while the tangential component is a *displacement with friction*, when the normal component is zero.

The fields of displacements $\{u_i\}$ and of stresses $\{\sigma_{ij}\}$ satisfy the equations and conditions

(5.111) $\quad \sigma_{ij,j}+f_i=0 \quad \text{in } \Omega,$

(5.112) $\quad \sigma_{ij}=a_{ijhk}\varepsilon_{kh}(u) \quad \text{in } \Omega,$

(5.113)
$$\left|\begin{array}{ll} \sigma_{ij}n_j=F_i & \text{on } \Gamma_2 \\ u_N\leqslant 0 & \text{on } \Gamma_1 \\ u_N<0 \Rightarrow \sigma_{ij}n_j=0 & \\ u_N=0 \Rightarrow \left|\begin{array}{l} \sigma_N\leqslant 0 \\ \text{and} \left|\begin{array}{l} |\sigma_T|<\mathscr{F}|\sigma_N| \Rightarrow u_T=0 \\ |\sigma_T|=\mathscr{F}|\sigma_N| \Rightarrow \exists\lambda\geqslant 0, \quad u_T=-\lambda\sigma_T \end{array}\right. \end{array}\right. & \text{on } \Gamma_1 . \end{array}\right.$$

If we set

$$\mathscr{U}_{\text{ad}}=\{v \mid \quad v=\{v_i\},\ v_i\in H^1(\Omega),\ v_in_i=v_N\leqslant 0 \text{ on } \Gamma_1\},$$

we can show that the solution u of Problem (5.111)–(5.113) satisfies

(5.114)
$$\begin{aligned} &u\in\mathscr{U}_{\text{ad}} \\ &a(u,v-u)+\int_{\Gamma_1}\mathscr{F}|\sigma_N(u)|(|v_T|-|u_T|)\,d\Gamma \\ &\qquad\qquad \geqslant(f,v-u)+(F,v-u)_{\Gamma_2} \quad \forall v\in\mathscr{U}_{\text{ad}}. \end{aligned}$$

The problem to find a solution u satisfying (5.114) is open. □

5.5. The Dynamic Cases

Orientation. We again take up Problem 5.1, Section 5.1.2. We will give a variational formulation, the "dynamic analogue" to the formulation of Problem 5.3. Then we will solve the thus formulated problem. Here, we will not touch upon the "dynamic analogues" of other problems considered in preceding sections; they give rise to analogous results and the open problems in the stationary cases again lead to open problems in the dynamic case.

5.5.1. Variational Formulation

We define

(5.115) $\quad \mathscr{U}_{\text{ad}}(t)=\{v \mid \quad v\in(H^1(\Omega))^3, \quad v=U'(t) \text{ on } \Gamma_U\}.$

Then, with the notation of Section 5.1.2, we are led to

Problem 5.5. Find a function u such that

(5.116) $\quad u'(t) \in \mathscr{U}_{ad}(t) \quad \forall t,$

(5.117)
$$(u''(t), v-u'(t)) + a(u(t), v-u'(t)) + j(v) - j(u'(t))$$
$$\geqslant (f(t), v-u'(t)) + \int_{\Gamma_{\mathscr{F}}} F_N(t)(v_N - u'_N(t))\,\mathrm{d}\Gamma \qquad \forall v \in \mathscr{U}_{ad}(t),$$

and with the initial conditions

(5.118) $\quad u(0) = u_0, \quad u'(0) = u_1. \quad \square$

Remark 5.14. From (5.116), (5.118), it follows that $u(t) = U(t) + (u_0 - U(0))$ on Γ_U, from which (5.6) follows if

(5.119) $\quad u_0 = U(0) \quad \text{on } \Gamma_U. \quad \square$

Remark 5.15. The preceding problem concerns an *inequality of evolution of second order in* t (the inequalities of evolution which occurred in Chapters I and II were *of first order in* t). □

Remark 5.16. If $\mathscr{F} = 0$ (the case *without friction*) and if $\Gamma_F = \Gamma - \Gamma_U = \Gamma_{\mathscr{F}}$, inequality (5.117) becomes an *equality* and we come back to (2.42). The solution of Problem 5.5 *will thus reproduce the results* of Section 4 *as a particular case.* For clarity of the discussion, we separated the results. □

Remark 5.17. In the preceding sections, we saw (static case) that the cases where

$$\text{measure } \Gamma_U > 0 \quad \text{and} \quad \Gamma_U = \emptyset$$

behave rather differently, the latter case requiring for the existence of a solution that the friction forces be sufficient to counterbalance the exterior forces which act on the system and produce positions of static equilibrium of the body.

In the dynamic case, we shall see that the results are obtained in the same way in both cases. Nevertheless, as already indicated in the case "without friction" (cf. Remark 4.4), there may occur in the case "$\Gamma_U = \emptyset$" a *global* motion of the elastic body which can be decomposed into a solid rigid motion and a deformation.

Such a solid rigid motion will generally destroy the imposed boundary conditions and, consequently, we will consider the solution as valid only if either measure $\Gamma_U > 0$ or else if $\Gamma_U = \emptyset$, and if for all $t > 0$:

(5.120)
$$\int_\Gamma g(t)|\rho_T|\,\mathrm{d}\Gamma - |(f(t), \rho) + \int_\Gamma F_N(t)\,\rho_N\,\mathrm{d}\Gamma| \geqslant 0$$
$$\forall \rho \in \mathscr{R}, \quad \rho \neq 0. \quad \square$$

5.5.2. Statement of Results

As in Section 4.1, we introduce $\Phi(t)$ satisfying

(5.121) $$\Phi(t) \in (H^1(\Omega))^3, \quad \Phi(t) = U(t) \text{ on } \Gamma_U.$$

We introduce V_0 as in 4.2:

(5.122) $$V_0 = \{v \mid v \in (H^1(\Omega))^3, \quad v = 0 \text{ on } \Gamma_U\}$$ [41].

Replacing $u(t)$ by $u(t) - \Phi(t)$, keeping the same notation $u(t)$, Problem 5.5 is equivalent to the following: *find a function $t \to u(t)$ of $[0, T] \to V_0$ such that*

(5.123) $$\begin{aligned}(u''(t), v - u'(t)) + a(u(t), v - u'(t)) + j(v + \Phi'(t)) - j(u'(t) + \Phi'(t))& \\ \geqslant (\Psi(t), v - u'(t)) \quad \forall v \in V_0,&\end{aligned}$$

where

(5.124) $$(\Psi(t), v) = (f(t), v) + \int_{\Gamma_{\mathscr{F}}} F_N(t) v_N \, d\Gamma - (\Phi''(t), v) - a(\Phi(t), v),$$

with the initial data

(5.125) $u(0) = u_0$, $u'(0) = u_1$ (where, comparing with (5.118), u_0 (resp. u_1) corresponds to $u_0 - \Phi(0)$ (resp. $u_1 - \Phi'(0)$)). □

We introduce $H = (L^2(\Omega))^3$ and the dual V_0' of V_0 with

$$V_0 \subset H \subset V_0'.$$

We will prove

Theorem 5.7. *We make the following assumptions:*

(5.126) $$f, f', f'' \in L^2(0, T; H); \quad F_N, F_N', F_N'' \in L^2(0, T; L^2(\Gamma_{\mathscr{F}})),$$

(5.127) $$\begin{aligned}&\Phi, \Phi', \Phi'' \in L^2(0, T; V) \quad (V = (H^1(\Omega))^3), \\ &\Phi^{(3)}, \Phi^{(4)} \in L^2(0, T; H), \quad \Phi(0) \in (H^2(\Omega))^3,\end{aligned}$$

(5.128) *g does not depend on t,*

(5.129) $$u_0 \subset (H^2(\Omega))^3,$$

(5.130) $$u_1 \subset (H^1(\Omega))^3, \quad u_{1T} + \Phi_T'(0) = 0 \text{ on } \Gamma_{\mathscr{F}},$$

(5.131) $$a(\Phi(0), v) = (A\Phi(0), v) - \int_{\Gamma_{\mathscr{F}}} \sigma_{0T} v_T \, d\Gamma \quad \forall v \in V_0.$$

---

[41] $V = V_0$ if $\Gamma_U = \emptyset$.

Under these conditions, there exists one and only one function u which is a solution of (5.123), (5.125) *with*

(5.132) $u, u' \in L^\infty(0, T; V_0)$,

(5.133) $u'' \in L^\infty(0, T; H)$.

5.5.3. Uniqueness Proof

Uniqueness is easy. Indeed, let u and u_1 be two possible solutions. Taking $v = u_1'(t)$ in (5.123) (resp. $v = u'(t)$ in the inequality with respect to u_1) and adding, we obtain, setting $w = u - u_1$,

$$-(w'', w') - a(w, w') \geqslant 0$$

i.e.,

$$\frac{d}{dt}\left[|w'(t)|^2 + a(w(t), w(t))\right] \leqslant 0 .$$

Therefore, since $w(0) = 0$, $w'(0) = 0$ and $a(v, v) \geqslant 0$ (in particular), we have:

$$w'(t) = 0, \quad \text{therefore} \quad w(t) = 0 .$$

5.5.4. Existence Proof

Outline. In order to see where the difficulties arise, in the absence of assumption (5.128) which a priori is very restrictive, we will *begin* the proof assuming that g may depend on t, with

$$g, \quad \partial g/\partial t, \quad \partial^2 g/\partial t^2 \in L^\infty(\Gamma_{\mathscr{F}} \times]0, T[) .$$

We regularize the functional j. We define

(5.134) $j_\varepsilon(v) = \int_{\Gamma_{\mathscr{F}}} g(x, t)\, \varphi_\varepsilon(|v_T|^2)\, d\Gamma$

where, for example,

$$\varphi_\varepsilon(\lambda) = \frac{1}{1+\varepsilon} |\lambda|^{(1+\varepsilon)}, \quad \varepsilon > 0 ;$$

j_ε is a *convex regularization* of j.

We consider the approximate equation:

(5.135) $(u_\varepsilon'', v) + a(u_\varepsilon, v) + (j_\varepsilon'(u_\varepsilon' + \Phi'), v) = (\Psi, v) \quad \forall v \in V_0$

with

(5.136) $u_\varepsilon(0) = u_0, \quad u_\varepsilon'(0) = u_1 .$

We first prove the *existence* of a solution u_ε of (5.135), (5.136), then we establish *a priori estimates independent of* ε. Then we pass to the limit in ε.

Of course, as always, the essential point is in the a priori estimate; the existence of u_ε can be proved, for example, by passing to the limit *with the dimension* in the Galerkin approximation, as in Section 4. Let us therefore establish the a priori estimate.

A priori estimate (I). In (5.135), we take $v = u'_\varepsilon + \Phi'$. Then we obtain, noting that

$$(j'_\varepsilon(w), w) \geq 0 \qquad \forall w \in V_0:$$

$$(u''_\varepsilon(t), u'_\varepsilon(t) + \Phi'(t)) + a(u_\varepsilon(t), u'_\varepsilon(t) + \Phi'(t)) \leq (\Psi(t), u'_\varepsilon(t) + \Phi'(t))$$

from which

$$\frac{1}{2}\frac{\mathrm{d}}{\mathrm{d}t}(|u'_\varepsilon(t)|^2 + a(u_\varepsilon(t), u_\varepsilon(t))) \leq (\Psi(t), u'_\varepsilon(t) + \Phi'(t)) + (u''_\varepsilon(t), \Phi'(t)) - a(u_\varepsilon(t), \Phi'(t)). \tag{5.137}$$

Since, for $\lambda > 0$, there exists an $\alpha > 0$ such that

$$a(v, v) + \lambda |v|^2 \geq \alpha \|v\|^2,$$

it follows from (5.137), after integration in t, that

$$\begin{aligned} &|u'_\varepsilon(t)|^2 + \alpha \|u_\varepsilon(t)\|^2 - \lambda |u_\varepsilon(t)|^2 \\ &\quad \leq |u_1|^2 + c\|u_0\|^2 + 2\textstyle\int_0^t (\Psi(\sigma), u'_\varepsilon(\sigma) + \Phi'(\sigma))\,\mathrm{d}\sigma \\ &\qquad - 2\textstyle\int_0^t (u''_\varepsilon(\sigma), \Phi'(\sigma))\,\mathrm{d}\sigma - \int_0^t a(u_\varepsilon(\sigma), \Phi'(\sigma))\,\mathrm{d}\sigma \\ &\quad = |u_1|^2 + c\|u_0\|^2 + 2\textstyle\int_0^t (\Psi(\sigma), u'_\varepsilon(\sigma) + \Phi'(\sigma))\,\mathrm{d}\sigma - 2(u'_\varepsilon(t), \Phi'(t)) \\ &\qquad + 2(u_1, \Phi'(0)) + 2\textstyle\int_0^t (u'_\varepsilon(\sigma), \Phi''(\sigma))\,\mathrm{d}\sigma - \int_0^t a(u_\varepsilon(\sigma), \Phi'(\sigma))\,\mathrm{d}\sigma. \end{aligned} \tag{5.138}$$

Using

$$|u_\varepsilon(t)|^2 \leq c\textstyle\int_0^t |u'_\varepsilon(\sigma)|^2\,\mathrm{d}\sigma + |u_0|^2$$

and observing that the assumptions made on Φ, f, F_N imply that

$$\Psi, \Psi' \in L^2(0, T; V'_0), \tag{5.139}$$

it follows from (5.138) that

$$\begin{aligned} |u'_\varepsilon(t)|^2 + \|u_\varepsilon(t)\|^2 &\leq c + c\textstyle\int_0^t |u'_\varepsilon(\sigma)|^2\,\mathrm{d}\sigma + c\int_0^t \|u_\varepsilon(\sigma)\|^2\,\mathrm{d}\sigma + 2\int_0^t (\Psi(\sigma), u'_\varepsilon(\sigma))\,\mathrm{d}\sigma \\ &= c(1 + \textstyle\int_0^t (|u'_\varepsilon(\sigma)|^2 + \|u_\varepsilon(\sigma)\|^2)\,\mathrm{d}\sigma) + 2(\Psi(t), u'(t)) \\ &\quad - 2(\Psi(0), u_1) - 2\textstyle\int_0^t \|\Psi'(\sigma)\|_* \|u_\varepsilon(\sigma)\|\,\mathrm{d}\sigma \end{aligned}$$

from which finally

$$|u'_\varepsilon(t)|^2 + \|u_\varepsilon(t)\|^2 \leq c(1 + \textstyle\int_0^t (|u'_\varepsilon(\sigma)|^2 + \|u_\varepsilon(\sigma)\|^2)\,\mathrm{d}\sigma). \tag{5.140}$$

Consequently, according to Gronwall's lemma:

$$(5.141) \qquad |u'_\varepsilon(t)| + \|u_\varepsilon(t)\| \leqslant c . \quad \square$$

A priori estimates (II). We differentiate (5.135) with respect to t and substitute $v = u''(t) + \Phi''(t)$ in the result. Setting

$$X(t) = \left(\frac{\mathrm{d}}{\mathrm{d}t} j'_\varepsilon(u'_\varepsilon(t) + \Phi'(t)) \, , \; u''_\varepsilon(t) + \Phi''(t) \right),$$

it follows that

$$(5.142) \qquad (u'''_\varepsilon(t), u''_\varepsilon(t) + \Phi''(t)) + a(u'_\varepsilon(t), u''_\varepsilon(t) + \Phi''(t)) + X(t) = (\Psi'(t), u''_\varepsilon(t) + \Phi''(t)) .$$

But

$$(j'_\varepsilon(w), v) = \int_{\Gamma_\mathscr{F}} g(x,t) \, \varphi'_\varepsilon(|w_T|^2) \, w_T v_T \, \mathrm{d}\Gamma = \int_{\Gamma_\mathscr{F}} g(x,t) \, \psi_\varepsilon(w_T) \, v_T \, \mathrm{d}\Gamma$$

therefore

$$\left(\frac{\mathrm{d}}{\mathrm{d}t} j'_\varepsilon(w(t)), v \right) = \int_{\Gamma_\mathscr{F}} \frac{\partial g}{\partial t} \psi_\varepsilon(w_T(t)) \, v_T \, \mathrm{d}\Gamma + \int_{\Gamma_\mathscr{F}} g(x,t) \lim_{h \to 0} \frac{\psi_\varepsilon(w_T(t+h)) - \psi_\varepsilon(w_T(t))}{h} v_T \, \mathrm{d}\Gamma .$$

Therefore

$$\left(\frac{\mathrm{d}}{\mathrm{d}t} j'_\varepsilon(w(t)), w'(t) \right) = \int_{\Gamma_\mathscr{F}} \frac{\partial g}{\partial t} \psi_\varepsilon(w_T) \, w'_T) \, \mathrm{d}\Gamma + \int_{\Gamma_\mathscr{F}} g \lim_{h \to 0} \frac{\psi_\varepsilon(w_L(t+h)) - \psi_\varepsilon(w_T(t))}{h} \cdot \frac{w_T(t+h) - w_T(t)}{h} \mathrm{d}\Gamma .$$

Since the operator is monotone, the last integral is $\geqslant 0$ and therefore

$$(5.143) \qquad \left(\frac{\mathrm{d}}{\mathrm{d}t} j'_\varepsilon(w(t)), w'(t) \right) \geqslant \int_{\Gamma_\mathscr{F}} \frac{\partial g}{\partial t} \psi_\varepsilon(w_T) \, w'_T \, \mathrm{d}\Gamma = \int_{\Gamma_\mathscr{F}} \frac{\partial g}{\partial t} \frac{\partial}{\partial t} \psi_\varepsilon(w_T) \, \mathrm{d}\Gamma .$$

Then, with this inequality, (5.142) gives

$$(5.144) \qquad \begin{aligned} &\frac{1}{2} \frac{\mathrm{d}}{\mathrm{d}t} [|u''_\varepsilon(t)|^2 + a(u'_\varepsilon(t), u'_\varepsilon(t))] + \int_{\Gamma_\mathscr{F}} \frac{\partial g}{\partial t} \frac{\partial}{\partial t} \psi_\varepsilon(w_T) \, \mathrm{d}\Gamma \\ &\leqslant (\psi'(t), u''_\varepsilon(t) + \Phi''(t)) - (u'''_\varepsilon(t), \Phi''(t)) - a(u'_\varepsilon(t), \Phi''(t)) \end{aligned}$$

where we replaced w *by* $u''_\varepsilon + \Phi''$.

Integrating from 0 to t, it follows that:

$$\begin{aligned}|u_\varepsilon''(t)|^2+\|u_\varepsilon'(t)\|^2 &\leqslant c(\|u_1\|^2+|u_\varepsilon''(0)|^2)+c\int_0^t |u_\varepsilon''(\sigma)|^2\,\mathrm{d}\sigma\\ &+2\int_0^t (\Psi'(\sigma),u_\varepsilon''(\sigma))\,\mathrm{d}\sigma-2\int_0^t (u_\varepsilon'''(\sigma),\Phi''(\sigma))\,\mathrm{d}\sigma\\ &-2\int_0^t\int_{\Gamma_{\mathscr{F}}}\frac{\partial g}{\partial t}\frac{\partial}{\partial t}\psi_\varepsilon(u_{\varepsilon T}''+\Phi_T'')\,\mathrm{d}\Gamma\,\mathrm{d}\sigma\,.\end{aligned}\tag{5.145}$$

But

$$\int_0^t (\Psi'(\sigma),u_\varepsilon''(\sigma))\,\mathrm{d}\sigma=(\Psi'(t),u_\varepsilon'(t))-(\Psi'(0),u_1)-\int_0^t(\Psi'',u_\varepsilon')\,\mathrm{d}\sigma$$

and the assumptions on f, F_N, Φ imply that

$$\Psi''\in L^2(0,T;V_0')\tag{5.146}$$

therefore

$$\int_0^t(\Psi'(\sigma),u_\varepsilon''(\sigma))\,\mathrm{d}\sigma\leqslant c\,\|\Psi'(t)\|_*\,\|u_\varepsilon'(t)\|+c+\int_0^t\|\Psi''(\sigma)\|_*\,\|u_\varepsilon'(\sigma)\|\,\mathrm{d}\sigma\,.\tag{5.147}$$

Then

$$\int_0^t (u_\varepsilon'''(\sigma),\Phi''(\sigma))\,\mathrm{d}\sigma=(u_\varepsilon''(t),\Phi''(t))-(u_\varepsilon''(0),\Phi''(0))-\int_0^t(u_\varepsilon'',\Phi''')\,\mathrm{d}\sigma$$

from which it follows, starting from (5.145), that

$$\begin{aligned}|u_\varepsilon''(t)|^2+\|u_\varepsilon'(t)\|^2 &\leqslant c(\|u_1\|^2+|u_\varepsilon''(0)|^2)+c\int_0^t(\|u_\varepsilon'(\sigma)\|^2+|u_\varepsilon''(\sigma)|^2)\,\mathrm{d}\sigma\\ &+2\left|\int_0^t\int_{\Gamma_{\mathscr{F}}}\frac{\partial g}{\partial t}\frac{\partial}{\partial t}\psi_\varepsilon(u_{\varepsilon T}''+\Phi_T'')\,\mathrm{d}\Gamma\,\mathrm{d}\sigma\right|.\end{aligned}\tag{5.148}$$

We now have to estimate $u''(0)$.

From (5.135) it follows that

$$(u_\varepsilon''(0),v)=(\Psi(0),v)-a(u_0,v)-(j'(u_1+\Phi'(0)),v)\,.$$

But since, according to (5.130), $u_{1T}+\Phi_T'(0)=0$, it follows from this that

$$\begin{aligned}(u_\varepsilon''(0),v)=(\Psi(0),v)-a(u_0,v)&=(f(0),v)+\int_{\Gamma_{\mathscr{F}}}F_N(0)\,v_N\,\mathrm{d}\Gamma-(\Phi''(0),v)\\ &-a(\Phi_0,v)-a(u_0,v)\,.\end{aligned}$$

But

$$a(u_0,v)=(Au_0,v)+\int_{\Gamma_{\mathscr{F}}}\sigma_{0N}v_N\,\mathrm{d}\Gamma+\int_{\Gamma_{\mathscr{F}}}\sigma_{0T}v_T\,\mathrm{d}\Gamma\,.$$

Now, $\sigma_{0N}=F_N(0)$ and therefore, using (5.131):

$$(u_\varepsilon''(0),v)=(f(0)-\Phi''(0)-Au_0-A\Phi(0),v)+\int_{\Gamma_{\mathscr{F}}}\sigma_{0T}v_T\,\mathrm{d}\Gamma-\int_{\Gamma_{\mathscr{F}}}\sigma_{0T}v_T\,\mathrm{d}\Gamma$$

from which

$$u_\varepsilon''(0)=f(0)-\Phi''(0)-A(u_0+\Phi(0))\,.\tag{5.149}$$

Then, in particular, in view of the assumptions made,

$$(5.150) \qquad |u_\varepsilon''(0)| \leqslant c.$$

We now take the last term of (5.148) which is equivalent to

$$\int_{\Gamma_{\mathscr{F}}} \frac{\partial g}{\partial t}(x,t)\psi_\varepsilon(u_{\varepsilon T}''(t)+\Phi_T''(t))\,\mathrm{d}\Gamma - \int_{\Gamma_{\mathscr{F}}} \frac{\partial g}{\partial t}(x,0)\psi_\varepsilon(u_{\varepsilon T}''(0) + \Phi_T''(0))\,\mathrm{d}\Gamma - \int_0^t \int_{\Gamma_{\mathscr{F}}} \frac{\partial^2 g}{\partial t^2}\psi_\varepsilon(u_{\varepsilon T}''+\Phi_T'')\,\mathrm{d}\Gamma\,\mathrm{d}\sigma.$$

In order to make use of these estimates in (5.148), we have to be able to majorize by $|u_\varepsilon''(t)|$. But then the term $\|u_\varepsilon''(t)\|$ appears which makes it impossible to draw the conclusion unless we assume $\partial g/\partial t = 0$, which is assumption (5.128).

Then (5.148) and (5.150) give:

$$(5.151) \qquad |u_\varepsilon''(t)|^2 + \|u_\varepsilon'(t)\|^2 \leqslant c\left(1 + \int_0^t (|u_\varepsilon''(\sigma)|^2 + \|u_\varepsilon'(\sigma)\|^2)\,\mathrm{d}\sigma\right),$$

from which it follows, with Gronwall's lemma, that:

$$(5.152) \qquad |u_\varepsilon''(t)| + \|u_\varepsilon'(t)\| \leqslant c.$$

Passage to the limit in ε. According to (5.141) and (5.152), we can select from the u_ε a sequence, again denoted by u_ε, such that

$$(5.153) \qquad \begin{aligned} &u_\varepsilon \to u, \quad u_\varepsilon' \to u' \quad (\text{resp. } u_\varepsilon'' \to u'') \quad \text{weakly star in } L^\infty(0,T;V_0) \\ &(\text{resp. } L^\infty(0,T;H)). \end{aligned}$$

From (5.135), it follows that

$$(5.154) \qquad \begin{aligned} &(u_\varepsilon'', v-u_\varepsilon') + a(u_\varepsilon, v-u_\varepsilon') + j_\varepsilon(v+\Phi') - j_\varepsilon(u_\varepsilon'+\Phi') - (\Psi, v-u_\varepsilon') \\ &\qquad = j_\varepsilon(v+\Phi') - j_\varepsilon(u_\varepsilon'+\Phi') - (j_\varepsilon'(u_\varepsilon'+\Phi'), v-u_\varepsilon') \geqslant 0. \end{aligned}$$

Taking in (5.154)

$$v = v(t), \qquad v \in L^2(0,T;V_0),$$

it follows that

$$(5.155) \qquad \begin{aligned} &\int_0^T [(u_\varepsilon'', v) + a(u_\varepsilon, v) + j_\varepsilon(v+\Phi') - (\Psi, v-u_\varepsilon')]\,\mathrm{d}t \\ &\qquad \geqslant \int_0^T [(u_\varepsilon'', u_\varepsilon') + a(u_\varepsilon, u_\varepsilon') + j_\varepsilon(u_\varepsilon'+\Phi')]\,\mathrm{d}t \\ &\qquad = \tfrac{1}{2}[|u_\varepsilon'(T)|^2 + a(u_\varepsilon(T), u_\varepsilon(T))] - \tfrac{1}{2}[|u_1|^2 + a(u_0,u_0)] \\ &\qquad\quad + \int_0^T j_\varepsilon(u_\varepsilon'+\Phi')\,\mathrm{d}t. \end{aligned}$$

But

$$\liminf_{\varepsilon\to 0}\left[\tfrac{1}{2}[|u'_\varepsilon(T)|^2+a(u_\varepsilon(T),u_\varepsilon(T))]+\int_0^T j_\varepsilon(u'_\varepsilon+\Phi')\,dt\right]$$
$$\geqslant \tfrac{1}{2}[|u'(T)|^2+a(u(T),u(T))+\int_0^T j(u'+\Phi')\,dt$$
$$=\int_0^T[(u'',u')+a(u,u')+j(u'+\Phi')]\,dt$$

and therefore (5.155) gives

$$\int_0^T[(u'',v-u')+a(u,v-u')+j(v+\Phi')-j(u'+\Phi') \tag{5.156}$$
$$-(\Psi,v-u')]\,dt\geqslant 0 \qquad \forall v\in L^2(0,T;V_0)\,.$$

From (5.156), we pass to the *pointwise* inequality (5.123)[42] by the same procedure as in Section I.5.6.3. □

6. Linear Visco-Elasticity. Material with Short Memory

6.1. Constituent Law and General Remarks

Visco-elastic materials are materials "endowed with memory" in the sense that the state of stress at the instant t depends on all the deformations undergone by the material in previous times.

They are called "*linear*", if in addition, the constituent law establishes a linear relation between the tensor of stress and the tensor of strain.

It is materials of this type which we will consider in Sections 6 and 7, in situations with or without friction.

A particularly important class of "visco-elastic constituent laws" concerns materials for which a linear relationship exists between the derivatives with respect to time (i.e., the rate of change with respect to t) of the strain and stress tensors. They are the materials of "rate type" (cf. A. M. Freudenthal et Geiringer [1]), for which we then have

$$\sigma_{ij}(t)+A^{(1)}_{ijkh}\partial\sigma_{kh}(t)/\partial t+\cdots+A^{(n_1)}_{ijkh}\partial^{n_1}\sigma_{kh}(t)/\partial t^{n_1} \tag{6.1}$$
$$=a^{(0)}_{ijkh}\varepsilon_{kh}(t)+a^{(1)}_{ijkh}\partial\varepsilon_{kh}(t)/\partial t+\cdots+a^{(n_2)}_{ijkh}\partial^{n_2}\varepsilon_{kh}(t)/\partial t^{n_2}\ ^{43}\,.$$

The derivatives with respect to t in (6.1) are to be understood in the sense of distributions on R_t, the tensors σ and ε being *zero for* $t<0$. (In order to conform with earlier sections, we write ε_{kh} instead of $\varepsilon_{kh}(u)$.) In the *one-dimensional* case, we can easily make an explicit calculation (for example, using Laplace transformations). *In the case where* $n_2\geqslant n_1$, we find:

$$\sigma(t)=a^{(0)}\varepsilon(t)+a^{(1)}\partial\varepsilon(t)/\partial t+\cdots+a^{(N)}\partial^N\varepsilon(t)/\partial t^N+\int_0^t b(t-s)\varepsilon(s)\,ds \tag{6.2}$$

---

[42] In the following precise sense: except possibly for t in a set of measure zero on $[0,T]$, we have (5.123) $\forall v\in V_0$.

[43] Cf. also W. A. Day [1].

where

$$N = n_2 - n_1,$$

the $a^{(0)}, \ldots, a^{(N)}$ are constants, b is a regular function for $t>0$, continuous for $t \geqslant 0$.

In the case where $n_2 < n_1$, we obtain

$$\sigma(t) = \int_0^t b(t-s)\,\varepsilon(s)\,\mathrm{d}s, \tag{6.3}$$

where b has the same properties as above.

The law (6.3) is generally not used because it implies behavior which does not agree well with experience. In Section 6, we will study the case $n_1 = 0$, $n_2 = 1$ and, in Section 7, the case $n_1 = n_2 = 1$.

When $n_1 = 0$, $n_2 = 1$, *the constituent law is*

$$\sigma_{ij}(t) = a^{(0)}_{ijkh}\,\varepsilon_{kh}(t) + a^{(1)}_{ijkh}\,\partial/\partial t\,\varepsilon_{kh}(t); \tag{6.4}$$

one then says that *the material* under consideration *has a short memory*, since the state of stress at the moment t depends only on the deformation *at the instant* t and *at the immediately preceding instants*. The coefficients $a^{(0)}_{ijkh}$ (resp. $a^{(1)}_{ijkh}$) play the role of coefficients of elasticity (resp. of viscosity[44]), and thus it is natural, and in agreement with experience, to assume that

$$\begin{aligned} a^{(0)}_{ijkh}\,\varepsilon_{ij}\,\varepsilon_{kh} &\geqslant \alpha\,\varepsilon_{ij}\,\varepsilon_{ij}, \\ a^{(1)}_{ijkh}\,\varepsilon_{ij}\,\varepsilon_{kh} &\geqslant \alpha\,\varepsilon_{ij}\,\varepsilon_{ij}. \end{aligned} \tag{6.5}$$

(The coefficients a_{ijkh} are the coefficients of the ε_{ij} constructed for the field of velocity vectors, hence coefficients of viscosity. The scalar $a^{1}_{ijkh}\,\varepsilon_{ij}(u')\,\varepsilon_{kh}(u')$ is then a term of dissipation of energy, necessarily positive, which accounts for the second condition (6.5).)

When $n_1 = n_2 = 1$, *the constituent law* is

$$\sigma_{ij}(t) = a^{(0)}_{ijkh}\,\varepsilon_{kh}(t) + \int_0^t b_{ijkh}(t-s)\,\varepsilon_{kh}(s)\,\mathrm{d}s. \tag{6.6}$$

A material with the constituent law (6.6) (in Section 7, we will spell out the assumptions for the functions $t \to b_{ijkh}(t)$) is said *to have a long memory*, because the state of stresses at the instant t depends on the deformation at the instant t and also on the deformations at the times preceding t. □

6.2. Dynamic Case. Formulation of the Problem

The equations are[45]

$$\frac{\partial^2 u_i}{\partial t^2} - \frac{\partial}{\partial x_j}\,a^{(0)}_{ijkh}\,\varepsilon_{kh}(u) - \frac{\partial}{\partial t}\,\frac{\partial}{\partial x_j}\,a^{(1)}_{ijkh}\,\varepsilon_{kh}(u) = f_i \quad \text{in } \Omega, \quad t \in \,]0, T[\,. \tag{6.7}$$

[44] Because they involve the tensor of strain velocities.

[45] We used *linearized* acceleration since it is the *linearized* strain tensor which occurs in the constituent law.

Boundary conditions. We consider the case *with friction* (with constant coefficient $\mathscr{F}$) on a part $\Gamma_{\mathscr{F}}$ of $\Gamma=\partial\Omega$, with a displacement imposed on $\Gamma_U=\Gamma-\Gamma_{\mathscr{F}}$.
The boundary condions are then analogous to those of Section 5.1.2:

(6.8) $$u_i=U_i \quad \text{on} \quad \Gamma_U\times]0,T[\,,$$

(6.9) $$\sigma_N=F_N \quad \text{on} \quad \Gamma_{\mathscr{F}}\times]0,T[\,,$$

(6.10) $$\begin{aligned}&|\sigma_T|<g=\mathscr{F}|F_N| \Rightarrow \partial u_T/\partial t=0\\ &|\sigma_T|=g \Rightarrow \text{there exists} \quad \lambda>0 \quad \text{such that} \quad \partial u_T/\partial t=-\lambda\sigma_T\,.\end{aligned}$$

The *initial conditions* are, naturally:

(6.11) $$u(x,0)=u_0(x)\,, \qquad \partial u(x,0)/\partial t=u_1(x)\,, \qquad x\in\Omega\,.$$

For the variational formulation, let us introduce the following notation:

(6.12) $$\begin{aligned}a^0(u,v)&=\int_\Omega a^{(0)}_{ijkh}\,\varepsilon_{kh}(u)\,\varepsilon_{ij}(v)\,\mathrm{d}x\,,\\ a^1(u,v)&=\int_\Omega a^{(1)}_{ijkh}\,\varepsilon_{kh}(u)\,\varepsilon_{ij}(v)\,\mathrm{d}x\,,\end{aligned}$$

(6.13) $$j(v)=\int_\Gamma g\,|v_T|\,\mathrm{d}\Gamma\,,$$

and (as in Section 5.5.1):

(6.14) $$\mathscr{U}_{\mathrm{ad}}(t)=\{v\mid v\in(H^1(\Omega))^3,\quad v=U'(t) \ \text{on} \ \Gamma_U\}\,.$$

Then the preceding problem can be formulated as in Section 5.5.1 with a now replaced by a^0 and a supplementary term a^1 (corresponding to the viscosity), say:

(6.15) $$\begin{aligned}&(u''(t),v-u'(t))+a^1(u'(t)),v-u'(t))+a^0(u(t),v-u'(t))+j(v)-j(u'(t))\\ &\qquad\geqslant(f(t),v-u'(t))+\int_{\Gamma_{\mathscr{F}}}F_N(t)(v_N-u'_N(t))\,\mathrm{d}\Gamma \qquad \forall v\in\mathscr{U}_{\mathrm{ad}}(t)\end{aligned}$$

with the boundary conditions (6.11). □

Remark 6.1. If $\mathscr{F}=0$ (case without friction), (6.15) reduces, if $U=0$, to the linear equation $(\Gamma_{\mathscr{F}}=\Gamma-\Gamma_U)$

(6.16) $$(u''(t),v)+a^1(u'(t),v)+a^0(u(t),v)=(f(t),v)+\int_{\Gamma_F}F_N v_N\,\mathrm{d}\Gamma\,, \qquad v=0 \quad \text{on} \ \Gamma_U$$

(otherwise, replace v by $v-u$ in (6.16)). □

Remark 6.2 (Analogous to Remarks 4.4 and 5.17). We will treat Problem (6.15) in the same manner whether Γ_U is empty or of positive measure. However, if $\Gamma_U=\emptyset$, a global motion of the visco-elastic body can occur which can be decomposed into a rigid body motion and a motion with deformation. We then (cf. Remark 4.4) come back to the case where, $\forall t$,

(6.17) $$|(f(t),\rho)+\int_\Gamma F_N(t)\,\rho_N\,\mathrm{d}\Gamma|\leqslant\int_\Gamma g\,|\rho_T|\,\mathrm{d}\Gamma \qquad \forall\rho\in\mathscr{R}\,. \quad □$$

6.3. Existence Theorem and Uniqueness in the Dynamic Case

As in Section 4.1, we introduce[46] $\Phi(t)$ satisfying

$$\Phi(t)\in(H^1(\Omega))^3\,, \quad \Phi(t)=U(t) \quad \text{on } \Gamma_U\,. \tag{6.18}$$

We introduce

$$V=(H^1(\Omega))^3\,, \tag{6.19}$$

$$V_0=\{v \mid v\in V,\ v=0 \text{ on } \Gamma_U\}\,^{47}\,. \tag{6.20}$$

Replacing u by $u-\Phi$ (keeping the notation u), the problem reduces to: find u such that

$$u(t)\in V_0\,, \tag{6.21}$$

$$\begin{aligned}&(u''(t), v-u'(t))+a^1(u'(t), v-u'(t))+a^0(u(t), v-u'(t))\\&\qquad +j(v+\Phi'(t))-j(u'(t)+\Phi'(t))\geqslant(L(t), v-u'(t))\,, \qquad \forall v\in V_0\,,\end{aligned} \tag{6.22}$$

where

$$(L(t),v)=(f(t),v)+\int_{\Gamma_{\mathscr{F}}} F_N(t)v_N\,\mathrm{d}\Gamma-(\Phi''(t),v)-a^1(\Phi'(t),v)-a^0(\Phi(t),v)\,, \tag{6.23}$$

with the initial data

$$\begin{aligned}&u(0)=u_0 \quad (\text{in fact } u_0-\Phi(0))\\&u'(0)=u_1 \quad (\text{in fact } u_1-\Phi(0))\,.\end{aligned} \tag{6.24}$$

We set $H=(L^2(\Omega))^3$ and $V_0\subset H\subset V_0'$. Then we have

Theorem 6.1. *We make the following assumptions*:

$$f,f',f''\in L^2(0,T;H)\,, \quad F_N,F_N',F_N''\in L^2(\Gamma_{\mathscr{F}}\times]0,T[)\,, \tag{6.25}$$

$$\Phi,\Phi',\Phi''\in L^2(0,T;V)\,, \quad \Phi''',\Phi^{(4)}\in L^2(0,T;H)\,, \quad \Phi(0)\in(H^2(\Omega))^3\,, \tag{6.26}$$

(6.27) *g does not depend on t*[48],

$$u_0\in(H^2(\Omega))^3\,, \tag{6.28}$$

$$u_1\in V_0\,, \quad u_{1T}+\Phi_T'(0)=0 \quad \text{on } \Gamma_{\mathscr{F}}\,, \tag{6.29}$$

$$(L(0),v)-a^0(u_0,v)-a^1(u_1,v)=(u_2,v)\,, \qquad u_2\in H\,. \tag{6.30}$$

---

[46] In the following, replace Φ by 0, when $\Gamma_U=\emptyset$.

[47] $V_0=V$ if $\Gamma_U=\emptyset$.

[48] When $\mathscr{F}$ does not depend on t, this amounts to saying that F_N does not depend on t, and the second part of asumption (6.25) becomes meaningless.

Under these conditions, there exists one and only one function u such that

(6.31) $u, u' \in L^\infty(0,T;V_0)$,

(6.32) $u'' \in L^2(0,T;V_0) \cap L^\infty(0,T;H)$,

and satisfying (6.22), (6.24).

Remark 6.3. Property (6.32) is stronger than (5.133); this is due to the viscosity term

$$a^1(u', v-u'). \quad \square$$

Proof. Uniqueness is proved as in 5.5.3.

For existence, we follow the method of Section 5.5.4. We give only an outline of the proof. We introduce j_ε as in (5.134).

The approximate regularized equation is then:

(6.33) $$(u_\varepsilon'', v) + a^1(u_\varepsilon', v) + a^0(u_\varepsilon, v) + (j_\varepsilon'(u_\varepsilon' + \Phi'), v) = (L(t), v)$$
$$\forall v \in V_0, \quad u_\varepsilon(0) = u_0, \quad u_\varepsilon'(0) = u_1.$$

Then we obtain *the first* a priori *estimates* by taking $v = u_\varepsilon' + \Phi'$.
We note that, $\forall \lambda > 0$, there exist $\alpha_0, \alpha_1 > 0$ with

(6.34) $$a^0(v,v) + \lambda|v|^2 \geqslant \alpha_0 \|v\|^2, \quad \alpha_0 > 0, \quad \forall v \in V_0$$
$$a^1(v,v) + \lambda|v|^2 \geqslant \alpha_1 \|v\|^2, \quad \alpha_1 > 0,$$

and

(6.35) we can take $\lambda = 0$ in (6.34) if Γ_U has positive measure.

Thus, we obtain an estimate analogous to (5.138), with the left hand side:

$$|u_\varepsilon'(t)|^2 + \alpha_1 \int_0^t \|u_\varepsilon(s)\|^2 \, ds + \alpha_0 \|u_\varepsilon(t)\|^2 - \lambda |u_\varepsilon(t)|^2 + \int_0^t |u_\varepsilon'(\sigma)|^2 \, d\sigma$$

from which follows (compare with (5.141)):

(6.36) $|u_\varepsilon'(t)| + \|u_\varepsilon(t)\| \leqslant c$,

(6.37) $\int_0^T \|u_\varepsilon'(t)\|^2 \, dt \leqslant c$.

For *the second* a priori *estimates*, we again proceed as in Section 5.5.4.
We need an estimate for $u_\varepsilon''(0)$; setting $t=0$ in (6.33), it follows, using (6.29), that

$$(u_\varepsilon''(0), v) = (L(0), v) - a^0(u_0, v) - a^1(u_1, v) = (\text{according to } (6.30)) = (u_2, v)$$

therefore

$$(6.38) \qquad u_\varepsilon''(0)=u_2 .$$

Then we differentiate (6.33) with respect to t, take $v=u_\varepsilon''(t)+\Phi''(t)$ and thus obtain (compare with (5.152))

$$(6.39) \qquad |u_\varepsilon''(t)|+\|u_\varepsilon'(t)\|\leqslant c ,$$

$$(6.40) \qquad \int_0^T \|u_\varepsilon''(t)\|^2 \, dt \leqslant c .$$

Then we pass to the limit with ε as in Section 5.5.4.

Remark 6.4. We can just as well introduce *other boundary conditions*, as for example in Section 5.4.1.

Remark 6.5. We can considerably weaken the assumptions in Theorem 6.1 and then obtain a "weaker" solution. Precisely, we have:

Theorem 6.1a. *We make the following assumptions*:

$$(6.25a) \qquad f\in L^2(0,T;H), \qquad F_N\in L^2(\Gamma_{\mathscr{F}}\times]0,T[),$$

$$(6.26a) \qquad \Phi,\Phi'\in L^2(0,T;V), \qquad \Phi''\in L^2(0,T;H),$$

$$(6.28a) \qquad u_0\in V_0 ,$$

$$(6.29a) \qquad u_1\in H .$$

Under these conditions, there exists one and only one function u such that

$$(6.31a) \qquad u\in L^\infty(0,T;V_0), \qquad u'\in L^2(0,T;V_0)\cap L^\infty(0,T;H)$$

$$(6.32a) \qquad u''\in L^2(0,T;V_0'),$$

and satisfying (6.22), (6.24).

Indeed, we again start from a solution u_ε of (6.33). With the estimates (6.36), (6.37), it follows from (6.33) that

$$(u_\varepsilon'',v)=(L(t),v)-a^1(u_\varepsilon',v)-a^0(u_\varepsilon,v)-(j_\varepsilon'(u_\varepsilon'+\Phi'),v)$$

hence, *in particular*:

$$u_\varepsilon'' \text{ remains in a bounded set of } L^2(0,T;V') .$$

The Theorem follows from this by the same type of technique as before. $\square$

6.4. Quasi-Static Problems. Variational Formulation

Since the constituent law (6.4) *depends on time* because of the viscosity term, it is not possible to consider *static problems* as we did in the case of elasticity (without viscosity).

However, when the given forces or displacements vary little for $t \geqslant t_0 > 0$ and if one is interested in the solution for $t \geqslant t_0$, is is customary[49] to neglect (in the equations of motion) the linearized acceleration terms $\partial^2 u_i/\partial t^2$ in the equations of motion: this is the *quasi-static* case. □

We can then write the equations of equilibrium *at each instant*, taking account of the constituent law (6.4):

(6.41) $$-\frac{\partial}{\partial t}\frac{\partial}{\partial x_j} a^{(1)}_{ijkh}\varepsilon_{kh}(u) - \frac{\partial}{\partial x_j} a^{(0)}_{ijkh}\varepsilon_{kh}(u) = f_i \quad \text{in} \quad \Omega \times]0, T[\,,$$

the boundary conditions (6.8), (6.9), (6.10) being unchanged and the *initial condition* being

(6.42) $$u(0) = u_0\,.$$

We are also led to the problem to find $u(t)$ with

(6.43) $$\begin{aligned} &u'(t) \in \mathcal{U}_{\mathrm{ad}}(t)\,,\\ &a^1(u'(t), v - u'(t)) + a^0(u(t), v - u'(t)) + j(v) - j(u'(t))\\ &\qquad \geqslant (f(t), v - u'(t)) + \textstyle\int_{\Gamma_{\mathcal{F}}} F_N(t)(v - u'(t))\,\mathrm{d}\Gamma \qquad \forall v \in \mathcal{U}_{\mathrm{ad}}(t)\,, \end{aligned}$$

and the initial condition (6.42).

Here we have to distinguish the cases carefully: measure $\Gamma_U > 0$ and $\Gamma_U = \emptyset$. We start with the case "measure $\Gamma_U > 0$".

6.5. Existence and Uniqueness Theorem for the Case when Γ_U has Measure >0

We again introduce $\Phi(t)$ satisfying (6.18) and replace u by $u - \Phi$. The problem becomes: find a u such that

(6.44) $$u(t) \in V_0\,,$$

(6.45) $$\begin{aligned} &a^1(u'(t), v - u'(t)) + a^0(u(t), v - u'(t)) + j(v + \Phi'(t)) - j(u'(t) + \Phi'(t))\\ &\qquad \geqslant (L_1(t), v - u'(t)) \qquad \forall v \in V_0\,, \end{aligned}$$

where

(6.46) $$(L_1(t), v) = (f(t), v) + \textstyle\int_{\Gamma_{\mathcal{F}}} F_N(t)\, v_N\,\mathrm{d}\Gamma - a^1(\Phi'(t), v) - a^0(\Phi(t), v)\,,$$

[49] In Section 6.7 below, we will see to what extent this is justified, at least for the case *without friction*.

and with the initial condition

$$u(0)=u_0(=u_0-\Phi(0))\,. \tag{6.47}$$

We will prove

Theorem 6.2. *We assume*

$$f\in L^2(0,T;H)\,,\qquad F_N\in L^2(\Gamma_{\mathscr{F}}\times]0,T[)\,, \tag{6.48}$$

$$\Phi,\Phi'\in L^2(0,T;V) \tag{6.49}$$

$$u_0\in V_0\,. \tag{6.50}$$

Under these conditions, there exists one and only one function u such that

$$u\in L^\infty(0,T;V_0)\,, \tag{6.51}$$

$$u'\in L^2(0,T;V_0)\,, \tag{6.52}$$

and such that we have (6.45)[50], (6.47).

Remark 6.6. The assumptions are analogous to those of Theorem 6.1a. ▯

Uniqueness proof. Let u and u_* be two possible solutions. Taking $v=u'_*(t)$ in (6.45) (resp. $v=u'(t)$ in the inequality with respect to u_*) and adding, we obtain, if $w=u-u'$,

$$-a^1(w'(t),w'(t))-a^0(w(t),w'(t))\geqslant 0$$

or again[51]

$$\frac{1}{2}\frac{\mathrm{d}}{\mathrm{d}t}a^0(w(t))+a^1(w'(t))\leqslant 0$$

whence

$$a^0(w(t))=0\,,\quad \text{therefore}\quad w(t)=0\,,\quad \text{since } a^0 \text{ and } a^1 \text{ satisfy}$$

$$a^i(v)\geqslant\alpha_i\|v\|^2\,,\qquad \alpha_i>0\,,\qquad \forall v\in V_0 \text{ [52]}\,. \tag{6.53}$$ ▯

Existence proof. We again start from a *regularized equation*.
First, with j_ε being defined as in (5.134), we look for a solution u_ε of

$$\begin{aligned} &a^1(u'_\varepsilon(t),v)+a^0(u_\varepsilon(t),v)+(j'_\varepsilon(u'_\varepsilon(t)+\Phi'(t)),v)=(L_1(t),v)\qquad \forall v\in V_0\,,\\ &u_\varepsilon(0)=u_0\,. \end{aligned} \tag{6.54}$$

[50] Precisely: except possibly for t in a set of measure zero, we have (6.45) for all v in V_0.
[51] We put: $a^i(v,v)=a^i(v)$.
[52] Since measure $\Gamma_U>0$.

We solve this problem by a new approximation (for example), introducing $\eta>0$ and a solution $u_{\varepsilon\eta}$ of

$$(6.55)\quad \begin{aligned}&\eta(u''_{\varepsilon\eta}(t),v)+a^1(u'_{\varepsilon\eta}(t),v)+a^0(u_{\varepsilon\eta}(t),v)+(j'_\varepsilon(u'_{\varepsilon\eta}(t)+\Phi'(t)),v)=(L_1(t),v) \qquad \forall v\in V_0,\\ &u_{\varepsilon\eta}(0)=u_0, \quad u'_{\varepsilon\eta}(0)=0 \quad \text{(for example)}.\end{aligned}$$

Letting η tend towards 0, we obtain a solution u_ε of (6.54).

Taking $v=u'_\varepsilon(t)+\Phi'(t)$ in (6.54), it follows, since $(j'_\varepsilon(\varphi),\varphi)\geqslant 0$, that

$$\begin{aligned}a^1(u'_\varepsilon(t))+a^0(u_\varepsilon(t),u'_\varepsilon(t)) &= -a^1_\varepsilon(u'_\varepsilon(t),\Phi'(t))-a^0(u_\varepsilon(t),\Phi'(t))\\ &\quad +(L_1(t),u'_\varepsilon(t)+\Phi'(t))\end{aligned}$$

from which

$$\begin{aligned}\int_0^t a_1(u'_\varepsilon(s))\,ds+\tfrac{1}{2}a^0(u_\varepsilon(t)) &\leqslant \tfrac{1}{2}a^0(u_0)-\int_0^t a^1(u'_\varepsilon,\Phi')\,ds\\ &\quad -\int_0^t a^0(u_\varepsilon,\Phi')\,ds+\int_0^t (L_1,u'_\varepsilon+\Phi')\,ds\end{aligned}$$

from which it follows that

$$\int_0^t \|u'_\varepsilon(s)\|^2\,ds+\|u_\varepsilon(t)\|^2\leqslant c\int_0^t (\|u'_\varepsilon\|+\|u_\varepsilon\|)\,M(s)\,ds, \qquad M\in L^2(0,T),$$

from which

$$\int_0^t \|u'_\varepsilon(s)\|^2\,ds+\|u_\varepsilon(t)\|^2\leqslant \tfrac{1}{2}\int_0^t \|u'_\varepsilon\|^2\,ds+\tfrac{1}{2}\int_0^t \|u_\varepsilon(s)\|^2\,ds+c\int_0^t M(s)^2\,ds$$

and therefore

$$\int_0^t \|u'_\varepsilon(s)\|^2\,ds+\|u_\varepsilon(t)\|^2\leqslant \int_0^t \|u_\varepsilon(s)\|^2\,ds+c\,.$$

Thus

$$(6.56)\qquad \|u_\varepsilon(t)\|\leqslant c, \qquad \int_0^T \|u'_\varepsilon(t)\|^2\,dt\leqslant c\,.$$

Then we can select from the u_ε a sequence, again denoted by u_ε, such that

$$(6.57)\quad \begin{aligned}&u_\varepsilon\to u \quad \text{weakly star in } L^\infty(0,T;V_0)\\ &u'_\varepsilon\to u' \quad \text{weakly in } L^2(0,T;V_0)\,.\end{aligned}$$

But since (the function $v\to j_\varepsilon(v)$ being convex)

$$(6.58)\qquad j_\varepsilon(v+\Phi'(t))-j_\varepsilon(u'_\varepsilon(t)+\Phi'(t))-(j'_\varepsilon(u'_\varepsilon(t)+\Phi'(t)),\,v-u'_\varepsilon(t))\geqslant 0$$

it follows from (6.54) that

$$(6.59)\quad \begin{aligned}&a^1(u'_\varepsilon(t),\,v-u'_\varepsilon(t))+a^0(u_\varepsilon(t),\,v-u'_\varepsilon(t))+j_\varepsilon(v+\Phi'(t))-j_\varepsilon(u'_\varepsilon(t)+\Phi'(t))\\ &\qquad \geqslant (L_1(t),\,v-u'_\varepsilon(t))\,.\end{aligned}$$

Let $v \in L^2(0,T;V_0)$. Taking $v=v(t)$ in (6.59) and integrating, we have:

$$\int_0^T [a^1(u'_\varepsilon, v-u'_\varepsilon)+a^0(u_\varepsilon, v-u'_\varepsilon)+j_\varepsilon(v+\Phi')-j_\varepsilon(u'_\varepsilon+\Phi')-L_1, v-u'_\varepsilon)]\,dt \geqslant 0, \tag{6.60}$$

or again

$$\int_0^T [a^1(u'_\varepsilon, v)+a^0(u_\varepsilon, v)+j_\varepsilon(v+\Phi')-(L_1, v-u'_\varepsilon)]\,dt \\ \geqslant \int_0^T a^1(u'_\varepsilon)\,dt+\tfrac{1}{2}a^0(u_\varepsilon(T))+\int_0^T j_\varepsilon(u'_\varepsilon+\Phi')\,dt-\tfrac{1}{2}a^0(u_0). \tag{6.61}$$

It follows, since

$$\liminf_{\varepsilon\to 0}\left[\int_0^T a^1(u'_\varepsilon)\,dt+\tfrac{1}{2}a^0(u_\varepsilon(T))+\int_0^T j_\varepsilon(u'_\varepsilon+\Phi')\,dt\right] \\ \geqslant \int_0^T a^1(u')\,dt+\tfrac{1}{2}a^0(u(T))+\int_0^T j(u'+\Phi')\,dt,$$

that

$$\int_0^T [a^1(u', v-u')+a^0(u, v-u')+j(v+\Phi')-j(u'+\Phi')-(L_1, v-u')]\,dt \geqslant 0 \\ \forall v \in L^2(0,T;V_0). \tag{6.62}$$

We pass from this to the pointwise inequality (6.45) by the same procedure as in Section I.5.6.3. □

6.6. Discussion of the Case when $\Gamma_U=\emptyset$

When $\Gamma_U=\emptyset$, we take $\Phi(t)=0$ in what preceded. The problem is then to find u with

$$u(t) \in V(=H^1(\Omega)^3), \tag{6.63}$$

$$a^1(u'(t), v-u'(t))+a^0(u(t), v-u'(t))+j(v)-j(u'(t)) \geqslant (L_2(t), v-u'(t)) \quad \forall v \in V \tag{6.64}$$

where

$$(L_2(t), v)=(f(t), v)+\int_\Gamma F_N(t)\, v_N\, d\Gamma, \tag{6.65}$$

and with the initial condition (6.47).

If in (6.64), we take $v=u'(t)+\rho$, $\rho \in \mathcal{R}$, and observe that

$$a^0(\varphi,\rho)=0, \quad a^1(\varphi,\rho)=0 \quad \forall \varphi \in V,$$

we have:

$$j(u'(t)+\rho)-j(u'(t)) \geqslant (L_2(t), \rho)$$

and, as

$$j(u'(t)+\rho)-j(u'(t)) \leqslant j(\rho)$$

it follows that

$$(L_2(t),\rho)\leqslant j(\rho),$$

or again (since $j(\rho)=j(-\rho)$):

$$|(L_2(t),\rho)|\leqslant j(\rho)\quad \forall\rho\in\mathcal{R},\quad \forall t\in[0,T]. \tag{6.66}$$

We will see that the *necessary* condition (6.66) is *sufficient* to enable us to solve the preceding problem in a certain sense.

We will pass to the quotient with $\mathcal{R}$. As in (3.47), (3.48), we introduce:

$$\dot V=V/\mathcal{R},\quad \dot H=H/\mathcal{R}, \tag{6.67}$$

$$a^i(\dot u,\dot v)=a^i(u,v),\quad u\in\dot u,\quad v\in\dot v,\quad i=0,1. \tag{6.68}$$

According to Theorem 3.4, we have:

$$a^i(\dot v)=a^i(\dot v,\dot v)\geqslant\alpha\|\dot v\|_{\dot V}^2,\quad \alpha>0,\quad i=0,1. \tag{6.69}$$

Moreover, we define:

$$\mathcal{H}(t;\dot v)=\inf_{\rho\in\mathcal{R}}[j(v+\rho)-(L_2(t),v+\rho)]. \tag{6.70}$$

We note that, if (6.66) holds, then

$$\mathcal{H}(t;\dot v)>-\infty. \tag{6.71}$$

Indeed, taking v arbitrary and *fixed* in $\dot v$, we have

$$j(v+\rho)\geqslant j(\rho)-j(v),$$

therefore

$$j(v+\rho)-(L_2(t),v+\rho)\geqslant j(\rho)-(L_2(t),\rho)+j(v)-(L_2(t),v)$$
$$\text{(according to (6.66))}\geqslant j(v)-(L_2(t),v)\ \text{ whence (6.71)}.$$

With this understood, we will prove

Theorem 6.3. *We assume that*

$$f\in L^2(0,T;H),\quad F_N\in L^2(\Gamma\times]0,T[), \tag{6.72}$$

$$u_0\in V \tag{6.73}$$

and that (6.66) holds a.e. in t, with L_2 defined by (6.65). Then there exists one and only one function $\dot u$ such that

(6.74) $u^{\bullet} \in L^{\infty}(0, T; V^{\bullet})$,

(6.75) $u^{\bullet\prime} \in L^{2}(0, T; V^{\bullet})$,

(6.76) $u^{\bullet}(0) = u_0^{\bullet}$,

and

(6.77) $a^1(u^{\bullet\prime}(t), v^{\bullet} - u^{\bullet\prime}(t)) + a^0(u^{\bullet}(t), v - u^{\bullet\prime}(t)) + \mathscr{H}(t; v^{\bullet}) - \mathscr{H}(t; u^{\bullet\prime}(t)) \geqslant 0 \quad \forall v^{\bullet} \in V^{\bullet}$.

Remark 6.7. In the case without friction, $j=0$. We have to make the assumption

(6.78) $(L_2(t), \rho) = 0 \qquad \forall \rho \in R$

and then

(6.79) $\mathscr{H}(t; v^{\bullet}) = -(L_2(t), v^{\bullet})$,

and (6.77) reduces to the equation:

(6.80) $a^1(u^{\bullet\prime}(t), v) + a^0(u^{\bullet}(t), v) = (L_2(t), v^{\bullet}) \qquad \forall v^{\bullet} \in V^{\bullet}$. □

Proof of Theorem 6.3. Uniqueness follows exactly as in Theorem 6.2, since (6.69) holds, due to the passage to the quotient with $\mathscr{R}$ (and Korn's inequality).

To prove existence, we start out by applying Theorem 6.2 with $a^1(u,v)$ replaced by $a^1(u,v) + \varepsilon(u,v)$, $\varepsilon > 0$, and $a^0(u,v)$ unchanged[53]. Then there exists a (unique) u_ε such that $u_\varepsilon \in L^{\infty}(0, T; V)$, $u'_\varepsilon \in L^2(0, T; V)$ and

(6.81) $a^1(u'_\varepsilon, v - u'_\varepsilon) + \varepsilon(u'_\varepsilon, v - u'_\varepsilon) + a^0(u_\varepsilon, v - u'_\varepsilon) + j(v) - j(u'_\varepsilon) \geqslant (L_2, v - u'_\varepsilon) \quad \forall v \in V$,

and $u_\varepsilon(0) = u_0$.

Setting $v=0$ in (6.81), it follows that

$$
\begin{aligned}
a^1(u'_\varepsilon(t)) + \varepsilon |u'_\varepsilon(t)|^2 + \frac{1}{2}\frac{\mathrm{d}}{\mathrm{d}t} a^0(u_\varepsilon(t)) &\leqslant (L_2(t), u'_\varepsilon(t)) - j(u'_\varepsilon(t)) \\
&= (L_2(t), u'_\varepsilon(t) + \rho) - (L_2(t), \rho) - j(u'_\varepsilon(t))) \\
&\leqslant (\text{according to } (6.66)) \leqslant (L_2(t), u'_\varepsilon(t) + \rho) + j(\rho) - j(u'_\varepsilon(t)) \\
&\leqslant (L_2(t), u'_\varepsilon(t) + \rho) + j(u'_\varepsilon(t) + \rho) \\
&\leqslant c(1 + \|L_2(t)\|_*) \|u'_\varepsilon(t) + \rho\| \qquad \forall \rho \\
&\leqslant c(1 + \|L_2(t)\|_*) \inf_\rho \|u'_\varepsilon(t) + \rho\| \\
&= c(1 + \|L_2(t)\|_*) \|u_\varepsilon^{\bullet\prime}(t)\|_{V^{\bullet}}.
\end{aligned}
$$

[53] Theorem 6.2 is valid when $a^1(v) \geqslant \alpha \|v\|^2$, $\alpha > 0$, and when $a^0(v) \geqslant 0$, $a^0(v) + \lambda |v|^2 \geqslant \|v\|^2$ when $\lambda > 0$.

whence, according to (6.69)

$$\alpha\int_0^t \|u_\varepsilon^{\cdot\prime}(s)\|_{V^\cdot}^2 ds + \tfrac{1}{2}\alpha\|u_\varepsilon^\cdot(t)\|_{V^\cdot}^2 \leqslant \tfrac{1}{2}a^0(u_0) + c\int_0^t (1+\|L_2(s)\|_*)\|u_\varepsilon^{\cdot\prime}(s)\|_{V^\cdot} ds$$
$$\leqslant \tfrac{1}{2}\alpha\int_0^t \|u_\varepsilon^{\cdot\prime}(s)\|_{V^\cdot}^2 ds + c$$

and consequently

$$\|u_\varepsilon^\cdot(t)\|_{V^\cdot}^2 + \int_0^t \|u_\varepsilon^{\cdot\prime}(s)\|_{V^\cdot}^2 ds \leqslant c\,. \tag{6.82}$$

Then we can select from the $u_\varepsilon^\cdot$ a sequence, again denoted by $u_\varepsilon^\cdot$, such that

$$\begin{aligned} &u_\varepsilon^\cdot \to u^\cdot \quad \text{weakly star in } L^\infty(0,T;V^\cdot)\\ &u_\varepsilon^{\cdot\prime} \to u^{\cdot\prime} \quad \text{weakly in } L^2(0,T;V^\cdot)\,. \end{aligned} \tag{6.83}$$

Moreover, we have

$$\varepsilon\int_0^T |u_\varepsilon'(t)|^2 dt \leqslant c\,. \tag{6.84}$$

We write (6.81) in the form

$$\varepsilon(u_\varepsilon', v-u_\varepsilon') + a^1(u_\varepsilon^{\cdot\prime}, v^\cdot - u_\varepsilon^{\cdot\prime}) + a^0(u_\varepsilon^\cdot, v^\cdot - u_\varepsilon^{\cdot\prime}) + j(v) - (L_2(t), v)$$
$$\geqslant [j(u_\varepsilon'(t)) - (L_2(t), u_\varepsilon'(t))]$$
$$\geqslant \inf[j(u_\varepsilon'(t)+\rho) - (L_2(t), u_\varepsilon'(t)+\rho)] = \mathscr{H}(t; u_\varepsilon^{\cdot\prime}(t)))$$

from which it follows, if $v\in L^2(0,T;V)$, that:

$$\begin{aligned} &\varepsilon\int_0^T (u_\varepsilon', v-u_\varepsilon')dt + \int_0^T [a^1(u_\varepsilon^{\cdot\prime}, v^\cdot - u_\varepsilon^{\cdot\prime}) + a^0(u_\varepsilon^\cdot, v^\cdot - u_\varepsilon^{\cdot\prime}) + j(v) - (L_2, v)]dt\\ &\qquad \geqslant \int_0^T \mathscr{H}(t; u_\varepsilon^{\cdot\prime})dt\,. \end{aligned} \tag{6.85}$$

But $v^\cdot \to \int_0^T \mathscr{H}(t; v^\cdot)dt$ is lower semi-continuous on $L^2(0,T;V^\cdot)$ for the weak topology, therefore

$$\liminf_{\varepsilon\to 0} \int_0^T \mathscr{H}(t; u_\varepsilon^{\cdot\prime})dt \geqslant \int_0^T \mathscr{H}(t; u^{\cdot\prime})dt\,. \tag{6.86}$$

We write (6.85) in the form

$$\int_0^T [\varepsilon(u_\varepsilon', v) + a^1(u_\varepsilon^{\cdot\prime}, v^\cdot) + a^0(u_\varepsilon^\cdot, v^\cdot) + j(v) - (L_2, v)]dt$$
$$\geqslant \int_0^T \mathscr{H}(t; u_\varepsilon^{\cdot\prime})dt + \varepsilon\int_0^T |u_\varepsilon'|^2 dt + \int_0^T a^1(u_\varepsilon^{\cdot\prime})dt + \tfrac{1}{2}a^0(u_\varepsilon^\cdot(t)) - \tfrac{1}{2}a^0(u_0)$$

whence

$$\int_0^T [a^1(u^{\cdot\prime}, v^\cdot) + a^0(u^\cdot, v^\cdot) + j(v) - (L_2, v)]dt$$
$$\geqslant \int_0^T \mathscr{H}(t; u^{\cdot\prime})dt + \int_0^T a^1(u^{\cdot\prime})dt + \tfrac{1}{2}a^0(u^\cdot(T)) - \tfrac{1}{2}a^0(u_0)$$

and consequently

$$\text{(6.87)} \qquad \int_0^T [j(v)-(L_2,v)]\,dt \leqslant \int_0^T [-a^1(u^{\cdot\prime}, v^{\cdot}-u^{\cdot\prime})-a^0(u^{\cdot}, v^{\cdot}-u^{\cdot\prime})+\mathscr{H}(t;u^{\cdot\prime})]\,dt\,.$$

By the same procedure as in Section I.5.6.3, it follows, except possibly for $t\in E$, *where E is a set of measure zero on* $[0,T]$, that

$$\text{(6.88)} \qquad \begin{aligned} j(v)-(L_2,v) \geqslant -a^1(u^{\cdot\prime}(t), v^{\cdot}-u^{\cdot\prime}(t))-a^0(u^{\cdot}(t), v^{\cdot}-u^{\cdot\prime}(t))+\mathscr{H}(t;u^{\cdot\prime}(t)) \\ \forall v\in V. \end{aligned}$$

If we substitute $v+\rho$ for v, the right-hand side is invariant, therefore (6.88) is *equivalent* to

$$\text{(6.89)} \qquad \begin{aligned} &\inf_\rho [j(v+\rho)-(L_2, v+\rho)] \\ &\qquad \geqslant -a^1(u^{\cdot\prime}(t), v^{\cdot}-u^{\cdot\prime}(t))-a^0(u^{\cdot}(t), v^{\cdot}-u^{\cdot\prime}(t))+\mathscr{H}(t;u^{\cdot\prime}(t))\,. \end{aligned}$$

The left-hand side of (6.89) is simply $\mathscr{H}(t;v^{\cdot})$, from which (6.77) follows. ☐

6.7. Justification of the Quasi-Static Case in the Problems without Friction

6.7.1. Statement of the Problem

We consider the dynamic case (discussed in Sec. 6.2 and 6.3) with the following assumptions:

(6.90) $\qquad U_i$, f_i and F_N *are not dependent on* $t\geqslant 0$ [54]

We assume that the initial data are *zero*:

$$\text{(6.91)} \qquad u_0=0\,, \qquad u_1=0\,,$$

then, for $t<0$, the material is in an undeformed state.

In the usual cases[55], the "*quasi-static hypothesis*" is that, under conditions (6.90), (6.91), the solutions of the dynamic problem and of the quasi static problem are "*close to each other*".

We are going to prove that, actually, this is the case **when there is no friction.** *The corresponding problem in the case "with friction" is an open question which seems interesting to us.* ☐

Since there is no friction, the solution u of the dynamic case satisfies:

$$\text{(6.92)} \qquad \begin{aligned} &(u''(t),v)+a^1(u'(t),v)+a^0(u(t),v)=(L,v) \qquad \forall v\in V_0\,, \\ &u(t)\in V_0\,, \end{aligned}$$

[54] An assumption which can be weakened. Cf. Remark 6.8.

[55] I. e. *without friction.*

where

(6.93) $(L,v)=(f,v)-a^0(\Phi,v)$. □

We will generalize this somewhat. We recall (cf. Sec. 5.1.2) that *in order to simplify the discussion*, we took $\Gamma_U=\Gamma-\Gamma_{\mathscr{F}}$. In fact, (with the notation of Sec. 2)

$$\Gamma-\Gamma_{\mathscr{F}}=\Gamma_U\cup\Gamma_{\mathscr{F}}$$

and here, since $\Gamma_{\mathscr{F}}=\emptyset$, we come back to

(6.94) $\Gamma=\Gamma_U\cup\Gamma_F$.

We then prescribe

(6.95) $u_i=U_i$ on Γ_U, U_i independent of t,

(6.96) $\sigma_{ij}n_j=F_i$ on Γ_F, F_i independent of t

from which finally

(6.97)
$$\begin{aligned}&(u''(t),v)+a^1(u'(t),v)+a^0(u(t),v)=(L,v)\,, \quad v\in V_0\,, \quad u(t)\in V_0\\ &u(0)=0\,, \quad u'(0)=0\,,\end{aligned}$$

where

(6.98) $(L,v)=(f,v)+\int_{\Gamma_{\mathscr{F}}}F_iv_i\,d\Gamma-a^0(\Phi,v)$.

By $\tilde{u}(t)$ we designate the solution of the associated quasi-static problem, say

(6.99)
$$\begin{aligned}&a^1(\tilde{u}'(t),v)+a^0(\tilde{u}(t),v)=(L,v)\,, \quad v\in V_0\,, \quad \tilde{u}(t)\in V_0\\ &\tilde{u}(0)=0\,.\end{aligned}$$

The aim is now to prove that u and $\tilde{u}$ are "close to each other", in a certain sense. We will distinguish two cases according as measure $\Gamma_U>0$ or $\Gamma_U=\emptyset$. □

6.7.2. The Case "Measure $\Gamma_U>0$"

We compare u and $\tilde{u}$ to the solution $\bar{u}$ of the following stationary problem:

(6.100) $a^0(\bar{u},v)=(L,v) \quad \forall v\in V_0\,, \quad \bar{u}\in V_0\,,$

a problem which certainly does have a unique solution when "measure $\Gamma_U>0$". We will prove the following results:

Theorem 6.4. *We assume that* measure $\Gamma_U>0$. *With the assumptions* (6.95), (6.96), *there then exist a* $\gamma>0$ *and a constant* c *such that*:

(6.101) $\|u(t)-\bar{u}\|\leqslant c\,e^{-\gamma t}$,

(6.102) $|u'(t)|\leqslant c\,e^{-\gamma t}$, $\quad e^{-\gamma t}u'\in L^2(0,\infty;V)$.

Theorem 6.5. *Under the assumptions of Theorem* 6.4, *there exist a* $\gamma>0$ *and a constant* c *such that*

$$\|\tilde{u}(t)-\bar{u}\| \leqslant c\,\mathrm{e}^{-\gamma t}, \tag{6.103}$$

$$\mathrm{e}^{\gamma t}\,\tilde{u}' \in L^2(0,\infty;V). \tag{6.104}$$

Corollary 6.1. *(Justification of the quasi static case.) Under the assumptions of Theorem* 6.4, *there exist a* $\gamma>0$ *and a* c *such that*

$$\begin{aligned}&\|u(t)-\tilde{u}(t)\| \leqslant c\,\mathrm{e}^{-\gamma t},\\ &\mathrm{e}^{\gamma t}(u'-\tilde{u}') \in L^2(0,\infty;V).\end{aligned} \tag{6.105}$$

Proof of Theorem 6.4. The function $w(t)=u(t)-\bar{u}$ satisfies

$$(w''(t),v)+a^1(w'(t),v)+a^0(w(t),v)=0, \tag{6.106}$$

with

$$w(0)=-\bar{u}, \quad w'(0)=0. \tag{6.107}$$

Let us set

$$w(t)=e^{-\lambda t}z(t), \quad \lambda>0 \quad \text{to be chosen.} \tag{6.108}$$

Then (6.106) gives

$$(z''(t),v)+a^1(z'(t),v)-2\lambda(z'(t),v)+a^0(z(t),v)+\lambda^2(z(t),v)-\lambda a^1(z(t),v)=0. \tag{6.109}$$

We choose λ small enough so that

$$a^0(v)-\lambda a^1(v)+\lambda|v|^2 \geqslant \beta\|v\|^2, \quad \beta>0, \quad \forall v\in V_0, \tag{6.110}$$

$$a^1(v)-2\lambda|v|^2 \geqslant 0, \quad \forall v\in V_0. \tag{6.111}$$

If we take $v=z'(t)$ in (6.109), we have:

$$\frac{1}{2}\frac{\mathrm{d}}{\mathrm{d}t}|z'(t)|^2+\frac{1}{2}\frac{\mathrm{d}}{\mathrm{d}t}\left[a^0(z(t))+\lambda^2(z(t)^2-\lambda a^1(z(t))\right]+a^1(z'(t))-2\lambda|z'(t)|^2=0$$

from which

$$\begin{aligned}|z'(t)|^2+a^0(z(t))+\lambda^2|z(t)|^2-\lambda a^1(z(t))+2\textstyle\int_0^t(a^1(z')-2\lambda|z'|^2)\,\mathrm{d}s&\\ =|z'(0)|^2+a^0(z(0))+\lambda^2|z(0)|^2-\lambda a^1(z(0))&\end{aligned}$$

from which, due to (6.110), (6.111):

(6.112) $$|z'(t)|^2+\beta\|z(t)\|^2\leqslant \text{constant}.$$

From this follow (6.101) and the first inequality in (6.102). But we can choose λ in such a way that, instead of (6.111), we have

(6.113) $$a^1(v)-2\lambda|v|^2\geqslant\beta\|v\|^2$$

($\beta>0$; the "best" β is not necessarily the same as in (6.110)).
Then it follows that

$$|z'(t)|^2+\|z(t)\|^2+\int_0^t\|z'(s)\|^2\,ds\leqslant C$$

from which, consequently,

(6.114) $$\int_0^\infty\|z'(s)\|^2\,ds\leqslant C.$$

But then

$$e^{\lambda t}w'=z'-\lambda z$$

gives, for $\gamma<1$:

$$e^{\gamma t}w'=e^{(\gamma-\lambda)t}z'-\lambda e^{(\gamma-\lambda)t}z\in L^2(0,\infty;V)$$

from which the second part of (6.102) follows. ▯

Proof of Theorem 6.5. The function $\tilde{w}(t)=\tilde{u}(t)-\bar{u}$ satisfies:

(6.115) $$a^1(\tilde{w}'(t),v)+a^0(\tilde{w}(t),v)=0.$$

The substitution (6.108) leads to

(6.116) $$a^1(z'(t),v)+a^0(z(t),v)-\lambda a^1(z(t),v)=0.$$

We choose λ in such a way that[56]

(6.117) $$a^0(v)-\lambda a^1(v)\geqslant\beta\|v\|^2.$$

Then, if we take $v=z'(t)$ in (6.114), we obtain

$$\|z(t)\|^2+\int_0^t\|z'(s)\|^2\,ds\leqslant C$$

from which the result follows as before. ▯

[56] The *optimal* choices for λ (and therefore for γ) are not necessarily the same in the two theorems.

Remark 6.8. We can prove analogous results, if U_i, f_i and F_N depend on t but such that we can choose $\Phi(t)$ in a manner that, for a sufficiently small $\lambda>0$, we have

(6.118)
$$\begin{aligned} & e^{\lambda t}\|L(t)-L\|_*\leqslant C, \\ & \int_0^\infty \left\|\frac{d}{dt}(e^{\lambda t}L(t))\right\|_* dt<\infty. \quad \square \end{aligned}$$

6.7.3. The case $\Gamma_U=\emptyset$

When $\Gamma_U=\emptyset$, the quasi static Problem (6.99) has a solution only if

(6.119) $$(L,\rho)=0 \qquad \forall \rho\in\mathcal{R}.$$

Under these conditions, we can "pass to the quotient with $\mathcal{R}$" in (6.99); *then the quasi-static problem has a unique solution* $\tilde{u}^\cdot$ *such that*

(6.120) $$\tilde{u}^\cdot\in L^\infty(0,T;V^\cdot), \qquad \tilde{u}^{\cdot\prime}\in L^2(0,T;V^\cdot)$$

(6.121) $$a^1(\tilde{u}^{\cdot\prime}(t),v^\cdot)+a^0(\tilde{u}^\cdot(t),v^\cdot)=(L,v^\cdot) \qquad \forall v^\cdot\in V^\cdot.$$

Furthermore, if we then take $v=\rho$ in (6.97), we find

$$(u''(t),\rho)=0$$

which, since $u(0)=0$, $u'(0)=0$, proves that

(6.122) $$(u(t),\rho)=0 \qquad \forall\rho\in\mathcal{R}.$$

Now we identify u with $u^\cdot$, the solution of

(6.123) $$(u^{\cdot\prime\prime}(t),v^\cdot)+a^1(u^{\cdot\prime}(t),v^\cdot)+a^0(u^\cdot(t),v^\cdot)=(L,v^\cdot) \qquad \forall v^\cdot\in V^\cdot.$$

But after this passage to the quotient, we have properties (6.69) and, consequently, we are in a situation analogous to the one in Section 6.7.2, but in the quotient space $V^\cdot$. We introduce $\overline{u}^\cdot$, the solution in $V^\cdot$ of

(6.124) $$a^0(\overline{u}^\cdot,v^\cdot)=(L,v^\cdot) \qquad \forall v^\cdot\in V^\cdot.$$

Then, transcribing Theorems 6.4 and 6.5, we have $(H^\cdot=H/\mathcal{R})$:

Theorem 6.6. *Assuming* $\Gamma_U=\emptyset$, (6.96) *with* $\Gamma_F=\Gamma$ and (6.119), *there then exist a* $\gamma>0$ *and a* c *such that*

(6.125) $$\|u^\cdot(t)-\overline{u}^\cdot\|_{V^\cdot}\leqslant c\,e^{-\gamma t},$$

(6.126) $$|u^{\cdot\prime}(t)|\leqslant c\,e^{-\gamma t}, \qquad e^{\gamma t}u^{\cdot\prime}\in L^2(0,\infty;V^\cdot).$$

Theorem 6.7. *Under the assumptions of Theorem 6.6, there exist a $\gamma>0$ and a c such that*

(6.127) $$\|\tilde{u}^{\cdot}(t)-\bar{u}^{\cdot}\|_{V^{\cdot}}\leq c\,e^{-\gamma t},$$

(6.128) $$e^{\gamma t}\tilde{u}^{\cdot\prime}\in L^2(0,\infty;V^{\cdot}).$$

We have a Corollary corresponding to Corollary 6.1. □

6.8. The Case without Viscosity as Limit of the Case with Viscosity

We consider a material for which the constituent law is (compare with (6.5)):

(6.129) $$\sigma_{ij}=a^{(0)}_{ijkh}\varepsilon_{kh}+\lambda a^{(1)}_{ijkh}\partial\varepsilon_{kh}/\partial t$$

where $\lambda>0$ and where we let λ tend towards zero; in other words: "the viscosity will tend towards zero".

Our aim is to prove that we obtain the "elastic" case discussed in Section 5 as limit of the "viscous" case when $\lambda\to 0$. □

With the notation of Theorem 6.1, we denote by u_λ *the solution* of

(6.130) $$u_\lambda(t)\in V_0,$$

(6.131) $$\begin{aligned}&(u''_\lambda(t),v-u'_\lambda(t))+\lambda a^1(u'_\lambda(t),v-u'_\lambda(t))+a^0(u_\lambda(t),v-u'_\lambda(t))\\&\qquad+j(v+\Phi'(t))-j(u'_\lambda(t)+\Phi'(t))\geq(L(t),v-u'_\lambda(t))\qquad\forall v\in V_0,\end{aligned}$$

with the initial data

(6.132) $$u_\lambda(0)=u_0,\qquad u'_\lambda=u_1.$$

We will prove

Theorem 6.8. *In addition to the assumptions of Theorem 6.1, we assume that*

(6.133) $$\left|\begin{array}{l}\text{there exist } \hat{u}_2,u_3\in H \text{ such that}\\(L(0),v)-a^0(u_0,v)=(\hat{u}_2,v),\\a^1(u_1,v)=(u_3,v)\qquad\forall v\in V_0\end{array}\right.$$ [57].

Then, when $\lambda\to 0$, we have:

(6.134) $$\begin{aligned}&u_\lambda\to u,\quad u'_\lambda\to u'\quad\text{weakly star in } L^\infty(0,T;V_0)\\&u''_\lambda\to u''\quad\text{weakly star in } L^\infty(0,T;H)\end{aligned}$$

[57] This assumption implies (6.30).

where u is the solution of the "elastic" case, given in Theorem 5.7. Furthermore,

(6.135) $\lambda^{1/2} u'_\lambda$ *stays in a bounded set in* $L^2(0,T;V_0)$.

Proof. As in (6.33), we consider the approximate regularized equation:

$$(u''_{\varepsilon\lambda}, v) + \lambda a^1(u'_{\varepsilon\lambda}, v) + a^0(u_{\varepsilon\lambda}, v) + (j'_{\varepsilon\lambda}(u'_{\varepsilon\lambda} + \Phi'), v) = (L, v)$$
$$u_{\varepsilon\lambda}(0) = u_0, \quad u'_{\varepsilon\lambda}(0) = u_1. \tag{6.136}$$

If we substitute $v = u'_{\varepsilon\lambda} + \Phi'$ in (6.136), we can derive estimates analogous to (6.36), (6.37), taking into account the presence of the factor λ, from which:

$$|u'_{\varepsilon\lambda}(t)| + \|u_{\varepsilon\lambda}(t)\| \leqslant c, \tag{6.137}$$

$$\lambda \textstyle\int_0^T \|u'_{\varepsilon\lambda}(t)\|^2 \, dt \leqslant c. \tag{6.138}$$

For the second a priori estimates, we proceed as in the proof of Theorem 6.1. First, setting $t=0$ in (6.136), it follows, if we make use of the assumptions of Theorem 6.1 and of (6.133) that:

$$u''_{\varepsilon\lambda}(0) = \hat{u}_2 - \lambda u_3 \quad \text{(therefore bounded in } H). \tag{6.139}$$

Then we differentiate (6.136) in t and, in the equation thus obtained, replace v by $u''_{\varepsilon\lambda}$; it follows (compare with (6.39), (6.40)) that:

$$|u''_{\varepsilon\lambda}(t)| + \|u'_{\varepsilon\lambda}(t)\| \leqslant c, \tag{6.140}$$

$$\lambda \textstyle\int_0^T \|u''_{\varepsilon\lambda}(t)\|^2 \, dt \leqslant c. \tag{6.141}$$

We pass to the limit *with* ε which shows that u_λ, the (unique) solution of (6.130), (6.131), (6.132) satisfies

$$u_\lambda, u'_\lambda \in \text{bounded set of } L^\infty(0,T;V_0)$$
$$u''_\lambda \in \text{bounded set of } L^\infty(0,T;H) \text{ when } \lambda \to 0. \tag{6.142}$$

and also satisfies (6.135).

Then we select a sequence from the λ, again denoted by $\lambda \to 0$, such that (6.134) holds, from which the Theorem follows provided that we prove that the u thus obtained is the "elastic" solution.

If, in (6.131), we set $v = v(t)$, where $v(.) \in L^2(0,T;V_0)$, it follows that

$$\begin{aligned}
&\textstyle\int_0^T [(u''_\lambda, v) + a^0(u_\lambda, v) + j(v+\Phi') - (L, v - u'_\lambda)] \, dt \\
&\qquad \geqslant \textstyle\int_0^T [(u''_\lambda, u'_\lambda) + \lambda a'(u'_\lambda, u'_\lambda) + a^0(u_\lambda, u'_\lambda) + j(u'_\lambda + \Phi')] \, dt \\
&\qquad \geqslant \tfrac{1}{2}|u'_\lambda(T)|^2 + \tfrac{1}{2}a^0(u_\lambda(T), u_\lambda(T)) + \textstyle\int_0^T j(u'_\lambda + \Phi') \, dt \\
&\qquad\quad - \tfrac{1}{2}|u_1|^2 - \tfrac{1}{2}a^0(u_0, u_0)
\end{aligned}$$

from which

$$\begin{aligned}&\int_0^T [(u'',v)+a^0(u,v)+j(v+\Phi')-(L,v-u')]\,dt\\&\quad\geq \liminf_{\lambda}\big[\tfrac{1}{2}|u'_\lambda(T)|^2+\tfrac{1}{2}a^0(u_\lambda(T),u_\lambda(T))\\&\qquad+\int_0^T j(u'_\lambda+\Phi')\,dt\big]-\tfrac{1}{2}|u_1|^2-\tfrac{1}{2}a^0(u_0,u_0)\\&\quad\geq \tfrac{1}{2}|u'(T)|^2+\tfrac{1}{2}a^0(u(T),u(T))+\int_0^T j(u'+\Phi')\,dt-\tfrac{1}{2}|u_1|^2-\tfrac{1}{2}a^0(u_0,v_0)\\&\quad=\int_0^T [(u'',u')+a^0(u,u')+j(u'+\Phi')]\,dt\end{aligned}$$

and consequently

$$\int_0^T [(u'',v-u')+a^0(u,v-u')+j(v+\Phi')-j(u'+\Phi')-(L,v-u')]\,dt\geq 0 \tag{6.143}$$

which corresponds to the integral form of inequality (5.123)—the form which is equivalent to the pointwise form. □

6.9. Interpretation of Viscous Problems as Parabolic Systems[58]

We consider the situation of Section 6.3 with, for the sake of simplicity, $\Gamma_U=\emptyset$. We will replace (6.22) by a *parabolic* (equivalent) system of inequalities.

We introduce

$$\mathscr{V}=V_0\times V_0\subset\mathscr{H}=V_0\times H. \tag{6.144}$$

If $u=\{u_1,u_2\}$, $v=\{v_1,v_2\}$ denote two elements of $\mathscr{V}$ or $\mathscr{H}$, we set:

$$[u,v]=a^0(u_1,v_1)+(u_2,v_2), \tag{6.145}$$

$$\pi(u,v)=-a^0(u_2,v_1)+a^1(u_2,v_2)+a^0(u_1,v_2); \tag{6.146}$$

$\mathscr{H}$ is a Hilbert space for the scalar product (6.145); the form $\pi(u,v)$ is continuous on $\mathscr{V}\times\mathscr{V}$ and we have, since the form a^0 is symmetric:

$$\pi(v,v)=a^1(v_2,v_2)\geq\alpha\|v_2\|^2. \tag{6.147}$$

Setting:

$$u(t)=\{u_1(t),u_2(t)\}, \tag{6.148}$$

we consider the "*parabolic inequality*":

$$\begin{aligned}&[u'(t),v-u'(t)]+\pi(u(t),v-u'(t))+j(v_2+\Phi'(t))-j(u_2(t)+\Phi'(t))\\&\qquad\qquad\geq(L(t),v_2-u_2(t))\qquad\forall v\in\mathscr{V}\end{aligned} \tag{6.149}$$

[58] This section may be passed over.

which is *equivalent* to (6.22) (with $u(t)=u_1(t)$); the proof of the equivalence follows immediately: it is sufficient to spell out (6.149) using notations (6.145), (6.146); then we obtain two inequalities, the first of which is

$$a^0(u_1'(t), v_1-u_1'(t))-a^0(u_2(t), v_1-u_1'(t))\geqslant 0 \qquad \forall v_1\in V_0$$

which is equivalent to

$$u_1'(t)-u_2(t)=0$$

from which the result follows. □

Remark 6.9. Thus we see that, in the form of the parabolic inequality (6.149), the problem is *of the type of the inequalities encountered in Chapter II.* □

Remark 6.10. Starting out from (6.149)—naturally, adding the initial condition $u(0)=\{u_0, u_1\}$—we can again obtain Theorem 6.1. □

7. Linear Visco-Elasticity. Material with Long Memory

7.1. Constituent Law and General Remarks

The constituent law is now given by (6.6), which we restate:

$$\sigma_{ij}(t)=a_{ijkh}\,\varepsilon_{kh}(u(t))+\int_0^t b_{ijkh}(t-s)\,\varepsilon_{kh}(u(t))\,ds\,. \tag{7.1}$$

The first term in (7.1) represents an instantaneous elastic effect; the coefficients a_{ijkh} (called: *coefficients of instantaneous elasticity*) may possibly depend on x[59] but are bounded in x and satisfy:

$$\begin{aligned} &a_{ijkh}=a_{jikh}=a_{khij}\\ &a_{ijkh}\,\varepsilon_{ij}\,\varepsilon_{kh}\geqslant\alpha\varepsilon_{ij}\,\varepsilon_{ij}\,, \qquad \alpha>0\,. \end{aligned} \tag{7.2}$$

The coefficients $b_{ijkh}(t)$ represent the *effects of the memory of the material.*

In the usual practical examples, these are linear combinations of exponentials with negative coefficients: the memory decreases exponentially with distance in time. Cf. Examples in P. Germain [2], W. Prager [1] and T.H. Lin [1].

The coefficients $b_{ijkh}=b_{ijkh}(x,t)$ which depend on t, possibly on x, are bounded in x, t and satisfy

$$b_{ijkh}=b_{jikh} \tag{7.3}$$

[59] In the same way, in a sufficiently regular manner, we can solve by the same kind of methods as below—except for Section 7.5—the case where the a_{ijkh} depend also on t.

and the *regularity assumption*:

(7.4) $$b_{ijkh}, \quad \partial b_{ijkh}/\partial t, \quad \partial^2 b_{ijkh}/\partial t^2 \in L^\infty(Q), \qquad (Q=\Omega\times]0,T[).$$

Visco-elastic material of "solid type". A visco-elastic material with long memory of type 7.1 is said to be of *solid type* if

(7.5) $$\int_0^\infty b_{ijkh}(t)\,dt = b_{ijkh}^\infty \quad \text{finite}$$

(7.6) the coefficients $a_{ijkh}^\infty = a_{ijkh} + b_{ijkh}^\infty$ satisfy assumptions analogous to (7.2).

Then the coefficients a_{ijkh}^∞ are called "*coefficients of retarded elasticity*".

We can prove that, under certain conditions, the solution $u(t)$ of the problems considered hereafter converges, when $t\to+\infty$, toward the solution of the problem of elasticity corresponding to the coefficients a_{ijkh}^∞. □

7.2. Dynamic Problems with Friction

The equations are

(7.7) $$\frac{\partial^2 u_i}{\partial t^2} - \frac{\partial}{\partial x_j} a_{ijkh}\,\varepsilon_{kh}(u) - \int_0^t \frac{\partial}{\partial x_j} b_{ijkh}(t-s)\,\varepsilon_{kh}(u(s))\,ds = f.$$

The boundary and initial conditions are identical with those of (6.8), (6.11).

Remark 7.1. We can equally well introduce other boundary conditions, as for example in Section 5.4.1. □

For the variational formulation, we introduce:

(7.8) $$a(u,v) = \int_\Omega a_{ijkh}\,\varepsilon_{kh}(u)\,\varepsilon_{ij}(v)\,dx,$$

(7.9) $$b(t;u,v) = \int_\Omega b_{ijkh}(t)\,\varepsilon_{kh}(u)\,\varepsilon_{ij}(v)\,dx,$$

(7.10) $$V=(H^1(\Omega))^3, \qquad H=(L^2(\Omega))^3,$$
$$V_0=\{v \mid v\in V,\ v=0 \text{ on } \Gamma_U\}.$$

The form $v\to b(t;u,v)$ is continuous on V_0, therefore

(7.11) $$b(t;u,v)=(B(t)u,v), \qquad B(t)u\in V_0', \qquad B(t)\in\mathscr{L}(V_0;V_0').$$

We introduce $j(v)$ and $\mathscr{U}_{ad}$ as in (6.13), (6.14).

Then the problem can be formulated in the following way (compare with Sections 5.5.1 and 6.2): find a function $u(t)$ such that

(7.12) $$(u''(t), v-u'(t)) + a(u(t), v-u'(t)) + \int_0^t b(t-s; u(s), v-u'(t))\,ds + j(v) - j(u'(t))$$
$$\geq (f(t), v-u'(t) + \int_{\Gamma_{\mathscr{F}}} F_N(v_N - u_N'(t))\,d\Gamma \qquad \forall v\in\mathscr{U}_{ad}(t)$$

with

$$(7.13)\qquad u'(t)\in\mathcal{U}_{\mathrm{ad}}(t)\,,$$

$$(7.14)\qquad u(0)=u_0\,,\qquad u'(0)=u_1\,.$$

Remark 7.2. If $\mathcal{F}=0$ (the case without friction), (7.12) reduces to the linear equation (if $U=0$, $\Gamma_F=\Gamma-\Gamma_U$):

$$(7.15)\qquad (u''(t),v)+a(u(t),v)+\int_0^t b(t-s;u(s),v)\,\mathrm{d}s=(f(t),v)+\int_{\Gamma_{\mathcal{F}}}F_N v\,\mathrm{d}\Gamma\qquad \forall v\in V_0.$$

If $U\neq 0$, replace v by $v-u$ in (7.15). □

Remark 7.3. We have analogous properties to those in Remarks 4.4, 5.17 and 6.2 concerning the distinction between the case "Γ_U has positive measure" and the case "$\Gamma_U=\emptyset$". □

7.3. Existence and Uniqueness Theorem in the Dynamic Case

We introduce $\Phi(t)$ as in (6.18) ($\Phi=0$, if $\Gamma_U=\emptyset$); substituting $u-\Phi$ for u, (but keeping the u notation), the problem reduces to: find a u such that

$$(7.16)\qquad u(t)\in V_0\,,$$

$$(7.17)\qquad \begin{aligned}&(u''(t),v-u'(t))+a(u(t),v-u'(t))+\int_0^t(b(t-s;u(s),v-u'(t))\,\mathrm{d}s\\&\qquad+j(v+\Phi'(t))-j(u'(t)+\Phi'(t))\geqslant(L(t),v-u'(t))\qquad\forall v\in V_0\end{aligned}$$

where

$$(7.18)\qquad \begin{aligned}(L(t),v)=&(f(t),v)+\int_{\Gamma_{\mathcal{F}}}F_N(t)v_N\,\mathrm{d}\Gamma-(\Phi''(t),v)-a(\Phi(t),v)\\&-\int_0^t(b(t-s;\Phi(s),v)\,\mathrm{d}s\,,\end{aligned}$$

with

$$(7.19)\qquad \begin{aligned}&u(0)=u_0\quad(\text{in fact, } u_0-\Phi(0))\\&u'(0)=u_1\quad(\text{in fact, } u_1-\Phi'(0))\,.\end{aligned}$$

Then we have

Theorem 7.1. *We make the following assumptions:*

$$(7.20)\qquad f,f',f''\in L^2(0,T;H)\,,\qquad F_N,F_N'F_N''\in L^2(\Gamma_{\mathcal{F}}\times]0,T[)\,,$$

$$(7.21)\qquad \Phi,\Phi',\Phi''\in L^2(0,T;V)\,,\qquad \Phi''',\Phi^{(4)}\in L^2(0,T;H)\,,\qquad \Phi(0)\in(H^2(\Omega))^3$$

$$(7.22)\qquad g \text{ does not depend on } t\,,$$

$$(7.23)\qquad u_0\in V_0\,,\qquad (L(0),v)-a(u_0,v)=(u_2,v)\,,\qquad u_2\in H\,,$$

$$(7.24)\qquad u_1\in V_0\,,\qquad u_{1T}+\Phi_T'(0)=0\quad\text{on } \Gamma_{\mathcal{F}}\,.$$

Under these conditions, there exists one and only one function u such that

(7.25) $$u, u' \in L^\infty(0, T; V_0),$$

(7.26) $$u'' \in L^\infty(0, T; H),$$

which satisfies (7.17), (7.19).

Uniqueness proof. Using the notation of (7.11), we will write

$$\left(\int_0^t B(t-s)u(s)\,\mathrm{d}s,\ v-u'(t)\right)$$

instead of

$$\int_0^t b(t-s; u(s), v-u'(t))\,\mathrm{d}s\,.$$

Let u and u_* be two possible solutions. If we take $v=u_*'(t)$ (resp. $v=u'(t)$) in inequality (7.17) (resp. in the inequality with respect to u_*) and add, it follows, setting $w=u-u_*$, that

$$-(w''(t), w'(t)) - a(w(t), w'(t)) - \left(\int_0^t B(t-s)\,w(s)\,\mathrm{d}s,\ w'(t)\right) \geqslant 0$$

from which[60]

$$\frac{\mathrm{d}}{\mathrm{d}t}\left[|w'(t)|^2 + a(w(t))\right] \leqslant -2\left(\int_0^t B(t-s)\,w(s)\,\mathrm{d}s,\ w'(t)\right)$$

and therefore

(7.27) $$|w'(t)|^2 + a(w(t)) \leqslant -2\int_0^t \mathrm{d}s\left(\int_0^s B(s-s_1)\,w(s_1)\,\mathrm{d}s_1,\ w'(s)\right).$$

But, generally, (for $t \leqslant T \leqslant \infty$):

(7.28) $$\left|\int_0^t \mathrm{d}s\left(\int_0^s B(s-s_1)\,\varphi(s_1)\,\mathrm{d}s_1,\ \varphi'(s)\right)\right| \leqslant c\left[\int_0^t \|\rho(s)\|^2\,\mathrm{d}s + \|\varphi(t)\| \int_0^t \|\varphi(s)\|\,\mathrm{d}s\right].$$

Indeed,

$$\begin{aligned}\int_0^t \mathrm{d}s\left(\int_0^s B(s-s_1)\,\varphi(s_1)\,\mathrm{d}s,\ \varphi'(s)\right) &= \int_0^t \mathrm{d}s_1 \int_s^t (B(s-s_1)\,\varphi(s_1), \varphi'(s))\,\mathrm{d}s\\ &= \int_0^t \mathrm{d}s_1\left[(B(t-s_1)\,\varphi(s_1), \varphi(t)) - (B(0)\,\varphi(s_1), \varphi(s_1))\right]\\ &\quad - \int_0^t \mathrm{d}s_1 \int_s^t (B'(s-s_1)\,\varphi(s_1), \varphi(s))\,\mathrm{d}s\end{aligned}$$

from which (7.28) follows.

Then (7.27) and (7.28) give:

(7.29) $$|w'(t)|^2 + a(w(t)) \leqslant c\int_0^t \|w(s)\|^2\,\mathrm{d}s + c\,\|w(t)\| \int_0^t \|w(s)\|\,\mathrm{d}s\,.$$

---

[60] We recall that $|f| = (f, f)^{1/2}$, $a(v) = a(v, v)$.

But for $\lambda>0$, there exists an $\alpha>0$ such that

$$a(v)+\lambda|v|^2 \geqslant \alpha\|v\|^2 \tag{7.30}$$

(and we can take $\lambda=0$ if Γ_U has measure >0), therefore (7.29) gives[61]

$$|w'(t)|^2+\alpha\|w(t)\|^2 \leqslant \lambda|w(t)|^2+c\int_0^t\|w(s)\|^2\,ds+\frac{\alpha}{2}\|w(t)\|^2 . \tag{7.31}$$

But

$$w(t)=\int_0^t w'(s)\,ds ,$$

therefore, it follows from (7.31) that

$$|w'(t)|^2+\|w(t)\|^2 \leqslant c\int_0^t(|w'(s)|^2+\|w(s)\|^2)\,ds \tag{7.32}$$

from which $w=0$ follows[62]. □

Existence proof. As in the proof of Theorem 6.1, we follow the method of Section 5.5.4.

The approximate regularized equation is

$$(u_\varepsilon'',v)+a(u_\varepsilon,v)+\left(\int_0^t B(t-s)u_\varepsilon(s)\,ds,v\right)+(j_\varepsilon'(u_\varepsilon'+\Phi'),v)=(L(t)v), \qquad \forall v\in V_0, \tag{7.33}$$

with

$$u_\varepsilon(0)=u_0 , \qquad u_\varepsilon'(0)=u_1 .$$

We obtain the *first estimates* by substituting $v=u_\varepsilon'+\Phi'$ in (7.33).

The only new fact to prove is that the term

$$X=\left(\int_0^t B(t-s)u_\varepsilon(s)\,ds,u_\varepsilon'(t)+\Phi'(t)\right)$$

can be suitably estimated. This is clear for the part

$$\left(\int_0^t B(t-s)u_\varepsilon(s)\,ds,\Phi'(t)\right) .$$

There remains

$$X_1=\left(\int_0^t B(t-s)u_\varepsilon(s)\,ds,u_\varepsilon'(t)\right)dt .$$

But according to (7.28)

$$\left|\int_0^t X_1(s)\,ds\right| \leqslant c\left(\int_0^t\|u_\varepsilon(s)\|^2\,ds+\|u_\varepsilon(t)\|\int_0^t\|u_\varepsilon(s)\|\,ds\right) \tag{7.34}$$

---

[61] The c's always denote various constants.

[62] Here we did not use all of assumption (7.4), but solely the fact that $\frac{\partial}{\partial t}b_{ijkh}(x,t)$ is in $L^\infty(\Omega\times[0,T])$.

and, consequently, we obtain (as in Section 5.5.4):

(7.35) $\|u_\varepsilon(t)\| + |u'_\varepsilon(t)| \leqslant c$.

If we substitute "$t=0$" in (7.33), we have, due to (7.24):

$$(u''_\varepsilon(0), v) + a(u_0, v) = (L(0), v)$$

from which, according to (7.23),

(7.36) $u''(0) = u_2$.

Differentiating (7.33) with respect to t, it follows that

(7.37) $(u'''_\varepsilon, v) + a(u'_\varepsilon, v) + (B(0)\,u_\varepsilon(t), v) + (\int_0^t B'(t-s)\,u_\varepsilon(s)\,\mathrm{d}s, v) + ((j'_\varepsilon(u'_\varepsilon + \Phi'))', v) = (L'(t), v)$

and we have the same estimates as in the preceding situations, provided that we can suitably majorize

a) $(B(0)\,u_\varepsilon(t), u''_\varepsilon(t) + \Phi''(t))$—which is clear,

b) $Y = (\int^t B'(t-s)\,u_\varepsilon(s)\,\mathrm{d}s, u''_\varepsilon(t) + \Phi''(t))$;—which is also clear for

$$(\int_0^t B'(t-s)\,u_\varepsilon(s)\,\mathrm{d}s, \Phi''(t))$$

then it only remains to majorize

$$Y_1(t) = (\int_0^t B'(t-s)\,u_\varepsilon(s)\,\mathrm{d}s, u''_\varepsilon(t)).$$

But

$$\begin{aligned}\int_0^t Y_1(s)\,\mathrm{d}s &= \int_0^t \mathrm{d}s(\int_0^s B'(s-s_1)\,u_\varepsilon(s_1)\,\mathrm{d}s, u''_\varepsilon(s)) \\ &= \int_0^t \mathrm{d}s_1 \int_{s_1}^t (B'(s-s_1)\,u_\varepsilon(s_1), u''_\varepsilon(s))\,\mathrm{d}s \\ &= \int_0^t [(B'(t-s_1)\,u_\varepsilon(s_1), u'_\varepsilon(t)) - (B'(0)\,u_\varepsilon(s_1), u'_\varepsilon(s_1))]\,\mathrm{d}s \\ &\quad - \int_0^t \mathrm{d}s_1 \int_{s_1}^t (B''(s-s_1)\,u_\varepsilon(s_1), u'_\varepsilon(s))\,\mathrm{d}s\end{aligned}$$

from which it follows—now we make use of (7.4)—that

$$|\int_0^t Y_1(s)\,\mathrm{d}s| \leqslant c\,\|u'_\varepsilon(t)\| + c + c\int_0^t \|u'_\varepsilon(s)\|\,\mathrm{d}s,$$

terms which do not represent any difficulty in the majorizations.

Thus we obtain

(7.38) $\|u'_\varepsilon(t)\| + |u''_\varepsilon(t)| \leqslant C$

and we complete the proof of the theorem as in Theorem 6.1.

7.4. The Quasi-Static Case

By considerations analogous to those in the preceding Section 6, we see that the quasi-static problem "associated" with Problem (7.17), (7.19) is the following (we assume U independent of t so that we can assume that Φ is independent of t): find $u(t)$ such that

(7.16) $$u(t) \in V_0,$$

(7.39) $$a(u(t), v-u'(t)) + (\textstyle\int_0^t B(t-s)u(s)\,ds, v-u'(t)) + j(v) - j(u') \geqslant (L(t), v-u'(t)) \quad \forall v \in V_0,$$

(7.40) $$u(0) = u_0.$$

7.4.1. Necessary Conditions for the Initial Data

Let us assume that the Problem (7.39), (7.40) has a solution, regular in t; we set: $u'(t) = u_1$ (u_1 is not given). Substituting $t=0$ in (7.39), we find

$$a(u_0, v-u_1) + j(v) - j(u_1) \geqslant (L(0), v-u_1)$$

i.e.,

(7.41) $$\inf_{v \in V_0} [a(u_0, v) - (L(0), v) + j(v)] = a(u_0, u_1) - (L(0), u_1) + j(u_1) > -\infty.$$

We can state (7.41) more precisely by assuming, for the sake of simplicity, that

(7.42) $$U = 0.$$

Then, using Green's formula, we have

$$a(u_0, v) - (L(0), v) + j(v) = (Au_0 - f(0), v) + \int_{\Gamma_{\mathscr{F}}} (\sigma_N^0 v_N F_N(0) v_N)\,d\Gamma + \int_{\Gamma_{\mathscr{F}}} (\sigma_T^0 v_T + g|v_T|)\,d\Gamma$$

and thus it follows that $\inf[a(u_0, v) - (L(0), v) + j(v)] = -\infty$ unless

(7.43) $$Au_0 = f(0), \quad \sigma_N^0 = F_N(0) \text{ on } \Gamma_{\mathscr{F}}, \quad |\sigma_T^0| \leqslant g \text{ on } \Gamma_{\mathscr{F}} \text{ (and } u_0 = 0 \text{ on } \Gamma_U).$$

Coming back to (7.43), we then have

(7.44) $$\inf_{v \in V_0} [a(u_0, v) - (L(0), v) + j(v)] = 0.$$

The necessary conditions (7.43) are also *sufficient* for the existence of a "strong" solution of (7.39), (7.40); cf. H. Brézis [2], Chapter 2 (where methods involving non linear semi-groups are used). From a mechanical point of view, the initial condition seems to be

$$u_0 = u_0^*,$$

where u_0^* is the solution of

$$\begin{aligned}
&Au_0^* = f(0) && \text{in } \Omega\\
&\sigma_N(u_0^*) = F_N(0) && \text{on } \Gamma_{\mathscr{F}} \text{ (and } u_0^* = 0 \text{ on } \Gamma_U)\\
&|\sigma_T(u_0^*)| < g \Rightarrow u_{0T}^* = 0 &&\\
&|\sigma_T(u_0^*)| = g \Rightarrow \exists \lambda \geqslant 0 && \text{such that} \quad u_{0T}^* = -\lambda \sigma_T(u_0^*),
\end{aligned}$$

that is to say that u_0^* is the *solution of the problem of static elasticity corresponding to the forces* $f(0), F_N(0)$ *and to the conditions for friction on* $\Gamma_{\mathscr{F}}$.

However, in order to simplify somewhat, we will only prove existence here under a more restrictive assumption

$$\begin{aligned}
&Au_0 = f_0 && \text{on } \Omega\\
\text{(7.45)} \qquad &\sigma_N^0 = F_N(0) && \text{on } \Gamma_{\mathscr{F}}\\
&\sigma_T^0 = 0 && \text{on } \Gamma_{\mathscr{F}} \text{ (and } u_0 = 0 \text{ on } \Gamma_U).
\end{aligned}$$

7.4.2. Discussion of the Case "Measure $\Gamma_U > 0$"

Theorem 7.2. *We assume that* measure $\Gamma_U > 0$ *and that* u_0 *is given in* $(H^2(\Omega))^3$, *satisfying* (7.45). *We assume that* $U = 0$, *therefore* $\Phi = 0$, *and that* (7.20), (7.22) *hold.*

Under these conditions, there exists one and only one function u such that

$$\text{(7.46)} \qquad u, u' \in L(0, T; V_0)$$

and satisfying (7.39), (7.40).

Uniqueness proof. Let u and u_* be two possible solutions, $w = u - u_*$. If we take $v = u_*'(t)$ in (7.39) (resp. $v = u'(t)$ in the corresponding inequality for u) and add, it follows after integration (compare with (7.29)), that

$$\text{(7.47)} \qquad a(w(t)) \leqslant c \int_0^t \|w(s)\|^2 \, ds + c \|w(t)\| \int_0^t \|w(s)\| \, ds.$$

But since Γ_U has positive measure, we have (7.30) with $\lambda = 0$ and therefore it follows from (7.47) that

$$\|w(t)\|^2 \leqslant c \int_0^t \|w(s)\|^2 \, ds$$

whence $w = 0$. □

Existence proof. We carry out a *double regularization*. We regularize j to j_ε and add a term in $\eta(u'',v)$, $\eta>0$. Then we "approximate" (7.39) by the *biregularized equation*:

(7.48) $$\eta(u''_{\varepsilon\eta},v)+a(u_{\varepsilon\eta},v)+(\textstyle\int_0^t B(t-s)u_{\varepsilon\eta}(s)\,\mathrm{d}s,v)+j'_\varepsilon(u'_{\varepsilon\eta}(t),v)=(L(t),v)$$

with

(7.49) $$u_{\varepsilon\eta}(0)=u_0\,,$$

(7.50) $$u'_{\varepsilon\eta}(0)=0\ ^{63}\,.$$

Then we obtain the first a priori estimates by taking $v=u'_{\varepsilon\eta}(t)$, which leads to

(7.51) $$u_{\varepsilon\eta}\leqslant c \quad \text{(the c's denote constants independent of ε and η)}.$$

(7.52) $$\sqrt{\eta}\,|u'_{\varepsilon\eta}|\leqslant c\,.$$

To go further, we differentiate (7.48) with respect to t, then substitute $u''_{\varepsilon\eta}(t)$ for v. Now we need an estimate for $u''_{\varepsilon\eta}(0)$, independent of ε and η.

Setting $t=0$ in (7.48), we find

(7.53) $$\eta(u''_{\varepsilon\eta}(0),v)+a(u_0,v)+(j'_\varepsilon(0),v)=(L(0),v)=(f(0),v)+\textstyle\int_{\Gamma_{\mathscr{F}}}F_N(0)\,v_N\,\mathrm{d}\Gamma\,.$$

But according to (7.45), $a(u_0,v)=(f(0),v)+\int_{\Gamma_{\mathscr{F}}}F_N(0)\,v_N\,\mathrm{d}\Gamma$ and therefore (7.53) gives

$$\eta(u''_{\varepsilon\eta}(0),v)=0\,,$$

therefore[64]

(7.54) $$u''_{\varepsilon\eta}=0\,.$$

Then it follows that

(7.55) $$\|u'_{\varepsilon\eta}(t)\|\leqslant c\,,$$

(7.56) $$\sqrt{\eta}\,|u''_{\varepsilon\eta}(t)|\leqslant c\,,$$

and we can pass to the limit in ε and η by the usual steps. □

Remark 7.4. In the case "without friction", we replace $\Gamma_{\mathscr{F}}$ by Γ_F (cf. Sec. 6.7). The problem is then to find a function u with (7.16) and

(7.57) $$a(u(t),v)+(\textstyle\int_0^t B(t-s)u(s)\,\mathrm{d}s,v)=(f(t),v)+\int_{\Gamma_F}F_i(t)\,v_i\,\mathrm{d}\Gamma\,.$$

---

[63] This is natural according to (7.44) and (7.41), since $u_1=0$ yields the minimum.

[64] It was for the purpose of obtaining this estimate, *independent of* η, that we made assumption (7.45).

Then u_0 *is imposed*: if we take $t=0$ in (7.57), we find

$$a(u_0, v) = (f(0), v) + \int_{\Gamma_F} F_i(0)\, v_i \,\mathrm{d}\Gamma \tag{7.58}$$

which is the analogue to (7.45) in this case. □

7.4.3. Discussion of the Case "$\Gamma_U = \emptyset$"

If $\Gamma_U = \emptyset$, the problem is to find u with

$$\begin{aligned} & u(t) \in V \\ & a(u(t), v - u'(t)) + \left(\int_0^t B(t-s)\, u(s)\,\mathrm{d}s,\, v - u'(t)\right) + j(v) - j(u'(t)) \\ & \qquad\qquad \geqslant (L(t), v - u'(t)) \qquad \forall v \in V \\ & u(0) = u_0\,. \end{aligned} \tag{7.59}$$

We obtain *necessary conditions* exactly as in Section 6.6: taking $v = u'(t) + \rho$ in (7.59), $\rho \in \mathscr{R}$, it follows, since

$$\left(\int_0^t B(t-s)\, u(s)\,\mathrm{d}s, \rho\right) = \int_0^t b(t-s; u(s), \rho)\,\mathrm{d}s = 0\,,$$

that

$$j(u'(t) + \rho) - j(u'(t)) \geqslant (L(t), \rho)$$

from which

$$j(\rho) \geqslant (L(t), \rho)$$

or again

$$|(L(t), \rho)| \leqslant j(\rho) \qquad \forall \rho \in \mathscr{R}\,. \tag{7.60}$$

We will see that the necessary condition (7.60) is *sufficient* in order to have a solution in a suitable sense. We pass to the quotient with $\mathscr{R}$ (cf. (6.67)); we set

$$\begin{aligned} & a(\dot{u}, \dot{v}) = a(u, v)\,, \\ & b(t; \dot{u}, \dot{v}) = b(t; u, v)\,, \qquad u \in \dot{u}\,, \qquad v \in \dot{v} \end{aligned} \tag{7.61}$$

and we define (compare with (6.70))

$$\mathscr{H}(t; \dot{v}) = \inf [j(v + \rho) - (L(t), v + \rho)]\,; \tag{7.62}$$

Due to (7.60), we have (cf. (6.71)) $\mathscr{H}(t; \dot{v}) > -\infty$.

The problem which we consider is then: to find $u^{\cdot}(t)$ such that

(7.63) $$u^{\cdot}(t) \in V^{\cdot},$$

(7.64) $$a(u^{\cdot}(t), v^{\cdot} - u^{\cdot\prime}(t)) + \int_0^t b(t-s; u^{\cdot}(s), v^{\cdot} - u^{\cdot\prime}(t))\,ds + \mathscr{H}(t; v^{\cdot}) - \mathscr{H}(t; u^{\cdot\prime}(t)) \geq 0 \quad \forall v^{\cdot} \in V^{\cdot},$$

(7.65) $$u^{\cdot}(0) = u_0^{\cdot}.$$

Then we have

Theorem 7.3. *We assume that* (7.60) *holds, as well as* (7.20) *and* (7.22), *and also that*

(7.66) $$a(u_0^{\cdot}, v^{\cdot}) + \mathscr{H}(0; v^{\cdot}) = 0.$$

Then there exists one and only one function $u^{\cdot}$ *such that*

(7.67) $$u^{\cdot}, u^{\cdot\prime} \in L^{\infty}(0, T; V^{\cdot})$$

and satisfying (7.63), (7.64), (7.65).

The proof uses the same methods as in Theorem 7.2 and in Theorem 6.3. □

Remark 7.5. In the case without friction, (7.60) gives

(7.68) $$(L(t), \rho) = 0 \quad \forall \rho \in \mathscr{R}$$

and therefore $\mathscr{H}(t; v^{\cdot}) = (L(t), v^{\cdot})$ so that (7.64) reduces to

(7.69) $$a(u^{\cdot}(t), v^{\cdot}) + \int_0^t b(t-s; u^{\cdot}(s), v^{\cdot})\,ds = (L(t), v^{\cdot}) \quad \forall v^{\cdot} \in V^{\cdot}.$$

7.5. Use of the Laplace Transformation in the Cases without Friction

Let us introduce the operator $A \in \mathscr{L}(V; V_0')$ by

(7.70) $$a(u, v) = (Au, v), \quad u, v \in V_0.$$

Then the dynamic problem can be stated by (cf. Remark 7.2)

(7.71) $$u'' + Au + B * u = L$$

where[65]

(7.72) $$B * u(t) = \int_0^t B(t-s)\,u(s)\,ds;$$

[65] Vector valued convolution in t.

In (7.71) we differentiate with respect to t on $]0, \infty[$; for what follows, it is preferable to continue u as 0 for $t<0$—again denoting this continuation by u; then, with the differentiations this time taken on $]-\infty, +\infty[$, we have:

$$(7.73)\qquad \begin{aligned} &u'' + Au + B*u = L + u_0\delta' + u_1\delta, \quad \text{an equation in } \mathscr{D}'(V_0'), \\ &u \text{ zero for } t<0. \quad \Box \end{aligned}$$

The *quasi-static problem* can be expressed by

$$(7.74)\qquad Au + B*u = L. \quad \Box$$

We assume (for simplicity) that L is independent of t.

The solution u of Problem (7.73) (resp. (7.74)) can be obtained by *Laplace transformation in t*. (This represents a special case of Lions [3].)
We introduce

$$(7.75)\qquad \hat{u}(p) = \int_0^\infty e^{-pt} u(t)\,dt, \quad p = \xi + i\eta, \quad \xi > 0,$$

(Laplace transform of u with respect to t)
We assume that

$$(7.76)\qquad b_{ijkh} \in L^1(0, \infty; L^\infty(\Omega))$$

which implies

$$(7.77)\qquad B \in L^1(0, \infty; \mathscr{L}(V_0; V_0'));$$

then we introduce[66]

$$(7.78)\qquad \hat{B}(p) = \int_0^\infty e^{-pt} B(t)\,dt \quad (\in \mathscr{L}(V_0; V_0')), \quad \xi \geqslant \xi_*.$$

By Laplace transformation, (7.73) (resp. (7.74)) is equivalent to

$$(7.79)\qquad (p^2 + A + \hat{B}(p))\hat{u}(p) = \frac{1}{p} L + p u_0 + u_1$$

(resp.

$$(7.80)\qquad (A + \hat{B}(p))\hat{u}(p) = \frac{1}{p} L).$$

One proves (cf. Lions, loc. cit.) that Problem (7.79) has a unique solution when ξ is large enough, say $\xi \leqslant \xi_0$, therefore

[66] We assume that $e^{-\delta t} B \in L^1(0, \infty; \mathscr{L}(V_0; V_0'))$ for $\xi \geqslant \xi_*$. Incidentally, this assumption can be eliminated. Cf. Lions [3].

(7.81) $$\hat{u}(p)=(p^2+A+\hat{B}(p))^{-1}\left(\frac{1}{p}L+\rho u_0+u_1\right)$$

and that the function $p\to(p^2+A+\hat{B}(p))^{-1}\left(\frac{1}{p}L+pu_0+u_1\right)$ has an inverse Laplace transform which is *the* solution u of the problem. ▯

For the *quasi-static* case, we prove an analogous result: for $\xi\geqslant\xi_1$, we can invert $A+\hat{B}(p)$ and have

(7.82) $$\hat{u}(p)=(A+\hat{B}(p))^{-1}\frac{1}{p}L;$$

the function $p\to(A+\hat{B}(p))^{-1}\frac{1}{p}L$ has an inverse Laplace transform which is *the* solution u of the problem. ▯

Remark 7.6. The method of Laplace transformation, which we briefly sketched, is just as applicable in the situation of Section 6 (short memory) in the case without friction. ▯

7.6. Elastic Case as Limit of the Case with Memory

We consider (compare with Sec. 6.8) the law

(7.83) $$\sigma_{ij}(t)=a_{ijkh}\varepsilon_{kh}(u(t))+\lambda\int_0^t b_{ijkh}(t-s)\varepsilon_{kh}(u(s))\,ds,$$

$\lambda>0$ is intended to tend toward zero.

By u_λ we denote the solution of the problem corresponding to the one solved in Theorem 7.1, namely

(7.84) $$u_\lambda(t)\in V_0,$$

(7.85) $$(u_\lambda''(t),v-u_\lambda'(t))+a(u_\lambda(t),v-u_\lambda'(t))+\lambda\int_0^t b(t-s;u_\lambda(s),v-u_\lambda'(t))\,ds$$
$$+j(v+\Phi'(t))-j(u_\lambda'(t)+\Phi'(t))\geqslant(L(t),v-u_\lambda'(t))\qquad\forall v\in V_0,$$

(7.86) $$u_\lambda(0)=u_0,\quad u_\lambda'(0)=u_1.$$

We have

Theorem 7.3. *Under the conditions of Theorem* 7.1, *we have, when* $\lambda\to0$:

(7.87) $u_\lambda\to u$, $u_\lambda'\to u'$ *weakly star in* $L^\infty(0,T;V_0)$

(7.88) $u_\lambda''\to u''$ *weakly star in* $L^\infty(0,T;H)$

where u is the solution of the "elastic" case given by Theorem 5.7.

For the proof, we consider the solution $u_{\varepsilon\lambda}$ (compare with (7.33)) of the regularized problem

$$(u''_{\varepsilon\lambda}, v) + a(u_{\varepsilon\lambda}, v) + \lambda\left(\int_0^t B(t-s)u_{\varepsilon\lambda}(s)\,\mathrm{d}s, v\right) + (j'_\varepsilon(u'_{\varepsilon\lambda} + \Phi'), v) = (L(t), v) \quad \forall v \in V_0 \tag{7.89}$$

with $u_{\varepsilon\lambda}(0) = u_0$, $u'_{\varepsilon\lambda}(0) = u_1$.

As in the proof of Theorem 7.1, we then obtain the estimates, *independent of ε and λ*:

$$\|u_{\varepsilon\lambda}(t)\| + \|u'_{\varepsilon\lambda}(t)\| + |u'_{\varepsilon\lambda}(t)| \leqslant c$$

from which the Theorem follows. □

Remark 7.7. More generally, the solution $u = u_{(b)}$, given in Theorem 7.1, "depends continuously on b". □

8. Comments

In Sections 2–4, we reviewed the essentials from the classical theory of linear elasticity which we needed for our essential goal which constitutes the subject of Sections 5–7. In particular, a proof of Korn's inequality can be found in Section 3 which plays a fundamental role in the classical theory as well as in situations "with friction" or in "unilateral" situations. A proof of this inequality requiring fewer assumptions on the boundary of Ω is due to J. Gobert [1]. Our proof is based on a fairly simple general theorem (Theorem 3.2) which, moreover, can be generalized: if $v \in H^{-m}(\Omega)$, $D^p v \in H^{-m}(\Omega)$ $\forall |p| = k$, then $v \in H^{-m+k}(\Omega)$. Problems of elasticity with conditions of friction at the boundary were introduced by the authors in Duvaut-Lions [6]. The first unilateral problem in elasticity is the problem of Signorini. It has been investigated by G. Fichera [1], and subsequently by Lions-Stampacchia [1].

The analogue to Signorini's problem in linear visco-elasticity with long memory has been investigated in G. Duvaut [4]. For results on classical problems of linear visco-elasticity, one may refer to L. Brun [1], C.M. Dafermos [1], J.N. Distefano [1] and G. Duvaut [3] and the bibliographies of these works.

In Section 3.5, we only gave introductory notions of the duality theory. For the general theory, we refer to Moreau [1], Rockafellar [1], Teman [2]. Examples of duality problems are discussed in M.J. Sewell [1] and K. Wasmizu [1].

Chapter IV

Unilateral Phenomena in the Theory of Flat Plates

1. Introduction

In this chapter, we will examine *unilateral* phenomena and phenomena with *friction* for flat elastic plates.

The theory of plates is *two-dimensional;* thus, it is an *approximate theory*, since a plate is a three-dimensional body, of which one dimension (the thickness) is small compared to the other two.

We briefly recapitulate the outline of the theory (obviously, we have no intention to undertake an exhaustive study of the subject), so that we may introduce the unilateral problems without ambiguity.

In the approximation, we limit ourselves to terms of first order, hence, to the "*linear theory*" (though, as we already stressed on several occasions, the unilateral problems are *non linear;* but the differential *operators* considered are linear). *The non linear theory* (or rather, the non linear *theories*, because, in the literature, one finds several types of non linear schemes)—for example, for the equations of Von Karman—leads to unilateral conditions *of the same type* as in the linear case; in order not to complicate the discussion disproportionately, we did not include this theory here, referring instead to a separate article by the authors.

2. General Theory of Plates

2.1. Definitions and Notation

In its non-deformed state, the flat plate, considered as a two-dimensional medium, occupies a region Ω such that

$$\Omega = \text{open bounded set of } \mathbb{R}^2, \text{ with a regular boundary } \Gamma.$$

Let:

n (resp. τ) be the exterior unit normal to Γ (resp. the unit tangent obtained from n through a rotation by $+\pi/2$).

The plane $\mathbb{R}^2$ is referred to the orthonormal axes Ox_1, Ox_2. The Ox_3-axis completes the orthonormal triple $Ox_1x_2x_3$.

Actually, the plate is a three-dimensional body occupying a volume $\mathscr{V}$ of $\mathbb{R}^3$ defined by

$$\mathscr{V}=\{x| \quad x=\{x_1,x_2,x_3\}\in\mathbb{R}^3, \quad x_1,x_2\in\Omega \quad \text{and} \quad -\tfrac{1}{2}h(x_1,x_2)<x_3<\tfrac{1}{2}h(x_1,x_2)\} \tag{2.1}$$

where the function $\{x_1,x_2\}\to h(x_1,x_2)$ denotes the *thickness* of the plate, $h(x_1,x_2)$ being "*small*".

Let

$$u(x_1,x_2,x_3,t)=u_i(x_1,x_2,x_3,t) \qquad i=1,2,3$$

be the displacement vector of the plate at the instant t. In what follows, we will determine the equations (and the inequalities) characterizing the displacement *of the points of* Ω, i. e., characterizing $u(x_1,x_2,0,t)$. □

Notational conventions

The Roman indices $i,j,\ldots$ range from 1 to 3.

The Greek indices $\alpha,\beta,\ldots$ range from 1 to 2.

Thus, the displacement vector has components u_i, the points of Ω have coordinates x_α.

2.2. Analysis of Forces

More precisely, we actually have a volume density $\tilde{f}=\{\tilde{f}_1,\tilde{f}_2,\tilde{f}_3\}$ of forces on $\mathscr{V}$, whose exact distribution we do not take into account, but only the elements of reduction per unit surface of

$$f_i(x_1,x_2)=\int_{-h/2}^{+h/2} \tilde{f}_i(x_1,x_2,x_3)\,dx_3,$$
$$m_i(x_1,x_2)=\int_{-h/2}^{+h/2} \varepsilon_{i3k}x_3\,\tilde{f}_k(x_1,x_2,x_3)\,dx_3.$$

Therefore, there act on an area element dx_1, dx_2 of Ω an elementary force and an elementary moment, given respectively by

$$\begin{aligned} &f_i(x_1,x_2)\,dx_1\,dx_2,\\ &m_i(x_1,x_2)\,dx_1\,dx_2. \end{aligned} \tag{2.2}$$

We observe that $m_3=0$.

Now, let Ω_1 be a portion of Ω, bounded by a contour Γ_1, and let $\mathscr{V}_1$ be the volume corresponding to Ω_1, so that

$$\mathscr{V}_1=\{x| \quad x=\{x_1,x_2,x_3\}, \quad \{x_1,x_2\}\in\Omega_1, \quad -h/2<x_3<h/2\}. \tag{2.3}$$

The portion $\mathscr{V}-\mathscr{V}_1$ exerts a system of forces on $\mathscr{V}$, which, in equilibrium, together with the density f of forces on Γ_1, form a system statically equivalent to zero. Accordingly, we examine the forces exerted by $\mathscr{V}-\mathscr{V}_1$ on $\mathscr{V}_1$.

These forces act across the interface

$$S_1=\Gamma_1\times]-h/2,+h/2[.$$

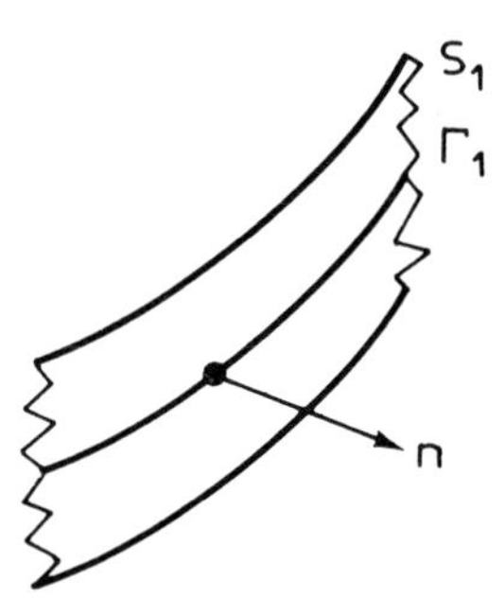

Fig. 16

Let σ_{ij} be the stress tensor in $\mathscr{V}$. Then, the portion $\mathscr{V}-\mathscr{V}_1$ exerts on $\mathscr{V}_1$ a force density $(\sigma_{ij}n_j)$ along S_1. Hence, per unit length of Γ_1, the reduction elements of these forces are given by

$$\begin{aligned} &R_i=\int_{-h/2}^{+h/2}\sigma_{ij}n_j\,\mathrm{d}x_3, && M_i=\int_{-h/2}^{+h/2}\varepsilon_{i3k}x_3\sigma_{kl}n_l\,\mathrm{d}x_3, \\ &(R_i)=\text{resultant}, && (M_i)=\text{resultant moment}. \end{aligned} \tag{2.4}$$

This allows us to introduce, in every point of Γ_1, and therefore in every point of Ω, a tensor (Σ_{ij}) of stress forces and a tensor (M_{ij}) of stress couples, defined by

$$\begin{aligned} &\Sigma_{ij}(x_1,x_2)=\int_{-h/2}^{+h/2}\sigma_{ij}(x_1,x_2,x_3)\,\mathrm{d}x_3, \qquad h=h(x_1,x_2) \\ &M_{ij}(x_1,x_2)=\int_{-h/2}^{+h/2}\varepsilon_{i3k}x_3\sigma_{kj}\,\mathrm{d}x_3. \end{aligned} \tag{2.5}$$

Then, the forces exerted by $\mathscr{V}-\mathscr{V}_1$ on $\mathscr{V}_1$ reduce to a density of forces (R_i) and of moments (M_i) on Γ_1 given by

$$\begin{aligned} &R_i=\Sigma_{ij}n_j=\Sigma_{i\alpha}n_\alpha \qquad (\text{since } n_3=0) \\ &M_i=M_{ij}n_j=M_{i\alpha}n_\alpha. \end{aligned} \tag{2.6}$$

The components of the stress forces and of the tensor of stress couples to be taken into consideration are therefore $\Sigma_{i\alpha}$, $M_{i\alpha}$.

One can verify that

$$\Sigma_{\alpha\beta}=\Sigma_{\beta\alpha}, \qquad M_{3\alpha}=0, \qquad M_{\beta\beta}=0. \tag{2.7}$$

Let us now write out that the portion Ω_1 of the plate Ω is in equilibrium under the action of the forces f (on Ω_1) and the forces exerted by $\mathscr{V}-\mathscr{V}_1$. Writing out that the resultant and the resultant moment at 0 vanish, we obtain

(2.8) $$\int_{\Omega_1} f_i \mathrm{d}x_1 \mathrm{d}x_2 + \int_{\Gamma_1} \Sigma_{i\alpha} n_\alpha \mathrm{d}\Gamma_1 = 0$$

and

(2.9) $$\int_{\Omega_1} \varepsilon_{ijk} x_j f_k \mathrm{d}x_1 \mathrm{d}x_2 + \int_{\Gamma_1} \varepsilon_{ijk} x_j \Sigma_{k\alpha} n_\alpha \mathrm{d}\Gamma_1 + \int_{\Omega_1} m_i \mathrm{d}x_1 \mathrm{d}x_2 + \int_{\Gamma_1} M_{i\alpha} n_\alpha \mathrm{d}\Gamma_1 = 0.$$

If we assume, which is permissible, that Γ_1 is regular, then (2.8) and (2.9) are, respectively, equivalent to

(2.10) $$\int_{\Omega_1} (f_i + \Sigma_{i\alpha,\alpha}) \mathrm{d}x_1 \mathrm{d}x_2 = 0,$$

(2.11) $$\int_{\Omega_1} [\varepsilon_{ijk} x_j f_k + (\varepsilon_{ijk} x_j \Sigma_{k\alpha})_{,\alpha}] \mathrm{d}x_1 \mathrm{d}x_2 + \int_{\Omega_1} (m_i + M_{i\alpha,\alpha}) \mathrm{d}x_1 \mathrm{d}x_2 = 0.$$

Since the region Ω_1 is arbitrary, (2.10) is equivalent to

(2.12) $$f_i + \Sigma_{i\alpha,\alpha} = 0.$$

Taking account of (2.12), we can reduce (2.11) to

$$\int_{\Omega_1} (\varepsilon_{i\alpha k} \Sigma_{k\alpha} + M_{i\alpha,\alpha} + m_i) \mathrm{d}x_1 \mathrm{d}x_2 = 0,$$

whence

(2.13) $$\varepsilon_{i\alpha k} \Sigma_{k\alpha} + M_{i\alpha,\alpha} + m_i = 0.$$

For $i=3$, the two sides of (2.13) are zero and, for $i=1,2$, we obtain

(2.14) $$\begin{aligned} &\Sigma_{32} + \mathrm{M}_{1\alpha,\alpha} + m_1 = 0, \\ &\Sigma_{31} - M_{2\alpha,\alpha} - m_2 = 0. \end{aligned}$$

Eliminating Σ_{31} and Σ_{32} between (2.12) and (2.14), we get

(2.15) $$\Sigma_{\alpha\beta,\beta} + f_\alpha = 0,$$

(2.16) $$M_{2\alpha,\alpha 1} - M_{1\alpha,\alpha 2} - m_{1,2} - m_{2,1} + f_3 = 0,$$

where only the $\Sigma_{\alpha\beta}$, $M_{\alpha\beta}$ enter. *This is the group of equations which we will retain in the following.* ▯

Remark 2.1. Up to this point, we did not make any simplification based on the fact that the thickness of the plate is "small". ▯

2.3. Linearized Theory

2.3.1. Hypotheses

In the following, we make the assumptions H_i):

H_1) The three-dimensional material which constitutes the plate, is isotropic and obeys the constituent law of Hooke, that is to say (cf. Sec. III.2.1):

$$\sigma_{ij} = \frac{E}{1+\nu}\left[\varepsilon_{ij} + \frac{\nu}{1-2\nu}\varepsilon_{kk}\delta_{ij}\right], \tag{2.17}$$

or equivalently

$$\varepsilon_{ij} = \frac{1}{E}\left[(1+\nu)\sigma_{ij} - \nu\sigma_{kk}\delta_{ij}\right] \tag{2.18}$$

where ν and E are respectively Poisson's ratio $(0<\nu<\frac{1}{2})$ and Young's Modulus $(E>0)$.

H_2) The forces $\{f_i\}$ applied to the plate are *normal* to it, that means[1]:

$$f_1 = f_2 = 0, \qquad m_1 = m_2 = 0. \tag{2.19}$$

H_3) The thickness h of the plate is "small".

H_4) The applied forces $\{0,0,f_3\}$ are sufficiently weak so that the induced displacements satisfy

$$u_1(x_1,x_2,0) = u_2(x_1,x_2,0) = 0 \tag{2.20}$$

$$\begin{aligned} u_1(x_1,x_2,x_3) &= x_3\frac{\partial u_1}{\partial x_3}(x_1,x_2,0) + O(h^2), \\ u_2(x_1,x_2,x_3) &= x_3\frac{\partial u_2}{\partial x_3}(x_1,x_2,0) + O(h^2), \end{aligned} \tag{2.21}$$

(where $O(h^2)$ signifies: $|O(h^2)| \leqslant Mh^2$, $h\to 0$).

H_5) The normal displacement u_3 permits a finite expansion of the form

$$\begin{aligned} u_3(x_1,x_2,x_3) = {} & u_3(x_1,x_2,0) + x_3\frac{\partial u_3}{\partial x_3}(x_1,x_2,0) \\ & + \frac{x_3^2}{2}\frac{\partial^2 u_3}{\partial x_3^2}(x_1,x_2,0) + O(h^3). \quad \square \end{aligned} \tag{2.22}$$

With these assumptions, we will determine the form of the strain and stress tensors in $\mathscr{V}$.

---

[1] At least, if $h = \text{const}$. In any case, we assume (2.19).

In the points of the upper and lower faces ($x_3 = \pm h/2$) of $\mathscr{V}$ no exterior force acts, and consequently

$$\sigma_{13}(x_1, x_2, \pm h/2) = \sigma_{23}(x_1, x_2, \pm h/2) = \sigma_{33}(x_1, x_2, \pm h/2) = 0. \tag{2.23}$$

Applying Hooke's law (2.17), we obtain for $x_3 = \pm h/2$

$$\frac{\partial u_1}{\partial x_3} + \frac{\partial u_3}{\partial x_1} = 0, \qquad \frac{\partial u_2}{\partial x_3} + \frac{\partial u_3}{\partial x_2} = 0, \qquad \frac{\partial u_3}{\partial x_3} + \frac{\nu}{1-2\nu}\varepsilon_{kk} = 0. \tag{2.24}$$

Taking account of (2.21), (2.22), it follows that

$$\begin{aligned} &\frac{\partial u_1}{\partial x_3}(x_1, x_2, 0) + \frac{\partial u_3}{\partial x_1}(x_1, x_2, 0) = O(h), \\ &\frac{\partial u_2}{\partial x_3}(x_1, x_2, 0) + \frac{\partial u_3}{\partial x_2}(x_1, x_2, 0) = O(h). \end{aligned} \tag{2.25}$$

Now, if we set

$$u_3(x_1, x_2, 0) = \zeta(x_1, x_2), \tag{2.26}$$ [2]

it follows, using (2.21), that

$$\begin{aligned} u_1(x_1, x_2, x_3) &= -x_3\, \partial\zeta/\partial x_1 + O(h^2), \\ u_2(x_1, x_2, x_3) &= -x_3\, \partial\zeta/\partial x_2 + O(h^2). \end{aligned} \tag{2.27}$$

Furthermore, the third relation (2.24) shows that

$$\frac{\partial u_3}{\partial x_3}\left(x_1, x_2, \pm\frac{h}{2}\right) = -\frac{\nu}{1-\nu}\left[\frac{\partial u_1}{\partial x_1}\left(x_1, x_2, \pm\frac{h}{2}\right) + \frac{\partial u_2}{\partial x_2}\left(x_1, x_2, \pm\frac{h}{2}\right)\right],$$

which implies, taking account of (2.27), that the left hand side is of order h and therefore that

$$\frac{\partial u_3}{\partial x_3}(x_1, x_2, 0) = 0. \tag{2.28}$$

Using (2.22), differentiated with respect to x_3, it follows that

$$\frac{\partial u_3}{\partial x_3}(x_1, x_2, x_3) = \frac{\nu x_3}{1-\nu}\left[\frac{\partial^2\zeta}{\partial x_1^2} + \frac{\partial^2\zeta}{\partial x_2^2}\right] + O(h^2), \tag{2.29}$$

---

[2] ζ is the vertical deflection.

from which we conclude:

$$\varepsilon_{kk}(x_1,x_2,x_3) = -x_3\frac{1-2\nu}{1-\nu}\left[\frac{\partial^2\zeta}{\partial x_1^2}+\frac{\partial^2\zeta}{\partial x_2^2}\right]+O(h^2). \tag{2.30}$$

From these relations, we can calculate the strain tensor at $x=(x_1,x_2,x_3)$,

$$\begin{aligned}
&\varepsilon_{11} = -x_3\,\partial^2\zeta/\partial x_1^2+O(h^2), \qquad \varepsilon_{22} = -x_3\,\partial^2\zeta/\partial x_2^2+O(h^2),\\
&\varepsilon_{33} = \frac{\nu x_3}{1-\nu}\Delta\zeta+O(h^2),\\
&\varepsilon_{12} = -x_3\,\partial^2\zeta/\partial x_1\,\partial x_2+O(h^2), \qquad \varepsilon_{23}=O(h^2), \qquad \varepsilon_{13}=O(h^2),
\end{aligned} \tag{2.31}$$

and then the stress tensor (σ_{ij})

$$\begin{aligned}
&\sigma_{11} = -x_3\frac{E}{1-\nu^2}\left(\frac{\partial^2\zeta}{\partial x_1^2}+\nu\frac{\partial^2\zeta}{\partial x_2^2}\right)+O(h^2),\\
&\sigma_{22} = -x_3\frac{E}{1-\nu^2}\left(\frac{\partial^2\zeta}{\partial x_2^2}+\nu\frac{\partial^2\zeta}{\partial x_1^2}\right)+O(h^2),\\
&\sigma_{12} = -\frac{E}{1+\nu}x_3\frac{\partial^2\zeta}{\partial x_1\,\partial x_2}+O(h^2),\\
&\sigma_{13}=O(h^2), \qquad \sigma_{23}=O(h^2), \qquad \sigma_{33}=O(h^2).
\end{aligned} \tag{2.32}$$

Two methods then present themselves for the formulation of equations.

2.3.2. Formulation of Equations. First Method

Starting out with relations (2.32), we calculate the $\Sigma_{i\alpha}$ and $M_{\alpha\beta}$ which were introduced in Section 2.2. We immediately obtain

$$\Sigma_{i\alpha}=O(h^3), \qquad (i=1,2,3,\ \alpha=1,2) \tag{2.33}$$

$$\left|\begin{aligned}
&M_{11} = -M_{22} = \frac{E}{1+\nu}\frac{h^3}{12}\frac{\partial^2\zeta}{\partial x_1\,\partial x_2}+O(h^4),\\
&M_{12} = \frac{E}{1-\nu^2}\frac{h^3}{12}\left(\frac{\partial^2\zeta}{\partial x_2^2}+\nu\frac{\partial^2\zeta}{\partial x_1^2}\right)+O(h^4),\\
&M_{21} = -\frac{E}{1-\nu^2}\frac{h^3}{12}\left(\frac{\partial^2\zeta}{\partial x_1^2}+\nu\frac{\partial^2\zeta}{\partial x_2^2}\right)+O(h^4).
\end{aligned}\right. \tag{2.34}$$

This shows that in Equations (2.14) it is not possible to neglect Σ_{13} and Σ_{23} compared to the moment terms, since all these terms are of the same order h^3.

Therefore, we have to eliminate Σ_{13} and Σ_{23} with the help of the third equation (2.15), which again leads to (2.16), i. e. here,

$$\frac{Eh^3}{12(1-\nu^2)}\Delta\Delta\zeta = f_3, \qquad \left(\Delta = \frac{\partial^2}{\partial x_1^2} + \frac{\partial^2}{\partial x_2^2}\right). \tag{2.35}$$

Frequently, one sets

$$D = Eh^3/12(1-\nu^2); \tag{2.36}$$

the coefficient D is known as the *stiffness coefficient of the plate (modulus of flexural rigidity)*. □

Green's formula (or *principle of virtual work*). The vertical deflection $\zeta(x_1, x_2)$, which is the unknown in this problem, satisfies the biharmonic equation (2.35) in Ω. In order to solve this last equation, it is obviously necessary to have conditions at the boundary Γ of Ω. These boundary conditions will involve either the displacement ζ itself or the given forces. We have in the x_1, x_2-plane two components of the line density of moments acting on Γ (the component in the direction Ox_3 of this moment density is identically zero). In the notation introduced in (2.6), we have

$$\begin{aligned} M_1 &= M_{1\alpha}n_\alpha = D\left\{(1-\nu)\frac{\partial^2\zeta}{\partial x_1\,\partial x_2}n_1 + \left(\frac{\partial^2\zeta}{\partial x_2^2} + \nu\frac{\partial^2\zeta}{\partial x_1^2}\right)n_2\right\}, \\ M_2 &= M_{2\alpha}n_\alpha = D\left\{-\left(\frac{\partial^2\zeta}{\partial x_1^2} + \nu\frac{\partial^2\zeta}{\partial x_2^2}\right)n_1 - (1-\nu)\frac{\partial^2\zeta}{\partial x_1\,\partial x_2}n_2\right\}. \end{aligned} \tag{2.37}$$

Naturally, we can form the normal and tangential components of this density of vector moments, say

$$M_m = M_1 n_1 + M_2 n_2, \qquad M_\tau = -M_1 n_2 + M_2 n_1. \tag{2.38}$$

Having introduced this notation, we can write down a *Green's formula*.

Let a function $v(x_1, x_2)$ be defined and sufficiently regular in Ω. Multiplying (2.16) with v and integrating over Ω, we obtain

$$\int_\Omega \varepsilon_{3\gamma\delta} M_{\gamma\alpha,\alpha\delta} v\,\mathrm{d}x_1\,\mathrm{d}x_2 = \int_\Omega f_3 v\,\mathrm{d}x_1\,\mathrm{d}x_2;$$

integrating by parts twice, it follows, setting $\mathrm{d}x = \mathrm{d}x_1\,\mathrm{d}x_2$ that

$$\int_\Omega \varepsilon_{3\gamma\delta} M_{\gamma\alpha} v_{,\alpha\delta}\,\mathrm{d}x - \int_\Gamma \varepsilon_{3\gamma\delta} M_{\gamma\alpha} v_{,\delta} n_\delta\,\mathrm{d}\Gamma + \int_\Gamma \varepsilon_{3\gamma\delta} M_{\gamma\alpha,\alpha} n_\delta v\,\mathrm{d}\Gamma = \int_\Omega f_3 v\,\mathrm{d}x. \tag{2.39}$$

We will successively transform each of the integrals on Γ. Using (2.13) and (2.6), we have

$$\int_\Gamma v\varepsilon_{3\gamma\delta} M_{\gamma\alpha,\alpha} n_\delta\,\mathrm{d}\Gamma = -\int_\Gamma \varepsilon_{3\gamma\delta}\varepsilon_{\gamma\alpha 3}\Sigma_{3\alpha} v n_\delta\,\mathrm{d}\Gamma = -\int_\Gamma v\Sigma_{3\alpha} n_\alpha\,\mathrm{d}\Gamma,$$

thus

$$\int_\Gamma v\varepsilon_{3\gamma\delta}M_{\gamma\alpha,\alpha}n_\delta \,\mathrm{d}\Gamma = -\int_\Gamma vR_3\,\mathrm{d}\Gamma. \tag{2.40}$$

Decomposing the vector $M_{\gamma\alpha}n_\alpha$ into normal and tangential components, we have:

$$\int_\Gamma \varepsilon_{3\gamma\delta}M_{\gamma\alpha}n_\alpha v_{,\delta}\,\mathrm{d}\Gamma = \int_\Gamma \varepsilon_{3\gamma\delta}(M_n n_\gamma + M_\tau \tau_\gamma)v_{,\delta}\,\mathrm{d}\Gamma. \tag{2.41}$$

Observing that

$$\tau_\delta = \varepsilon_{\delta 3\gamma}n_\gamma, \qquad n_\delta = -\varepsilon_{\delta 3\gamma}\tau_\gamma, \tag{2.42}$$

we obtain

$$\int_\Gamma \varepsilon_{3\gamma\delta}M_n n_\gamma v_{,\delta}\,\mathrm{d}\Gamma = \int_\Gamma M_n v_{,\delta}\tau_\delta\,\mathrm{d}\Gamma = -\int_\Gamma v(\partial M_n/\partial\tau)\,\mathrm{d}\Gamma, \tag{2.43}$$

since Γ is a closed curve. The other term is easily transformed into

$$\int_\Gamma \varepsilon_{3\gamma\delta}M_\tau \tau_\gamma v_{,\delta}\,\mathrm{d}\Gamma = -\int_\Gamma M_\tau v_{,\delta}n_\delta\,\mathrm{d}\Gamma = -\int_\Gamma M_\tau(\partial v/\partial n)\,\mathrm{d}\Gamma. \tag{2.44}$$

Regrouping the different terms, we obtain

$$\int_\Omega \varepsilon_{3\gamma\delta}M_{\gamma\alpha}v_{,\alpha\delta}\,\mathrm{d}x = \int_\Omega f_3 v\,\mathrm{d}x - \int_\Gamma M_\tau(\partial v/\partial n)\,\mathrm{d}\Gamma + \int_\Gamma (R_3 - \partial M_n/\partial\tau)v\,\mathrm{d}\Gamma. \tag{2.45}$$

We write the bilinear form on the left hand side explicitly

$$a(\zeta, v) = \int_\Omega \varepsilon_{3\gamma\delta}M_{\gamma\alpha}v_{,\alpha\delta}\,\mathrm{d}x = \int_\Omega (M_{1\alpha}v_{,\alpha 2} - M_{2\alpha}v_{,\alpha 1})\,\mathrm{d}x. \tag{2.46}$$

Using the expressions for $M_{\gamma\alpha}$ given in (2.34), it follows that

$$\begin{aligned} a(\zeta, v) = D\int_\Omega \Bigg[& \frac{\partial^2\zeta}{\partial x_1^2}\frac{\partial^2 v}{\partial x_1^2} + \frac{\partial^2\zeta}{\partial x_2^2}\frac{\partial^2 v}{\partial x_2^2} + \nu\left(\frac{\partial^2\zeta}{\partial x_1^2}\frac{\partial^2 v}{\partial x_2^2} + \frac{\partial^2\zeta}{\partial x_2^2}\frac{\partial^2 v}{\partial x_1^2}\right) \\ & + 2(1-\nu)\frac{\partial^2\zeta}{\partial x_1\partial x_2}\frac{\partial^2 v}{\partial x_1\partial x_2}\Bigg]\mathrm{d}x. \end{aligned} \tag{2.47}$$

We see that the bilinear form $a(\zeta, v)$ is *symmetric.*

Let us further set

$$(\varphi, \psi) = \int_\Omega \varphi\psi\,\mathrm{d}x, \qquad (\varphi, \psi)_\Gamma = \int_\Gamma \varphi\psi\,\mathrm{d}\Gamma. \tag{2.48}$$

Green's formula can then be written

$$a(\zeta, v) = (f_3, v) + (R_3 - \partial M_n/\partial\tau, v)_\Gamma - (M_\tau, \partial v/\partial n)_\Gamma \tag{2.49}$$

where the expressions for M_τ *and* M_n *are given in terms of* ζ *by* (2.37), (2.38) *and* (2.34).

This formula suggests that *certain* "well posed" problems, unilateral or not, are going to involve relations on Γ between M_τ, $R_3 - \partial M/\partial\tau$, ζ and $\partial\zeta/\partial n$. *In the following, we will put*

$$F_3 = R_3 - \partial M_n/\partial\tau\,.$$

2.3.3. Formulation of Equations. Second Method (due to Landau and Lifshitz [1])

We can approximately calculate the strain energy of the three-dimensional body $\mathscr{V}$ by using the expressions (2.31) and (2.32) for the tensors of strains and stresses.

Now, in Section III.3.2, we saw that the solution of a problem of linear elasticity can be obtained by minimizing the potential energy of the fields of kinematically admissible displacements. Since this potential energy is a convex functional, it is sufficient to make it stationary. We saw that we know the strain energy $E(\zeta)$ of the body $\mathscr{V}$ as a function of $\zeta(x_1, x_2)$, say

$$E(\zeta) = \tfrac{1}{2}\int_{-h/2}^{+h/2} \mathrm{d}x_3 \int_\Omega \sigma_{ij}\varepsilon_{ij}\,\mathrm{d}x_1\,\mathrm{d}x_2\,, \tag{2.50}$$

that is to say, after all calculations have been carried out,

$$E(\zeta) = \frac{h^3 E}{24(1-\nu^2)} \int_\Omega \left\{(\Delta\zeta)^2 + 2(1-\nu)\left[\left(\frac{\partial^2\zeta}{\partial x_1\,\partial x_2}\right)^2 - \frac{\partial^2\zeta}{\partial x_1}\frac{\partial^2\zeta}{\partial x_2}\right]\right\}\mathrm{d}x\,. \tag{2.51}$$

Since $E(\zeta)$ is the only non linear term of the potential energy of the kinematically admissible fields, we form its differential

$$(E'(\zeta), v) = \lim_{\lambda\to 0} \frac{E(\zeta+\lambda v) - E(\zeta)}{\lambda}\,, \qquad (v = v(x_1, x_2))\,. \tag{2.52}$$

We easily obtain

$$(E'(\zeta), v) = a(\zeta, v) \tag{2.53}$$

where $a(\zeta, v)$ is defined as in (2.47). Furthermore, we see that $E(\zeta) = \frac{1}{2}a(\zeta,\zeta)$. After integration by parts, $a(\zeta, v)$ can be written

$$\begin{aligned} a(\zeta, v) &= D\int_\Omega \Delta^2\zeta\, v\,\mathrm{d}x \\ &\quad - D\int_\Gamma \left\{\frac{\partial\Delta\zeta}{\partial n} + (1-\nu)\frac{\partial}{\partial\tau}\left[n_1 n_2\left(\frac{\partial^2\zeta}{\partial x_1^2} - \frac{\partial^2\zeta}{\partial x_2^2}\right) + (n_1^2 - n_2^2)\frac{\partial^2\zeta}{\partial x_1\,\partial x_2}\right]\right\} v\,\mathrm{d}\Gamma \\ &\quad + D\int_\Gamma \left\{\Delta\zeta + (1-\nu)\left(2n_1 n_2\frac{\partial^2\zeta}{\partial x_1\,\partial x_2} - n_2^2\frac{\partial^2\zeta}{\partial x_1^2} - n_1^2\frac{\partial^2\zeta}{\partial x_2^2}\right)\right\}\frac{\partial v}{\partial n}\,\mathrm{d}\Gamma\,. \end{aligned} \tag{2.54}$$

In order that the potential energy is stationary, it is necessary and sufficient that the work done by the exterior forces in the displacement v is given by $a(\zeta, v)$.

This implies that this work has the form

(2.55) $\int_\Omega f_3 v \,\mathrm{d}x + \int_\Gamma F_3 v \,\mathrm{d}\Gamma - \int_\Gamma \tilde{M}(\partial v/\partial n)\,\mathrm{d}\Gamma$

with the relations

(2.56) $$D\Delta^2\zeta = f_3 .$$

(2.57) $$-D\left\{\frac{\partial \Delta\zeta}{\partial n} + (1-\nu)\frac{\partial}{\partial \tau}\left[n_1 n_2\left(\frac{\partial^2\zeta}{\partial x_2^2} - \frac{\partial^2\zeta}{\partial x_1^2}\right) + (n_1^2 - n_2^2)\frac{\partial^2\zeta}{\partial x_1\,\partial x_2}\right]\right\} = \tilde{F}_3,$$

(2.58) $$D\left\{\Delta\zeta + (1-\nu)\left(2n_1 n_2\frac{\partial^2\zeta}{\partial x_1\,\partial x_2} - n_2^2\frac{\partial^2\zeta}{\partial x_1^2} - n_2^2\frac{\partial^2\zeta}{\partial x_2^2}\right)\right\} = -\tilde{M}.$$

This again gives us the relations obtained by the first method if we verify that

(2.59) $$\begin{aligned}&\tilde{F}_3 = F_3(=R_3 - \partial M_n/\partial \tau),\\ &\tilde{M} = M_\tau,\end{aligned}$$

which is easily done by using (2.14) and (2.34).

2.3.4. Summary

In the previously developed linearized theory, which is valid when the displacements of the points of the plate are small, the equation which governs the deformation is (2.56), with which we associate boundary conditions. If these latter conditions involve the displacement, they can be directly expressed in terms of ζ; if they involve tractions, they can be expressed by (2.57) if they concern forces, and by (2.58) if they concern a moment about the axis formed by the tangent to Γ.

The strain energy is given by $\frac{1}{2}a(\zeta,\zeta)$, calculated from the bilinear form $a(\zeta,v)$, defined in (2.46) and written out explicitly in (2.47). Moreover, we have Green's formula or the principle of virtual work, (2.49). □

Dynamic case. The displacement then is a function of x and t, say $\zeta=\zeta(x,t)$. In (2.56), we have to add to the forces f_3 the inertial force $-\rho h\partial^2\zeta/\partial t^2$ and obtain

(2.60) $\rho h\partial^2\zeta/\partial t^2 + D\Delta^2\zeta = f_3$, ($\rho$ = surface density in the undeformed state).

Obviously, we have to prescribe initial conditions

(2.61) $$\begin{aligned}&\zeta(x,0) = \zeta_0(x)\\ &\partial\zeta(x,0)/\partial t = \zeta_1(x),\end{aligned}$$

to which we add boundary conditions (involving equations or *inequalities*) analogous to those in the stationary case. □

3. Problems to be Considered

The problems which we intend to consider can be either static or dynamic. With every static problem, we can associate a dynamic problem by adding to the acting surface forces the forces of inertia and by prescribing initial conditions. Here, however, we will explicitly formulate only *static problems.* They can be divided into two groups: classical problems whose variational formulation leads to equations, and "unilateral" problems whose formulation leads to inequalities.

3.1. Classical Problems

In every point of Γ we prescribe either F_3 or ζ and either M_τ or $\partial_\zeta/\partial n$ which physically corresponds to

*) a density of forces F_3 on Γ or the displacement ζ of the points of Γ,

**) the couple M_τ whose moment is about the tangent τ at Γ or $\partial_\zeta/\partial n$, that is to say the slope of the plate counted in the direction of the normal at the points of Γ.

Examples. 1) *Clamped Plate.* We prescribe

$$\zeta = \partial_\zeta/\partial n = 0 \quad \text{on} \quad \Gamma. \tag{3.1}$$

2) Plate with free boundary on a portion Γ_1 of Γ and with fixed boundary on Γ_2, free to rotate around the tangent to Γ_2. One has then

$$F_3 = M_\tau = 0 \quad \text{on} \quad \Gamma_1 \tag{3.2}$$

$$\zeta = M_\tau = 0 \quad \text{on} \quad \Gamma_2. \tag{3.3}$$

3.2. Unilateral Problems

1) *Unilateral displacement of the points of Ω*

For example

$$\begin{aligned} &\zeta(x) \geqslant 0 \quad \text{on} \quad \Omega \\ &\left.\begin{aligned} \zeta(x) > 0 \Rightarrow f_3 = 0 \\ \zeta(x) = 0 \Rightarrow f_3 \geqslant 0 \end{aligned}\right\} \quad \text{on} \quad \Omega, \end{aligned} \tag{3.4}$$

while the conditions on the boundary Γ are of classical type.

2) *Unilateral displacement of the points of Γ*

$$\left.\begin{aligned} &\zeta(x) \geqslant 0 \\ &\zeta(x) > 0 \Rightarrow F_3 = 0 \\ &\zeta(x) = 0 \Rightarrow F_3 \geqslant 0 \end{aligned}\right\} \quad \text{on} \quad \Gamma, \tag{3.5}$$

the conditions for M_τ or $\partial_\zeta/\partial n$ on Γ are of classical type and f_3 is given.

3) *Unilateral rotation of the points of Γ*

$$\begin{aligned} &\partial\zeta/\partial n \geqslant 0 \quad \text{on } \Gamma \text{ with} \\ &\partial\zeta/\partial n > 0 \Rightarrow M_\tau = 0, \\ &\partial\zeta/\partial n = 0 \Rightarrow M_\tau \leqslant 0, \end{aligned} \tag{3.6}$$

while the conditions for F_3 or ζ on Γ are of classical type and f_3 is given.

4) *Displacement with friction of the points of Ω*

$$\begin{aligned} &|f_3| < \mathscr{F} \quad \text{(given constant or positive function)} \quad \Rightarrow \zeta = 0, \\ &|f_3| = \mathscr{F} \Rightarrow \text{there exists a } \lambda \geqslant 0 \text{ such that} \quad \zeta = -\lambda f_3, \end{aligned} \tag{3.7}$$

while the conditions for Γ are of classical type.

5) *Displacement with friction of the points of Γ*

$$\begin{aligned} &|F_3| < \mathscr{F} \Rightarrow \zeta = 0 \quad \text{on} \quad \Gamma \\ &|F_3| = \mathscr{F} \Rightarrow \quad \text{there exists a } \lambda \geqslant 0 \text{ such that} \quad \zeta = -\lambda F_3, \end{aligned} \tag{3.8}$$

while the conditions for the rotation of points of Γ are of classical type.

6) *Rotation with friction of the points of Γ*

$$\begin{aligned} &|M_\tau| < \mathscr{M} \quad \text{(given positive constant)} \quad \Rightarrow \partial\zeta/\partial n = 0 \quad \text{on } \Gamma, \\ &|M_\tau| = \mathscr{M} \Rightarrow \quad \text{there exists a } \lambda \geqslant 0 \text{ with} \quad \partial\zeta/\partial n = \lambda M_\tau \text{ on } \Gamma, \end{aligned} \tag{3.9}$$

while the conditions for F_3 or ζ on Γ are of classical type.

7) There can also be conditions of one of the preceding types on one part of Γ and conditions of other types on other parts of Γ (where the parts Γ_i of Γ considered satisfy $\bigcup \Gamma_i = \Gamma$). ▯

Remark 3.1. Problems 1) and 4) will not be examined explicitly in what follows; they involve identical methods and do not give rise to any particular difficulty; we prefer to treat problems in the following which present specific difficulties. ▯

4. Stationary Unilateral Problems

4.1. Notation

By putting the problem into a non-dimensional form, we can effect a change of scale on the x_i, reducing the problem to the case $D = 1$.

Then, for $u, v \in H^2(\Omega)$, we set

(4.1)
$$a(u,v) = \int_\Omega \frac{\partial^2 u}{\partial x_\alpha^2}\frac{\partial^2 v}{\partial x_\alpha^2}\,dx + \nu \int_\Omega \left(\frac{\partial^2 u}{\partial x_1^2}\frac{\partial^2 v}{\partial x_2^2} + \frac{\partial^2 u}{\partial x_2^2}\frac{\partial^2 v}{\partial x_1^2}\right) dx$$
$$+2(1-\nu)\int_\Omega \frac{\partial^2 u}{\partial x_1 \partial x_2}\frac{\partial^2 v}{\partial x_1 \partial x_2}\,dx .$$

Green's formula is

(4.2)
$$a(u,v) = (\Delta^2 u, v) + (F_3, v) - (M_\tau, \partial v/\partial n)_\Gamma .$$

We set

(4.3)
$$j_0(v) = \int_\Gamma (g_2 v^+ - g_1 v^-)\,d\Gamma, \quad g_1 < 0 < g_2, \text{ [3]}$$

(4.4)
$$j_1(v) = \int_\Gamma [k_2(\partial v/\partial n)^+ - k_1(\partial v/\partial n)^-]\,d\Gamma, \quad k_1 < 0 < k_2, \text{ [4]}$$

4.2. Problems (stationary)

Problem 4.1. We want to find $u \in H^2(\Omega)$ with

(4.5)
$$a(u, v-u) + j_0(v) - j_0(u) \geq (f, v-u) \qquad \forall v \in H^2(\Omega),$$

(we set $f_3 = f$).

Problem 4.1*a*. We want to find $u \in H^2(\Omega)$ with $\partial u/\partial n = 0$ on Γ, and

(4.5*a*)
$$a(u, v-u) + j_0(v) - j_0(u) \geq (f, v-u) \qquad \forall v \in H^2(\Omega) \text{ with } \partial v/\partial n = 0 \text{ on } \Gamma .$$

Problem 4.2. We want to find $u \in H^2(\Omega)$ with

(4.6)
$$a(u, v-u) + j_1(v) - j_1(u) \geq (f, v-u) \qquad \forall v \in H^2(\Omega) \cap H_0^1(\Omega) .$$

Problem 4.2*a*. We want to find $u \in H^2(\Omega) \cap H_0^1(\Omega)$, with

(4.6*a*)
$$a(u, v-u) + j_1(v) - j_1(u) \geq (f, v-u), \qquad \forall v \in H^2(\Omega)\ \ H_0^1(\Omega) .$$
$$v_n + q_n \to w \quad \text{in} \quad H^2(\Omega) .$$

We will now *interpret* these problems and verify that they include (possibly after passage to the limit) the problems of Section 3 [5]. □

---

[3] For the g_i we take constants. We could also take functions which are measurable and bounded on Γ.
[4] Remark analogous to (3).
[5] In analogous fashion, we treat the problem with unilateral displacement or friction in the points of Ω.

If we take $v=u\pm\varphi$, $\varphi\in\mathscr{D}(\Omega)$, in all the above inequalities, we can conclude that, in all cases,

$$\Delta^2 u = f. \tag{4.7}$$

Therefore, with Green's formula (4.2), we conclude that

$$a(u,v-u)-(F_3,v-u)_\Gamma+(M_\tau,\partial(v-u)/\partial n)_\Gamma=(f,v-u) \tag{4.8}$$

whence: *for Problems* 4.1 *and* 4.1*a*:

$$\begin{aligned}&(F_3,v-u)_\Gamma+j_0(v)-j_0(u)-(M_\tau,\partial(v-u)/\partial n)_\Gamma\geqslant 0\\ &\forall v\in H^2(\Omega)\quad (\text{resp. } \forall v \text{ with } \partial v/\partial n=0 \text{ on } \Gamma \text{ in Problem 4.1}a);\end{aligned} \tag{4.9}$$

and for Problems 4.2 *and* 4.2*a*

$$\begin{aligned}&-(M_\tau,\partial(v-u)/\partial n)_\Gamma+j_1(v)-j_1(u)+(F_3,v-u)_\Gamma\geqslant 0\\ &\forall v\in H^2(\Omega)\quad (\text{resp. } \forall v \text{ with } v=0 \text{ on } \Gamma \text{ in Problem 4.2}a).\end{aligned} \tag{4.10}$$

Boundary conditions in Problem 4.1. Now, (4.9) implies (and is equivalent to)

$$\begin{aligned}&M_\tau=0 \quad\text{and}\\ &\int_\Gamma [F_3(v^+-v^-)+g_2v^+-g_1v^-]\,d\Gamma-(F_3,u)_\Gamma-j_0(u)\geqslant 0.\end{aligned} \tag{4.11}$$

Taking $\varphi\geqslant 0$ on Γ and $v=\pm\lambda\varphi$, $\lambda>0$, we conclude that

$$F_3+g_2\geqslant 0, \qquad F_3+g_1\leqslant 0, \qquad (F_3,u)_\Gamma+j_0(u)=0$$

whence

$$\begin{aligned}&-g_2\leqslant F_3\leqslant g_1\\ &F_3u+g_2u^+-g_1u^-=0 \quad\text{i.e.}\quad (F_3+g_2)u^++(-F_3-g_1)u^-=0,\end{aligned} \tag{4.12}$$

or again

$$\begin{aligned}-g_2<F_3<-g_1 &\Rightarrow u=0\\ F_3=-g_1 &\Rightarrow u\leqslant 0\\ F_3=-g_2 &\Rightarrow u\geqslant 0. \quad \square\end{aligned} \tag{4.13}$$

Special case:

$g_2=-g_1=\mathscr{F}$; *this is Problem* 5 *of* Section 3. □

Limit case. We take $g_2=0$ and let g_1 tend toward $-\infty$. In the limit (which exists), u satisfies

$$\begin{aligned} &0<F_3 \Rightarrow u=0 \\ &F_3=0 \Rightarrow u\geqslant 0 . \end{aligned} \tag{4.14}$$

This is Problem 2 *of* Section 3.2. The *direct* variational formulation of this problem is the following:

Problem 4.3. We introduce

$$K_1=\{v| \quad v\in H^2(\Omega), \quad v\geqslant 0 \text{ on } \Gamma\}, \tag{4.15}$$

which is a convex closed set of $H^2(\Omega)$; we want to find a solution u of

$$\begin{aligned} &u\in K_1, \\ &a(u,v-u)\geqslant (f,v-u) \quad \forall v\in K_1 . \quad \Box \end{aligned} \tag{4.16}$$

Boundary conditions in Problem 4.1*a*. The unilateral boundary conditions (4.12) (or (4.13)) are unchanged, condition (4.11) being replaced by

$$\partial u/\partial n=0 \quad \text{on} \quad \Gamma . \quad \Box \tag{4.17}$$

Boundary conditions in Problem 4.2. Condition (4.10) implies (and is equivalent to)

$$F_3=0 \tag{4.18}$$

and

$$\int_\Gamma [-M_\tau((\partial v/\partial n)^+ -(\partial v/\partial n)^-)+k_2(\partial v/\partial n)^+ -k_1(\partial v/\partial n)^-]\,d\Gamma$$
$$+(M_\tau, \partial u/\partial n)_\Gamma - j_1(u)\geqslant 0,$$

whence

$$-M_\tau+k_2\geqslant 0, \qquad -M_\tau+k_1\leqslant 0, \qquad (M_\tau, \partial u/\partial n)-j_1(u)=0$$

whence

$$\begin{aligned} &k_1\leqslant M_\tau\leqslant k_2, \\ &(k_2-M_\tau)(\partial u/\partial n)^+ +(M_\tau-k_1)(\partial u/\partial n)^- =0, \end{aligned} \tag{4.19}$$

or again

$$\begin{aligned} &k_1<M_\tau<k_2 \Rightarrow \partial u/\partial n=0, \\ &M_\tau=k_2 \Rightarrow \partial u/\partial n\geqslant 0, \\ &M_\tau=k_1 \Rightarrow \partial u/\partial n\leqslant 0. \end{aligned} \tag{4.20}$$

Special case

$$k_2 = -k_1 = \mathcal{M}.$$

This is Problem 6 of Section 3.2. □

Limit case. We take $k_2=0$ and let $k_1 \to -\infty$. We find

$$M_\tau > 0 \Rightarrow \partial u/\partial n = 0,$$
$$M_\tau = 0 \Rightarrow \partial u/\partial n \geqslant 0,$$

this is Problem 3 of Section 3.2.

The direct variational formulation of this problem is the following:

Problem 4.4. We introduce

$$K_2 = \{v \mid \quad v \in H^2(\Omega), \quad \partial v/\partial n \geqslant 0 \text{ on } \Gamma\}, \tag{4.21}$$

a convex closed set of $H^2(\Omega)$; we seek a solution u of

$$\begin{aligned} & u \in K_2, \\ & a(u, v-u) \geqslant (f, v-u) \quad \forall v \in K_2. \quad □ \end{aligned} \tag{4.22}$$

Boundary conditions in Problem 4.2*a.* The unilateral boundary conditions (4.19) are unchanged, condition (4.18) is replaced by (4.23):

$$u = 0 \quad \text{on} \quad \Gamma. \quad □ \tag{4.23}$$

Remark 4.1. If $j_0(v)=0$, $g_1=g_2=0$, Problem 4.1 is equivalent to

$$a(u,v) = (f,v) \quad \forall v \in H^2(\Omega),$$

i. e., (4.7) with boundary conditions of classical type:

$$\partial u/\partial n = 0, \qquad F_3 = 0 \quad \text{on} \quad \Gamma. \tag{4.24}$$

Problem 4.1*a* is equivalent to the boundary conditions

$$\partial u/\partial n = 0, \qquad F_3 = 0. \tag{4.25}$$

If $j_1(v)=0$, i. e., $k_1=k_2=0$, Problem 4.2 is equivalent to (4.7) with the boundary conditions (4.24), and Problem 4.2*a* is equivalent to (4.7) with the conditions

$$u = 0, \qquad M_\tau = 0. \quad □ \tag{4.26}$$

The problem with the conditions (clamped plate)

$$u = 0, \qquad \partial u/\partial n = 0 \quad \text{on} \quad \Gamma \tag{4.27}$$

is obvious; it is equivalent to the minimization on $H_0^2(\Omega)$ of

$$\tfrac{1}{2}a(v,v)-(f,v),$$

a problem which has a unique solution, since

$$a(v,v)\geqslant\alpha\,\|v\|_{H^2}^2, \qquad \alpha>0, \quad \forall v\in H_0^2(\Omega). \quad \square$$

4.3. Solution of Problem 4.1. Necessary Conditions for the Existence of a Solution

To simplify writing, we set

$$\psi(\lambda)=g_2\lambda^+-g_1\lambda^-, \tag{4.28}$$

so that

$$j_0(v)=\int_\Gamma \psi(v)\,\mathrm{d}\Gamma.$$

The problem is equivalent to the minimization on $H^2(\Omega)$ of the functional

$$a(v,v)+j_0(v)-(f,v)=H(v). \tag{4.29}$$

If we designate by $\mathscr{P}$ the space of polynomials of degree $\leqslant 1$, we have:

$$\mathscr{P}\subset H^2(\Omega) \text{ (since } \Omega \text{ is bounded) and } a(u,p)=0 \quad \forall p\in\mathscr{P}.$$

In order that a solution exists, we must have:

(4.30) $\forall p\in\mathscr{P}$, $H(\lambda p)$ is *bounded from below* when $\lambda\to\pm\infty$.

Therefore

$$H(\lambda p)=j_0(\lambda p)-\lambda(f,p)=\lambda\left[\int_\Gamma p\frac{\psi(\lambda p)}{\lambda p}\mathrm{d}\Gamma-(f,p)\right].$$

But when $\lambda\to\infty$

$$\int_\Gamma p\frac{\psi(\lambda p)}{\lambda p}\mathrm{d}\Gamma\to\int_\Gamma(g_2p^+-g_1p^-)\mathrm{d}\Gamma$$

and when $\lambda\to-\infty$

$$\int_\Gamma p\frac{\psi(\lambda p)}{\lambda p}\mathrm{d}\Gamma\to\int_\Gamma(g_1p^+-g_2p^-)\mathrm{d}\Gamma$$

so that condition (4.30) implies $(\lambda\to+\infty)$

$$\int_\Gamma(g_2p^+-g_1p^-)\mathrm{d}\Gamma-(f,p)\geqslant 0$$

i. e.,

$$(f,p)\leqslant j_0(p) \quad \forall p\in\mathscr{P}, \tag{4.31}$$

and ($\lambda\to-\infty$)

$$\int_\Gamma (g_1 p^+ - g_2 p^-)\,\mathrm{d}\Gamma - (f,p)\leqslant 0$$

i. e.

$$-j_0(-p)-(f,p)\leqslant 0 \quad \forall p\in\mathscr{P}.$$

Changing p to $-p$, this is equivalent to (4.31). Thus we proved

Theorem 4.1. *In order that Problem* 4.1 *has a solution, it is necessary that f satisfies* (4.31).

Special case. Let us assume $j_0=0$ (which corresponds—cf. Remark 4.1—to a problem of classical type). Then, (4.31) is equivalent to

$$(f,p)=0 \quad \forall p\in\mathscr{P},$$

i. e., to

$$(f,1)=0, \qquad (f,x_\alpha)=0, \qquad \alpha=1,2. \quad \square \tag{4.32}$$

Limit case. For $g_2=0$, $g_1=-\infty$, (4.31) yields

$$(f,p)\leqslant 0 \quad \forall p\in\mathscr{P}, \qquad p\geqslant 0 \quad \text{on} \quad \Gamma. \tag{4.33}$$

Moreover, the necessary condition (4.33) can be established directly from relation (4.16); we note that (4.16) is equivalent to

$$\begin{aligned} &u\in K_1,\\ &a(u,v)\geqslant(f,v) \quad \forall v\in K_1,\\ &a(u,u)=(f,u). \end{aligned} \tag{4.34}$$

Taking $v=p\in K_1$, we find the necessary condition (4.33).

4.4. Solution of Problem 4.1. Sufficient Conditions

We shall prove

Theorem 4.2. *We assume f to be given in* $L^2(\Omega)$ (which can be extended). *We assume $g_1<0<g_2$, g_1, g_2 finite, and we assume that*[6]

$$(f,p)<j_0(p) \quad \forall p\in\mathscr{P}, \qquad p\neq 0. \tag{4.35}$$

Then Problem 4.1 *has at least one solution.*

[6] "Strong" analogue of (4.31).

Remark 4.2. In the case "$j_0=0$", the conditions (4.32) are sufficient; this is the classical Fredholm alternative. Then there are *infinitely many* solutions: if u is a solution, then all solutions are given by

$$u+p, \qquad p\in\mathscr{P}.$$

Remark 4.3. For $g_2=0$, $g_1=-\infty$, we have an analogous result:

Theorem 4.2a. *We assume f to be given in $L^2(\Omega)$* (which can be extended). *We assume that*

$$(f,p)<0 \quad \forall p\in\mathscr{P}, \qquad p\geqslant 0 \quad on \quad \Gamma, \qquad p\neq 0. \tag{4.36}$$

Then Problem 4.3 *has at least one solution.* □

Proof of Theorem 4.2. From (4.35) and the fact that $\mathscr{P}$ is of finite dimension, it follows that there exists an $\varepsilon>0$ such that

$$(f,p)\leqslant j_0(p)-\varepsilon|p| \tag{4.37}$$

where $|\varphi|=$norm of φ in $L^2(\Omega)$ (obviously, we could take some other norm in (4.37) instead of $|\varphi|$, possibly changing ε). We decompose $v\in H^2(\Omega)$ as

$$\begin{aligned} &v=\tilde{v}+p, && p\in\mathscr{P},\\ &(\tilde{v},q)=0 && \forall q\in\mathscr{P}. \end{aligned} \tag{4.38}$$

Then, writing $a(v)$ instead of $a(v,v)$, the functional (4.29) becomes:

$$H(v)=H(\tilde{v}+p)=\tfrac{1}{2}a(v)+j_0(\tilde{v}+p)-(f,\tilde{v}+p). \tag{4.39}$$

But

$$\begin{aligned} &j_0(\tilde{v}+p)\geqslant j_0(p)-c_1\|\tilde{v}\|, \quad \text{where} \quad \|\ \|=\text{norm in } H^2(\Omega),\\ &|(f,\tilde{v})|\leqslant c_2\|\tilde{v}\|. \end{aligned}$$

Assuming for a moment the

Lemma 4.1. *There exists an $\alpha>0$ such that*

$$a(\tilde{v})\geqslant\alpha\|\tilde{v}\|^2, \qquad \tilde{v}\in H^2(\Omega), \qquad (\tilde{v},q)=0 \quad \forall q\in\mathscr{P}, \tag{4.40}$$

we then have

$$H(v)\geqslant\alpha\|\tilde{v}\|^2-c_3\|\tilde{v}\|+j_0(p)-(f,p)$$

and therefore, according to (4.37),

$$H(v)\geqslant\alpha\|\tilde{v}\|^2-c_3\|\tilde{v}\|+\varepsilon|p|. \tag{4.41}$$

But then $H(v) \to +\infty$ as $\|v\| \to +\infty$, according to

Lemma 4.2. *On $H^2(\Omega)$, the norms $\|v\|$ and $(a(\tilde{v}))^{1/2} + |p|$ are equivalent.*

Thus the theorem is proved, subject to the verification of the lemmas. Naturally, Lemma 4.1 follows from Lemma 4.2. It remains therefore to prove Lemma 4.2. The principle of the proof is the same as in Lemma 7.1, Chapter I. Let us set

$$\|\|v\|\| = (a(\tilde{v}))^{1/2} + |p|. \tag{4.42}$$

The function $v \to \|\|v\|\|$ is a norm. Indeed, the only non trivial point is to see that $\|\|v\|\| = 0$ implies $v = 0$. Now, we have $a(\tilde{v}) = 0$, $|p| = 0$, therefore $p = 0$ and $\tilde{v} \in \mathscr{P}$ which leads to $\tilde{v} = 0$ (since $(\tilde{v}, q) = 0$ $\forall q \in \mathscr{P}$). According to the closed graph theorem (as in Chap. I, Lemma 7.1), it is then sufficient to show that $H^2(\Omega)$ is *complete* for the norm (4.42). Let then $v_n = \tilde{v}_n + p_n$ be a Cauchy sequence for this norm. Then the $\partial^2 v_n / \partial x_\alpha \partial x_\beta$ converge in $L^2(\Omega)$ and the p_n converge in $\mathscr{P}$. According to Deny-Lions [1], we then can find a sequence $q_n \in \mathscr{P}$ such that

$$\tilde{v}_n + q_n \to w \quad \text{in} \quad H^2(\Omega).$$

But then

$$(v_n + q_n, q) \to (w, q) \qquad \forall q \in \mathscr{P}$$

i. e.,

$$(q_n, q) \to w, q) \qquad \forall q \in \mathscr{P}.$$

Thus, $q_n \to q_0$ in $\mathscr{P}$, and therefore $v_n = (v_n + q_n) - q_n \to w - q_0$ in $H^2(\Omega)$, therefore $v_n = \tilde{v}_n + p_n$ converge in $H^2(\Omega)$.

Proof of Theorem 4.2*a*. Theorem 4.2*a* is a consequence of Theorem 5.1 of Lions-Stampacchia [1] (in the notation of this theorem, take $V = H^2(\Omega)$, $p_0(v) = |v|$, $p_1(v) = a(v))^{1/2}$; then $Y = \mathscr{P}$, and the assumption (5.3) of Theorem 5.1 of Lions-Stampacchia is equivalent to Lemma 4.1).

4.5. The Question of Uniqueness in Problems 4.1 and 4.3

The question of uniqueness has not been solved in Problem 4.1; it can be seen immediately that if u_1 and u_2 are two possible solutions, then $u_1 - u_2 \in \mathscr{P}$; but we have not been able to settle the question whether one can or cannot have several solutions u and $u + p$, $p \in \mathscr{P}$.

Here is a *partial* result relating to the case of Problem 4.3.

Theorem 4.3. *We make the same assumptions as in Theorem* 4.2*a and, in addition, assume that Γ contains no rectilinear portion. Then Theorem* 4.3 *has a unique solution.*

Proof. Let u and $u+p$ be two possible solutions. The unilateral conditions on Γ are

$$\begin{array}{lll} u\geqslant 0, & F_3\geqslant 0, & uF_3=0, \\ u+p\geqslant 0, & F_3\geqslant 0, & (u+p)F_3=0. \end{array}$$

But it is impossible to have $F_3=0$ almost everywhere. Indeed, Green's formula then shows that

$$\int_\Omega \Delta^2 u\,dx=(f,1)=-\int_\Gamma F_3\,d\Gamma=0,$$

which is impossible according to (4.36) for $p=1$. Therefore, $F_3>0$ on a set E of Γ of measure >0, and, on E, we have $u=0$ and $u+p=0$, therefore $p=0$ on E, which is impossible according to the assumption made on Γ, except when $p\equiv 0$.

4.6. Solution of Problem 4.1*a*

We have the following results:

Theorem 4.4. *In order that Problem* 4.1*a has a solution, it is necessary that*

(4.43) $$g_1 \text{ measure } \Gamma\leqslant(f,1)\leqslant g_2 \text{ measure } \Gamma.$$

Proof. We have to minimize $H(v)$ on the space V of $v\in H^2(\Omega)$ with $\partial v/\partial n=0$ on Γ. Then, in particular, $H(\mu)$, $\mu\in\mathbb{R}$, is bounded from below, if the problem has a solution from which (in a manner analogous to Theorem 4.1)

$$(f,\mu)\leqslant j_0(\mu) \qquad \forall\mu\in\mathbb{R},$$

which, if we take $\mu=1$ and $\mu=-1$, is equivalent to (4.43). ▯

We next prove, in a manner analogous to Theorem 4.2,

Theorem 4.5. *We assume that*

(4.44) $$g_1 \text{ measure } \Gamma<(f,1)<g_2 \text{ measure } \Gamma.$$

Then Problem 4.1*a has at least one solution.*

This time, we can settle the problem of uniqueness.

Theorem 4.6. *Under the assumptions of Theorem* 4.5, *Problem* 4.1*a has a unique solution.*

Proof. Let u_1 and u_2 be two possible solutions. Then

$$a(u_1-u_2, u_1-u_2)=0,$$

therefore $u_1 - u_2 = p$, $p \in \mathscr{P}$. But we must have $\partial u_1/\partial n = \partial u_2/\partial n = 0$, therefore $\partial p/\partial n = 0$ on Γ which is equivalent to $p = \text{constant} = c \in \mathbb{R}$. Thus, two possible solutions are of the form u and $u+c$, the unilateral boundary conditions being:

$$\begin{aligned} -g_2 < F_3 < -g_1 &\Rightarrow u = 0 \quad \text{and} \quad u+c=0 \\ F_3 = -g_1 &\Rightarrow u \leqslant 0 \quad \text{and} \quad u+c \leqslant 0 \\ F_3 = -g_2 &\Rightarrow u \geqslant 0 \quad \text{and} \quad u+c \geqslant 0. \end{aligned}$$

It is impossible to have $F_3 = -g_1$ (or $F_3 = -g_2$) almost everywhere on Γ, since then, according to Green's formula,

$$(f,1) = -\int_\Gamma F_3 \, d\Gamma = \int_\Gamma g_1 \, d\Gamma,$$

which is impossible according to (4.44). If $-g_2 < F_3 < -g_1$ on a set E of positive measure, then $u = u + c = 0$ on E, therefore $c = 0$. The only remaining case is thus $F_3 = -g_1$ on Γ_1, $F_3 = -g_2$ on Γ_2, $\Gamma_1 \cup \Gamma_2 = \Gamma$, except for sets of measure zero. We conclude as in Theorem 7.5, Chapter I.

4.7. Solution of Problem 4.2

Problem 4.2 is equivalent to minimizing on $H^2(\Omega)$ the functional

$$J(v) = \tfrac{1}{2} a(v) + j_1(v) - (f, v). \tag{4.45}$$

We have

Theorem 4.7. *In order that Problem* 4.2 *has a solution, it is necessary that*

$$(f, p) \leqslant j_1(p), \quad \forall p \in \mathscr{P}. \tag{4.46}$$

Remark 4.4. Since $j_1(c) = 0$, $c = \text{constant}$, (4.46) *implies*

$$(f, 1) = 0. \tag{4.47}$$

Proof of Theorem 4.7. In order that a solution exists, the function $\lambda \to J(\lambda v)$ must be bounded from below, which leads to (4.46) as in Theorem 4.1.

Next, we prove, in a manner analogous to that of Theorem 4.2,

Theorem 4.8. *If we assume that*

$$(f, p) < j_1(p) \quad \forall p \in \mathscr{P}, \quad p \neq 0, \tag{4.48}$$

then Problem 4.2 *has a solution.*

It is easy to see that if u is a solution, *then* $u+c$ *is a solution;* therefore, the solution is not unique. We did not completely settle the question of finding *all* solutions.

Remark 4.5. *The solution of Problem* 4.5 *is unique. Indeed, the form $a(u,v)$ is coercive on* $H^2(\Omega)\cap H_0^1(\Omega)$: there exists a constant $\alpha_1>0$ such that

$$a(v)\geqslant\alpha_1\|v\|_{H^2(\Omega)}^2 \qquad \forall v\in H^2(\Omega)\cap H_0^1(\Omega).$$

More generally, we get the same result if we require that u vanishes *on a non-straight part* Γ_0 of Γ. This observation extends to other types of problems discussed in this chapter. ▯

Remark 4.6. In order that Problem 4.4 has a solution, it is necessary that

(4.49) $\qquad (f,p)\leqslant 0 \quad \forall p\in\mathscr{P}\cap K_2 \quad (K_2$ defined in (4.21)).

In fact,

(4.50) $\qquad$ if Γ is a regular curve, $\mathscr{P}\cap K_2=\mathbb{R}$.

Indeed, if $p=a_0+a_1x_1+a_2x_2\in\mathscr{P}$, then $\partial p/\partial n=a_1n_1+a_2n_2$ and, therefore, $p\in\mathscr{P}\cap K_2$, if $a_1n_1+a_2n_2\geqslant 0$ on Γ. But if we take points on Γ *with opposite normals*, we conclude that $a_1n_1+a_2n_2=0$ and $a_1=a_2=0$. Thus, (4.50) follows, and then (4.49) is *equivalent* to

(4.51) $\qquad (f,1)=0.$

We do not know whether this condition is sufficient.

We will give a sufficient condition, decidedly much too restrictive, in order that Problem 4.4 has a solution.

We make the following assumptions:

(4.52) $\qquad (f,x_\alpha)=0, \qquad \alpha=1,1$ *and* (4.51);

(4.53) $\qquad \Omega$ is an open bounded *convex* set with a *regular* boundary Γ.

Then Problem 4.4 *has a solution.*

Indeed, the problem is equivalent to minimizing the functional

$$\tfrac{1}{2}a(v)-(f,v)$$

on K_2.

But, according to (4.52) and the Fredholm alternative, there exists an $F\in H^2(\Omega)$ such that

(4.54) $\qquad (f,v)=a(F,v) \quad \forall v\in H^2(\Omega);$

(F is defined up to an additive element p of $\mathscr{P}$; we choose F). Then the problem is equivalent to minimizing

(4.55) $$a(v-F)=\mathcal{J}(v)$$

on $H^2(\Omega)$.

We pass to the quotient with $\mathcal{P}$; we introduce

(4.56) $$\mathcal{H}=H^2(\Omega)/\mathcal{P}$$

and let $\varphi\to\varphi^{\boldsymbol{\cdot}}$ be the canonical mapping of $H^2(\Omega)\to\mathcal{H}$. On $\mathcal{H}$, we define

$$a(\varphi^{\boldsymbol{\cdot}},\psi^{\boldsymbol{\cdot}})=a(\varphi,\psi),\qquad \varphi\in\varphi^{\boldsymbol{\cdot}},\qquad \psi\in\psi^{\boldsymbol{\cdot}},$$

and the functional

(4.57) $$\mathcal{J}(v^{\boldsymbol{\cdot}})=a(v^{\boldsymbol{\cdot}}-F^{\boldsymbol{\cdot}}).$$

By $K_2^{\boldsymbol{\cdot}}$ we designate the image of K_2 by $\varphi\to\varphi^{\boldsymbol{\cdot}}$. Then the problem is equivalent to the minimization of $\mathcal{J}(v^{\boldsymbol{\cdot}})$ on $K_2^{\boldsymbol{\cdot}}$; but $(a(v^{\boldsymbol{\cdot}},v^{\boldsymbol{\cdot}}))$ is equivalent to the quotient norm on $\mathcal{H}$, so that we will have existence of a solution if we show that

(4.58) $K_2^{\boldsymbol{\cdot}}$ *is closed in* $\mathcal{H}$.

Let $v_j^{\boldsymbol{\cdot}}$ be a sequence of $K_2^{\boldsymbol{\cdot}}$ with $v_j^{\boldsymbol{\cdot}}\to v^{\boldsymbol{\cdot}}$ in $\mathcal{H}$. There exists a $v_j\in v_j^{\boldsymbol{\cdot}}$ such that $\partial v_j/\partial n\geqslant 0$ on Γ, and we can find a sequence $p_j\in\mathcal{P}$ such that

(4.59) $$v_j+p_j\to w \quad\text{in}\quad H^2(\Omega).$$

It follows from (4.59) that

(4.60) $$\partial v_j/\partial n+\partial p_j/\partial n\to\partial w/\partial n \quad\text{in}\quad L^2(\Gamma)\ \text{(and even in } H^{1/2}(\Gamma)).$$

Setting $\partial v_j/\partial n+\partial p_j/\partial n=a_j$, we have:

(4.61) $$\partial p_j/\partial n=a_j-\partial v_j/\partial n\leqslant a_j \quad\text{on}\quad \Gamma.$$

But since Ω is a convex set with regular boundary, there exists a mapping $x\to\psi(x)$ which is a diffeomorphism of $\Gamma\to\Gamma$, and such that, denoting the normal in $y\in\Gamma$ by n_y, we have:

(4.62) $$n_{\psi(x)}=-n_x.$$

Setting

$$p_j(x)=c_{0j}+c_{1j}x_1+c_{2j}x_2,$$

we have

$$\partial p_j(\psi(x))/\partial n=-\partial p_j(x)/\partial n,$$

so that (4.61) gives:

$$\partial p_j(x)/\partial n\geqslant -a_j(\psi^{-1}(x))=b_j(x)$$

and consequently

$$a_j(x) \leqslant \partial p_j(x)/\partial n \leqslant b_j(x), \tag{4.63}$$

where a_j and b_j remain in bounded sets of $L^2(\Gamma)$.

From this we conclude, since

$$|\partial p_j(x)/\partial n| \leqslant |a_j(x)| + |b_j(x)|,$$

that

$$\partial p_j/\partial n \quad \text{remains in a bounded set of} \quad L^2(\Gamma). \tag{4.64}$$

But then $|c_{1j}|+|c_{2j}| \leqslant$ constant, and therefore we can extract a sequence, again called c_{1j}, c_{2j}, such that

$$q_j = c_{1j}x_1 + c_{2j}x_2 \to q = c_1 x_1 + c_2 x_2. \tag{4.65}$$

But then (4.59), which can be written $v_j + c_{0j} + q_j \to w$ in $H^2(\Omega)$, together with (4.65), implies

$$v_j + c_{0j} \to \psi = w - q \quad \text{in} \quad H^2(\Omega) \tag{4.66}$$

and therefore

$$\frac{\partial}{\partial n}(v_j + c_{0j}) = \frac{\partial v_j}{\partial n} \to \frac{\partial \psi}{\partial n} \quad \text{in} \quad L^2(\Gamma)$$

which proves that $\partial\psi/\partial n \geqslant 0$. Therefore $\psi \in v^{\boldsymbol{\cdot}}$ and $\partial\psi/\partial n \geqslant 0$ which proves that $v^{\boldsymbol{\cdot}} \in K_2^{\boldsymbol{\cdot}}$.

This proves the *existence* of a solution.

The solution $u^{\boldsymbol{\cdot}}$ is unique in $\mathscr{H}$. The set of solutions is thus given by the $u \in u^{\boldsymbol{\cdot}}$ such that $\partial u/\partial n \geqslant 0$ on Γ.

Remark 4.7. Naturally, we can introduce *dual formulations* of the preceding problems according to the principles of Section III.3.5. ▯

5. Unilateral Problems of Evolution

5.1. Formulation of the Problems

The notation is the same as in Section 4. In addition, we set

$$H = L^2(\Omega). \tag{5.1}$$

As usual, we denote by $u(t)$ the function $x \to (u(x,t))$, and we set

$$u'(t) = \partial u(.,t)/\partial t, \qquad u''(t) = \partial^2 u(.,t)/\partial t^2, \ldots$$

Problem 5.1. We seek a function $t \to (u(t))$ of $[0,T] \to H^2(\Omega)$ such that

$$u'(t) \in H^2(\Omega), \qquad u''(t) \in H$$

and such that

$$(u''(t), v - u'(t)) + a(u(t), v - u'(t)) + j_0(v) - j_0(u'(t)) \geqslant (f(t), v - u'(t)) \quad \forall v \in H^2(\Omega), \tag{5.2}$$

with the initial conditions

$$u(0) = u_0, \qquad u'(0) = u_1. \tag{5.3}$$

This problem is the "evolution analogue" of Problem 4.1. It can be interpreted in the following way: First of all, if u satisfies (5.2), then

$$\partial^2 u/\partial t^2 + \Delta^2 u = f \quad \text{in} \quad Q = \Omega \times]0, T[. \tag{5.4}$$

The boundary conditions on $\Sigma = \Gamma \times]0, T[$ are obtained as in Section 4. They are:

$$M_\tau = 0 \quad \text{on} \quad \Sigma \tag{5.5}$$

and the *unilateral conditions:*

$$\begin{aligned} -g_2 < F_3 < -g_1 &\Rightarrow \partial u/\partial t = 0 \quad \text{on} \quad \Sigma, \\ F_3 = -g_1 &\Rightarrow \partial u/\partial t \leqslant 0, \\ F_3 = -g_2 &\Rightarrow \partial u/\partial t \geqslant 0. \quad \square \end{aligned} \tag{5.6}$$

Problem 5.1*a*. We introduce the space

$$V = \{v \mid \quad v \in H^2(\Omega), \quad \partial v/\partial n = 0 \text{ on } \Gamma\}. \tag{5.7}$$

We seek a function $t \to u(t)$ of $[0,T] \to V$ such that $u'(t) \in V$, $u''(t) \in H$ and such that (5.2) is satisfied $\forall v \in V$, with the initial conditions (5.3).

The boundary conditions are then

$$\partial u/\partial n = 0 \quad \text{(which replaces (5.5))}, \tag{5.8}$$

and the unilateral conditions (5.6) are unchanged.

Problem 5.2. We seek a function which satisfies the conditions of Problem 5.1, but *with* j_0 *replaced by* j_1, i. e.,

(5.9) $$(u''(t), v-u'(t)) + a(u(t), v-u'(t)) + j_1(v) - j_1(u'(t)) \geqslant (f(t), v-u'(t)), \quad \forall v \in H^2(\Omega),$$

with the initial conditions (5.3).

This problem is the "evolution analogue" of Problem 5.1. It can be interpreted in the following way: u satisfies (5.4) with the initial conditions (5.3) and the boundary conditions

(5.10) $$F_3 = 0 \quad \text{on} \quad \Sigma$$

and

(5.11) $$\begin{cases} k_1 < M_\tau < k_2 \Rightarrow \dfrac{\partial}{\partial n} \dfrac{\partial u}{\partial t} = 0 \quad \text{on} \quad \Sigma \\ M_\tau = k_2 \Rightarrow \dfrac{\partial}{\partial n} \dfrac{\partial u}{\partial t} \geqslant 0, \\ M_\tau = k_1 \Rightarrow \dfrac{\partial}{\partial n} \dfrac{\partial u}{\partial t} \leqslant 0. \quad \square \end{cases}$$

Problem 5.2*a*. In Problem 5.2, we replace $H^2(\Omega)$ by $H^2(\Omega) \cap H_0^1(\Omega)$; thus, we seek a function u with values in $H^2(\Omega) \cap H_0^1(\Omega)$, satisfying (5.9) with

$$v \in H^2(\Omega) \cap H_0^1(\Omega)$$

and the initial conditions (5.3).

The problem can be interpreted thus: u satisfies (5.4), (5.3) and the boundary conditions

(5.12) $u = 0$ on Σ (which "replaces" (5.10)),
conditions (5.11) remaining unchanged. □

Limiting cases. If, in Problem 5.1, we (formally) take $g_2 = 0$, $g_1 = -\infty$, we are led to the following problem, the "evolution analogue" of Problem 4.3:

Problem 5.3. If the convex set K_1 is defined by (4.15), we seek a function $t \to u(t)$ such that:

(5.13) $$\begin{aligned} &u'(t) \in K_1, \\ &(u''(t), v-u'(t)) + a(u(t), v-u'(t)) \geqslant (f(t), v-u'(t)) \quad \forall v \in K_1, \end{aligned}$$

with the initial conditions (5.3). □

The problem can be interpreted thus: u is a solution of (5.4) with (5.3) and the following conditions on Σ:

(5.14) $$\begin{aligned} &M_\tau = 0, \\ &\frac{\partial u}{\partial t} \geqslant 0, \qquad F_3 \geqslant 0, \qquad \frac{\partial u}{\partial t} F_3 = 0. \quad \square \end{aligned}$$

In the same way, if we (formally) set $k_2=0$, $k_1=-\infty$ in Problem 5.2, we are led to

Problem 5.4. If K_2 is the convex set defined by (4.21), we seek a function $t\to u(t)$ which satisfies the analogue of (5.13), *with* K_1 *being replaced by* K_2.

While the other conditions remain unchanged, the boundary conditions (5.14) are here replaced by

$$F_3=0,$$

$$\frac{\partial}{\partial t}\frac{\partial u}{\partial n}\geqslant 0, \qquad M_\tau\geqslant 0, \qquad \left(\frac{\partial}{\partial t}\frac{\partial u}{\partial n}\right)M_\tau=0. \quad \square \tag{5.15}$$

Remark 5.1. In an analogous fashion, we could state and solve dynamic problems with unilateral displacement of the points of Ω (cf. Sec. 3.2.1) or friction at points of Ω (cf. Sec. 3.2.4). □

5.2. Solution of Unilateral Problems of Evolution

The methods are *identical* with those of Section III.5.5. We merely state some of the results without giving proofs.

Theorem 5.1. *We assume that*

$$f, f'\in L^2(0,T;H)=L^2(Q), \tag{5.16}$$

$$u_0\in H^4(\Omega), \qquad M_\tau(u_0)=0, \qquad F_3(u_0)=0 \quad \text{on} \quad \Gamma, \tag{5.17}$$

$$u_1\in H^2(\Omega)\cap H_0^1(\Omega). \tag{5.18}$$

Then Problem 5.1 *has a unique solution satisfying*

$$u, u'\in L^\infty(0,T;H^2(\Omega)), \tag{5.19}$$

$$u''\in L^\infty(0,T;H). \tag{5.20}$$

The principle of the proof consists *in regularizing* $j_0(v)$ into $j_{0\varepsilon}(v)$, *for example* by

$$j_{0\varepsilon}(v)=\int_\Gamma\left[g_2\frac{(v^+)^{1+\varepsilon}}{1+\varepsilon}-g_1\frac{(v^-)^{1+\varepsilon}}{1+\varepsilon}\right]d\Gamma, \qquad \varepsilon>0, \tag{5.21}$$

and then considering the *regularized equation*

$$(u_\varepsilon'',v)+a(u_\varepsilon,v)+(j_{0\varepsilon}'(u_\varepsilon'),v)=(f,v), \qquad \forall v\in H^2(\Omega), \tag{5.22}$$

with

$$u_\varepsilon(0)=u_0, \qquad u_\varepsilon'(0)=u_1. \tag{5.23}$$

Due to the assumptions made for u_0, u_1, we conclude from (5.22) that

$$u_\varepsilon''(0)=f(0)-\Delta^2 u_0, \tag{5.24}$$

which, with (5.16), permits us to differentiate (5.22) once with respect to t and to obtain a priori estimates corresponding to (5.19), (5.20). ☐

Remark 5.2. We can weaken the assumptions on f, u_0, u_1 by introducing *weak solutions* of inequalities, cf. H. Brézis et J. L. Lions [1] and Brézis [2]. ☐

We can take $g_2=0$ and set: $g_1=-g$, $g>0$ and call u_g the solution corresponding to Problem 5.1. Then we have the following result:

Theorem 5.2. *We assume that* (5.16), (5.17), (5.18) *hold. Then Problem* 5.3 *has a unique solution u satisfying* (5.19), (5.20) *and, when* $g\to+\infty$, *we have:*

$$u_g\to u,\ u_g'\to u' \quad \text{weakly star in} \quad L^\infty(0,T;H^2(\Omega)), \tag{5.25}$$

$$u_g''\to u'' \quad \text{weakly star in} \quad L^\infty(0,T;L^2(\Omega)). \tag{5.26}$$

For the proof, we observe that, from (5.22), we can derive a priori estimates, that are independent of ε *and of* g, for u_ε, u_ε' in $L^\infty(0,T;H^2(\Omega))$ and for u_ε'' in $L^\infty(0,T;L^2(\Omega))$.

Remark 5.3. We can solve Problem 5.3 directly by penalization, if we approximate (5.13) by

$$\begin{aligned}&(u_\eta'',v)+a(u_\eta,v)+\eta^{-1}\int_\Gamma(-(u_\eta')^-)v\,d\Gamma=(f,v) \qquad \forall v\in H^2(\Omega),\\ &u_\eta(0)=u_0, \qquad u_\eta'(0)=u_1.\end{aligned} \tag{5.27}$$

The assumption "$u_1=0$ on Γ" can be replaced by "$u_1\geqslant 0$ on Γ". ☐

Remark 5.4. We have a result relative to Problem 5.1*a* which, in all points, is analogous to that for Theorem 5.1.

With regard to Problem 5.2, we have the

Theorem 5.3. *We assume that* (5.16), (5.17) *hold and that*

$$u_1\in H^2(\Omega), \qquad \partial u_1/\partial n=0. \tag{5.28}$$

Then Problem 5.2 *has a unique solution satisfying* (5.19), (5.20).

The principle of the proof is to *regularize* $j_1(v)$ into $j_{1\varepsilon}(v)$, *for example* by

$$j_{1\varepsilon}(v)=\int_\Gamma\left\{\frac{k_2}{1+\varepsilon}\left(\left(\frac{\partial v}{\partial n}\right)^+\right)^{1+\varepsilon}-\frac{k_1}{1+\varepsilon}\left(\left(\frac{\partial v}{\partial n}\right)^-\right)^{1+\varepsilon}\right\}d\Gamma, \qquad \varepsilon>0. \tag{5.29}$$ ☐

Let us take:

$$k_2=0, \qquad k_1=-k, \qquad k>0$$

and let u_k be the solution corresponding to Problem 5.2. Then:

Theorem 5.4. *We assume that* (5.16), (5.17), (5.28) *hold. Then Problem* 5.4 *has a unique solution* u *satisfying* (5.19), (5.20) *and, when* $k \to +\infty$, *we have:*

(5.30) $\quad u_k \to u,\ u'_k \to u'$ *weakly star in* $L^\infty(0, T; H^2(\Omega))$,

(5.31) $\quad u''_k \to u''$ *weakly star in* $L^\infty(0, T; L^2(\Omega))$. □

6. Comments

The linearized theory of plates—which has been used here—is the one named after Kirchhoff. There are other linearized theories named after Reissner and after Hencky, whose formulations can be found in the thesis of Sander [1]. One could also consult A. E. H. Love [1], S. Timoshenko and S. Woinowsky-Krieger [1], G. Kirchhoff [1], E. Reissner [1] and [2], H. Hencky [1], [2], A. E. Green [1]. Non linear theories have not been treated here. In another, separate work by the authors, one can find the analysis of unilateral problems for the non linear theory which leads to the equations of von Karman.

The problems of "plastic" plates also lead to inequalities; results for this question can be found in N. Coutris [1].

From the mathematical point of view, we saw that while the dynamic problems have been solved satisfactorily, the same does not hold for the static problems. In the text, we mentioned several open questions concerning uniqueness, or not, of solutions. In the same way, *problems of regularity* of solutions seem to be largely open; we point out, however, that for Problems 4.3 and 4.4, one can prove that *for* $f \in L^2(\Omega)$ *there is a* $u \in H^4(\Omega)$ (by the method of "translations parallel to the boundary").

Chapter V

Introduction to Plasticity

1. Introduction

The term "plastic material" covers materials with various types of behavior, such as "elasto-visco-plastic", "rigid-perfectly plastic", "plastic with work hardening", etc.

In all these situations, there exists a threshold separating two types of behavior. This threshold can be fixed, or it can depend on the history of the deformations; in the latter case we say: the material undergoes work hardening, That type of behavior will not be discussed in this book. For the case of a fixed threshold, we will state the elasto-visco-plastic constituent law and, by passage to the limit, we will obtain rigid vicso-plastic materials, elastic perfectly plastic materials and, finally, rigid perfectly plastic materials. In the last two sections, we will discuss two constituent laws applicable in special situations of static or quasi static problems: we are here concerned with Hencky's law and with so-called "locking materials". In the latter case, the threshold involves the strains.

2. The Elastic Perfectly Plastic Case (Prandtl-Reuss Law) and the Elasto-Visco-Plastic Case

2.1. Constituent Law of Prandtl-Reuss

Using the notation introduced in Chapter I, we recall the relations deriving from the conservation laws

(2.1) $\partial\rho/\partial t+\operatorname{div}(\rho v)=0$ (conservation of mass)

(where ρ and v are, respectively, the density and the velocity vector);

(2.2) $\sigma_{ij,j}+f_i=\rho(\mathrm{d}v_i/\mathrm{d}t)$ (conservation of momentum)

(the indices i and j take the values 1 to 3),

(2.3) $\rho(\mathrm{d}e/\mathrm{d}t)=\sigma_{ij}\varepsilon_{ij}(v)$ (conservation of energy)

(flux and sources of heat are assumed to be negligible).

As in Chapter III, we linearize these equations, which leads us to consider the density as a constant ρ_0 *in equations* (2.2) and (2.3); equation (2.1) then furnishes the small variations of ρ about ρ_0 as a function of $\operatorname{div} v$, that is determined by the single equation (2.2) that then takes the form

(2.4) $\rho_0(\partial^2 u_i/\partial t^2)=\sigma_{ij,j}+f_i$.

The vector $\{u_i\}$, $i=1,2,3$, is the displacement vector and $\{f_i\}$ designates a volume density of given forces (independent of x and t). Equation (2.3) gives the internal energy of the material at every moment. In the following, we take $\rho_0=1$, thus fixing the unit of specific mass. □

2.1.1. Preliminary Observation (Tresca [1], St. Venant [1, 2], Levy [1, 2])

We consider a metal rod (mild steel, for example) that is subjected to a traction force σ and that in consequence undergoes a relative elongation ε.

In a system of two orthogonal axes, we mark off ε as abscissa and σ as ordinate and trace the graph of the relation (ε, σ) (Fig. 17). When ε increases, starting from zero, σ increases and the point (ε, σ) describes a straight line segment from the origin O. If we continue to increase ε, the curve described by the point $(\varepsilon. \sigma)$ curves, starting from the point S, and becomes progressively closer to a parallel to $O\varepsilon$. Thus, *when* ε *increases from* 0 *to* $+\infty$, the graph for the relation (ε, σ) is composed of a straight line segment OS and a curved arc Sz.

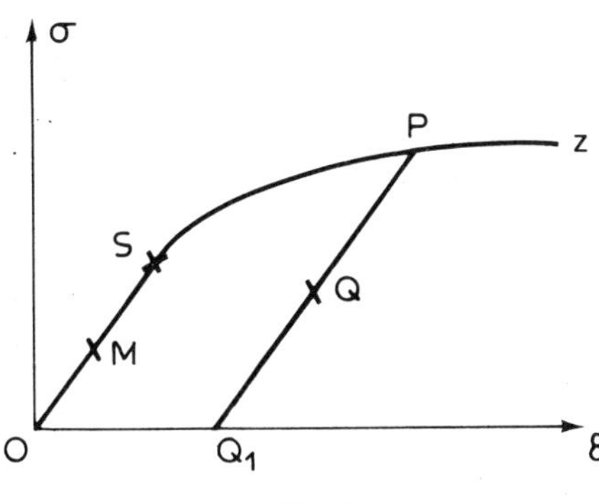

Fig. 17

Assume first that, starting with a point M of the open segment OS, we let ε undergo small changes of arbitrary sign $(0<\varepsilon<\varepsilon_S)$: we observe that the representative point moves in a neighborhood of M while staying on the segment OS. The latter then represents a region where the behavior of the material is linear and reversible, that is to say, *elastic*.

Now, we place ourselves at a point P of the arc Sz and let ε decrease. We observe that the representative point (ε, σ) describes a line segment starting from P, essentially parallel to OS, say PQ. At the point P, the behavior of the material is then no longer reversible; the arc Sz represents a *plastic* region.

If we continue the segment PQ to Q_1 on the $O\varepsilon$-axis, we again find an open segment PQ_1 on which the behavior of the material is reversible. Furthermore, $PQ_1 > OS$ as long as the arc Sz is not a half-line parallel to $O\varepsilon$: this is the phenomenon of *work hardening* that is very interesting in practice because it permits to obtain a zone of linear reversible behavior starting from Q_1 *of greater amplitude* than that obtained starting from the natural state represented by the origin.

If on the contrary, Sz is a half-line parallel to $O\varepsilon$ (or is sufficiently close to being considered one), we say that the material under consideration is *perfectly plastic*. We retain this assumption in the remainder of this section. In this case, the stress σ never passes a certain threshold g independent of the amount of strain. □

We formulate the constituent law for such a perfectly plastic metal rod (Fig. 18). Since every point Q of the region between the $O\varepsilon$-axis and the graph OSz can be reached by the point (ε, σ), *it is impossible* to give to this rod a constituent law in the form ε = function of σ or σ = function of ε. On the contrary, if starting from a state (ε, σ), $\sigma \leqslant g$, we increase ε by a "small" $d\varepsilon$, then σ changes by $d\sigma$ such that

$$d\varepsilon = A\, d\sigma + \tilde{\lambda} \tag{2.5}$$

where

$$\begin{aligned} &\tilde{\lambda} = 0 \quad \text{when} \quad \sigma < g \quad \text{or when} \quad \sigma = g, \quad d\sigma < 0, \\ &\tilde{\lambda} \geqslant 0 \quad \text{when} \quad \sigma = g \quad \text{and} \quad d\sigma = 0. \end{aligned}$$

($1/A$ is the slope of the segment OS).

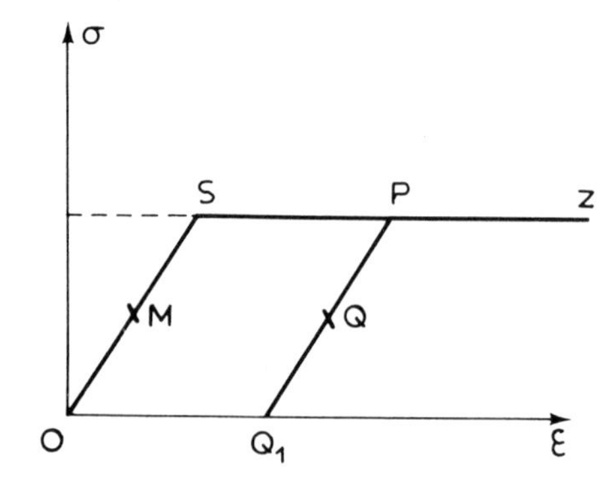

Fig. 18

If the increase $d\varepsilon$ takes place during the time dt, the constituent law (2.5) can be written:

(2.6) $$\dot{\varepsilon}=A\dot{\sigma}+\lambda \quad (\dot{x}=\partial x/\partial t)$$ [1]

where

(2.7) $$\begin{aligned}&\lambda=0 \quad \text{when} \quad \sigma<g \quad \text{or when} \quad \sigma=g,\ \dot{\sigma}<0\\ &\lambda\geqslant 0 \quad \text{when} \quad \sigma=g \quad \text{and} \quad \dot{\sigma}=0.\end{aligned}$$

Relation (2.7) can be written equivalently

(2.8) $$\begin{aligned}&\lambda(\tau-\sigma)\leqslant 0 \qquad \forall\tau\leqslant g\\ &\lambda\dot{\sigma}=0. \quad \square\end{aligned}$$

Remark 2.1. If we let the coefficient A tend toward zero, that is to say if the slope $1/A$ of OS tends toward infinite, the constituent law (2.6), (2.7) becomes

$$\dot{\varepsilon}=\lambda$$

where λ satisfies (2.7). This is the constituent law of a *rigid perfectly plastic* material: the material undergoes no deformation if $\sigma<g$, and it deforms according to the indicated law if $\sigma=g$. $\square$

2.1.2. Generalization

One can generalize the obtained constituent law to *three-dimensional* perfectly plastic bodies.

We assume that the elastic zone is defined by

(2.9) $$\mathscr{F}(\sigma_{ij})<0$$

and the plastic zone by

(2.10) $$\mathscr{F}(\sigma_{ij})=0.$$

The scalars σ_{ij} are components of the stress tensor and satisfy $\sigma_{ij}=\sigma_{ji}$. The function $\mathscr{F}$ is *convex* and *continuous* with respect to σ_{ij}, so that the domain of $\mathbb{R}^6$, where $\mathscr{F}(\sigma_{ij})\leqslant 0$, is a closed convex set. Actually, $\mathscr{F}(\sigma_{ij})$ involves only the deviator $\{\sigma_{ij}^D\}$ of the tensor $\{\sigma_{ij}\}$. $\square$

We mention *two classical examples* of functions $\mathscr{F}$:

i) *Von Mises' model*

(2.11) $$\mathscr{F}(\sigma_{ij})=\tfrac{1}{2}\sigma_{ij}^D\sigma_{ij}^D-k^2,$$

where k is a given constant.

[1] This notation, which is classical in mechanics, is retained here for the formulation of the equations. It will be abandoned in the mathematical development.

ii) *Tresca's model*

$$\mathscr{F}(\sigma_{ij}) = \sup_{I,J;I \neq J} |\sigma_I - \sigma_J| - g \tag{2.12}$$

where the σ_T are the eigenvalues of the stress tensor $\{\sigma_{ij}\}$ and where g is a positive constant. We show that the function $\mathscr{F}(\sigma_{ij})$ defined by (2.12) is convex. For every unit vector $n=(n_1,n_2,n_3)$ we introduce the stress vector Σ by its components

$$\Sigma_i = \sigma_{ij} n_j;$$

we decompose Σ into a vector in the direction of n, say Σ_n, and a perpendicular vector Σ_T. One easily verifies—by passing to principal axes for $\{\sigma_{ij}\}$—that, for $\mathscr{F}$ given by (2.12), $\mathscr{F}(\sigma_{ij}) \leqslant 0$ is equivalent to

$$\Sigma_T \leqslant g, \qquad \forall n \text{ with } |n|=1.$$

The mapping $\{\sigma_{ij}\} \to \Sigma_T$ is linear for each n, thus, if $\{\sigma_{ij}^{(1)}\}$ and $\{\sigma_{ij}^{(2)}\}$ are two stress tensors to which $\Sigma_T^{(1)}$ and $\Sigma_T^{(2)}$ correspond and if $\alpha \in]0,1[$, then

$$|\alpha \Sigma_T^{(1)} + (1-\alpha)\Sigma_T^{(2)}| \leqslant \alpha |\Sigma_T^{(1)}| + (1-\alpha)\Sigma_T^{(2)}| \leqslant g$$

if $|\sigma_T^{(1)}| \leqslant g$ and $|\sigma_T^{(2)}| \leqslant g$ which implies that $\mathscr{F}$, given by (2.12), is convex. □

We observe that, in the two examples given above, the function $\mathscr{F}$ is independent of the deformation. This characterizes *perfectly plastic* materials. □

The analogue to (2.6), in the present situation, is

$$\dot{\varepsilon}_{ij}(u) = A_{ijkh}\dot{\sigma}_{kh} + \lambda_{ij}. \tag{2.13}$$

The coefficients A_{ijkh} are the elasticity coefficients; they have the usual properties

$$\begin{aligned} &A_{ijkh} = A_{ijhk} = A_{khij} \\ &A_{ijkh}\sigma_{ij}\sigma_{kh} \geqslant \alpha\sigma_{ij}\sigma_{ij}, \qquad \alpha = \text{constant} > 0. \end{aligned} \tag{2.14}$$

The quantity $A_{ijkh}\sigma_{kh}$ is the *elastic* part of the tensor of strain velocities $\dot{\varepsilon}_{ij}(u)$, while, by definition, λ_{ij} is the *velocity of plastic deformation.*

In this preliminary observation, the part λ satisfies (2.7) or (2.8) which here are generalized to

$$\lambda_{ij}(\tau_{ij} - \sigma_{ij}) \leqslant 0 \qquad \forall \tau \text{ with } \mathscr{F}(\tau_{ij}) \leqslant 0, \tag{2.15}$$

$$\lambda_{ij}\dot{\sigma}_{ij} = 0. \tag{2.16}$$

(Property (2.15) also expresses the principle of maximum plastic work formulated by Hill [3].)

If, moreover, the functions $t \to \sigma_{ij}(t)$ *are differentiable in* t—which we assume—*we can derive* (2.16) *from* (2.15).

Indeed: let $\Delta t > 0$; we take in (2.15)

$$\tau_{ij} = \sigma_{ij}(t + \Delta t) \quad (\text{resp. } \tau_{ij} = \sigma_{ij}(t - \Delta t)),$$

divide by Δt and let Δt tend toward zero; we get

$$\lambda_{ij} \dot{\sigma}_{ij}(t) \leqslant 0 \quad (\text{resp. } \lambda_{ij} \dot{\sigma}_{ij}(t) \geqslant 0)$$

from which (2.16) follows.

Summarizing: *the constituent law of elastic perfectly plastic materials is expressed by*

$$(2.17) \qquad \mathscr{F}(\sigma_{ij}) \leqslant 0$$

and (2.13), (2.15). ▯

Remark 2.2. As in Remark 2.1 for the one-dimensional case, we can set $A_{ijkh} = 0$, and the corresponding constituent law becomes that for rigid perfectly plastic materials. ▯

2.2. Elasto-Visco-Plastic Constituent Law

In the Prandtl-Reuss (or elastic perfectly plastic) constituent law, the speed of plastic deformation *at constant stress* can be very large in absolute value; in other words, the material shows no effect of viscosity: this is an idealized situation and we can legitimately think that *plastic deformations, often of large amplitude, in practice show a viscosity effect, that can be more or less important.* For this reason, we will describe a model of a law of elasto-visco-plastic behavior, where the material shall be elastic below the threshold of plasticity and of visco-plastic type beyond. This rather comprehensive law permits to obtain, or to recover by passage to the limit, the following special cases:

i) Rigid visco-plastic ($A_{ijkh} = 0$). This case will be discussed completely, and by different methods, in Chapter VI.

ii) Elastic perfectly plastic (viscosity tends toward zero) (the Prandtl-Reuss law given in the preceding section).

iii) Rigid perfectly plastic ($A_{ijkh} = 0$ and the viscosity tends toward zero). This case will be discussed in Section 5.2. ▯

We introduce the closed convex set of

$$(2.18) \qquad \tilde{K} = \{\sigma \mid \quad \sigma = \{\sigma_{ij}\} \in \mathbb{R}^6, \quad \sigma_{ij} = \sigma_{ij}, \quad \mathscr{F}(\sigma_{ij}) \leqslant 0\}$$

where $\mathscr{F}$ is a convex continuous function of the six scalar variables σ_{ij}, as in the preceding Section 2.1.2, and the mapping $\sigma \to P_{\tilde{K}}(\sigma) =$ orthogonal projection, in the usual Euclidian structure of $\sigma \in \mathbb{R}^6$ on the convex set $\tilde{K}$.

The announced constituent law can then be written

$$(2.19)\qquad \begin{aligned} &\varepsilon_{ij}(\dot{u}) = A_{ijkh}\dot{\sigma}_{kh} + \lambda_{ij},\\ &\{\lambda_{ij}\} = 0 \quad \text{if} \quad \mathscr{F}(\sigma) < 0,\\ &\lambda_{ij} = \frac{1}{2\mu}[\sigma_{ij} - (P_{\tilde{K}}\sigma)_{ij}] \quad \text{if} \quad \mathscr{F}(\sigma) \geq 0, \end{aligned}$$

where μ is a positive scalar, which can be considered as a *viscosity* coefficient; for, if $\{\sigma_{ij}\}$ is constant, $(\dot{\sigma}_{ij} = 0)$, the law (2.19) becomes

$$\sigma_{ij} = (P_{\tilde{K}}\sigma)_{ij} + 2\mu\varepsilon_{ij}(\dot{u}),$$

a relation in which μ and $\varepsilon_{ij}(u)$ (tensor of strain velocities) play the same role as, in the Navier Stokes constituent law for fluids

$$\sigma_{ij} = -p\delta_{ij} + 2\mu\varepsilon_{ij}(\dot{u}),$$

the term $-p\delta_{ij} = (P_{\tilde{K}}\sigma)_{ij}$ when $\tilde{K} = \{\sigma \mid \{\sigma_{ij}\} = \text{sperical tensor}\}$.

In the particular case of Von Mises' condition, i. e., $\mathscr{F}$ given by (2.11), the law (2.19) takes the form

$$(2.20)\qquad \begin{aligned} &\varepsilon_{ij}(\dot{u}) = A_{ijkh}\dot{\sigma}_{kh} + \lambda_{ij}\\ &\lambda_{ij} = 0 \quad \text{if} \quad \mathscr{F}(\sigma) < 0\\ &\lambda_{ij} = \frac{1}{2\mu}\,\frac{\sigma_{\text{II}}^{1/2} - k}{\sigma_{\text{II}}^{1/2}}\,\sigma_{ij}^{D} \end{aligned}$$

where $\{\sigma_{ij}^{D}\}$ represents the deviator of the stress tensor and

$$\sigma_{\text{II}} = \tfrac{1}{2}\sigma_{ij}^{D}\sigma_{ij}^{D}. \quad \square$$

We introduce the mapping of $\mathbb{R}^6$ into $\mathbb{R}$:

$$\mathscr{J}_{\mu}(\tau) = \frac{1}{4\mu}[\tau_{ij} - (P_{\tilde{K}}\tau)_{ij}][\tau_{ij} - (P_{\tilde{K}}\tau)_{ij}].$$

The law (2.19) can then be written *equivalently*

$$(2.21)\qquad \begin{aligned} &\varepsilon_{ij}(\dot{u}) = A_{ijkh}\dot{\sigma}_{kh} + \lambda_{ij}\\ &\text{where } \lambda_{ij} \text{ is characterized by}^{2}\\ &\mathscr{J}_{\mu}(\tau) - \mathscr{J}_{\mu}(\sigma) \geq \lambda_{ij}(\tau_{ij} - \sigma_{ij}), \qquad \forall \tau \in \mathbb{R}^6. \quad \square \end{aligned}$$

[2] We can also say $\{\lambda_{ij}\} \in \partial\mathscr{J}_{\mu}(\sigma)$, where $\partial\mathscr{J}_{\mu}(\sigma)$ represents the subgradient of $\mathscr{J}_{\mu}$ in σ (in fact, the derivative when $\mu > 0$)

(2.21a) $\varepsilon_{ij}(\dot{u}) = A_{ijkh}\dot{\sigma}_{kh} + (\mathscr{J}_{\mu}'(\sigma))_{ij}$.

The case where the viscosity tends toward zero. When μ tends toward zero, the function $\mathscr{J}_\mu(\tau)$ tends toward the indicator function of the convex set $\tilde{K}$ of $\mathbb{R}^6$, i. e., toward the "proper" convex function, zero on $\tilde{K}$ and equal to $+\infty$ outside of $\tilde{K}$ ($\tilde{K}\neq\emptyset$). Formula (2.21) takes the limiting form

$$\begin{aligned}&\varepsilon_{ij}(\dot{u})=A_{ijkh}\dot{\sigma}_{kh}+\lambda_{ij}\\ &\text{where } \{\lambda_{ij}\} \text{ is characterized by}\\ &\lambda_{ij}(\tau_{ij}-\sigma_{ij})\leqslant 0 \quad \forall\tau\in\tilde{K}, \quad \sigma\in\tilde{K},\end{aligned} \tag{2.22}$$

that is simply the *Prandtl-Reuss law*, stated in the preceding section.

"Rigid" case as limit of the "elastic" case. If all the A_{ijkh} are zero, (2.19) is the constituent law of rigid-visco-plastic materials and (2.22) is the constituent law of rigid-perfectly plastic materials. □

Moreover, we point out that, if $\tilde{K}=\mathbb{R}^6$, the elasto-visco-plastic constituent law simply yields

$$\varepsilon_{ij}(\dot{u})=A_{ijkh}\dot{\sigma}_{kh}$$

that, if we assume that there was an instant t_0 when the material was in a stress free state[3] and that the displacements are measured from this state, yields by integration from t_0 to t:

$$\varepsilon_{ij}(u)=A_{ijkh}\sigma_{kh}$$

that is to say the elastic constituent law. □

Orientation. In the present chapter, we will investigate the *deformations* of a medium which, *in its non-deformed state, occupies a domain Ω of* $\mathbb{R}^6$, *while the displacements of the points of Ω are small.* If $u(x,t)$ represents the displacement vector at the instant t of the point which accupied the position x of Ω in the non-deformed state, then the quantity $\dot{u}(x,t)=\partial u(x,t)/\partial t$ represents the velocity at the time t at the point with the coordinate $x+u(x,t)$, or also the velocity at the instant t of the material particle which was located at x in the non-deformed state. In fact, *since $u(x,t)$ is small,* we take $\dot{u}(x,t)=\partial u(x,t)/\partial t$ to be the velocity at the point x. For the same reason, $\partial^2 u(x,t)/\partial t^2$ represents, strictly speaking, the acceleration at the instant t at the point $x+u(x,t)$; since $u(x,t)$ is small, $\partial^2 u(x,t)/\partial t^2$ is taken to be the acceleration at the point x. The approximations correspond to a linearization which amounts to not distinguishing x, the initial position of a material particle, from the position of that particle in the deformed state. □

In Chapter VI, we will investigate problems that by boundary conditions and by the nature of the constituent law are *flow problems*, that is to say problems where *the displacements of material points are large.* The open set Ω of $\mathbb{R}^3$ then

[3] Which is implicit in all constituent laws stated in this book which involve the stress tensor.

represents the domain in which we investigate the flow, and $\dot{u}(x,t)$ *is the velocity at the instant* t *of the material particle which is located at the point* x *of* Ω *at this instant* (the dot "$\cdot$" does not indicate partial differentiation with respect to t, but $\dot{u}(x,t)$ always denotes the velocity in x at the instant t). In order to avoid confusion and to conform to the customary notation, we will not write $\varepsilon_{ij}(\dot{u})$ but $D_{ij}(v)$, where $v=v(x,t)$ represents the velocity of the particle located at x at the instant t. The preceding description is known as the *Euler description* of the flow.

We still add that the constituent law to be used is

$$\begin{aligned} &D_{ij}(v)=0 \quad \text{when} \quad \sigma_{\mathrm{II}}^{1/2}<g \\ &D_{ij}(v)=\frac{1}{2\mu}\frac{\sigma_{\mathrm{II}}^{1/2}-g}{\sigma_{\mathrm{II}}^{1/2}}\sigma_{ij}^{D} \quad \text{when} \quad \sigma_{\mathrm{II}}^{1/2}\geqslant g \end{aligned} \tag{2.23}$$

(g is a given positive constant).

From this follows that

$$D_{kk}(v)=\operatorname{div} v=0$$

that is to say that the material which obeys such a constituent law is *incompressible.* □

2.3. Problems to be Discussed

We intend to investigate, in *dynamic* and *quasi-static* situations, the fields of displacements $\{u_i\}$ and of stresses $\{\sigma_{ij}\}$ in the interior of a medium that, in its non-deformed state, occupies a region Ω of $\mathbb{R}^3$. We set, as we did in the preceding chapters:

$$\begin{aligned} &\partial\Omega=\Gamma, \quad n=\text{exterior unit normal to } \Gamma \\ &\Gamma=\Gamma_U\cup\Gamma_F, \qquad \Gamma_U\cap\Gamma_F=\emptyset. \end{aligned}$$

The material obeys the constituent law (2.21). Then we formulate the two following problems:

Problem 2.1 (Dynamic case). *To find* $\{u_i\}$ *and* $\{\sigma_{ij}\}$ *in* $\Omega\times[0,T]$ *which satisfy* (2.21), *the linearized equations of motion*

$$\partial^2 u_i/\partial t^2=\sigma_{ij,j}+f_i, \tag{2.24}$$

the boundary conditions

$$u_i=U_i \quad \text{on} \quad \Gamma_U, \tag{2.25}$$

$$\sigma_{ij}n_j=F_i \quad \text{on} \quad \Gamma_F, \tag{2.26}$$

and the initial conditions

(2.27) $u(0)=u_0, \qquad \partial u(0)/\partial t=u_1,$

(2.28) $\sigma(0)=\sigma_0.$

The functions $f_i(x,t)$, $F_i(x,t)$, $U_i(x,t)$, $u_0(x)$, $u_1(x)$, $\sigma_0(x)$ *are prescribed for* $t\geqslant 0$. □

Problem 2.2 (Quasi-static case). This is a problem in which the data vary sufficiently slowly with respect to t so that we can neglect the terms of acceleration in the equations of motion:

To find $\{u_i\}$ *and* $\{\sigma_{ij}\}$ *in* $\Omega\times[0,T]$ *that satisfy* (2.21), *the equations of equilibrium*

(2.29) $\sigma_{ij,j}+f_i=0$

the boundary conditions (2.25) *and* (2.26) *and initial conditions*

(2.30) $u(0)=u_0,$

(2.31) $\sigma(0)=\sigma_0.$

The functions $f_i(x,t)$, $F_i(x,t)$, $U_i(x,t)$, $u_0(x)$, $\sigma_0(x)$ *are given for* $t\geqslant 0$. □

In the following Section 3, we will discuss Problems 2.1 and 2.2 and then, in Sections 4 and 5, the passages to the limit when $\mu\to 0$ or when, in a suitable sense, $A_{ijkh}\to 0$.

3. Discussion of Elasto-Visco-Plastic, Dynamic and Quasi-Static Problems

3.1. Variational Formulation of the Problems

First, we examine the situation in Problem 2.1. We set

(3.1) $\partial u/\partial t=v;$

once v is known, u will then be defined by

(3.2) $u(t)=\int_0^t v(s)\,ds+u_0.$

With this notation and making use of (2.21a), the equations of the problem are:

(3.3)
$$A_{ijkh}\sigma'_{kh}+(\mathcal{J}'_\mu(\sigma))_{ij}-\varepsilon_{ij}(v)=0,$$
$$v'_i-\sigma_{ij,j}=f_i, \quad \text{in} \quad \Omega\times]0,T[,^4$$

with the boundary conditions

(3.4)
$$v_i=U'_i \quad \text{on} \quad \Gamma_U\times]0,T[$$
$$\sigma_{ij}n_j=F_i \quad \text{on} \quad \Gamma_F\times]0,T[$$

and the initial conditions

(3.5)
$$v(0)=u_1, \qquad \sigma(0)=\sigma_0.$$

This is a *non linear* system of partial differential equations. The boundary conditions are *non homogeneous* in (3.4), and we start by reducing them by translation to the homogeneous case.

For that purpose, we introduce functions $\sigma^0=\{\sigma_{ij}\}$ and $v^0=\{v_i^0\}$ [5] such that

(3.6)
$$\sigma_{ij}^0 n_j=F_i \quad \text{on} \quad \Gamma_F\times]0,T[,$$
$$\sigma^0(0)=\sigma_0$$

and

(3.7)
$$v_i^0=U'_i \quad \text{on} \quad \Gamma_U\times]0,T[,$$
$$v^0(0)=u_1.$$

If we now introduce the new unknown functions

(3.8)
$$v^1=v-v^0, \qquad \sigma^1=\sigma-\sigma^0,$$

Equations (3.3) become:

(3.9)
$$A_{ijkh}\sigma_{kh}^{1'}+(\mathcal{J}'_\mu(\sigma^1+\sigma^0))_{ij}-\varepsilon_{ij}(v^1)=g_{ij},$$
$$v_i^{1'}-\sigma_{ij,j}^1=h_i,$$

where

(3.10)
$$g_{ij}=\varepsilon_{ij}(v^0)-A_{ijkh}\sigma_{kh}^{0'}$$
$$h_i=f_i-(v_i^0)'+\sigma_{ij,j}^0.$$

The boundary and initial conditions become *zero* (we just arranged things that way!)

To simplify the writing, we omit the index "1".

[4] T some arbitrary fixed finite quantity.

[5] The necessary *regularity assumptions* are stated later on.

The formulation is now:

$$(3.11)\qquad \begin{aligned} &A_{ijkh}\sigma'_{kh}+(\mathscr{J}'_\mu(\sigma+\sigma^0))_{ij}-\varepsilon_{ij}(v)=g_{ij},\\ &v'_i-\sigma_{ij,j}=f_i, \end{aligned}$$

with the boundary conditions

$$(3.12)\qquad \begin{aligned} &\sigma_{ij}n_j=0 \quad \text{on} \quad \Gamma_F\times]0,T[,\\ &v_i=0 \quad \text{on} \quad \Gamma_U\times]0,T[, \end{aligned}$$

and the initial conditions

$$(3.13)\qquad \sigma(0)=0, \qquad v(0)=0. \quad \square$$

We now introduce (as in (3.69), Chap. III):

$$(3.14)\qquad \mathscr{A}(\sigma,\tau)=\int_\Omega A_{ijkh}\sigma_{kh}\tau_{ij}\,dx$$

and the spaces

$$(3.15)\qquad \begin{aligned} \mathscr{H}&=\{\tau=\{\tau_{ij}\} \mid \tau_{ij}=\tau_{ji}, \quad \tau_{ij}\in L^2(\Omega)\},\\ H&=\{v=v_i\} \mid v_i\in L^2(\Omega)\} \end{aligned}$$

equipped with the scalar products[6]

$$\begin{aligned} &(\sigma,\tau)=\int_\Omega \sigma_{ij}\tau_{ij}\,dx,\\ &(v,w)=\int_\Omega v_iw_i\,dx. \end{aligned}$$

Next, we introduce:

$$(3.16)\qquad \begin{aligned} \mathscr{V}&=\{\tau \mid \tau\in\mathscr{H},\ \tau_{ij,j}\in L^2(\Omega),\ \tau_{ij}n_j=0 \quad \text{on } \Gamma_F\},\\ V&=\{v \mid v\in H,\ v_{i,j}\in L^2(\Omega), \quad v_i=0 \quad \text{on } \Gamma_U\}. \end{aligned}$$

We note that

$$(3.17)\qquad (\varepsilon_{ij}(v),\tau_{ij})+(v_i,\tau_{ij,j})=0 \qquad \forall v\in V,\ \tau\in\mathscr{V}.$$

Then, if σ, v is a regular solution of (3.11), (3.12), we have:

$$(3.18)\qquad \begin{aligned} &\mathscr{A}(\sigma',\tau)+(\mathscr{J}'_\mu(\sigma+\sigma^0),\tau)+\int_\Omega v_i\tau_{ij,j}\,dx=(g,T) \qquad \forall\tau\in\mathscr{V},\\ &(v',w)+\int_\Omega \sigma_{ij}w_{i,j}\,dx=(h,w) \qquad \forall w\in V. \end{aligned}$$

[6] Although denoted in the same way, this inconsistency of notation cannot possibly cause confusion.

Conversely, if σ is a "regular" solution of (3.18), we can deduce (3.11) as well as

$$\int_\Gamma v_i \tau_{ij} n_j \, d\Gamma = 0, \qquad \int_\Gamma \sigma_{ij} n_j w_i \, d\Gamma = 0$$

i.e.

$$\int_{\Gamma_U} v_i \tau_{ij} n_j \, d\Gamma = 0, \qquad \int_{\Gamma_F} (\sigma_{ij} n_j) w_i \, d\Gamma = 0$$

from which (3.12) follows.

Therefore, we now consider the problem in the form (3.18), (3.13).

In the same notation, the *quasi static problem* can be stated:

$$\text{(3.19)} \quad \begin{aligned} &\mathscr{A}(\sigma', \tau) + (\mathscr{J}'_\mu(\sigma + \sigma^0), \tau) + \int_\Omega v_i \tau_{ij,j} \, dx = (g, \tau) \quad \forall \tau \in \mathscr{V}, \\ &\int_\Omega \sigma_{ij} w_{i,j} \, dx = (h, w) \quad \forall w \in V, \end{aligned}$$

with

$$\text{(3.20)} \quad \sigma(0) = 0 . \quad \square$$

3.2. Statement of Results

The following *assumptions* will be made:

$$\text{(3.21)} \quad g, g' \in L^2(0, T; \mathscr{H}),$$

$$\text{(3.22)} \quad h, h' \in L^2(0, T; H)$$ [7]

and

(3.23) σ^0 *is independent of* t.

Remark 3.1. Assumption (3.23) *appears* to be indispensable for obtaining *strong* solutions, such as we shall obtain subsequently. It is *probable* (but has not been proved) that this condition can be avoided by introducing *weak solutions;* the essential difficulty then is the (possible) *uniqueness* of the solution. □

In the following sections, we will prove:

Theorem 3.1. *We assume that* (3.21), (3.22), (3.23) *hold. Then there exists one and only one pair of functions* σ *and* v *such that*

$$\text{(3.24)} \quad \sigma, \sigma' \in L^\infty(0, T; \mathscr{H}),$$

$$\text{(3.25)} \quad v, v' \in L^\infty(0, T; H),$$

$$\text{(3.26)} \quad \sigma_{ij,j} \in L^\infty(0, T; L^2(\Omega)),$$

$$\text{(3.27)} \quad v_{i,j} \in L^\infty(0, T; L^2(\Omega))$$

and that satisfy (3.18), (3.13).

[7] It is not difficult to spell out sufficient conditions for the data U_i, u_1, σ_0 in order that these assumptions are satisfied.

Remark 3.2. According to (3.26) and (3.27), we can define $\sigma_{ij}n_j$ and v_i on Γ. The calculations of integration by parts, carried out according to (3.18), are legitimate. Thus, we have the existence of *strong* solutions of the problem[8].

We now proceed to the *quasi-static* case. We distinguish two cases according as Γ_U has positive measure or is empty.

Theorem 3.2. *In addition to the assumptions of Theorem* 3.1, *we assume that* Γ_U *has positive measure and that* $h(0)=0$.

Then there exists one and only one pair of functions σ, v *such that*

$$\sigma \in L^\infty(0,T;\mathscr{V}), \tag{3.28}$$

$$\sigma' \in L^\infty(0,T;\mathscr{H}), \tag{3.29}$$

$$v \in L^\infty(0,T;V), \tag{3.30}$$

and satisfying (3.19), (3.20). □

When $\Gamma_U=\emptyset$ (and only in this case), we can take $w=\rho\in\mathscr{R}$ in the second equation of (3.19), from which the *necessary* condition follows (since $\int_\Omega \sigma_{ij}\rho_{i,j}\,dx=0$):

$$(h,\rho)=0 \qquad \forall \rho\in\mathscr{R};$$

but when $\Gamma_U=\emptyset$, $v^0=0$ and according to (3.10), we then have the condition

$$(f,\rho)+(\sigma^0_{ij,j},\rho_i)=0$$

or again, according to (3.6)

$$(f,\rho)+\int_\Gamma F_N\rho_N\,d\Gamma=0 \qquad \forall\rho\in\mathscr{R}\ ^{9} \tag{3.31}$$

Thus, we have

Theorem 3.3. *We make the same assumptions as in Theorem* 3.1 *with* $\Gamma_U=\emptyset$ *and we postulate that* (3.31) *holds and that* $h(0)=0$.

Then there exists one and only one function σ, *and a function* v *defined up to addition of an element of* $\mathscr{R}$, *such that* (3.28), (3.29) *hold and* $\varepsilon_{ij}(v)\in L^\infty(0,T;L^2(\Omega))$ *and such that* (3.19), (3.20) *hold.* □

3.3. Uniqueness Proof in the Theorems

Uniqueness in Theorem 3.1. Let σ, v and σ_*, v_* be two pairs of possible solutions. Setting

$$\hat\sigma=\sigma-\sigma_*, \qquad \hat v=v-v_*,$$

---

[8] We can also state:

$\sigma\in L^\infty(0,T;\mathscr{V})$, $\qquad v\in L^\infty(0,T;V)$.

[9] Therefore, the system of forces $f_i F_i$ must be statically equivalent to 0.

we conclude from (3.18) and from the analogous equations for σ_*, v_* that

$$\mathscr{A}(\hat{\sigma}',\tau)+(\hat{v}',w)+(\mathscr{J}'_\mu(\sigma+\sigma_0)-\mathscr{J}'_\mu(\sigma_*+\sigma_0),\tau)+\int_\Omega \hat{v}_i\tau_{ij,j}dx + \int_\Omega \hat{\sigma}_{ij}w_{i,j}dx=0. \tag{3.32}$$

If we take $\tau=\hat{\sigma}$ and $w=\hat{v}$ in (3.32), it follows, applying (3.17) and the fact that

$$\sigma \to \mathscr{J}'_\mu(\sigma+\sigma_0) \quad \textit{is monotone},$$

that

$$\mathscr{A}(\hat{\sigma}',\hat{\sigma})+(\hat{v}',\hat{v})\leqslant 0 \tag{3.33}$$

which, together with the fact that $\hat{\sigma}(0)=0$, $\hat{v}(0)=0$, leads to $\hat{\sigma}=0$, $\hat{v}=0$. ▯

Uniqueness in Theorem 3.2 *and uniqueness modulo* $\mathscr{R}$ *in Theorem* 3.3. With the same notation as before, we obtain

$$(\hat{\sigma}',\hat{\sigma})\leqslant 0 \tag{3.34}$$

from which we conclude that $\hat{\sigma}=0$.

We now make use of the first equation (3.19) in the form

$$\varepsilon_{ij}(v)=g_{ij}-A_{ijkh}\sigma'_{kh}-(\mathscr{J}'_\mu(\sigma+\sigma^0))_{ij}$$

from which, since $\sigma=\sigma_*$:

$$\varepsilon_{ij}(v)=\varepsilon_{ij}(v_*) \tag{3.35}$$

from which the result follows: if Γ_U has positive measure, (3.35) implies that $v=v_*$, and if $\Gamma_U=\emptyset$, we have equality up to an element of $\mathscr{R}$ (depending on t). ▯

3.4. Existence Proof in the Dynamic Case

We now "regularize" Equations (3.18) *in the space variables.*

For this purpose, we introduce

$$[\sigma,\tau]=\int_\Omega \sigma_{ij,j}\tau_{ij,j}dx, \tag{3.36}$$

$$((v,w))=\int_\Omega v_{i,j}w_{i,j}dx \tag{3.37}$$

and, η being a parameter >0 (intended to tend toward 0), we consider the *regularized problem*

$$\begin{aligned}&\mathscr{A}(\sigma',\tau)+(\mathscr{J}'_\mu(\sigma+\sigma^0),\tau)+\eta[\sigma,\tau]+\int_\Omega v_i\tau_{ij,j}dx=(g,\tau),\\ &(v',w)+\eta((v,w))+\int_\Omega \sigma_{ij}w_{i,j}dx=(h,w)\end{aligned} \tag{3.38}$$

with

$$(3.39) \qquad \sigma(0)=0\,, \qquad v(0)=0\,.$$

For every $\eta>0$, this problem has a unique solution[10].

We now obtain a priori estimates for these solutions $\{\sigma, v\}$ (or, more precisely, $\{\sigma_\eta, v_\eta\}$, to indicate the dependence on η) and then let $\eta\to 0$.

A priori estimates (I). In (3.38), we replace τ by $\sigma+\sigma^0$ and w by v. It follows, (since

$$(\mathscr{J}_\mu'(\sigma+\sigma^0), \sigma+\sigma^0)\geqslant 0\,, \quad \text{and since} \quad \int_\Omega v_i\sigma_{ij,j}\mathrm{d}x+\int_\Omega \sigma_{ij}v_{i,j}\mathrm{d}x=0):$$

that

$$(3.40) \qquad \mathscr{A}(\sigma', \sigma+\sigma^0)+\eta[\sigma, \sigma+\sigma^0]+(v',v)+\eta(v,v)\leqslant (g,\sigma+\sigma^0)+(h,v)\,.$$

From this, we conclude[11] that when $\eta\to 0$:

$$(3.41) \qquad \sigma=\sigma_\eta \quad \text{remains in a bounded set of} \quad L^\infty(0,T;\mathscr{H})\,,$$

$$(3.42) \qquad v=v_\eta \quad \text{remains in a bounded set of} \quad L^\infty(0,T;H)\,,$$

$$(3.43) \qquad \begin{array}{l} \eta^{1/2}\sigma_\eta \text{ (resp. } \eta^{1/2}v_\eta) \quad \text{remains in a bounded set of} \quad L^2(0,T;\mathscr{V}) \\ \text{(resp. of } L^2(0,T;V)). \end{array}$$

Moreover, setting $t=0$ in (3.38), we conclude, taking account of (3.39), that

$$(3.44) \qquad \begin{array}{l} \mathscr{A}(\sigma'(0),\tau)=(g(0)-\mathscr{J}_\mu'(\sigma^0),\tau) \\ (v'(0),w)=(h(0),w) \end{array}$$

so that, when $\eta\to 0$:

$$(3.45) \qquad \sigma'(0) \quad \text{(resp. } v'(0)) \text{ remain in a bounded set of } \mathscr{H} \text{ (resp. } H).$$ □

A priori estimates (II). We differentiate (3.38) with respect to t[12]; it follows that:

$$(3.46) \qquad \begin{array}{l} \mathscr{A}(\sigma'',\tau)+((\mathscr{J}_\mu'(\sigma+\sigma^0))',\tau)+\eta[\sigma',\tau]+\int_\Omega v_i'\tau_{ij,j}\mathrm{d}x=(g',\tau)\,, \\ (v'',w)+\eta((v',w))+\int_\Omega \sigma_{ij}'w_{i,j}\mathrm{d}x=(h',w)\,. \end{array}$$

Taking $\tau=\sigma'$ and $w=v'$ in (3.46), we obtain by adding:

$$(3.47) \qquad \mathscr{A}(\sigma'',\sigma')+\eta[\sigma',\sigma']+(v'',v')+\eta((v',v'))+((\mathscr{J}_\mu'(\sigma+\sigma^0))',\sigma')=(g',\sigma')+(h',v')\,.$$

---

[10] This follows from the general theory of monotone parabolic problems (cf. Lions [1]). Besides, we can also prove this, starting from the a priori estimates which follow.

[11] Note that *in this first estimate the hypothesis* (3.23) *is not involved.*

[12] This is legitimate if we approximate (for example) (3.38) by the Galerkin method.

But

$$((\mathscr{J}_\mu'(\sigma+\sigma^0))',\sigma')$$

(3.48)
$$= \lim_{\Delta t\to 0} \frac{1}{(\Delta t)^2}(\mathscr{J}_\mu'(\sigma(t+\Delta t)+\sigma^0)-\mathscr{J}_\mu'(\sigma(t)+\sigma^0),\sigma(t+\Delta t)-\sigma(t))$$

$$\geqslant 0 \quad \text{(due to the monotonicity of } \sigma\to\mathscr{J}_\mu'(\sigma+\sigma^0)\text{)}.$$ [13]

Consequently, we conclude from (3.47) that

(3.49)
$$\mathscr{A}(\sigma'',\sigma')+\eta[\sigma',\sigma']+(v'',v')+\eta((v',v'))\leqslant(g',\sigma')+(h',v')$$

from which it follows, taking account of (3.45), that, when $\eta\to 0$,

(3.50) $\sigma'=\sigma_\eta'$ remains in a bounded set of $L^\infty(0,T;\mathscr{H})$,

(3.51) $v'=v_\eta'$ remains in a bounded set of $L^\infty(0,T;H)$,

(3.52) $\eta^{1/2}\sigma_\eta'$ (resp. $\eta^{1/2}v_\eta'$) remains in a bounded set of $L^2(0,T;\mathscr{V})$ (resp. of $L^2(0,T;V)$). □

Passage to the limit in η. From the preceding estimates, it follows that

(3.53) $\mathscr{J}_\mu'(\sigma_\eta+\sigma^0)$ remains in a bounded set of $L^\infty(0,T;\mathscr{H})$.

Then we can select a sequence, again denoted by σ_η, v_η, such that we have, when $\eta\to 0$:

(3.54) $\sigma_\eta\sigma_\eta'\to\sigma,\sigma'$ weakly star in $L^\infty(0,T;\mathscr{H})$,

(3.55) $v_\eta, v_\eta'\to v,v'$ weakly star in $L^\infty(0,T;H)$,

(3.56) $\mathscr{J}_\mu'(\sigma_\eta+\sigma^0)\to\chi$ weakly star in $L^\infty(0,T;\mathscr{H})$.

and due to (3.43) we can then pass to the limit in (3.38).

Thus, we have produced σ, v satisfying

(3.57)
$$\mathscr{A}(\sigma',\tau)+(\chi,\tau)+\int_\Omega v_i\tau_{ij,j}\,dx=(g,\tau),$$
$$(v',w)+\int_\Omega \sigma_{ij}w_{i,j}\,dx=(h,w).$$

But, by a type of "monotonicity" argument (which we already employed several times—cf. also Chapter 2 of Lions [1]), we show that

(3.58)
$$\chi=\mathscr{J}_\mu'(\sigma+\sigma^0).$$

Thus, we proved the existence of σ, v that satisfy (3.18), (3.13) and (3.24), (3.25).

---

[13] *Here we made use of the fact that σ^0 does not depend on t.*

But from (3.18) it follows that (3.11) holds, in the sense of distributions in the cylinder $\Omega\times]0,T[$, from which (3.26), (3.27) follow, since

$$\varepsilon_{ij}(v)=A_{ijkh}\sigma'_{kh}+(\mathscr{J}'_\mu(\sigma+\sigma^0))_{ij}-g_{ij},$$
$$\sigma_{ij,j}=v'_i-f_i.$$

(From this, (3.12) follows immediately, as we already saw). ☐

3.5. Existence Proof in the Quasi-Static Case

We will obtain the solution of quasi-static problems as *limit* of the solutions of the following problems: for $\xi\to 0$ (intended to tend toward 0), we denote by $\{\sigma_\xi, v_\xi\}=\{\sigma,v\}$ *the solution of*

(3.59) $$\begin{aligned}&\mathscr{A}(\sigma',\tau)+(\mathscr{J}'_\mu(\sigma+\sigma^0),\tau)+\textstyle\int_\Omega v_i\tau_{ij,j}\,dx=(g,\tau), \qquad \forall\tau\in\mathscr{V},\\ &\xi(v',w)+\textstyle\int_\Omega \sigma_{ij}w_{i,j}\,dx=(h,w) \qquad \forall w\in V,\end{aligned}$$

with

(3.60) $$\sigma(0)=0, \qquad v(0)=0.$$

According to Theorem 3.1, we know the existence and uniqueness of a pair σ_ξ, v_ξ solving this problem.

Let us now examine *how the a priori estimates of* 3.4 *depend on* ξ.

The analogue to (3.41), (3.42) is:

(3.61) σ_ξ remains in a bounded set of $L^\infty(0,T;\mathscr{H})$ as $\xi\to 0$,

(3.62) $\xi^{1/2}v_\xi$ remains in a bounded set of $L^\infty(0,T;H)$.

Setting $t=0$ in (3.59), we can conclude that

(3.63) $$\mathscr{A}(\sigma'(0),\tau)=(g(0),\tau)-(\mathscr{J}^i_\mu(\sigma^0),\tau),$$

(3.64) $$\xi v'(0)=h(0)$$

and therefore, since we assume in Theorems 3.2 and 3.3 that $h(0)=0$:

(3.65) $$v'_\xi(0)=v'(0)=0.$$

The analogue to (3.50), (3.51) now is:

(3.66) σ'_ξ remains in a bounded set of $L^\infty(0,T;\mathscr{H})$,

(3.67) $\xi^{1/2}v'_\xi$ remains in a bounded set of $L^\infty(0,T;H)$. ☐

Furthermore, *it follows from the first equation of* (3.59) *that*

(3.68) $\varepsilon_{ij}(v_\xi) = A_{ijkh}\sigma'_{kh} + \mathscr{J}'_\mu(\sigma + \sigma^0) - g$ (where $\sigma = \sigma_\xi$)

which, with (3.61) and (3.66) implies that

(3.69) $\varepsilon_{ij}(v_\xi)$ remains in a bounded set of $L^\infty(0, T; L^2(\Omega))$.

The use made of (3.69) differs according as Γ_U has positive measure or $\Gamma_U = \emptyset$.
If Γ_U has positive measure, it follows from (3.69) and from Theorem 3.3, Chapter III (a consequence of Korn's inequality), that

(3.70) v_ξ remains in a bounded set of $L^\infty(0, T; V)$.

If $\Gamma_U = \emptyset$, we have to introduce (as we did in Chap. III) the quotient space

(3.71) $\dot{V} = V/\mathscr{R}$

and then

(3.72) $\dot{v}_\xi \in$ bounded set of $L^\infty(0, T; \dot{V})$. □

As a consequence, we can select a sequence, again denoted by σ_ξ, v_ξ such that, when $\xi \to 0$:

(3.73) $\sigma_\xi, \sigma'_\xi \to \sigma, \sigma'$ weakly star in $L^\infty(0, T; \mathscr{H})$,

(3.74)
(i) $v_\xi \to V$ weakly star in $L^\infty(0, T; V)$ when $\Gamma_U \neq \emptyset$,
(ii) $\dot{v}_\xi \to V$ weakly star in $L^\infty(0, T; V)$ when $\Gamma_U = \emptyset$,

and, by a monotonicity argument, we can pass to the limit in the term $\mathscr{J}'_\mu(\sigma_\xi + \sigma^0)$. Thus it follows that σ and v satisfy the conditions of Theorem 3.2 and 3.3[14], provided we check that σ satisfies (3.26). But this is a consequence of the second equation of (3.19) which can be written

$$-\sigma_{ij,j} = h_i. \quad □$$

Remark 3.3 At the same time, we have obtained the fact that for $\xi \to 0$ the quasi-static problem is the limit of the solution of problem (3.59), (3.60). □

[14] In the case $\Gamma_U = \emptyset$, we can pass to the limit in the second equation of (3.59) due to the fact that, by assumption, $(h, \rho) = 0$.

4. Discussion of Elastic Perfectly Plastic Problems

4.1. Statement of the Problems

In the preceding results, we now let the viscosity coefficient μ tend to zero. On $\mathscr{H}$, we introduce the functional

$$(4.1)\qquad \mathscr{J}(\tau)=\tfrac{1}{4}\int_\Omega (\tau_{ij}-(P_{\tilde{K}}\tau)_{ij})(\tau_{ij}-(P_{\tilde{K}}\tau)_{ij})\,\mathrm{d}x\,.$$

Then the first equation of (3.18) can be written in the form of an *inequality*

$$(4.2)\qquad \begin{aligned}\mathscr{A}(\sigma',\tau-\sigma)+\mu^{-1}\mathscr{J}(\tau+\sigma^0)-\mu^{-1}\mathscr{J}(\sigma+\sigma^0)+\int_\Omega v_i(\tau_{ij,j}-\sigma_{ij,j})\,\mathrm{d}x&\\ \geqslant(g,\tau-\sigma)\qquad \forall\tau\in\mathscr{V}&,\end{aligned}$$

everything else remaining unchanged.

When $\mu\to 0$, the term $\mu^{-1}\mathscr{J}(\sigma+\sigma^0)$ represents a *penalizing term* attached to the convex set K of $\mathscr{H}$ defined by

$$(4.3)\qquad K=\{\tau|\quad \tau\in\mathscr{H},\quad \tau(x)\in K \text{ a.e. in } \Omega\}\,.$$

We proceed purely formally; this will be justified in the following sections. We will show that the solution σ_μ, v_μ *converges* when $\mu\to 0$, the limit satisfying

$$(4.4)\qquad \sigma(t)+\sigma^0\in K\,,$$

$$(4.5)\qquad \mathscr{A}(\sigma',\tau-\sigma)+\int_\Omega v_i(\tau_{ij,j}-\sigma_{ij,j})\,\mathrm{d}x\geqslant(g,\tau-\sigma)\qquad \forall\tau\in(K-\sigma^0)\cap\mathscr{V}\,,$$

$$(4.6)\qquad (v',w)+\int_\Omega \sigma_{ij}w_{i,j}\,\mathrm{d}x=(h,w)\qquad \forall w\in V\,,$$

$$(4.7)\qquad \sigma(0)=0\,,\qquad v(0)=0\,.$$

We now take (4.4)...(4.7) as *definition of the dynamic elastic-perfectly plastic problem.* □

Remark 4.1. Naturally, we arrive at the *same statement* using the considerations of Section 2.1.2. □

Remark 4.2. We can *eliminate* v in the preceding system[15]. Indeed, (4.6) is equivalent to

$$(4.8)\qquad v_i'-\sigma_{ij,j}=h_i\,.$$

If we introduce

$$(4.9)\qquad S(t)=\int_0^t \sigma(t_1)\,\mathrm{d}t_1\,,$$

---

[15] We could also have applied an elimination calculation in Section 3, but there it would not have been particularly useful.

then (4.8) yields

(4.10) $$v_i(t)=S_{ij,j}+\int_0^t h_i(t_1)\,dt_1$$

and (4.5) becomes, using the notation of (3.36),

(4.11) $$\mathscr{A}(S'',\tau-S')+[S+\int_0^t h(t_1)\,dt_1,\tau-S']\geqslant(g,\tau-S').$$

If we set

(4.12) $$G(t)=g(t)-\int_0^t h(t_1)\,dt_1$$

the problem becomes:
we seek an S such that

(4.13) $$S'(t)\in(K-\sigma^0)\cap\mathscr{V},$$

(4.14) $$\mathscr{A}(S'',\tau-S')+[S,\tau-S']\geqslant(G,\tau-S')\qquad\forall\tau\in(K-\sigma^0)\cap\mathscr{V},$$

(4.15) $$S(0)=0,\qquad S'(0)=0.$$

Once S is determined by (4.13), (4.14), (4.15), (see Theorem 4.1 below), v_i is defined by (4.10) and $\sigma=S'$. □

The quasi-static problem is obtained by omitting the term with v' in (4.6), Then the problem is to find σ and v such that (4.4) and (4.5) are satisfied as well as

(4.16) $$\int_\Omega \sigma_{ij}w_{i,j}\,dx=(h,w)\qquad\forall w\in V,$$

and

(4.17) $$\sigma(0)=0.$$

Again, we can eliminate v. In fact, (4.16) is equivalent to

(4.18) $$-\sigma_{ij,j}=h_i.$$

We then introduce

(4.19) $$L(t)=\{\tau\mid\quad \tau\in\mathscr{H},\ -\tau_{ij,j}=h_i,\ \tau_{ij}n_j=0\ \text{ on }\Gamma_F\}.$$

If we take $\tau\in L(t)\cap(K-\sigma^0)$ in (4.5), the term $\int_\Omega v_i(\tau_{ij,j}-\sigma_{ij,j})\,dx$ vanishes and the problem becomes: to determine a solution σ of

(4.20) $$\sigma(t)\in(K-\sigma^0)\cap L(t),$$

(4.21) $$\mathscr{A}(\sigma',\tau-\sigma)\geqslant(g,\tau-\sigma)\qquad\forall\tau\in(K-\sigma^0)\cap L(t),$$

(4.22) $$\sigma(0)=0.$$

Once σ is determined, obtaining v runs into difficulties, to which we shall return in Remark 4.4.

4.2. Formulation of the Results

Theorem 4.1. *We make the same assumptions as in Theorem* 3.1 *with the addition of*

(4.23) $$\sigma^0 \in K.$$

Then there exists one and only one function S such that

(4.24) $$S, S', S'' \in L^\infty(0, T; \mathscr{H}),$$

(4.25) $$S, S' \in L^\infty(0, T; \mathscr{V}),$$

which satisfies (4.13), (4.14), (4.15). *We have: $\sigma = S'$ and v is given by* (4.10).

This solves the *dynamic* elastic-perfectly plastic problem.

Furthermore, the elastic-perfectly plastic problem *is the limit of the cases with viscosity when $\mu \to 0$.* If we denote the solution obtained in Theorem 3.1 for $\mu > 0$ by σ_μ, v_μ, we have:

Theorem 4.2. *With the assumptions of Theorem* 4.1, *let σ, v be the solution defined by Theorem* 4.1. *Then, as $\mu \to 0$:*

(4.26) $$\sigma_\mu, \sigma'_\mu \to \sigma, \sigma' \quad \text{weakly star in} \quad L^\infty(0, T; \mathscr{H}),$$

(4.27) $$v_\mu, v'_\mu \to v, v' \quad \text{weakly star in} \quad L^\infty(0, T; H),$$

(4.28) $$\sigma_{\mu ij,j} \to \sigma_{ij,j} \quad \text{weakly star in} \quad L^\infty(0, T; L^2(\Omega)). \quad \square$$

Concerning the quasi-static case, we have

Theorem 4.3. *We make the same assumptions as in Theorem* 4.1 *and the additional one:*

(4.29) $$\sigma^0 = 0$$

and, as in Theorems 3.2, 3.3, $h(0) = 0$.

Then there exists one and only one function σ such that

(4.30) $$\begin{aligned} &\sigma, \sigma' \in L^\infty(0, T; \mathscr{H}), \\ &\sigma \in L^\infty(0, T; \mathscr{V}), \end{aligned}$$

which satisfies (4.20) (a. e. *in t*) *and* (4.21), (4.22).

Furthermore, we have

Theorem 4.4. *We make the same assumptions as in Theorem* 4.3. *Then, if* σ_μ, v_μ *denote the solution obtained in Theorem* 3.2 (resp. σ_μ, v'_μ *obtained in Theorem* 3.3), *we have*

$$(4.31)\qquad \begin{aligned} &\sigma_\mu, \sigma'_\mu \to \sigma, \sigma' \quad \textit{weakly star in} \quad L^\infty(0,T;\mathscr{H}) \\ &\sigma_{\mu ij,j} \to \sigma_{ij,j} \quad \textit{weakly star in} \quad L^\infty(0,T;L^2(\Omega)). \end{aligned}$$

Remark 4.3. We did not obtain information on the $\varepsilon_{ij}(v_\mu)$. ▯

4.3. Proof of the Uniqueness Results

Let S and S_* be two possible solutions of (4.13), (4.14), (4.15). We take $\tau = S'_*$ (resp. $\tau = S'$) in inequality (4.14) (resp. the analogous inequality with respect to S_*) and add; if $\hat{S} = S - S_*$, we obtain

$$\mathscr{A}(\hat{S}'', \hat{S}') + [\hat{S}, \hat{S}'] \leqslant 0$$

from which it follows that

$$\hat{S} = 0. \quad ▯$$

In the same way, if σ, σ_* are possible solutions of (4.20), (4.21), (4.22), we obtain, if $\hat{\sigma} = \sigma - \sigma_*$:

$$\mathscr{A}(\hat{\sigma}', \hat{\sigma}) \leqslant 0$$

whence: $\hat{\sigma} = 0$. ▯

4.4. Proof of Theorems 4.1 and 4.2

We consider the solution σ_μ, v_μ, given by Theorem 3.1. Then

$$(4.32)\qquad \begin{aligned} &\mathscr{A}(\sigma'_\mu, \tau) + (\mathscr{J}'_\mu(\sigma_\mu + \sigma^0), \tau) + \textstyle\int_\Omega v_{\mu i} \tau_{ij,j} \mathrm{d}x = (g, \tau), \qquad \tau \in \mathscr{V}, \\ &(v'_\mu, w) + \textstyle\int_\Omega \sigma_{\mu ij} w_{i,j} \mathrm{d}x = (h, w) \qquad \forall w \in V \end{aligned}$$

and we can verify *the dependence on* μ in the a priori estimates of Section 3.

Replacing τ by σ_μ and w by v_μ in (4.32), we obtain:

(4.33) $\qquad \sigma_\mu$ remains (when $\mu \to 0$) in a bounded set of $L^\infty(0,T;\mathscr{H})$

(4.34) $\qquad v_\mu$ remains in a bounded set of $L^\infty(0,T;H)$.

Setting $t=0$ in (4.32), it follows, since $\mathscr{J}'_\mu(\sigma^0)=0$, that

$$\begin{aligned}&\mathscr{A}(\sigma'_\mu(0),\tau)=(g(0),\tau),\\&v'_\mu(0)=h(0).\end{aligned}\tag{4.35}$$

Differentiating (4.32) with respect to t and replacing τ by σ'_μ and w by v'_μ in the result, we obtain, using (4.35):

(4.36) σ'_μ (resp. v'_μ) remains in a bounded set of $L^\infty(0,T;\mathscr{H})$ (resp. of $L^\infty(0,T;H)$).

Furthermore, if we take $\tau=\sigma_\mu+\sigma^0$, $w=v_\mu$ in (4.32) and add, it follows that

$$\int_0^T(\mathscr{J}'_\mu(\sigma_\mu+\sigma^0),\sigma_\mu+\sigma^0)\,dt\leqslant\text{constant}=C$$

or again, in the notation of (4.1)

$$\int_0^T(\mathscr{J}(\sigma_\mu+\sigma^0)\,dt\leqslant C\mu.\tag{4.37}$$

Then we can extract a sequence, again denoted by σ_μ, v_μ, such that (4.26), (4.27) are satisfied. The second equation of (4.32) then shows that

$$\sigma_{\mu ij,j}=v'_{\mu i}-h_i$$

from which (4.28) follows.

We deduce from (4.37) that

$$\mathscr{J}(\sigma+\sigma^0)=0$$

whence (4.4).

As we saw in (4.2), we can write the first equation of (4.32) in the form

$$\begin{aligned}\mathscr{A}(\sigma'_\mu,\tau-\sigma_\mu)+\int_\Omega v_{\mu i}(\tau_{ij,j}-\sigma_{\mu ij,j})\,dx+\mu^{-1}\mathscr{J}(\tau+\sigma^0)&\\-\mu^{-1}\mathscr{J}(\sigma_\mu+\sigma^0)\geqslant(g,\tau-\sigma_\mu).&\end{aligned}\tag{4.38}$$

If, in (4.38), we take $\tau\in K-\sigma^0$, we have: $\mathscr{J}(\tau+\sigma^0)=0$ and, since $\mathscr{J}(\sigma_\mu+\sigma^0)\geqslant0$, it follows that:

$$\mathscr{A}(\sigma'_\mu,\tau-\sigma_\mu)+\int_\Omega v_{\mu i}(\tau_{ij,j}-\sigma_{\mu ij,j})\,dx\geqslant(g,\tau-\sigma_\mu),\qquad\tau\in K-\sigma^0.\tag{4.39}$$

By analogy with (4.9), we introduce

$$S_\mu(t)=\int_0^t\sigma_\mu(t_1)\,dt,\tag{4.40}$$

which then satisfies—eliminating v_μ as above—

$$\mathscr{A}(S''_\mu,\tau-S'_\mu)+[S_\mu,\tau-S'_\mu]\geqslant(G,\tau-S'_\mu)\qquad\forall\tau\in(K-\sigma^0)\cap\mathscr{V}.\tag{4.41}$$

It remains to show that the S, given by (4.9), satisfies (4.14); thus, we have to pass to the limit in (4.41) which we do by the usual procedure; we take $\tau=\tau(t)$ in (4.41) where

$$\tau, \tau' \in L^2(0,T;\mathscr{V}), \qquad \tau(t) \in K-\sigma^0 \quad \text{a. e.} \tag{4.42}$$

and integrate with respect to t; it follows that

$$\begin{aligned}\int_0^T \mathscr{A}(S''_\mu,\tau)\,dt + \int_0^T [S_\mu,\tau]\,dt - \int_0^T (G,\tau-S'_\mu)\,dt \\ \geqslant \int_0^T (\mathscr{A}(S''_\mu,S'_\mu)+[S_\mu,S'_\mu])\,dt \\ = \tfrac{1}{2}\mathscr{A}(S'_\mu(T),S'_\mu(T)) + \tfrac{1}{2}[S_\mu(T),S_\mu(T)]\end{aligned}$$

whence

$$\begin{aligned}\int_0^T \mathscr{A}(S'',\tau)\,dt + \int_0^T \{[S,\tau]-(G,\tau-S')\}\,dt \\ \geqslant \liminf_{\mu\to 0}\{\tfrac{1}{2}\mathscr{A}(S'_\mu(T),S'_\mu(T)) + \tfrac{1}{2}[S_\mu(T),S_\mu(T)]\} \\ \geqslant \tfrac{1}{2}\mathscr{A}(S'(T),S'(T)) + \tfrac{1}{2}[S(T),S(T)] \\ = \int_0^T \{\mathscr{A}(S'',S')+[S,S']\}\,dt\end{aligned}$$

and consequently

$$\int_0^T \{\mathscr{A}(S'',\tau-S')+[S,\tau-S']-(G,\tau-S')\}\,dt \geqslant 0 \tag{4.43}$$

for all τ satisfying (4.42).

From this, we pass to (4.14) by the usual argument (we take $\tau(t)=S'(t)$ in (4.43) except for a neighborhood $\mathcal{O}$ of a point t, where we take $\tau(t)=\tau$ fixed in $K-\sigma^0$; dividing the result by measure $(\mathcal{O}_j)$ and letting measure $(\mathcal{O}_j)$ tend toward 0, we deduce (4.14)). □

4.5. Proof of Theorems 4.3 and 4.4

By $\{\sigma, v\} = \{\sigma_{\mu\xi}, v_{\mu\xi}\}$, we denote the solution of

$$\begin{aligned}&\mathscr{A}(\sigma',\tau) + (\mathscr{J}'_\mu(\sigma),\tau) + \int_\Omega v_i \tau_{ij,j}\,dx = (g,\tau), \\ &\xi(v',w) + \int_\Omega \sigma_{ij} w_{i,j}\,dx = (h,w) \qquad (\xi>0),\end{aligned} \tag{4.44}$$

$$\sigma(0)=0, \qquad v(0)=0. \tag{4.45}$$

If we set $\tau=\sigma$, $w=v$ in (4.44), it follows that

(4.46) $\sigma_{\mu\xi} \in$ bounded set of $L^\infty(0,T;\mathscr{H})$, when $\mu,\xi\to 0$,

(4.47) $\xi^{1/2} v_{\mu\xi} \in$ bounded set of $L^\infty(0,T;H)$,

and also that

$$\int_0^T (\mathscr{J}'_\mu(\sigma_{\mu\xi}),\sigma_{\mu\xi})\,dt \leqslant \text{constant} = C$$

and therefore

$$\int_0^T \mathscr{J}(\sigma_{\mu\xi})\,dt \leqslant C\mu\,. \tag{4.48}$$

Furthermore, setting $t=0$ in (4.44):

$$\begin{aligned}&\sigma_{\mu\xi}(0) \quad \text{remains in a bounded set of } \mathscr{H},\\ &v_{\mu\xi}(0)=0\,.\end{aligned} \tag{4.49}$$

From this, it follows, after differentiating (4.44) with respect to t, that:

$$\begin{aligned}&\sigma'_{\mu\xi} \text{ (resp. } \xi^{1/2} v'_{\mu\xi}) \quad \text{remains in a bounded set of} \quad L^\infty(0,T;\mathscr{H})\\ &\text{(resp. of } L^\infty(0,T;H)).\end{aligned} \tag{4.50}$$

We can pass to the limit in ξ as we already saw in Theorems 3.2 and 3.3; thus we obtain the solution $\{\sigma_\mu, v_\mu\}$ or, $\{\sigma_\mu, \dot{v}_\mu\}$ given in Theorems 3.2, 3.3 with the estimates

$$\sigma_\mu, \sigma'_\mu \quad \text{remains in a bounded set of} \quad L^\infty(0,T;\mathscr{H}), \tag{4.51}$$

$$\int_0^T \mathscr{J}(\sigma_\mu)\,dt \leqslant C\mu\,, \tag{4.52}$$

$$\sigma_{\mu ij,j} \in \quad \text{bounded set of} \quad L^\infty(0,T;L^2(\Omega))\,. \tag{4.53}$$

But the second equation of (3.19) yields

$$-\sigma_{\mu ij,j} = h_i\,, \tag{4.54}$$

and the first equation of (3.19) implies ($\sigma^0=0$)

$$\mathscr{A}(\sigma'_\mu, \tau-\sigma_\mu) + \mu^{-1}\mathscr{J}(\tau) - \mu^{-1}\mathscr{J}(\sigma_\mu) + \int_\Omega v_i(\tau_{ij,j} - \sigma_{\mu ij,j})\,dx \geqslant (g, \tau-\sigma_\mu)\,. \tag{4.55}$$

Then we take τ in $L(t)$ (cf. (4.19)), from which

$$\mathscr{A}(\sigma'_\mu, \tau-\sigma_\mu) + \mu^{-1}\mathscr{J}(\tau) - \mu^{-1}\mathscr{J}(\sigma_\mu) \geqslant (g, \tau-\sigma_\mu)\,. \tag{4.56}$$

If *in addition* $\tau\in K$, then $\mathscr{J}(\tau)=0$ and, since $\mathscr{J}(\sigma_\mu)\geqslant 0$, it follows that

$$\mathscr{A}(\sigma'_\mu, \tau-\sigma_\mu) \geqslant (g, \tau-\sigma_\mu)\,, \qquad \tau\in K\cap L(t)\,. \tag{4.57}$$

From (4.51), (4.52), (4.53), it follows that we can select a subsequence, again denoted by σ_μ, such that (4.31) is satisfied. Due to (4.52), we can show that $\sigma\in K$, therefore $\sigma\in K\cap L(t)$ and from (4.57) it follows by the usual arguments that (4.21) is satisfied. ☐

Remark 4.4. *Determination of the velocity field.* Since the field of stresses, the solution $\sigma(x,t)$, is known, the field of displacement velocities $\partial u(x,t)/\partial t$ satisfies

$$\begin{aligned}&\varepsilon_{ij}(\partial u/\partial t)=A_{ijkh}\dot{\sigma}_{kh}+\lambda_{ij}\\&\lambda_{ij}(\varphi_{ij}-\sigma_{ij})\leqslant 0\qquad\forall\varphi\in K.\end{aligned}$$

The problem to determine $\partial u(x,t)/\partial t$ from these relations, *in the general case*, is unsolved as far as we know.

We encounter an analogous situation in the problem of elasto-plastic torsion of a cylindrical tree. There, the problem can be solved almost explicitly. Cf. Section 6.6, Remark 6.3. □

Remark 4.5. In the quasi-static cases with the constituent law of Prandtl-Reuss, one can obtain a variational formulation directly involving $\partial\{\sigma_{ij}\}/\partial t$ (cf. W. I. Koiter [1], H. D. Bui and K. Dangvan [1], C. T. Herakovitch and P. G. Hodge [1]). This property can be utilized in numerical analysis, but it seems difficult to derive global existence on $[0,T]$ from it. □

5. Discussion of Rigid-Visco-Plastic and Rigid Perfectly Plastic Problems

5.1. Rigid Visco-Plastic Problems

We shall let "the coefficients A_{ijkh} tend toward zero" in the following sense[16]:

(5.1) we replace A_{ijkh} by θA_{ijkh}, $\theta>0$, $\theta\to 0$.

Then we consider the solution $\{\sigma_\theta, v_\theta\}=\{\sigma,v\}$ of

$$\begin{aligned}&\theta\mathscr{A}(\sigma',\tau)+(\mathscr{J}'_\mu(\sigma+\sigma^0),\tau)+\int_\Omega v_i\tau_{ij,j}\mathrm{d}x=(g,\tau)\\&(v',w)+\int_\Omega\sigma_{ij}w_{i,j}\mathrm{d}x=(h,w),\end{aligned}\tag{5.2}$$

with

$$\sigma(0)=0,\qquad v(0)=0.\tag{5.3}$$

If we *formally* pass to the limit in θ, so that, in a suitable topology, $\sigma_\theta\to\sigma$, $v_\theta\to v$ (this will be stated more precisely later on), we obtain:

$$\begin{aligned}&(\mathscr{J}'_\mu(\sigma+\sigma^0),\tau)+\int_\Omega v_i\tau_{ij,j}\mathrm{d}x=(g,\tau),\\&(v',w)+\int_\Omega\sigma_{ij}w_{i,j}\mathrm{d}x=(h,w),\end{aligned}\tag{5.4}$$

[16] One might consider a much more general situation, even with coefficients A_{ijkh} *dependent on* x, but tending strongly toward 0 in $L^\infty(\Omega)$. To simplify the discussion, we limit ourselves to (5.1).

with

$$v(0)=0. \tag{5.5}$$

But then we eliminate σ.[17] The first equation of (5.4) is equivalent to

$$\mathscr{J}_\mu(\sigma+\sigma^0)=g+\varepsilon(v) \quad (\text{or } g+D(v)), \tag{5.6}$$

or again

$$g+\varepsilon(v)\in\partial\mathscr{J}_\mu(\sigma+\sigma^0). \tag{5.6a}$$

We now introduce the dual functional $\mathscr{J}_\mu^*$ of $\mathscr{J}_\mu$, defined on $\mathscr{H}$ by

$$\mathscr{J}_\mu^*(\sigma)=\sup_\tau\,[(\sigma,\tau)-\mathscr{J}_\mu(\tau)]. \tag{5.7}$$

Here, we can be more precise since $\mathscr{J}_\mu(\sigma)=\mu^{-1}\mathscr{J}(\sigma)$. Then we easily verify that

$$\begin{aligned}&\mathscr{J}_\mu^*(\sigma)=\mu^{-1}\mathscr{J}^*(\mu\sigma),\\&\mathscr{J}^*=\text{functional dual to } \mathscr{J}.\end{aligned} \tag{5.8}$$

Then (5.6a) *is equivalent* (cf. Moreau [1]) to

$$\sigma+\sigma^0\in\partial\mathscr{J}_\mu^*(\varepsilon(v)+g) \tag{5.9}$$

that is

$$\mathscr{J}_\mu^*(\varepsilon(w)+g)-\mathscr{J}_\mu^*(\varepsilon(v)+g)-(\sigma+\sigma^0,\varepsilon(w)-\varepsilon(v))\geqslant 0. \tag{5.10}$$

The second equation of (5.4) can be written

$$(v',w-v)+(\sigma,\varepsilon(w)-\varepsilon(v))-(h,w-v)=0$$

and, using (5.10), it follows that

$$(v',w-v)+\mathscr{J}_\mu^*(\varepsilon(w)+g)-\mathscr{J}_\mu^*(\varepsilon(v)+g)\geqslant(-\sigma_{ij,j}^0+h_i,w_i-v_i),\quad v,w\in V. \tag{5.11}$$ □

We now prove the following result, always by methods of the same type:

Theorem 5.1. *We make the same assumptions as in Theorem* 3.1 *with* (4.23) *and*

$$g(0)=0. \tag{5.12}$$

---

[17] There is a *duality* between the rigid-visco-plastic case (where one eliminates σ) and the elastic perfectly plastic case (where one eliminates v). We will state this precisely in the following.

Let σ_θ, v_θ be the solution of problem (5.2), (5.3). *We have:*

$$(5.13)\qquad \begin{aligned} &v_\theta, v'_\theta \to v, v' \quad \text{weakly star in} \quad L^\infty(0,T;H),\\ &v_\theta \to v \quad \text{weakly star in} \quad L^\infty(0,T;V), \end{aligned}$$

where v is the solution of (5.11) *with* $v(0)=0$.

As always, the proof is based on the a priori estimates. From (5.2), it follows that

$$(5.14)\qquad v_\theta \quad \text{remains in a bounded set of} \quad L^\infty(0,T;H),$$

$$(5.15)\qquad \theta^{1/2}\sigma_\theta \quad \text{remains in a bounded set of} \quad L^\infty(0,T;\mathscr{H}).$$

Setting $t=0$ in (5.2), it follows, due to (5.12), that

$$(5.16)\qquad \sigma'_\theta(0)=0,$$

and furthermore

$$(5.17)\qquad v'_\theta(0)=h(0).$$

If we now differentiate equations (5.2) with respect to t, we obtain:

$$(5.18)\qquad \begin{aligned} &v'_\theta \text{ (resp. } \theta^{1/2}\sigma'_\theta) \quad \text{remains in a bounded set of} \quad L^\infty(0,T;H)\\ &\text{(resp. of } L^\infty(0,T;\mathscr{H})). \end{aligned}$$

Then the second equation of (5.2) gives

$$(5.19)\qquad \sigma_{\theta ij,j} \in \quad \text{bounded set of} \quad L^\infty(0,T;L^2(\Omega))$$

and furthermore

$$(5.20)\qquad \int_0^T (\mathscr{I}'_\mu(\sigma_\theta+\sigma^0), \sigma_\theta+\sigma^0)\,dt \leqslant C$$

so that the first equation of (5.2) yields:

$$(5.21)\qquad v_{\theta ij,j} \in \quad \text{bounded set of} \quad L^\infty(0,T;L^2(\Omega)).$$

From this, the result follows. ☐

Remark 5.1. Once v is known, the determination of σ_{ij} runs into difficulties analogous to those encountered in Remark 4.4 in a "dual" case. ☐

Remark 5.2. The associated "quasi-static" problem is actually a simple elliptic problem, obtained by omitting the term with v' in (5.11). ☐

Remark 5.3. The problem discussed here will be taken up again in Chapter VI; it will be stated more precisely in a nonlinear context. ☐

5.2. Rigid Perfectly Plastic Problems

Orientation. The rigid perfectly plastic case corresponds to

$$A_{ijkh}=0, \qquad \mu=0.$$

One can be led to this problem *by passage to the limit* in A_{ijkh} and in μ which can be done in two different ways according as one first passes to the limit in A (resp. μ) and then in μ (resp. A).[18] □

Let us first consider the problem obtained by letting the viscosity μ tend toward 0, starting out from the "rigid-visco-plastic problem".

We denote by v_μ the solution of Problem (5.11) with $v_\mu(0)=0$. Here we will obtain only a *formal* result.

First we observe, starting from the a priori estimates (5.14) and (5.18), that

(5.22) $\quad v_\mu, v'_\mu$ remain in a bounded set of $L^\infty(0,T;H)$.

Furthermore, we can show that

(5.23)
$$\lim_{\mu\to 0} \mu^{-1}\mathcal{J}^*(\mu\sigma)=j^*(\sigma),$$
j a continuous functional on $\mathcal{H}$, which is convex, non differentiable and $\geqslant 0$.

Then, according to (5.22), we can select a sequence, again denoted by v_μ, such that

(5.24) $\quad v_\mu, v'_\mu \to v, v'$ weakly star in $L^\infty(0,T;H)$.

Using (5.8), (5.23) and *formally* passing to the limit in (5.11), we obtain

(5.25)
$$(v', w-v)+j^*(\varepsilon(w)+g)-j^*(\varepsilon(v)+g)\geqslant(h_i-\sigma^0_{ij,j}, w_i-v_i) \qquad \forall w\in V,$$

with

(5.26) $\quad v(0)=0.$ □

If we *first* pass to the limit in μ, we obtain, according to Section 4, a solution S_θ of

(5.27) $\quad \theta\mathscr{A}(S''_\theta, \tau-S'_\theta)+[S_\theta, \tau-S'_\theta]\geqslant(G,\tau-S'_\theta) \quad \forall\tau$ with $\tau\in(K-\sigma^0)\cap\mathscr{V}$,

(5.28) $\quad S'_\theta(t)\in(K-\sigma^0)\cap\mathscr{V},$

(5.29) $\quad S_\theta(0)=0, \qquad S'_\theta(0)=0.$

[18] The problem of equivalence of the two passages to the limit has not been entirely settled.

Now we let $\theta \to 0$. We verify that, if $G(0)=0$ (i. e., $g(0)=0$):

(5.30) $\quad S_\theta, S'_\theta$ remains in a bounded set of $L^\infty(0,T;\mathscr{V})$,

(5.31) $\quad \theta^{1/2} S''_\theta$ remains in a bounded set of $L^\infty(0,T;\mathscr{H})$.

This is sufficient for passage to the limit.

The rigid perfectly plastic problem consists in finding a function S such that

(5.32) $\quad S'(t) \in (K-\sigma^0) \cap \mathscr{V}$,

(5.33) $\quad [S, \tau - S'] \geqslant (G, \tau - S') \quad \forall \tau \in (K-\sigma^0) \cap \mathscr{V}$,

(5.34) $\quad S(0)=0$.

Thus we have

Theorem 5.2. *Under the assumptions of Theorem* 4.1 *and when* $g(0)=0$, *Problem* (5.32), (5.33), (5.34) *has a unique solution which satisfies*

(5.35) $\quad S, S' \in L^\infty(0,T;\mathscr{V})$.

Furthermore, if S_θ *denotes the solution of* (5.27), (5.28), (5.29), *we have*

(5.36) $\quad S_\theta, S'_\theta \to S, S'$ *weakly star in* $L^\infty(0,T;\mathscr{V})$.

Thus we have

(5.37) $\quad \sigma = S'$.

Remark 5.4. The terms "in S" and "in v" are dual. □

Remark 5.5. We can give *a different dual formulation* of the problem "in S", according to an idea by Brézis [2]. We introduce the functional ψ, the indicator function of $K-\sigma^0$; then

(5.38) $\quad \psi(\sigma)=0$ if $\sigma \in K-\sigma_0$, $\qquad \psi(\sigma)=+\infty$ otherwise.

Furthermore, since $\tau \to (G,\tau)$ is continuous on $\mathscr{H}$, and therefore on $\mathscr{V}$, we have:

(5.39) $\quad (G,\tau)=[\tilde{G},\tau]$.

Then (5.33) is equivalent to

(5.40) $\quad [S-\tilde{G}, \tau - S'] + \psi(\tau) - \psi(S') \geqslant 0 \quad \forall \tau$

i. e., to

(5.41) $\quad -(S-\tilde{G}) \in \partial\psi(S')$.

We introduce the dual (or conjugate) function to ψ on $\mathscr{V}$ by

$$\psi^*(\sigma) = \sup_{\tau \in \mathscr{V}} \{[\sigma, \tau] - \psi(\tau)\} \tag{5.42}$$

and set

$$R = -S + \tilde{G}. \tag{5.43}$$

Then (5.41) is equivalent to $R \in \partial\psi(S')$ which is equivalent to

$$S' \in \partial\psi^*(R)$$

i. e.

$$-[S', \tau - R] + \psi^*(\tau) - \psi^*(R) \geqslant 0$$

therefore

$$[R', \tau - R] - [\tilde{G}', \tau - R] + \psi^*(\tau) - \psi^*(R) \geqslant 0$$

and finally:

$$[R', \tau - R] + \psi^*(\tau) - \psi^*(R) \geqslant (G', \tau - R) \tag{5.44}$$

with

$$R(0) = \tilde{G}(0) = 0. \tag{5.45}$$

Problem (5.44), (5.45) has a unique solution such that

$$R, R' \in L^\infty(0, T; \mathscr{V}). \quad \square \tag{5.46}$$

6. Hencky's Law. The Problem of Elasto-Plastic Torsion

6.1. Constituent Law

We consider the following stress-strain relation

$$\begin{aligned} &\varepsilon_{ij}(u) = A_{ijkh}\sigma_{kh} + \lambda_{ij} \\ &\text{with} \quad \mathscr{F}(\sigma) \leqslant 0 \\ &\text{and} \quad \lambda_{ij}(\varphi_{ij} - \sigma_{ij}) \leqslant 0 \qquad \forall \varphi, \qquad \mathscr{F}(\varphi) \leqslant 0. \end{aligned} \tag{6.1}$$

This law was introduced by Hencky [1], essentially with the idea to solve *static or quasi-static* problems without taking into consideration the history of prior deformations. □

In practice, this law can lead to solutions which coincide with those obtained by the Prandtl-Reuss law. In particular, this is the case in the problem of elasto-plastic torsion of a cylindrical tree (cf. H. Brézis [2] and H. Lanchon [1, 2, 4].

6.2. Problems to be Considered

We seek a field of displacements $\{u_i(x)\}$ and of stresses $\{\sigma_{ij}(x)\}$ in Ω such that

$$\sigma_{ij,j} + f_i = 0 \quad \text{in} \quad \Omega, \tag{6.2}$$

$$\sigma_{ij} n_j = F_i \quad \text{on} \quad \Gamma_F, \tag{6.3}$$

$$u_i = U_i \quad \text{on} \quad \Gamma_U, \tag{6.4}$$

The notation and the geometry are the same as in the preceding sections of this chapter and the data f_i, F_i, U_i are functions of x alone.

6.3. Variational Formulation for the Stresses

As in Section 2, we introduce the Hilbert space $\mathscr{H}$, the closed convex set $K \subset \mathscr{H}$ and the set of admissible displacements

$$\mathscr{U}_{\text{ad}} = \{v \mid v = \{v_i\},\ v_i \in H^1(\Omega),\ v_i = U_i \text{ on } \Gamma_U\}, \tag{6.5}$$

where it is assumed implicitly, in order that $\mathscr{U}_{\text{ad}}$ is not empty, that

$$U_i \quad \text{is a restriction to } \Gamma_U \text{ of an element of} \quad H^{1/2}(\Gamma). \tag{6.6}$$

In addition, we assume

$$f_i \in L^2(\Omega), \qquad F_i \in L^2(\Gamma_F) \tag{6.7}$$

and we define

$$M = \{\varphi \mid \varphi \in \mathscr{H},\ \varphi_{ij,j} + f_i = 0 \text{ in } \Omega,\ \varphi_{ij} n_j = F_i \text{ on } \Gamma_F\} \tag{6.8}$$

and set, as in the preceding sections,

$$\mathscr{A}(\varphi, \psi) = \int_\Omega A_{ijkh} \varphi_{kh} \psi_{ij} \, dx, \tag{6.9}$$

which represents a norm equivalent to the classical norm for $\mathscr{H}$.

Then, if $\{u_i, \sigma_{ij}\}$ *is a solution of the problem* (6.1)...(6.4), *the field of stresses* $\{\sigma_{ij}\}$ *minimizes on* $K \cap M$ (assumed not to be empty) *the functional*

$$\tfrac{1}{2}\mathscr{A}(\varphi,\varphi) - \int_{\Gamma_U} U_i \varphi_{ij} n_j \, d\Gamma .$$

Proof. The pointwise constituent law (6.1) is equivalent to

$$
\begin{aligned}
&\varepsilon_{ij}(u) = A_{ijkh}\sigma_{kh} + \lambda_{ij}\\
(6.10) \qquad &\sigma \in K\\
&\int_\Omega \lambda_{ij}(\varphi_{ij} - \sigma_{ij})\,dx \leqslant 0 \qquad \forall \varphi \in K,
\end{aligned}
$$

from which it follows immediately that

$$\int_\Omega \varepsilon_{ij}(u)\,(\varphi_{ij} - \sigma_{ij})\,dx \leqslant \int_\Omega A_{ijkh}\sigma_{kh}(\varphi_{ij} - \sigma_{ij})\,dx \qquad \forall \varphi \in K,$$

and then, by integration by parts,

$$(6.11) \qquad \mathscr{A}(\sigma, \varphi - \sigma) \geqslant \int_{\Gamma_U} U_i(\varphi_{ij} - \sigma_{ij}) n_j \, d\Gamma, \qquad \forall \varphi \in K \cap M,$$

which is equivalent to saying that $\{\sigma_{ij}\}$ minimizes the functional

$$(6.12) \qquad I(\varphi) = \tfrac{1}{2}\mathscr{A}(\varphi,\varphi) - \int_{\Gamma_U} U_i \varphi_{ij} n_j \, d\Gamma$$

on $K \cap M$. □

From this immediately follows:

Theorem 6.1. *Under the assumption* $K \cap M \neq \emptyset$, *there exists a unique* σ *which minimizes* $I(\varphi)$ on $K \cap M$. □

6.4. Determination of the Field of Displacements

There is still the problem:

"*Does there exist a u corresponding to* σ?"

This question will be dealt with by a duality method.

Recapitulation of facts on duality (Moreau [1], Rockafellar [1], R. Témam [1], [2]).

Let X and Y be two Banach spaces, reflexive or not, and let $L \in \mathscr{L}(X, Y)$.

The functions F and G, defined respectively on X and Y, with values in $]-\infty, +\infty]$, are convex lower semi-continuous and proper (that is to say, not identically $+\infty$).

Let F^* and G^* be the convex functions, conjugate to F and G and defined on X^* and Y^*, respectively, by

$$F^*(x^*) = \sup_{x \in X} \{(x, x^*) - F(x)\}$$

where (x, x^*) represents the duality between X and $X^*(F^{**} \equiv F)$.

Furthermore, let L^* be the mapping conjugate to L; we have

$$L^* \in \mathscr{L}(Y^*; X^*).$$

Under the *assumptions*

$$\exists x_0 \in X \quad \text{such that}$$

1) $F(x_0) < \infty$,
2) G is finite and continuous in Lx_0,

we have:
there exists a y_0^* such that

$$\inf_{y^* \in Y^*} [F^*(L^* y^*) + G^*(-y^*)] = F^*(L^* y_0^*) + G^*(-y_0^*) \tag{6.13}$$

and

$$\inf_{y^* \in Y^*} [F^*(L^* y^*) + G^*(-y^*)] + \inf[F(x) + G(Lx)] = 0. \tag{6.14}$$

If moreover, there exists a solution x_0 of $\inf[F(x)+G(Lx)]$, then

$$F(x_0) + F^*(L^* y_0^*) = (L^* y_0^*, x_0) \tag{6.15}$$

and

$$G(Lx_0) + G^*(y_0^*) = (y_0^*, Lx_0). \quad \square \tag{6.16}$$

Application

$$\left| \begin{array}{l} X = Y = (L^2(\Omega))^6 = \mathscr{H} \\ L = \text{identity}. \end{array} \right.$$

Then we define

$$F(e) = -\inf_{\tau \in K} \{\tfrac{1}{2} \mathscr{A}(\tau, \tau) - (e, \tau)\} = \sup_{\tau \in K} \{(e, \tau) - \tfrac{1}{2} \mathscr{A}(\tau, \tau)\}. \tag{6.17}$$

For e given in X, let τ_e be the solution of

$$\mathscr{A}(\tau_e, \tau - \tau_e) - (e, \tau - \tau_e) \geqslant 0 \quad \forall \tau \in K; \quad \tau_e \in K. \tag{6.18}$$

We immediately verify that

(6.19) the mapping $e \to \tau_e$ is Lipschitz continuous from X into X, and that

(6.20) $F(e) = -\frac{1}{2} \mathscr{A}(\tau_e, \tau_e) + (e, \tau_e)$ is continuous on X,
F is convex (as upper envelope of linear functions).

We set

$$(6.21)\qquad G(e)=\begin{cases}-(v,f)-\int_{\Gamma_F}F_iv_i\,\mathrm{d}\Gamma & \text{when}\quad e=\varepsilon_{ij}(v),\ v\in\mathcal{U}_{\mathrm{ad}},\\ +\infty & \text{otherwise.}\end{cases}$$

The set of $e=\{\varepsilon_{ij}(v)\}$, $v\in\mathcal{U}_{\mathrm{ad}}$ *is a convex closed set of* X *if* Γ_U *has positive measure.*

Indeed, if $\varepsilon_{ij}(v_n)$ is convergent in $L^2(\Omega)$, then, taking v_0 in $\mathcal{U}_{\mathrm{ad}}$, $\varepsilon_{ij}(v_n-v_0)$ converges in $L^2(\Omega)$, $v_n-v_0=0$ on Γ_U and therefore v_n-v_0 converges in $(H^1(\Omega))^3$, therefore $v_n\to v$ in $(H^1(\Omega))^3$ and $\varepsilon_{ij}(v_n)\to\varepsilon_{ij}(v)$.

From this it follows that G *is a proper convex function on* $\mathcal{H}$, *and according to* (6.20) *we have all the conditions for applying duality.*

Then we consider the problem

$$(6.22)\qquad \inf_{e\in\mathcal{H}}\,[F(e)+G(e)],$$

that means

$$(6.22\text{a})\qquad \inf_{v\in\mathcal{U}_{\mathrm{ad}}}\,\Big[\sup_{\tau\in K}\,[(\varepsilon(v),\tau)-\tfrac{1}{2}\mathcal{A}(\tau,\tau)]-(v,f)-\int_{\Gamma_F}v_iF_i\,\mathrm{d}\Gamma\Big].$$

We ignore whether or not this problem generally has a solution, but we shall prove directly, without reference to a general theory:

Theorem 6.2. *The problem dual to* (6.22a) *is the initial value problem for the stresses.*

Proof. According to (6.14), it is sufficient to prove that

i) $F^*(\tau)=\frac{1}{2}\mathcal{A}(\tau,\tau)+\psi_K(\tau)$, where ψ_K is the indicator function of K, which is obvious since F is defined as conjugate function to

$$\tfrac{1}{2}\mathcal{A}(\tau,\tau)+\psi_K(\tau).$$

ii) $G^*(\tau)=\int_{\Gamma_U}U_i\tau_{ij}n_j\,\mathrm{d}\Gamma+\psi_M(-\tau)$

(because then the problem dual to (6.22a) is

$$\inf_{\tau\in\mathcal{H}}\,[F^*(\tau)+G^*(-\tau)]=\inf_{\tau\in K\cap M}\,[\tfrac{1}{2}\mathcal{A}(\tau,\tau)-\int_{\Gamma_U}u_i\tau_{ij}n_j\,\mathrm{d}\Gamma]).$$

We calculate $G^*(\tau)$:

$$\begin{aligned}G^*(\tau)&=\sup_{v\in\mathcal{U}_{\mathrm{ad}}}\,[(\tau,\varepsilon(v))+(f,v)+\int_{\Gamma_F}v_iF_i\,\mathrm{d}\Gamma]\\ &=\sup_{v\in\mathcal{U}_{\mathrm{ad}}}\,\Big[\int_\Omega(\tau_{ij}\varepsilon_{ij}(v)+f_iv_i)\,\mathrm{d}x+\int_{\Gamma_F}v_iF_i\,\mathrm{d}\Gamma\Big].\end{aligned}$$

Then the supremum is $+\infty$ except possibly when

$$(6.23)\qquad -\tau_{ij,j}+f_i=0$$

and then

$$G^*(\tau) = \sup_{v \in \mathscr{U}_{ad}} \int_{\Gamma_F} (\tau_{ij} n_j + F_i) v_i \, d\Gamma + \int_{\Gamma_U} \tau_{ij} n_j U_i \, d\Gamma] = +\infty$$

except when

$$(6.24) \qquad -\tau_{ij} n_j = F_i \quad \text{on} \quad \Gamma_F$$

and then we see that

$$G^*(\tau) = \int_{\Gamma_U} \tau_{ij} n_j U_i \, d\Gamma + \psi_M(-\tau) . \quad \square$$

Theorem 6.3. *If problem* (4.22a) *has a solution u and if σ is the solution of the initial value problem* (*Theorem* 6.1), *then u and σ are linked by* (6.1)...(6.4).

Proof. We have indeed

$$u = U \quad \text{on} \quad \Gamma_U$$
$$\sigma \in K \cap M ,$$

therefore (6.2), (6.3), (6.4) are satisfied. From (6.15) we have

$$F(\varepsilon(u)) + F^*(\sigma) = (\sigma, \varepsilon(u)) .$$

But $F^*(\sigma) = \frac{1}{2} \mathscr{A}(\sigma, \sigma)$ according to the proof of Theorem 6.2i). Therefore

$$\tfrac{1}{2} \mathscr{A}(\sigma, \sigma) - (\sigma, \varepsilon(u)) = -F(\varepsilon(u)) = \inf_{\tau \in K} [\tfrac{1}{2} \mathscr{A}(\tau, \tau) - (\varepsilon(u), \tau)]$$

therefore

$$\tfrac{1}{2} \mathscr{A}(\sigma, \sigma) - (\sigma, \varepsilon(u)) \leqslant \tfrac{1}{2} \mathscr{A}(\tau, \tau) - (\varepsilon(u), \tau) \qquad \forall \tau \in K ,$$

which is equivalent to

$$\mathscr{A}(\sigma, \tau - \sigma) - (\varepsilon(u), \tau - \sigma) \geqslant 0 \qquad \forall \tau \in K$$

whence (6.1).

6.5. Isotropic Material with the Von Mises Condition

Let $\mathscr{F}$ be given by (2.11) and let the coefficients A_{ijkh} correspond to the elasticity law given in (2.4) of Chaper III. We calculate $\mathscr{A}(\sigma, \tau)$ and find

$$\mathscr{A}(\sigma, \tau) = \frac{1}{9K_0} \int_\Omega \sigma_{hk} \delta_{ij} \tau_{ij} \, dx + \frac{1}{2\mu} \int_\Omega \sigma_{ij}^D \tau_{ij} \, dx$$

(by K_0 we denote the modul of volume compressibility in order not to confuse it with the convex set K).

But

$$\sigma_{ij}^D \tau_{ij} = \sigma_{ij}^D \tau_{ij}^D, \qquad \delta_{ij}\tau_{ij} = \tau_{jj},$$

from which

$$\mathscr{A}(\sigma,\tau) = \frac{1}{9K_0} \int_\Omega \sigma_{kk}\tau_{jj}\,dx + \frac{1}{2\mu}\int_\Omega \sigma_{ij}^D \tau_{ij}^D\,dx.$$

Then

$$\sup_{\tau\in K} \left[(\varepsilon(v),\tau) - \tfrac{1}{2}\mathscr{A}(\tau,\tau)\right]$$
$$= \sup_{\tau\in K}\left[\int_\Omega \tau_{ij}^D \varepsilon_{ij}^D(v) + \frac{1}{3}\tau_{kk}\varepsilon_{kk}(v) - \frac{1}{18K_0}(\tau_{kk})^2 - \frac{1}{4\mu}\tau_{ij}^D\tau_{ij}^D\right].$$

We easily obtain

$$\sup_{\tau\in K}\left[(\varepsilon(v),\tau) - \tfrac{1}{2}\mathscr{A}(\tau,\tau)\right] = \int_\Omega \left[\tfrac{1}{2}K_0(\operatorname{div} v)^2 + \Phi(\varepsilon^D(v))\right]dx$$

where Φ is given by

$$\Phi(\varepsilon^D) = \begin{cases} \mu\varepsilon_{ij}^D\varepsilon_{ij}^D & \text{if } \varepsilon_{ij}^D\varepsilon_{ij}^D \le k^2/2\mu^2 \\ k((2\varepsilon_{ij}^D\varepsilon_{ij}^D)^{1/2} - k/2\mu) & \text{if } \varepsilon_{ij}^D\varepsilon_{ij}^D > k^2/2\mu^2. \end{cases}$$

Problem (4.22a) is thus

(6.25) $$\inf_{v\in\mathscr{U}} \left\{\int_\Omega \left[\tfrac{1}{2}K_0(\operatorname{div} v)^2 + \Phi(\varepsilon^D(v))\right]dx - (f,v) - \int_{\Gamma_F} F_i v_i\,d\Gamma\right\}.$$

The functional to be minimized is not necessarily infinite at infinity on $(H^1(\Omega))^3$; by contrast, the functional is infinite on $(W^{1,1}(\Omega))^3$ [19, 20] at least in the following important cases:

i) $f_i = F_i = 0$ (a situation which occurs in the problem of elasto-plastic torsion of a cylindrical tree. (Cf. Sec. 6.6 below)).

ii) $F_i = 0$ and $f_i = \partial q/\partial x_i$, $q \in H_0^1(\Omega)$, because then $(f,v) = -(q, \operatorname{div} v)$, and

$$\tfrac{1}{2}K_0 \int_\Omega (\operatorname{div} v)^2\,dx - c\,|\operatorname{div} v|_{L^2(\Omega)} \to +\infty \quad \text{if} \quad |\operatorname{div} v|_{L^2(\Omega)} \to +\infty.$$

iii) There exists a $\sigma_1 \in K \cap M$ such that

$$C_1 = \sup_{x\in\Omega} \tfrac{1}{2}\sigma_{1ij}^D \sigma_{1ij}^D < k^2.$$

[19] $W^{1,1}(\Omega) = \{\varphi \mid \varphi, \partial\varphi/\partial x_1, \ldots, \partial\varphi/\partial x_n \in L^1(\Omega)\}$.

[20] Note the analogy with the theory of non-parametric minimal surfaces, where we have the functional $\int_\Omega (1 + |\operatorname{grad} v|^2)^{1/2}\,dx$ which is infinite at infinity on $W^{1,1}(\Omega)$.

Indeed, we then have

$$(f,v)+\int_{\Gamma_F} F_i v_i \, d\Gamma = \int_\Omega \sigma_{1ij}\varepsilon_{ij}(v)\,dx - \int_{\Gamma_U} U_i \sigma_{1ij} n_j \, d\Gamma$$

and

$$\int_\Omega \sigma_{1ij}\varepsilon_{ij}(v)\,dx = \tfrac{1}{3}\int_\Omega \sigma_{1jj} \operatorname{div} v \, dx + \int_\Omega \sigma^D_{1ij}\varepsilon^D_{ij}(v)\,dx$$

where the last term satisfies

$$\left|\int_\Omega \sigma^D_{1ij}\varepsilon^D_{ij}(v)\,dx\right| \leqslant (2C_0)^{1/2}\int_\Omega (\varepsilon^D_{ij}\varepsilon^D_{ij})^{1/2}\,dx. \quad \square$$

We can then define a "very weak" solution of (6.25) in the space $((L^\infty(\Omega))')^6$. To do this, we consider the functional that occurs in (6.25), not on $(H^1(\Omega))^3$ but on $(W^{1,1}(\Omega))^3$; the dual problem is again the problem of stresses

$$\inf\left[\tfrac{1}{2}\mathscr{A}(\tau,\tau) - \int_{\Gamma_U} U_i \tau_{ij} n_j \, d\Gamma\right], \quad \tau \in K \cap M, \tag{6.26}$$

observing that $K \subset (L^\infty(\Omega))^6$.

Then, due to the fact that

$$\int_\Omega \left[\tfrac{1}{2}K_0(\operatorname{div} v)^2 + \Phi(\varepsilon^D(v))\right]dx - (f,v) - \int_{\Gamma_F} F_i v_i \, d\Gamma \to +\infty \quad \text{if} \quad \|v\|_{W^{1,1}(\Omega)} \to +\infty$$

(the analog on $(H^1(\Omega))^3$ being incorrect), it follows from R. Téman [2] that the problem dual to (6.26) has a solution in $((L^\infty(\Omega))')^6$. Moreover, according to (6.16), Hencky's constituent law is satisfied in the sense of duality between $(L^\infty(\Omega))'$ and $L^\infty(\Omega)$. $\square$

6.6. Torsion of a Cylindrical Tree (Fig. 19)
(cf. Annin [1], Lauchon [1–4], Ting [1], [2])

Here the open set Ω of $\mathbb{R}^3$ is a cylinder *without holes*, bounded by two plane sections Γ_0 and Γ_1 with respective equations $x_3=0$ and $x_3=h$ (h: the given positive length) and the lateral surface Γ_2. We always denote by n the exterior unit normal to $\partial\Omega$. We impose boundary conditions of the type contemplated in Section 6.2, more precisely, with

$$\begin{aligned} &f_i = 0 \quad \text{in} \quad \Omega \\ &\sigma_{ij} n_j = 0 \quad \text{on} \quad \Gamma_2 \quad (i=1,2,3), \end{aligned} \tag{6.27}$$

$$\sigma_3 = 0 \quad \text{on} \quad \Gamma_0 \quad \text{and} \quad \Gamma_1, \tag{6.28}$$

$$u_i = \alpha \varepsilon_{i3j} x_j x_3 \quad \text{on} \quad \Gamma_0 \quad \text{and} \quad \Gamma_1 \tag{6.29}$$

(in other words

$$\begin{aligned} &u_1 = u_2 = 0 \quad \text{on} \quad \Gamma_0 \\ &u_1 = -\alpha h x_2, \qquad u_2 = \alpha h x_1 \quad \text{on} \quad \Gamma_1). \end{aligned}$$

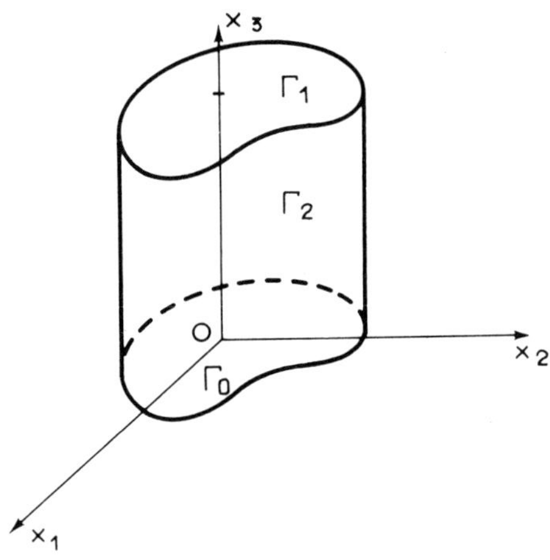

Fig. 19

The scalar constant α is *the angle of unit torsion.*

Moreover, the material is assumed to be isotropic, so that the elasticity coefficients A_{ijkh} are those of the preceding Section 6.5, and the plasticity condition is that of Von Mises (2.11).

Under these conditions

$$M = \{\tau \mid \tau \in H, \quad \tau_{ij} = 0 \text{ in } \Omega, \quad \tau_{ij} n_j = 0 \text{ on } \Gamma_2, \quad \tau_3 = 0 \text{ on } \Gamma_0 \text{ and } \Gamma_1\}. \tag{6.30}$$

According to Section 6.3, the stress field solution minimizes the functional

$$\tfrac{1}{2}\mathscr{A}(\tau,\tau) - \alpha h \int_{\Gamma_1} (x_1 \tau_{23} - x_2 \tau_{13})\,\mathrm{d}\Gamma. \tag{6.31}$$

on $M \cap K$. □

Now, let $\{\sigma_{ij}\}$ be the stress field solution. It was proved by H. Lauchon [4] that

$$\begin{aligned} &\sigma_{ij} = 0 \quad \text{except for } \sigma_{13} \text{ and } \sigma_{23}, \\ &\sigma_{13} = \sigma_{13}(x_1, x_2), \qquad \sigma_{23} = \sigma_{23}(x_1, x_2). \end{aligned} \tag{6.32}$$

Under these conditions, the equilibrium equations $\sigma_{ij,j} = 0$ reduce to

$$\sigma_{13,1} + \sigma_{23,2} = 0, \tag{6.33}$$

that is to say, there exists[21] a $\theta = \theta(x_1, x_2)$ such that

$$\sigma_{13} = \partial\theta/\partial x_2, \qquad \sigma_{23} = -\partial\theta/\partial x_1. \tag{6.34}$$

[21] In the multiple connected case, we have to modify the reasoning slightly; cf. H. Lauchon [2].

The equations $\sigma_{ij}n_j=0$ on Γ_2 can then simply be written as

$$d\theta/ds=0$$

where s is a curvilinear abscissa on Γ_0, that is to say, $\theta=$constant on the boundary of Γ_0 in $\mathbb{R}^2$. Since (6.34) defines θ within a constant, we can take $\theta\in H_0^1(\Gamma_0)$, since we must have $\sigma_{i3}\in L^2(\Omega)$.

The condition $\{\sigma_{ij}\}\in K$ can then be written with reference to

$$\theta\in K_1=\{v|\ v\in H_0^1(\Gamma_0),\ |\text{grad}\, v|\leqslant g \ \text{ a. e. on } \Gamma_0\}. \tag{6.35}$$

The field θ associated with $\{\sigma_{ij}\}$ by (6.34) minimizes on K_1 the functional analogous to (6.31)

$$\tfrac{1}{2}\int_{\Gamma_0}|\text{grad}\, v|^2 dx_1\, dx_2+\mu\alpha\int_{\Gamma_0}(x_1\,\partial v/\partial x_1+x_2\,\partial v/\partial x_2)dx_1\, dx_2,$$

(μ: the shear modulus)

that is to say, after transformation of the second integral,

$$\tfrac{1}{2}\int_{\Gamma_0}|\text{grad}\, v|^2 dx_1\, dx_2-2\mu\alpha\int_{\Gamma_0} v\, dx_1\, dx_2. \tag{6.36}$$

If we set

$$a(\theta,v)=\int_{\Gamma_0}\text{grad}\,\theta\,\text{grad}\, v\, dx_1\, dx_2, \tag{6.37}$$

$$f=2\mu\alpha, \qquad (f,v)=\int_{\Gamma_0} fv\, dx_1\, dx_2, \tag{6.38}$$

we have:

the field θ, associated by (6.34) with the stress field solution $\{\sigma_{ij}\}$ minimizes on K_1 the functional

$$\tfrac{1}{2}\alpha(v,v)-(f,v); \tag{6.39}$$

or again:

the field θ is characterized by

$$\begin{aligned}&\theta\in K_1\\ &a(\theta,v-\theta)\geqslant(f,v-\theta)\qquad \forall v\in K_1.\end{aligned} \tag{6.30}$$

Each of these statements naturally implies the existence of a unique field of stresses that is the solution of the elasto-plastic problem of torsion of a cylindrical tree with a Hencky type constituent law. □

Remark 6.1. We can assume that the torsion angle α is a known function of t, say $\alpha(t)$. Then we obtain, by the method we described here, a solution $\theta_{\alpha(t)}$ for each value of t. □

Suppose now that we examine the same torsion problem, taking the *Prandtl-Reuss law* as constituent law, with $\alpha(t)$ given as continuously differentiable and with $\alpha(0)=0$.

Then we find that $\tilde{\theta}(t)$, always associated by (6.34) with the stress field solution, satisfies

$$\begin{aligned}&\tilde{\theta}(t)\in K_1\\ &a(\tilde{\theta}'(t), v-\tilde{\theta}(t))\geqslant 2\mu\alpha'(t)\int_{\Gamma_0}(v-\tilde{\theta}(t))\,\mathrm{d}x_1\,\mathrm{d}x_2 \qquad \forall v\in K_1\\ &\theta(0)=0.\end{aligned} \tag{6.41}$$

Moreover, it has been established by H. Brézis that if $\alpha(t)$ is a non-decreasing function of t, then

$$\tilde{\theta}(t)=\theta_{\alpha(t)}. \quad \square \tag{6.42}$$

Remark 6.2. The extension of property (6.42) to the general problem of Section 6.2 is an open problem. $\square$

Remark 6.3. Knowing the stress field solution

$$\sigma_{13}=\partial\theta/\partial x_2, \qquad \sigma_{23}=-\partial\theta/\partial x_1, \tag{6.43}$$

it is legitimate to look for the field of associated displacements, say $u=(u_1,u_2,u_3)$, in the form

$$\begin{aligned}&u_1=\alpha x_3 x_2\\ &u_2=\alpha x_3 x_1\\ &u_3=\alpha\psi(x_1,x_2).\end{aligned} \tag{6.44}$$

Thus the strain tensor is

$$\begin{aligned}&\varepsilon_{11}(u)=\varepsilon_{22}(u)=\varepsilon_{33}(u)=0\\ &\varepsilon_{12}(u)=0, \qquad \varepsilon_{13}(u)=\alpha[-x_2+\partial\psi/\partial x_1]\\ &\qquad\qquad\qquad\;\; \varepsilon_{23}(u)=\alpha[x_1+\partial\psi/\partial x_2].\end{aligned} \tag{6.45}$$

Since the material is isotropic and μ denotes one of the Lamé constants, we have

$$\begin{aligned}&\varepsilon_{13}(u)=\frac{1}{2\mu}\sigma_{13}+\lambda_{13}\\ &\varepsilon_{23}(u)=\frac{1}{2\mu}\sigma_{23}+\lambda_{23}\end{aligned} \tag{6.46}$$

where

$$\lambda_{13}(\tau_{13}-\sigma_{13})+\lambda_{23}(\tau_{23}-\sigma_{23})\leqslant 0 \qquad \forall\tau_{13},\,\tau_{23} \text{ with } \tau_{13}^2+\tau_{23}^2\leqslant g^2. \tag{6.47}$$

We shall indicate two possible methods to determine u, the first one *actually* leading to an impasse.

1) If the convex set K_1 is defined by (6.35) and χ_{K_1} is its indicator function, the above constituent law can also be written

$$(G+\chi_{K_1})(\sigma)+(G+\chi_{K_1})^*(\varepsilon(u))=(\sigma,\varepsilon) \tag{6.48}$$

where we put

$$G(\sigma)=\frac{1}{4\mu}\int_{\Gamma_0}(\sigma_{13}^2+\sigma_{23}^2)\,dx_1\,dx_2. \tag{6.49}$$

From this it then follows that the displacement field solution u minimizes the functional

$$(G+\chi_{K_1})^*(\varepsilon(\tilde{u})) \tag{6.50}$$

among the displacement fields $\tilde{u}$ of the form (6.44).

We can calculate the functional (6.50) explicitly; we find

$$(G+\chi_{K_1})^*(\varepsilon(\tilde{u}))=\int_\Omega \Phi(\varepsilon_{13},\varepsilon_{23})\,dx_1\,dx_2$$

with

$$\Phi(\varepsilon_{13},\varepsilon_{23})=\begin{vmatrix}\mu|\varepsilon|^2 & \text{if} \quad |\varepsilon|\leqslant g/2\mu \\ g|\varepsilon|-g^2/4\mu & \text{if} \quad |\varepsilon|\geqslant g/2\mu\end{vmatrix} \tag{6.51}$$

where we put

$$|\varepsilon|=(\varepsilon_{13}^2+\varepsilon_{23}^2)^{1/2}$$

and

$$\varepsilon_{13}=\alpha[-x_2+\partial\psi/\partial x_1]$$
$$\varepsilon_{23}=\alpha[x_1+\partial\psi/\partial x_2].$$

This method, which expresses that u is the solution of a variational problem, does not lead to a solution, because the functional to be minimized is not coercive, for example on the set of functions $\psi(x_1,x_2)$ which belong to $H^1(\Gamma_0)$. □

2) Thus we again take the constituent law (6.47) and rewrite it (H. Brézis [3]):

$$\begin{vmatrix} -x_2+\partial\psi/\partial x_1=\dfrac{1}{\mu\alpha}\partial\theta/\partial x_2+\lambda\partial\theta/\partial x_2 \\ +x_1+\partial\psi/\partial x_2=-\dfrac{1}{\mu\alpha}\partial\theta/\partial x_1-\lambda\partial\theta/\partial x_1 \\ \text{with} \quad \begin{matrix}\lambda=0 & \text{if} & |\text{grad}\,\theta|<g \\ \lambda\geqslant 0 & \text{if} & |\text{grad}\,\theta|=g.\end{matrix}\end{vmatrix} \tag{6.52}$$

The existence of a function $\psi(x_1, x_2)$ implies $\partial^2\psi/\partial x_1 \partial x_2 = \partial^2\psi/\partial x_2 \partial x_1$, from which it follows that

$$-2 = \frac{1}{\mu\alpha}\Delta\theta + \frac{\partial}{\partial x_2}\left(\lambda \frac{\partial\theta}{\partial x_2}\right) + \frac{\partial}{\partial x_1}\left(\lambda \frac{\partial\theta}{\partial x_1}\right)$$

or again

$$\frac{\partial\lambda}{\partial x_1}\frac{\partial\theta}{\partial x_1} + \frac{\partial\lambda}{\partial x_2}\frac{\partial\theta}{\partial x_2} + \lambda\Delta\theta = -2 - \frac{1}{\mu\alpha}\Delta\theta. \tag{6.53}$$

In the elastic domain ($|\text{grad}\,\theta| < g$), we have $\lambda = 0$ (which certainly is a solution of (6.53), because the right hand side is then zero). In the plastic domain ($|\text{grad}\,\theta| = g$) where $\theta(x_1, x_2) = g \times \text{distance}\,(\{x_1, x_2, \partial\Gamma_0)$ and, consequently, Equation (6.53) becomes a differential equation on the normals at the points of $\partial\Gamma_0$, say

$$-g\,\partial\lambda/\partial n + \lambda\Delta\theta = -2 - \Delta\theta/\mu\alpha. \tag{6.54}$$

Then we can calculate $\lambda(x_1, x_2)$ explicitly by observing that $\lambda = 0$ in the elastic domain. The condition $\lambda(x_1, x_2) \geqslant 0$ in the plastic domain follows from $\Delta\theta + 2\mu\alpha \geqslant 0$ in the plastic domain. Thus, in the particular case of torsion, we find the displacement field associated with the stress solution. □

7. Locking Material

7.1. Constituent Law

The notion of "locking material" was introduced by W. Prager [3], who formulated a graphic constituent law for it. Intuitively, this material has linear elastic behavior as long as the stresses do not reach a certain threshold. When that is attained, the stresses can increase without causing greater strains. □

Precisely, we will formulate this law in the following way: let $f(\varepsilon_{ij})$ be a continuous convex function of the strain tensor ε, such that $f(0) < 0$. The domain of possible strains is defined by

$$f(\varepsilon) \leqslant 0, \tag{7.1}$$

and the constituent law can be written

$$\begin{aligned} &\sigma_{ij} = a_{ijkh}\varepsilon_{kh}(u) + \mu_{ij} \\ &\mu_{ij}(e_{ij} - \varepsilon_{ij}(u)) \leqslant 0 \qquad \forall e = \{e_{ij}\}, \qquad f(e) \leqslant 0. \end{aligned} \tag{7.2}$$

In practice, this law applies only in certain *static* or *quasi-static* situations.

As usual, the elasticity coefficients a_{ijkh} satisfy

$$(7.3)\qquad \begin{aligned} &a_{ijkh} = a_{ijhk} = a_{khij} \\ &a_{ijkh} e_{ij} e_{kh} \geqslant \alpha e_{ij} e_{ij}, \qquad \alpha = \text{constant} > 0. \quad \square \end{aligned}$$

Example of a law with threshold. By e_I, $I=1,2,3$, we denote the eigenvalues of the tensor $\{e_{ij}\}$, and with a given positive constant a we define

$$f(e) = -\Big(\inf_{I=1,2,3} e_I + a\Big),$$

which means that $f(e) \leqslant 0$ if and only if $e_I \leqslant -a \ \forall I = 1,2,3$.

We shall prove that

(7.4) *the set of $e \in \mathbb{R}^6$ such that $f(e) \leqslant 0$ is convex.*

Indeed, let n be the unit vector and E the vector with the components $E_i = e_{ij} n_j$; we set $E(n) = e_{ij} n_i n_j$. It is easy to prove, for example in principal axes for the $\{e_{ij}\}$, that $f(e) \leqslant 0$ is equivalent to

$$\inf_{|n|=1} E(n) \geqslant = a.$$

From the linearity of $E(n)$ with respect to $\{e_{ij}\}$, the property then follows. $\square$

As in the case of perfect plasticity, this constituent law can be written in integral form. We introduce the Hilbert space H

$$H = \{e \mid e = \{e_{ij}\},\ e_{ij} \in L^2(\Omega),\ e_{ij} = e_{ji}\},$$

with the scalar product

$$(7.5)\qquad (\varphi, \psi) = \int_\Omega \varphi_{ij} \psi_{ij} \, dx,$$

where Ω denotes the open bounded set in $\mathbb{R}^3$, with regular boundary Γ, occupied by the material under consideration.

We denote by k the convex set in H defined by

$$(7.6)\qquad k = \{e \mid e \in H,\ f(e) \leqslant 0 \ \text{a.e. in } \Omega\}.$$

Condition (7.2) is *equivalent* to

$$(7.7)\qquad \begin{aligned} &\sigma_{ij} = a_{ijkh} \varepsilon_{kh}(u) + \mu_{ij} \\ &\int_\Omega \mu_{ij}(e_{ij} - \varepsilon_{ij}(u)) \, dx \leqslant 0 \qquad \forall e \in k \end{aligned}$$

with

$$(7.8)\qquad \varepsilon(u) \in k, \qquad \mu_{ij} \in L^2(\Omega).$$

(Indeed, (7.2) implies (7.7) and one can prove that (7.7) and (7.8) imply (7.2) almost everywhere in Ω by an argument similar to the one in Section 2.1.2. □

7.2. Problem to be Considered

Following W. Prager, loc. cit., we examine the *stationary* problem of the deformation of a body subjected to a volume density of forces $f(x)$ with components $f_i(x)$, to a surface density of forces $F(x)$ with components $F_i(x)$ on Γ_F, and to imposed displacements $U(x)$ with components $U_i(x)$ on Γ_U, using the constituent law introduced above. The regions Γ_U and Γ_F constitute a partition of Γ.

The solution fields of displacements $u=\{u_i\}$ and of stresses $\sigma=\{\sigma_{ij}\}$ must satisfy the equations

$$\begin{aligned} &\sigma_{ij,j}+f_i=0 \quad &&\text{in} \quad \Omega \\ (7.9)\qquad &\sigma_{ij}n_j=F_i &&\text{on} \quad \Gamma_F \\ &u_i=U_i &&\text{on} \quad \Gamma_U, \end{aligned}$$

and the constituent law (7.7), (7.8). □

7.3. Double Variational Formulation of the Problem

The constituent law (7.7) can also be written as

$$(7.10)\qquad \int_\Omega \mu_{ij}\varepsilon_{ij}(u)\,\mathrm{d}x=\sup_{e\in k}\int_\Omega \mu_{ij}e_{ij}\,\mathrm{d}x=\psi_k^*(\mu)$$

where ψ_k is the indicator function of the convex set $k\subset H$, and ψ_k^* the conjugate function to ψ_k (cf. Sec. 6.3 for the definition). From this follows

Theorem 7.1. *The constituent law* (7.7), (7.8) *is equivalent to*

$$(7.11)\qquad (g+\psi_k)(\varepsilon(u))+(g+\psi_k)^*(\sigma)=(\varepsilon,\sigma)$$

where

$$(7.12)\qquad g(e)=\tfrac{1}{2}\int_\Omega a_{ijkh}e_{ij}e_{kh}\,\mathrm{d}x,$$

and ψ_k denotes the indicator function of the set k of H and ψ_k^* its conjugate function.

Proof. i) (7.7), (7.8) $\Rightarrow$ (7.11). We calculate the left hand side of (7.11) for $\varepsilon(u)\in k$,

$$
\begin{aligned}
g(\varepsilon) + \sup_{e\in k} \int_\Omega (e_{ij}\sigma_{ij} - \tfrac{1}{2} a_{ijkh} e_{ij} e_{kh})\,dx
&= \sup_{e\in k} \int_\Omega (e_{ij}\sigma_{ij} - \tfrac{1}{2} a_{ijkh} e_{ij} e_{kh} + \tfrac{1}{2} a_{ijkh}\varepsilon_{ij}\varepsilon_{kh})\,dx \\
&= \sup_{e\in k} \int_\Omega [e_{ij}\sigma_{ij} - \tfrac{1}{2} a_{ijkh}(e_{ij}-\varepsilon_{ij})(e_{kh}-\varepsilon_{kh}) + a_{ijkh}\varepsilon_{ij}(\varepsilon_{kh}-e_{kh})]\,dx \\
&\leqslant \sup_{e\in k} \int_\Omega [e_{ij}(\sigma_{ij} - a_{ijkh}\varepsilon_{kh}) + a_{ijkh}\varepsilon_{ij}\varepsilon_{kh}]\,dx \\
&= \sup_{e\in k} \int_\Omega (e_{ij}\mu_{ij} + a_{ijkh}\varepsilon_{ij}\varepsilon_{kh})\,dx = \int_\Omega (\varepsilon_{ij}\mu_{ij} + a_{ijkh}\varepsilon_{ij}\varepsilon_{kh})\,dx = (\sigma,\varepsilon).
\end{aligned}
$$

Then

$$(g+\psi_k)(\varepsilon) + (g+\psi_k)^*(\sigma) \leqslant (\sigma,\varepsilon), \tag{7.13}$$

and, since by definition of the dual function

$$(g+\psi_k)(\varepsilon) + (g+\psi_k)^*(\sigma) \geqslant (\sigma,\varepsilon) \tag{7.14}$$

(7.11) follows.

ii) (7.11) ⇒ (7.7), (7.8)

Equation (7.11) implies $\varepsilon(u)\in k$ and can thus be written

$$g(\varepsilon) + \sup_{e\in k} \int_\Omega (e_{ij}\sigma_{ij} - \tfrac{1}{2}\alpha_{ijkh} e_{ij} e_{kh})\,dx = (\varepsilon,\sigma),$$

that is to say, after regrouping,

$$(\varepsilon,\sigma) = \sup_{e\in k} \int_\Omega [e_{ij}\sigma_{ij} - \tfrac{1}{2} a_{ijkh}(\varepsilon_{ij}-e_{ij})(\varepsilon_{kh}-e_{kh}) - a_{ijkh}\varepsilon_{kh}(e_{ij}-\varepsilon_{ij})]\,dx. \tag{7.15}$$

We introduce μ by (7.7); (7.14) can then be written

$$\sup_{e\in k} \int_\Omega [e_{ij}\mu_{ij} - \tfrac{1}{2} a_{ijkh}(e_{ij}-\varepsilon_{ij})(e_{kh}-\varepsilon_{kh})]\,dx = \int_\Omega \varepsilon_{ij}\mu_{ij}\,dx. \tag{7.16}$$

Let e be an arbitrary element of k; according to (7.16), we have

$$\int_\Omega [(e_{ij}-\varepsilon_{ij})\mu_{ij} - \tfrac{1}{2} a_{ijkh}(e_{ij}-\varepsilon_{ij})(e_{kh}-\varepsilon_{kh})]\,dx \leqslant 0. \tag{7.17}$$

Then we apply (7.17) for the element $e^{(\alpha)}$ of k defined by

$$e^{(\alpha)} \pm \alpha e + (1-\alpha)\varepsilon. \tag{7.18}$$

After division by α, it follows that

$$\int_\Omega (e_{ij}-\varepsilon_{ij})\mu_{ij}\,dx - \tfrac{1}{2}\alpha \int_\Omega g(e-\varepsilon)\,dx \leqslant 0 \tag{7.19}$$

from which inequality (7.7) follows in the limit, when α tends toward zero. □

Consequences: Variational formulations. If e and τ are two arbitrary elements of H, we have

$$(g+\psi_k)(e)+(g+\psi_k)^*(\tau)\geqslant(e,\tau)$$

and, consequently, the possible solution $(u_i, \sigma_{ij}\}$ of the posed problem minimizes the functional

$$(g+\psi_k)(e)+(g+\psi_k)^*(\tau)-(e,\tau) \tag{7.20}$$

among the fields of stresses τ, that *are statically admissible*, that is to say, such that

$$\begin{aligned} &\tau\in H, \\ &\tau_{ij,j}+f_i=0 \quad \text{in} \quad \Omega, \\ &\tau_{ij}n_j=F_i \quad \text{on} \quad \Gamma_F, \end{aligned} \tag{7.21}$$

and the fields of strains e *that are kinematically admissible*, that is to say, such that

$$\begin{aligned} &e=\varepsilon(v) \\ &v=\{v_i\}, \quad v_i\in H^1(\Omega) \\ &v_i=U_i \quad \text{on} \quad \Gamma_U. \end{aligned} \tag{7.22}$$ [22]

But then we have

$$(e,\tau)=\int_\Omega f_i v_i \mathrm{d}x+\int_{\Gamma_F} F_i v_i \mathrm{d}\Gamma+\int_{\Gamma_U} u_i\tau_{ij}n_j \mathrm{d}\Gamma, \tag{7.23}$$

and, instead of minimizing the functional given by (7.20), *we can, equivalently, minimize the functionals separately*

$$I_1(v)=(g+\psi_k)(\varepsilon(v))-\int_\Omega f_i v_i \mathrm{d}\Gamma-\int_{\Gamma_F} F_i v_i \mathrm{d}\Gamma$$

on the set of kinematically admissible fields, and

$$I_2(\tau)=(g+\psi_k)^*(\tau)-\int_{\Gamma_U} u_i\tau_{ij}n_j \mathrm{d}\Gamma$$

on the set of statically admissible fields.

7.4. Existence and Uniqueness of a Displacement Field Solution

The functional $I_1(v)$, defined in (7.24), is strictly convex and lower semi-continuous on the set of kinematically admissible fields if measure $\Gamma_U>0$ which we assume.

---

[22] We assume, as we already did several times, that $f_i\in L^2(\Omega)$, $F_i\in H^{-1/2}(\Gamma_F)$, $u_i\in H^{1/2}(\Gamma)$ (that is to say, it is the restriction to Γ_U of an element of $H^{1/2}(\Gamma)$).

Furthermore, its value tends toward $+\infty$ when the norm of v in $(H^1(\Omega))^3$ tends toward $+\infty$.

(In the case where $\Gamma_U = \emptyset$, it is necessary that the set of forces f and F is statically equivalent to zero. The functional $I_1(v)$ is then defined on the quotient space $(H^1(\Omega))^3/\mathcal{R}$, where $\mathcal{R}$ is the set of rigid displacements of $\mathbb{R}^3$. On this quotient space, the functional $I_1(v)$ possesses the properties stated above. We already met with considerations of this type in Section III.3.) From this follows

Theorem 7.2. *There exists a unique field of kinematically admissible displacements u that minimizes the functional $I_1(v)$ when Γ_U has positive measure.*

If Γ_U is empty, and if the set of applied forces is statically equivalent to zero, there exists an element u of $(H^1(\Omega))^3$ that minimizes $I_1(v)$ among the kinematically admissible fields. This element is unique except for a field of rigid displacements.

7.5. The Associated Field of Stresses

Making use of the results on duality which we recapitulated in Section 6, we can state

Theorem 7.3. *If there exists a field of stresses σ that minimizes the functional $I_2(\tau)$ among the statically admissible fields defined by* (7.21), *then σ and $\varepsilon(u)$ are connected by the constituent law* (7.11). □

Remark 7.1. The problem of the existence of a field of stresses σ is open as far as we know. The situation here is similar to the one encountered in Section 6 if we interchange the roles of stresses and strains. □

8. Comments

For the classical work on plasticity, we refer to W. I. Koiter [1], J. Mandel [1, 2], W. Prager [1, 2], W. Prager and P. G. Hodge [1] and the bibliographies in these works.

Here we adopted a presentation where the different constituent laws of plasticity—except for plasticity with work hardening which we did not treat—appear as special cases or as limit cases of the elasto-visco-plastic law.

The problems touched upon in this chapter are problems with small deformations, the space variable being a Lagrange variable.

It will be different in Chapter VI, where we will discuss a flow problem, the space variable then being an Euler variable. (For a presentation of these two types of variables we refer to P. Germain [1], Chapter 3.)

In Sections 3 and 4, the results are obtained under the assumption that the convex set K is independent of time. This condition is realized in important special cases, such as the case of torsion of a cylindrical tree (H. Lauchon

[1–4], W. Ting [1, 2], B. A. Annin [1]). Nevertheless, the case of a convex set which depends on time occurs frequently and, to our knowledge, presents an unsolved problem. This chapter contains other important open problems, such as, in Section 6, the determination of a displacement field solution or, in Section 7, the determination of a stress field solution. □

We did not touch upon problems of plasticity involving large deformations; for the pertinent formulation of equations, we refer to M. M. Balaban, A. E. Green, P. M. Naghdi [1].

Chapter VI

Rigid Visco-Plastic Bingham Fluid

In this chapter, we assume familiarity with Sections I.1 to I.3.

1. Introduction and Problems to be Considered

A rigid visco-plastic fluid is a continuous medium which obeys the general conservation laws stated in Section I.1, and in addition obeys special constituent laws.

Here we state precisely the set of corresponding equations as well as the types of problems which we propose to examine.

1.1. Constituent Law of a Rigid Visco-Plastic, Incompressible Fluid

The hypothesis of *incompressibility*, physically very realistic, can be expressed by

$$\operatorname{div} v = 0 \tag{1.1}$$

where v is the field of velocity vectors. Applying the equation of conservation of mass (Sec. 1.1.3), it follows that

$$\frac{d\rho}{dt} = \frac{\partial \rho}{\partial t} + \frac{\partial \rho}{\partial x_i} v_i = 0 . \tag{1.2}$$

Consequently, the specific mass of a fluid particle remains constant in the course of the flow; hence, it is independent of the space variables as well, if it is independent of them at any particular instant—which we assume. Under these conditions, we have

$$\rho = \text{constant} = \rho_0 . \tag{1.3}$$

It is even permissible to set $\rho_0 = 1$ which we will do in the following; this simply amounts to choosing the unit of specific mass.

Having fixed ρ determines one unknown function of the problem. Moreover, it uncouples the equations of conservation of momentum from the energy equation, always under the condition that the constituent laws which express the stress tensor *do not involve the temperature*; we will assume here that this is the case. Thermal phenomena (field of temperatures, heat flow) will not come up in this chapter; we will solely study the properties of the flow (velocity and stress fields).

1.2. The Dissipation Function

In Equation (1.27) of Chapter I, the term $\sigma_{ij}D_{ij}$ [1] is sometimes called the *dissipation function* $\mathscr{D}$. Prescribing this function, as suggested for example by experimental results, permits us to formulate the constituent laws of the material under consideration. Thus, we will proceed by postulating that $\mathscr{D}$ depends only on tensor of strain rates (P. Germain [2], W. Prager [1]), namely

$$\sigma_{ij}D_{ij}=\mathscr{D}_1(\mathbb{D})+\mathscr{D}_2(\mathbb{D}) \tag{1.4}$$

where $\mathscr{D}_1$ and $\mathscr{D}_2$ are positive homogeneous functions of order 1 and 2, respectively, in the components of the tensor $\mathbb{D}$. It follows that

$$\begin{aligned} \mathscr{D}_1 &= \frac{\partial \mathscr{D}_1}{\partial D_{kl}} D_{kl} \\ \mathscr{D}_2 &= \frac{1}{2}\frac{\partial \mathscr{D}_2}{\partial D_{kl}} D_{kl} \, . \end{aligned} \tag{1.5}$$

Since the identity (1.4) must hold whatever the components D_{ij} are, subject to the constraint

$$D_{kk}=0 \quad \text{(another form of (1.1))}\,, \tag{1.6}$$

the components σ_{ij} of the stress tensor are given by

$$\sigma_{ij}=-p\delta_{ij}+\partial\mathscr{D}_1/\partial D_{ij}+\tfrac{1}{2}\partial\mathscr{D}_2/\partial D_{ij}\,, \tag{1.7}$$

where p is a scalar, independent of the D_{ij}. If, moreover, the fluid is assumed to be isotropic, the scalars $\mathscr{D}_1$ and $\mathscr{D}_2$ are functions of only the invariants of the tensor D. □

We call *Bingham fluid* the material for which the functions $\mathscr{D}_1$ and $\mathscr{D}_2$ are given (W. Prager [1]) by

$$\mathscr{D}_1=2g(D_{\mathrm{II}})^{1/2}\,, \qquad \mathscr{D}_2=4\mu D_{\mathrm{II}}\,, \tag{1.8}$$

[1] D_{ij} is defined in Chapter I by (1.23).

where D_{II} is the invariant of the tensor $\mathbb{D}$, given by

(1.9) $$D_{II} = \tfrac{1}{2} D_{ij} D_{ij}.$$

The positive scalars g and μ are respectively the *yield limit* (threshold of plasticity) and the *viscosity* of the Bingham fluid.

The constituent law (1.7) can then be written

(1.10) $$\sigma_{ij} = -p\delta_{ij} + g D_{ij}/(D_{II})^{1/2} + 2\mu D_{ij},$$

an expression which makes sense only when $D_{II} \neq 0$.

If $D_{II} \neq 0$, the stress tensor is indeterminate. In (1.10), the scalar $-p$ represents the spherical part[2] of the stress tensor; one can identify p with the *pressure*. ▯

In order to invert relations (1.10), we set

(1.11) $$\sigma_{II} = \tfrac{1}{2} \sigma_{ij}^D \sigma_{ij}^D$$

where the σ_{ij}^D represent the components of the deviation of the stresses. Then we have, with (1.10),

(1.12) $$\sigma_{II} = (g + 2\mu D_{II}^{1/2})^2,$$

which implies $\sigma_{II}^{1/2} \geqslant g$, in which case (1.10) is inverted to

(1.13) $$D_{ij} = \frac{1}{2\mu}(1 - g/\sigma_{II}^{1/2})\sigma_{ij}^D.$$

Returning to the case $D_{II} = 0$, we should add that, while the stress tensor is then indeterminate, it is still so that $\sigma_{II}^{1/2} \leqslant g$, because, in the opposite case, (1.13) furnishes the tensor $\mathbb{D}$ with $D_{II} > 0$.

To sum up, the constituent laws for the Bingham fluid can be written

(1.14) $$\begin{aligned} &\sigma_{II}^{1/2} < g \Leftrightarrow D_{ij} = 0 \\ &\sigma_{II}^{1/2} \geqslant g \Leftrightarrow D_{ij} = \frac{1}{2\mu}(1 - g/\sigma_{II}^{1/2})\sigma_{ij}^D. \end{aligned}$$ ▯

Remark 1.1. If in Equation (1.10) $g = 0$, one recovers the constituent law for a classical viscous incompressible fluid (Newtonian fluid). Thus, for small g, we can consider the Bingham fluid as a model close to the classical viscous fluid[3]. It has the additional particularity, made apparent in (1.14), that it moves like a rigid medium, if a certain function of the stresses—here $\sigma_{II}^{1/2}$—does not reach the yield limit g. ▯

---

[2] If a second order tensor is given with components T_{ij}, we again recall its decomposition into a spherical part and a deviation by

$T_{ij} = s\delta_{ij} + T_{ij}^D$, where $s = \tfrac{1}{3} T_{kk}$, and $T_{kk}^D = 0$.

[3] This will be more precisely stated in Section 5.

This type of behavior can be observed in the case of certain oils or certain sediments which are used in the process of oil drilling, as well as in concrete.

If g is strictly positive, one can observe rigid zones in the interior of the flow.

As g increases, these rigid zones become larger and may completely block the flow when g is sufficiently large. ▯

Remark 1.2. If we put $\mu=0$ in (1.10), the relations (1.14) are no longer valid; more precisely, the first one of the relations (1.14) remains in force, while the second one, which is the inverse of

$$\sigma_{ij}^D = g D_{ij}/D_{\mathrm{II}}^{1/2}$$

expresses that $\frac{1}{2}\sigma_{ij}^D\sigma_{ij}^D = g^2$ and that then the tensor D_{ij} is proportional to the tensor σ_{ij}^D; the relations (1.14) must therefore be replaced by

$$\begin{aligned} &\sigma_{\mathrm{II}} < g \Leftrightarrow D_{ij}=0 \\ &\sigma_{\mathrm{II}} = g \Rightarrow \exists \lambda \geqslant 0, \quad \text{such that} \quad D_{ij} = \lambda \sigma_{ij}^D. \end{aligned} \tag{1.15}$$

The materials which are subject to this type of constituent law are known as "*rigid perfectly plastic*", or also "with plastic Von Mises potential". We see that the representative point in $\mathbb{R}^6$ of the deviation of the stress tensor remains in the interior or on the boundary of the sphere with the equation

$$\sigma_{\mathrm{II}} \leqslant g^2. \tag{1.16}$$

When this point is in the interior, the material is in a rigid state, when it is on the boundary, the material may undergo a plastic deformation. ▯

1.3. Problems to be Considered and Recapitulation of the Equations

We consider three types of problems: flow in *the interior* of an open bounded set of $\mathbb{R}^3$, flow in *the exterior* of an open bounded set of $\mathbb{R}^3$ with given uniform velocity at infinity, laminar flow in the interior of an infinitely long cylindrical tube.

1) *Interior problem.* Let Ω be an open set of $\mathbb{R}^3$ with the boundary Γ. We seek the field of velocities $v(x,t)$ and the field of pressures $p(x,t)$ in a flow of a Bingham fluid in the interior of Ω.

The equations and boundary conditions are the following:

$$\gamma_i = \sigma_{ij,j} + f_i \quad \text{in } \Omega, \quad i=1,2,3, \quad \text{(equations of movement)} \tag{1.17}$$

where

$$\gamma_i = \partial v_i/\partial t + v_{i,j} v_j, \tag{1.18}$$

$$v_{i,j} = 0 \quad \text{(incompressibility)}. \tag{1.19}$$

The scalars f_i are components of a volume density $f(x,t)$ of given forces.

The components σ_{ij} of the stress tensor are related to the D_{ij} by the constituent law for Bingham fluids explicitly stated in Section 1.2.

The *boundary conditions* are the same as for classical viscous fluids, namely those of adhesion to the wall, i.e.,

(1.20) $$v_i = 0 \quad \text{on } \Gamma .$$

This problem can be treated in the stationary and non stationary case: in the first case, the unknown functions v_i and p are independent of time, in the second case, we have the initial conditions

(1.21) $$v_i|_{t=0} = v_{0i} ,$$

where the v_{0i} are given functions of the x_i satisfying (1.19) and (1.20).

2) *Exterior problem.* This is the problem of flow of a Bingham fluid with given constant velocity at infinity around a rigid fixed body.

The rigid body occupies a bounded domain $\tilde{\Omega}$ of $\mathbb{R}^3$ with a regular boundary. In $\mathbb{R}^3 - \tilde{\Omega} = \Omega$ the velocity field and the pressure satisfy Equations (1.17) to (1.20) to which we add the condition at infinity

(1.22) $$\lim_{|x| \to \infty} v(x,t) = (U_1, 0, 0) ,$$

where U_1 is a given positive constant, as well as an initial condition of type (1.21).

3) *Flow in a pipe.* This problem has already been treated by Mosolov and Miasnikov [1] by a method somewhat different from the one given here.

We are concerned with the stationary and *laminar*[4] flow of a Bingham fluid in a cylindrical pipe under the effect of a drop in pressure. ▯

The generators of the pipe are parallel to Ox_3 in the orthonormal system of axes $Ox_1x_2x_3$ (cf. Fig. 20).

Let Ω be the domain of $\mathbb{R}^2$ represented by a cross section of the cylinder. We study the flow between the cross sections

(1.23) $$x_3 = 0 \quad \text{and} \quad x_3 = L \quad (L\text{: given length})$$

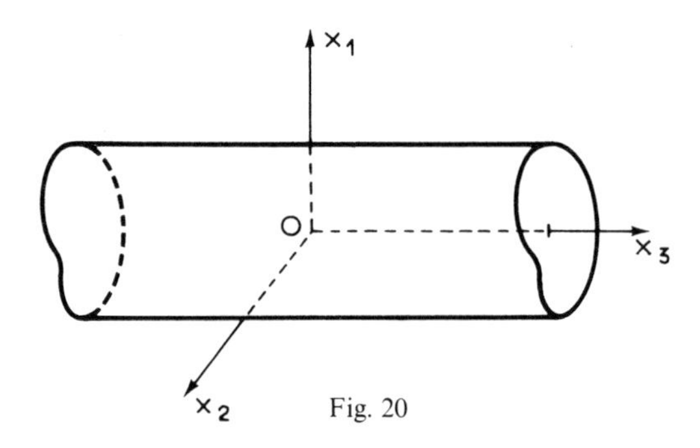

Fig. 20

[4] Which means here that the velocities are parallel to the generators of the cylinder.

on which we impose certain pressures say

$$(1.24)\qquad p(x_3)|_{x_3=0}=0\,, \qquad p(x_3)|_{x_3=L}=-cL\,.$$

The positive scalar c is the drop in pressure per length. ▯

If we seek the field of velocities and of pressures which satisfy (1.17), (1.18), (1.19), the constituent laws for Bingham fluids, the adhesion condition

$$(1.25)\qquad v=0 \quad \text{on} \quad \partial\Omega\times[0,L]\,,$$

and the conditions (1.24), we are led to a problem which is generally considered to be ill posed[5]. We restrict the class of admissible solutions by imposing on the flow the condition to be *laminar*.

This property is experimentally well confirmed, provided that the drop in pressure "c" is not too large, that is to say, viewed from the end effect, that the *Reynolds number*[6] of the flow is not too large.

Under these conditions, the velocity field is $(0,0,v)$ and Equation (1.19) implies that v depends only on x_1 and x_2. The tensor of strain rates $\mathbb{D}$ can then be written

$$(1.26)\qquad \mathbb{D}=\begin{pmatrix} 0 & 0 & \frac{1}{2}\partial v/\partial x_1 \\ 0 & 0 & \frac{1}{2}\partial v/\partial x_2 \\ \frac{1}{2}\partial v/\partial x_1 & \frac{1}{2}\partial v/\partial x_2 & 0 \end{pmatrix};$$

the stress deviator depends only on x_1 and x_2 and has the form

$$(1.27)\qquad \sigma+p\,\mathbb{1}=\begin{pmatrix} 0 & 0 & \sigma_{13}^D \\ 0 & 0 & \sigma_{23}^D \\ \sigma_{13}^D & \sigma_{23}^D & 0 \end{pmatrix}.$$

The equations of motion (1.17) then give:

$$(1.28)\qquad \begin{aligned} &\partial p/\partial x_1=0 \\ &\partial p/\partial x_2=0 \\ &\partial p/\partial x_3=\sigma_{31,1}^D+\sigma_{32,2}^D \qquad (f=0)\,. \end{aligned}$$

---

[5] The reason for this—but this is not a proof—is that the *restricted* problem which we are going to consider is *well posed*.

[6] One calls Reynolds number of the flow of a viscous fluid, the scalar

$\mathcal{R}=Ul\rho/\mu$,

where

U: is a characteristic velocity of the flow,
l: a characteristic length of the flow,
ρ: the specific mass of the fluid,
μ: the viscosity of the fluid.

In the third equation of (1.28), the left hand side depends only on x_3 and the right hand side only on x_1 and x_2. The two sides are therefore equal to the opposite of the drop in pressure c per unit length, from which

(1.29) $$p=-cx_3\,.$$

To sum up, we are led to look for a scalar function $v(x_1,x_2)$ in Ω which satisfies

(1.30) $$\sigma^D_{31,1}+\sigma^D_{32,2}+c=0 \quad \text{in } \Omega$$

with

(1.31) $$\begin{aligned}&\sigma^D_{3i}=gD_{3i}/D_{\mathrm{II}}^{1/2}+2\mu D_{3i}\,, \qquad i=1,2\\ &\text{when} \quad D_{\mathrm{II}}\neq 0\,, \quad (D_{\mathrm{II}}=(D_{31})^2+(D_{32})^2)\,,\end{aligned}$$

and the boundary condition

(1.32) $$v=0 \quad \text{on } \Gamma \quad (=\partial\Omega)\,. \quad \square$$

Remark 1.3. We set $f=0$ in (1.28) which gave us interesting simplifications.
The solution, if it exists, also permits us to solve a somewhat more general problem, which, in particular, includes the case of the volume forces of gravity. Let

(1.33) $$f=-\operatorname{grad}\Phi$$

with

(1.34) $$\Phi=\varphi(x_3)+\psi(x_1,x_2)\,.$$

If we set

(1.35) $$\tilde{p}=p+\Phi\,,$$

this quantity must satisfy Equations (1.28) where p has been replaced by $\tilde{p}$ which implies

$$\tilde{p}=p+\Phi=C_1x_3+C_2$$

where C_1 and C_2 are constants. Two cases must be distinguished:
i) $\psi\equiv 0$: conditions (1.24) can be satisfied by taking

$$p(x_3)=-cx_3+x_3L^{-1}[\varphi(L)-\varphi(0)]-[\varphi(x_3)-\varphi(0)]\,.$$

ii) $\psi\not\equiv 0$: conditions (1.24) can no longer be satisfied by a laminar flow. Moreover, these are no longer "natural" conditions: then we simply have to impose

(1.36) $$p(x_1^0, x_2^0, 0) = 0\,, \quad p(x_1^L, x_2^L, L) = -cL$$

where

$$(x_1^0, x_2^0) \in \Omega\,, \quad (x_1^L, x_2^L) \in \Omega\,,$$

and we obtain the solution in p by taking

(1.37) $$p(x_1, x_2, x_3) = C_1 x_3 + C_2 - \varphi(x_3) - \psi(x_1, x_2)$$

where the constants C_1 and C_2 are given by

(1.38) $$\begin{aligned} C_1 &= -c + \frac{\varphi(L) - \varphi(0)}{L} + \frac{\psi(x_1^L, x_2^L) - \psi(x_1^0, x_2^0)}{L} \\ C_2 &= \varphi(0) + \psi(x_1^0, x_2^0)\,. \end{aligned}$$

2. Flow in the Interior of a Reservoir. Formulation in the Form of a Variational Inequality

Now we *formally* derive a *variational inequality* "equivalent" to Problem 1 presented in the preceding Section 1.2. This variational inequality—which will be taken as the precise formulation of the problem—will subsequently be studied in Sections 3 to 6.

2.1. Preliminary Notation

We denote by $u = u(x,t)$ (or $u(x)$ in the stationary case) the field of vectors which solves Problem 1.

For an arbitrary field of vectors v, we set

(2.1) $$D_{\mathrm{II}}(v) = \tfrac{1}{2} D_{ij}(v) D_{ij}(v)\,,$$

(2.2) $$j(v) = 2 \int_\Omega (D_{\mathrm{II}}(v))^{1/2}\, \mathrm{d}x\,.$$

If w is a second vector field, we set

(2.3) $$a(v,w) = 2 \int_\Omega D_{ij}(v) D_{ij}(w)\, \mathrm{d}x\,.$$

The acceleration $\gamma = \gamma(v)$ associated with the velocity field v is the vector field with components $\gamma_i = \gamma_i(v)$:

(2.4) $$\gamma_i(v) = \mathrm{d}v_i/\mathrm{d}t = \partial v_i/\partial t + v_j v_{i,j}\,. \quad \square$$

2.2. Variational Inequality

We now prove

Theorem 2.1. *If* $u = u(x,t)$ *is a* (regular) *solution of Problem* 1, *then*

$$\operatorname{div} u = 0 \quad \forall t \in]0,T[, \tag{2.5}$$

$$u = 0 \quad on \quad \Gamma \times]0,T[, \tag{2.6}$$

$$a(u(t), v - u(t)) + gj(v) - gj(u(t)) \geq \int_\Omega (f - \gamma)(v - u(t))\,dx \tag{2.7}$$

$\forall$ *test vectors* v *such that*

$$\operatorname{div} v = 0, \quad v = 0 \quad \text{on } \Gamma \tag{2.8}$$

to which, obviously, one adds the initial condition (cf. (1.21)):

$$u(0) = u_0. \tag{2.9}$$

Remark 2.1. Naturally, for the stationary case, it is sufficient to set

$$\text{“}\partial u_i / \partial t = 0\text{”} \quad \text{in (2.7)}. \quad \square$$

Proof of Theorem 2.1. The only thing that has to be proved is inequality (2.7).

Let $\sigma_{ij}(u)$ be the components of the stress tensor associated with the field of velocities u by the constituent laws (1.11). Then, if v is a “test” field of velocities satisfying (2.8), we have

$$\int_\Omega \sigma_{ij}(u) D_{ij}(v - u)\,dx = \int_\Omega (f - \gamma)(v - u(t))\,dx \tag{2.10}$$

(where $u - u(t)$ on the left hand side of (2.10)). Writing out the left hand side of (2.10), we obtain as its value, using (1.10),

$$g \int_\Omega D_{II}(u)^{-1/2} D_{ij}(u)(D_{ij}(v - u))\,dx + \mu a(u, v - u). \tag{2.11}$$

In the first term of (2.11), we use the inequality

$$D_{ij}(u) D_{ij}(v) \leq 2 D_{II}(u)^{1/2} D_{II}(v)^{1/2} \tag{2.12}$$

so that the first term of (2.11) is majorized by

$$2g \int_\Omega (D_{II}(v)^{1/2} - D_{II}(u)^{1/2})\,dx = gj(v) - gj(u)$$

and (2.10) then implies (2.7). $\square$

We now prove, *purely formally*, that if u is a solution of (2.5), (2.6), (2.7), (2.9), then u is a solution of Problem 1.

Remark 2.2. This proof can only be a formal one, since the formulation of Problem 1 in "ordinary" terms is not perfectly precise. ▯

If the (or a) function $u=u(t)$ is a solution of (2.7) such that the functional $v \to j(v)$ is *differentiable* in u, i.e., if

$$D_{II}(u) \neq 0 \quad \text{a.e. in } \Omega, \tag{2.13}$$

then (2.7) is "equivalent" to

$$a(u(t),v)+g(j'(u(t)),v)=(f-\gamma,v)\ ^{7} \tag{2.14}$$

where

$$(j'(u),v) = \frac{d}{d\lambda} j(u+\lambda v)|_{\lambda=0} = \int_\Omega D_{II}(u)^{-1/2} D_{ij}(u) D_{ij}(v)\,dx\,. \tag{2.15}$$

Indeed, we replace v in (2.7) by

$$u(t)+\lambda v\,, \qquad \lambda>0$$

from which

$$\mu a(u(t),v)+g\lambda^{-1}[j(u(t)+\lambda v)-j(u(t))] \geqslant (f-\gamma,v)$$

and, letting λ tend towards 0:

$$\mu a(u(t),v)+g(j'(u),v) \geqslant (f-\gamma,v)$$

from which, changing v to $-v$, (2.14) follows.

But (2.14) *is equivalent* to

$$-\int_\Omega 2\mu \frac{\partial}{\partial x_j} D_{ij}(u) v_i\,dx + \int_\Omega g\left(\frac{\partial}{\partial x_j}(D_{II}(u)^{-1/2} D_{ij}(u))\right) v_i\,dx = \int_\Omega (f_i-\gamma_i) v_i\,dx \tag{2.16}$$

$\forall v=\{v_i\}$ with $\operatorname{div} v=0$.

Now, if $F=\{F_i\}$ is a vector orthogonal to the vectors of divergence zero:

$$\int_\Omega F_i v_i\,dx=0$$

then there exists a p (defined except for an additive constant) such that

$$F=\operatorname{grad} p \qquad (F_i=\partial p/\partial x_i)\,. \tag{2.17}$$

[7] Where $(f,\varphi)=\int_\Omega f_i \varphi_i\,dx$.

Consequently, (2.16) is equivalent to:

$$-2\mu(D_{ij}(u))_{,j} - g(D_{II}(u)^{-1/2} D_{ij}(u))_{,j} = f_i - \gamma_i - p_{,i}\,. \tag{2.18}$$

We again obtain (1.16) with (1.11). □

In view of this remark, it now appears reasonable[8] to take as our **definition of Problem 1**[9] **the problem of solving the variational inequality** (2.7) (with the conditions (2.5), (2.6), (2.9)). □

Remark 2.3. As we already observed repeatedly in connection with problems of variational inequalities, formulation (2.7) "automatically" takes into account **"free surfaces"**, here the "surface" *separating the region where the flow is regulated by* (2.18) *and the region where the fluid moves like a rigid medium.* □

Unsolved problem. No results seem to be known concerning the nature of this free surface.

3. Solution of the Variational Inequality, Characteristic for the Flow of a Bingham Fluid in the Interior of a Reservoir

3.1. Tools from Functional Analysis

Let Ω be a reservoir; Ω is an open bounded set of $\mathbb{R}^n$ [10] with a boundary Γ which we assume to be "sufficiently regular".

In Section I.3 we defined the space $H^s(\Omega)$ for an integer $s \geqslant 0$ and the space $H^s(\mathbb{R}^n)$ for an arbitrary real s.

For an arbitrary real $s \geqslant 0$, we can define

(3.1) $H^s(\Omega)$ = space of restrictions to Ω of the elements of $H^s(\mathbb{R}^n)$.
More precisely, let the mapping

$$f \to \pi f = \text{restriction of } f \text{ to } \Omega$$

of $H^s(\mathbb{R}^n) \to L^2(\Omega)$ (for example); then we define $H^s(\Omega)$ as the image of $H^s(\mathbb{R}^n)$ in the mapping π and we assign to $H^s(\Omega)$ the *Hilbert norm*:

$$\|\varphi\|_{H^s(\Omega)} = \inf_{f \in H^s(\mathbb{R}^n),\, \pi f = \varphi} \|f\|_{H^s(\mathbb{R}^n)}\,.$$

---

[8] This will be fully justified, in particular for two space dimensions, by the *existence and uniqueness* Theorem 3.1 (cf. Sec. 3 below).

[9] *We emphasize the fact that* (to our knowledge) *there is no other precise formulation of the problem once the function spaces where we look for a solution have been spelled out*; cf. also Remark 2.3 below.

[10] Obviously, in the applications, $n = 2$ or 3.

This definition *coincides* (within an equivalence of norms) with the usual definition when s is an integer[11]. By $(\varphi,\psi)_{H^s(\Omega)}$ we will denote the scalar product in $H^s(\Omega)$.

Next, we introduce:

$$\mathscr{V}=\{\varphi\,|\,\varphi\in(\mathscr{D}(\Omega))^n,\ \operatorname{div}\varphi=0\}, \tag{3.2}$$

$$V_s=\text{closure of } \mathscr{V} \text{ in } (H^s(\Omega))^n; \tag{3.3}$$

V_s is a Hilbert space for the norm

$$\|v\|_s=((v,v))_s^{1/2} \tag{3.4}$$

if

$$((v,w))_s=(v_i,w_i)_{H^s(\Omega)}\,. \tag{3.5}$$

We set:

$$V_1=V, \qquad \|v\|_1=\|v\|, \tag{3.6}$$

$$V_0=H, \qquad \|v\|_0=|v|, \qquad ((v,w))_0=(v,w)\,. \tag{3.7}$$

Lemma 3.1. *If* $s=n/2$, *we have*

$$v_{i,j}\in L^n(\Omega) \qquad \forall v\in V_s\,. \tag{3.8}$$

Proof. If $v\in V_s$, then $v_{i,j}\in H^{s-1}(\Omega)$. In general, (*J*. Peetre [1]), if $\varphi\in H^r(\Omega)$, then $\varphi\in L^p(\Omega)$, $1/p=1/2-r/n$ (if $1/2-r/n>0$); whence (3.8) follows. ◻

For a vector field u,v,w on Ω, we put (whenever this makes sense):

$$b(u,v,w)=\int_\Omega u_i v_{j,i} w_j\,dx\,. \tag{3.9}$$

In order to simplify the writing, we set

$$\|\varphi\|_{(L^p(\Omega))^n}=\|\varphi\|_{L^p(\Omega)}\,;$$

then the Hölder inequality, applied to (3.9), gives

$$\begin{gathered}|b(u,v,w)|\leqslant c_1\|u\|_{L^p(\Omega)}\|w\|_{L^p(\Omega)}\textstyle\sum_{i,j}\|D_i v_j\|_{L^n(\Omega)},\\ \dfrac{2}{p}+\dfrac{1}{n}=1\,.\end{gathered} \tag{3.10}$$

Using Lemma 3.1, it follows that:

$$|b(u,v,w)|\leqslant c_2\|u\|_{L^p(\Omega)}\|w\|_{L^p(\Omega)}\|v\|_s\,, \qquad s=n/2\,. \quad ◻ \tag{3.11}$$

[11] See for example Lions-Magenes [1], Vol. 1, Chap. 1.

Moreover, we make use of the *following inequality of convexity*[12]

$$\|v\|_{L^p(\Omega)} \leqslant c_3 \|v\|^{1/2}_{H^1(\Omega)} \|v\|^{1/2}_{L^2(\Omega)} \quad \forall v \in H^1_0(\Omega), \qquad (3.12)$$
$$\frac{1}{p} = \frac{1}{2} - \frac{1}{2n}.$$

Remark 3.1. An elementary proof of (3.12) is easy when $n=2$, in which case $p=4$. It is sufficient to prove (3.12) with $v \in \mathscr{D}(\Omega)$, and by extending v as 0 outside of Ω, it is sufficient to prove (3.12) for $v \in \mathscr{D}(\mathbb{R}^2)$. Now

$$v^2(x) = 2\int_{-\infty}^{x_1} v(\partial v/\partial x_1)\,dx_1 \leqslant 2\int_{-\infty}^{+\infty} |v|\,|\partial v/\partial x_1|\,dx_1 = 2v_1(x_2)$$

and by exchanging the indices 1 and 2, $v^2(x) \leqslant 2v_2(x_1)$ from which

$$\int_{\mathbb{R}^2} v^4(x)\,dx \leqslant 4\int_{-\infty}^{+\infty} v_1(x_2)\,dx_2 \int_{-\infty}^{+\infty} v_2(x_1)\,dx_1$$
$$\leqslant 4\|v\|_{L^2(\mathbb{R}^2)} \|\partial v/\partial x_1\|_{L^2(\mathbb{R}^2)} \|v\|_{L^2(\mathbb{R}^2)} \|\partial v/\partial x_2\|_{L^2(\mathbb{R}^2)}$$

from which, in particular, (3.12) follows (when $n=2$). □

From (3.11), (3.12) it follows that

$$|b(u,v,w)| \leqslant c_4 \|u\|^{1/2} |u|^{1/2} \|w\|^{1/2} \|v\|_s, \quad s=n/2. \qquad (3.13)$$

Henceforth, we will choose

$$s=n/2. \qquad (3.14)$$

We note that

$$V_s = V \quad \text{if} \quad n=2 \qquad (3.15)$$

and that, in all cases (with the possibility that $V_s - V$):

$$V_s \subset V \subset H \subset V' \subset V'_s \qquad (3.16)$$

where V' (resp. V'_s) denotes the dual of V (resp. V_s) *when H is identified with its dual.*

By w_j we denote the *eigenfunctions* of the canonical isomorphism Λ_s of $V_s \to V'_s$, i.e.,

$$((w_j, v))_s = \lambda_j (w_j, v)\ ^{13} \quad \forall v \in V_s, \quad |w_j| = 1. \quad □ \qquad (3.17)$$

---

[12] A consequence of the *theory of interpolation* which shows (see Lions-Magenes [1], Chap. 1) that $\|v\|_{H^{1/2}(\Omega)} \leqslant c\|v\|^{1/2}_{H^1(\Omega)} \|v\|^{1/2}_{L^2(\Omega)}$
and of the result of J. Peetre [1] (already used in Lemma 3.1) which shows that
$$H^{1/2}(\Omega) \subset L^p(\Omega), \quad \frac{1}{p} = \frac{1}{2} - \frac{1}{2n}.$$

[13] No summation in j on the right hand side of (3.17)!

We easily verify that, for $u, v, w \in \mathscr{V}$, we have:

$$b(u,v,w)+b(u,w,v)=0 \tag{3.18}$$

from which analogous relations follow by extension by continuity, applying (3.11).

3.2. Functional Formulation of the Variational Inequalities

If in (2.7) we replace $\gamma=\gamma(u)$ by its value taken from (2.4), we obtain, using the notation of (3.9) (and $u'=\partial u/\partial t$):

$$\begin{aligned}(u'(t),v-u(t))+\mu a(u(t),v-u(t))+b(u(t),u(t),v-u(t))+gj(v)-gj(u(t))\\ \geqslant (f(t),v-u(t))\,.\end{aligned} \tag{3.19}$$

But—here we continue the *formal* transformations of the problem so as to arrive at a definitive formulation—according to (3.18):

$$b(u(t),u(t),u(t))=0 \quad \text{and} \quad b(u(t),u(t),v)=-b(u(t),v,u(t))\,,$$

whence

$$\begin{aligned}(u'(t),v-u(t))+\mu a(u(t),v-u(t))-b(u(t),v,u(t))+gj(v)-gj(u(t))\\ \geqslant (f(t),v-u(t))\,.\end{aligned} \tag{3.20}$$ □

We now distinguish *two cases* according to the dimension n.

First case: Dimension $n=2$. We will prove further on

Theorem 3.1. *We assume that* $n=2$ *and that* f *and* u^0 *are given with*

$$f\in L^2(0,T;V')\,, \tag{3.21}$$

$$u^0\in H\,. \tag{3.22}$$

Then there exists a unique function u *such that*

$$u\in L^2(0,T;V)\,, \tag{3.23}$$

$$\partial u/\partial t\in L^2(0,T;V')\,, \tag{3.24}$$

u *satisfies* (3.19) (or (3.20)) $\forall v\in V$ *and*

$$u(0)=u^0\,. \tag{3.25}$$

Second case: Dimension $n\geqslant 3$. When the dimension is $\geqslant 3$, it is necessary (at least with the methods at our disposal) to introduce "*weaker*" solutions of (3.20).

In order to simplify the exposition, we make the assumption, for this case,

$$u^0=0\,. \tag{3.26}$$

We introduce the set

(3.27) $$W=\{v| \quad v\in L^2(0,T;V_s), \quad v'\in L^2(0,T;H), \quad v(0)=0\}.$$

We can take $v=v(t)$ (almost everywhere) in (3.20); then we show that, if u is a solution of (3.20), we have:

(3.28) $$\int_0^T \{(v',v-u)+\mu a(u,v-u)-b(u,v,u)+gj(v)-gj(u)-(f,v-u)\}\,\mathrm{d}t \geqslant 0 \quad \forall v\in W.$$

Indeed, taking account of (3.20) (with $v=v(t)$), we have:

$$(v',v-u)+\mu a(u,v-u)-b(u,v,u)+gj(v)-gj(u)-(f,v-u)=(v'-u',v-u)$$

so that the integral on the left hand side of (3.28) equals

$$\tfrac{1}{2}|v(T)-u(T)|^2$$

from which (3.28) follows.

We now take *inequality* (3.28) as the **definition** *of the problem.* Below, we shall prove

Theorem 3.2. *For an arbitrary n, let f be given with* (3.21). *There exists a function u such that*

(3.29) $$u\in L^2(0,T;V)\cap L^\infty(0,T;H), \quad \partial u/\partial t\in L^2(0,T;V_s'), \quad u(0)=0,$$

and u satisfies (3.28). ▯

Remark 3.2. The question of possible uniqueness of the solution for Theorem 3.2 is open for $n\geqslant 3$. ▯

Remark 3.3. For $n=2$, we will prove *uniqueness* of the solution for which (3.29) holds and also that u is then a solution in the sense of Theorem 3.1 (with $u^0=0$). ▯

Remark 3.4. We can also treat the case "$u^0\neq 0$" in the framework of Theorem 3.2, but at the price of additional technical difficulties—which we wanted to avoid. ▯

Remark 3.5. The case "$g=0$" corresponds to formulations, which have become classical since the work of Leray [1–3], of weak (or turbulent) solutions of the Navier-Stokes equations, at least if $n=2$. ▯

Remark 3.6. If $n=2$, it follows from (3.23) and (3.24), after possible modification on a set of measure zero, that $t\to u(t)$ is continuous from $[0,T]\to H$. ▯

We now prove the preceding Theorems, starting with Theorem 3.2.

3.3. Proof of Theorem 3.2

Outline of the proof. We *approximate* j by a *differentiable* functional: we choose[14]

$$j_\varepsilon(v) = \frac{2}{1+\varepsilon} \int_\Omega (D_{\mathrm{II}}(v))^{(1+\varepsilon)/2} \mathrm{d}x\,, \qquad \varepsilon>0\,. \tag{3.30}$$

Then we have

$$(j'_\varepsilon(u), v) = \int_\Omega D_{\mathrm{II}}(u)^{(\varepsilon-1)/2} D_{ij}(u) D_{ij}(v) \mathrm{d}x\,. \tag{3.31}$$

The natural idea is then to "approximate" the inequality by the equation

$$(u'_\varepsilon, v) + \mu a(u_\varepsilon, v) + b(u_\varepsilon, u_\varepsilon, v) + g(j'_\varepsilon(u_\varepsilon), v) = (f, v)\,, \tag{3.32}$$

with

$$u_\varepsilon(0) = 0 \;\text{[15]}. \tag{3.33}$$

However, there exists a technical difficulty in the solution of (3.32), (3.33) if $n>2$, which leads to the introduction of *a second regularization* through the addition of a "viscosity term"

$$\eta((u,v))_s \qquad (s=n/2)\,, \qquad \eta>0\,.$$

We thus introduce *the bi-regularized equation*:

$$(u'_{\varepsilon\eta}, v) + \mu a(u_{\varepsilon\eta}, v) + b(u_{\varepsilon\eta}, u_{\varepsilon\eta}, v) + \eta((u_{\varepsilon\eta}, v))_s + g(j'_\varepsilon(u_{\varepsilon\eta}), v) = (f, v)\,, \tag{3.34}$$

$$u_{\varepsilon\eta}(0) = 0\,, \tag{3.35}$$

the plan being the following:
  i) solution of (3.34), (3.35) and derivation of a priori estimates;
  ii) passage to the limit in ε and η. □

Step i). We use a "base" $w_1, \ldots, w_m, \ldots$ of the space V_s, more precisely, the "*special*" *base* of the eigenfunctions (3.17). We define $u_m (= u_{\varepsilon\eta m})$ as the solution of

$$\begin{aligned}(u'_m(t), w_j) + \mu a(u_m(t), w_j) + b(u_m(t), u_m(t), w_j) + \eta((u_m(t), w_j))_s + g(j'_\varepsilon(u_m(t)), w_j)\\ = (f(t), w_j)\,, \quad 1 \leqslant j \leqslant m\,,\end{aligned} \tag{3.36}$$

$$u_m(0) = 0 \tag{3.37}$$

which defines u_m in an interval $[0, t_m]$; but the a priori estimates which follow prove that $t_m = T$.

[14] The following proof is valid also *for other regularizations for* j, which can be of importance in *numerical* problems.

[15] In the case $n=2$ (Theorem 3.1), we use this method with $u_\varepsilon(0) = u^0 (\neq 0)$.

Making use of the fact that $b(u_m(t), u_m(t), u_m(t)) = 0$, we conclude from (3.36)[16]

$$\frac{1}{2}\frac{\mathrm{d}}{\mathrm{d}t}|u_m(t)|^2 + \mu a(u_m(t)) + \eta\,\|u_m(t)\|_s^2 + g(j'_\varepsilon(u_m(t)), u_m(t)) = (f(t), u_m(t))\,. \tag{3.38}$$

But there exists an $\alpha > 0$ such that

$$a(v) \geqslant \alpha\,\|v\|^2 \qquad \forall v \in V \tag{3.39}$$ [17]

and

$$(j'_\varepsilon(v), v) \geqslant 0\,. \tag{3.40}$$

Then it follows from (3.38) that:

$$|u_m(t)|^2 + 2\alpha\mu \int_0^t \|u_m(\sigma)\|^2\,\mathrm{d}\sigma + 2\eta \int_0^t \|u_m(\sigma)\|_s^2\,\mathrm{d}\sigma$$ [18]

$$\leqslant 2\int_0^t \|f(\sigma)\|_*\,\|u_m(\sigma)\|\,\mathrm{d}\sigma$$

$$\leqslant \alpha\mu \int_0^t \|u_m(\sigma)\|^2\,\mathrm{d}\sigma + \frac{1}{\alpha\mu}\int_0^t \|f(\sigma)\|_*^2\,\mathrm{d}\sigma\,,$$

and therefore

$$|u_m(t)|^2 + \alpha\mu \int_0^t \|u_m(\sigma)\|^2\,\mathrm{d}\sigma + 2\eta \int_0^t \|u_m(\sigma)\|_s^2\,\mathrm{d}\sigma \leqslant \frac{1}{\alpha\mu}\int_0^t \|f(\sigma)\|_*^2\,\mathrm{d}\sigma\,. \tag{3.41}$$

It follows that:

(3.42) u_m remains in a bounded set of $L^2(0,T;V) \cap L^\infty(0,T;H)$, *independent of* m, ε, η,

(3.43) $\eta^{1/2} u_m$ remains in a bounded set (a set independent of m, ε, η) of $L^2(0,T;V_s)$. □

We now prove that

(3.44) u'_m remains in a bounded set (independent of m, ε, η) of $L^2(0,T;V'_s)$.

To begin with, we observe that, for $v \in V_s$,

$$b(u_m(t), u_m(t), v) = -b(u_m(t), v, u_m(t))$$

satisfies, according to (3.13):

$$|b(u_m(t), u_m(t), v)| \leqslant c_4\,\|u_m(t)\|\,|u_m(t)|\,\|v\|_s$$

[16] We set: $a(v,v) = a(v)$.
[17] Because $v \in V \Rightarrow v_i \in H_0^1(\Omega)$.
[18] We set: $\|f\|_* =$ norm in V' dual to $\|\ \ \| = \sup[|(f,v)|/\|v\|]$, $v \in V$.

and since, according to (3.42), $u_m(t) \leqslant C$, we have (the c's denoting different constants, independent of m, ε, η):

$$(3.45) \qquad |b(u_m(t), u_m(t), v)| \leqslant c \|u_m(t)\| \|v\|_s$$

whence (applying (3.42)):

$$(3.46) \qquad \begin{aligned} &b(u_m(t), u_m(t), v) = (h_m(t), v), \quad v \in V_s, \\ &h_m \text{ remaining in a bounded set of } L^2(0,T; V_s'). \end{aligned}$$

We note that the form $v \to a(u,v)$ is continuous on V, therefore

$$(3.47) \qquad a(u,v) = (Au, v), \quad A \in \mathscr{L}(V; V')$$

so that (3.36) can be written in the (equivalent) form

$$(3.48) \qquad (u_m' + \mu A u_m + h_m + \eta \Lambda_s u_m + g j'(u_m) - f, w_j) = 0, \quad 1 \leqslant j \leqslant m.$$

If P_m denotes the orthogonal projection operator of $H \to [w_1, \dots, w_m]$, therefore

$$(3.49) \qquad P_m h = (h, w_j) w_j \quad (j \text{ varying from 1 to } m),$$

it follows from (3.48) (since $P_m u_m' = u_m'$) that

$$(3.50) \qquad u_m' = P_m(f - \mu A u_m - h_m - \eta \Lambda_s u_m - g j_\varepsilon'(u_m)).$$

But because of (3.42) and (3.47), $A u_m$ remains in a bounded set of

$$L^2(0,T; V') \subset L^2(0,T; V_s');$$

because of (3.43)

$$\|\eta \Lambda_s u_m\|_{L^2(0,T; V')} = 0(\eta^{1/2});$$

finally, because of (3.31)

$$\|j_\varepsilon'(u)\|_* \leqslant c\left(\int_\Omega D_{II}(u)^\varepsilon dx\right)^{1/2}$$

thus, *in particular*, $j_\varepsilon'(u_m)$ remains in a bounded set of $L^2(0,T; V')$. Therefore (3.50) implies

$$(3.51) \qquad \begin{aligned} &u_m' = P_m k_m, \\ &k_m \in \text{bounded set of } L^2(0,T; V_s'). \end{aligned}$$

Thus, we will have proved (3.44), if we verify that

$$(3.52) \qquad \|P_m \varphi\|_{V_s'} \leqslant c \|\varphi\|_{V'_s}.$$

But $\lambda_j^{1/2} w_j$ forms a complete orthogonal system of V_s' (for the norm

$$\|\varphi\|_{V_s'} = \|\Lambda_s^{-1}\varphi\|_{V_s})$$

so that

$$\|\varphi\|_{V_s'}^2 = \sum_{j=1}^{\infty} (\varphi, w_j \lambda_j^{1/2}),$$

$$\|P_m\varphi\|_{V_s'}^2 = \sum_{j=1}^{m} (\varphi, w_j \lambda_j^{1/2})$$

whence (3.52), with $c=1$. □

Passage to the limit in m. Our arguments will make use of both, compactness and monotonicity[19]. Because of (3.42), (3.44), we can select from the u_m a subsequence u_μ such that

$$(3.53) \qquad \begin{aligned} &u_\mu \to u_\varepsilon \text{ weakly star in } L^2(0,T;H) \text{ and weakly in } L^2(0,T;V),\\ &u'_\mu \to u'_{\varepsilon\eta} \text{ weakly in } L^2(0,T;V_s'), \end{aligned}$$

$u_\mu \to u_{\varepsilon\eta}$ strongly in $L^2(0,T;H)$ and, possibly after further selection,

$$(3.54) \qquad \begin{aligned} &u_{i\mu} \ (i\text{-th component of } u_\mu) \to u_{i\varepsilon\eta} \ (i\text{-th component of } u_{\varepsilon\eta})\\ &\text{a.e. in } \Omega \times]0,T[\,; \end{aligned}$$

furthermore, $j_\varepsilon'(u_m)$ (resp. $u_{i\mu}u_{j\mu}$) remains in a bounded set of $L^2(0,T;V')$ (resp. of $L^2(0,T;L^{p/2}(\Omega))$, according to (3.12) and (3.42); therefore, we can assume equally well that

$$(3.55) \qquad j'(u_\mu) \to \chi \quad \text{weakly in } L^2(0,T;V'),$$

$$(3.56) \qquad u_{i\mu}u_{j\mu} \to \theta_{ij} \quad \text{weakly in } L^2(0,T;L^{p/2}(\Omega)).$$

But, according to (3.45), $u_{i\mu}u_{j\mu} \to u_{i\varepsilon\eta}u_{j\varepsilon\eta}$, for example in the sense of distributions in Q, whence, by comparison with (3.56):

$$(3.57) \qquad \theta_{ij} = u_{i\varepsilon\eta}u_{j\varepsilon\eta}.$$

From this, it follows that

$$(3.58) \qquad \begin{aligned} &b(u_\mu, u_\mu, w_j) = -b(u_\mu, w_j, u_\mu) \to -b(u_{\varepsilon\eta}, w_j, u_{\varepsilon\eta})\\ &(\text{weakly in } L^2(0,T), \ \forall w_j). \end{aligned}$$

We then derive from (3.36) (for $m=\mu$) that

$$(3.59) \qquad (u'_{\varepsilon\eta}, w_j) + \mu a(u_{\varepsilon\eta}, w_j) = b(u_{\varepsilon\eta}, w_j, u_{\varepsilon\eta}) + \eta((u_{\varepsilon\eta}, w_j))_s + g(\chi, w_j) = (f, w_j), \quad \forall j.$$

[19] Cf. other examples in Lions [1], Chap. 2.

Thus, since the system of the w_j is complete in V_s, it follows from (3.59) that

$$(3.60)\qquad (u'_{\varepsilon\eta}, v) + \mu a(u_{\varepsilon\eta}, v) - b(u_{\varepsilon\eta}, v, u_{\varepsilon\eta}) + \eta((u_{\varepsilon\eta}, v))_s + g(\chi, v) = (f, v) \qquad \forall v \in V_s.$$

Since, obviously, (3.35) holds, we will thus have concluded step i), if we prove that

$$(3.61)\qquad \chi = j'_\varepsilon(u_{\varepsilon\eta}).$$

To this end, we apply a "monotonicity argument". □

Let φ be a function $\in L^2(0, T; V_s)$, with $\varphi' \in L^2(0, T; V'_s)$, $\varphi(0) = 0$; we set

$$X_\mu = g \int_0^T (j'_\varepsilon(u_\mu) - j'_\varepsilon(\varphi), u_\mu - \varphi)\,dt + \mu \int_0^T a(u_\mu - \varphi)\,dt$$
$$+ \eta \int_0^T \|u_\mu - \varphi\|_s^2\,dt + \int_0^T (u'_\mu - \varphi', u_\mu - \varphi)\,dt.$$

Using (3.36), we have

$$X_\mu = \int_0^T (f, u_\mu)\,dt - g \int_0^T [(j'_\varepsilon(u_\mu), \varphi) + (j'_\varepsilon(\varphi), u_\mu - \varphi)]\,dt$$
$$- \mu \int_0^T [a(u_\mu, \varphi) + a(\varphi, u_\mu - \varphi)]\,dt - \eta \int_0^T [((u_\mu, \varphi))_s$$
$$+ ((\varphi, u_\mu - \varphi))_s]\,dt - \int_0^T [(u'_\mu, \varphi) + (\varphi', u_\mu - \varphi)]\,dt$$

whence follows: $X_\mu \to X$ with

$$X = \int_0^T \{(f, u_{\varepsilon\eta}) - g(\chi, \varphi) - g(j'_\varepsilon(\varphi), u_{\varepsilon\eta} - \varphi) - \mu a(u_{\varepsilon\eta}, \varphi)$$
$$- \mu a(\varphi, u_{\varepsilon\eta} - \varphi) - \eta((u_{\varepsilon\eta}, \varphi))_s - \eta((\varphi, u_{\varepsilon\eta} - \varphi))_s$$
$$- (u'_{\varepsilon\eta}, \varphi) - (\varphi', u_{\varepsilon\eta} - \varphi)\}\,dt.$$

But, taking $v = u_{\varepsilon\eta}(t)$ (almost everywhere) in (3.60), which is permissible, we deduce that

$$X = \int_0^T \{g(\chi - j'_\varepsilon(\varphi), u_{\varepsilon\eta} - \varphi) + \mu a(u_{\varepsilon\eta} - \varphi) + \eta \|u_{\varepsilon\eta} - \varphi\|_s^2$$
$$+ (u'_{\varepsilon\eta} - \varphi', u_{\varepsilon\eta} - \varphi)\}\,dt.$$

Since $X_\mu \geqslant 0$ $\forall \mu$, we have:

$$X \geqslant 0.$$

Taking $\varphi = u_{\varepsilon\eta} - \lambda\psi$, $\psi \in L^2(0, T; V_s)$, $\psi' \in L^2(0, T; V'_s)$, $\psi(0) = 0$, $\lambda > 0$, we conclude (after division by λ) that:

$$g \int_0^T (\chi - j'_\varepsilon(u_{\varepsilon\eta} - \lambda\psi), \psi)\,dt + \lambda \int_0^T \{\mu a(\psi) + \eta \|\psi\|_s^2 + (\psi', \psi)\}\,dt \geqslant 0.$$

Letting $\lambda \to 0$, we obtain

$$g \int_0^T (\chi - j'_\varepsilon(u_{\varepsilon\eta}), \psi)\,dt \geqslant 0 \qquad \forall \psi$$

from which (3.61) follows. □

Thus we completed step i) and proved the existence of $u_{\varepsilon\eta}$ satisfying (3.34), (3.35) and:

$$\begin{aligned}&u_{\varepsilon\eta} \text{ remains in a bounded set of } L^2(0,T;V)\cap L^\infty(0,T;H),\\ &u'_{\varepsilon\eta} \text{ remains in a bounded set of } L^2(0,T;V'_s),\\ &\eta^{1/2}u_{\varepsilon\eta} \text{ remains in a bounded set of } L^2(0,T;V_s). \quad \square\end{aligned} \tag{3.62}$$

Step ii). For a fixed v in W (cf. (3.28)), we introduce the expression:

$$\begin{aligned}Y_{\varepsilon\eta} = \int_0^T \{&(v', v-u_{\varepsilon\eta}) + \mu a(u_{\varepsilon\eta}, v-u_{\varepsilon\eta}) + b(u_{\varepsilon\eta}, u_{\varepsilon\eta}, v-u_{\varepsilon\eta})\\ &+\eta((u_{\varepsilon\eta}, v-u_{\varepsilon\eta}))_s + gj_\varepsilon(v) - gj_\varepsilon(u_{\varepsilon\eta}) - (f, v-u_{\varepsilon\eta})\}\,\mathrm{d}t.\end{aligned} \tag{3.63}$$

According to (3.34), we have

$$Y_{\varepsilon\eta} = \int_0^T (v'-u'_{\varepsilon\eta}, v-u_{\varepsilon\eta})\,\mathrm{d}t + g\int_0^T \{j_\varepsilon(v) - j_\varepsilon(u_{\varepsilon\eta}) - (j'_\varepsilon(u_{\varepsilon\eta}), v-u_{\varepsilon\eta})\}\,\mathrm{d}t. \tag{3.64}$$

But the first term in the expression (3.64) for $Y_{\varepsilon\eta}$ equals $\frac{1}{2}|v(T)-u_{\varepsilon\eta}(T)|^2$ and the second term is $\geqslant 0$ because of the convexity of $v\to j'_\varepsilon(v)$. Therefore $Y_{\varepsilon\eta}\geqslant 0$, or again

$$\begin{aligned}\int_0^T \{&(v', v-u_{\varepsilon\eta}) + \mu a(u_{\varepsilon\eta}, v) - b(u_{\varepsilon\eta}, v, u_{\varepsilon\eta}) + \eta((u_{\varepsilon\eta}, v))_s\\ &+ gj_\varepsilon(v) - (f, v-u_{\varepsilon\eta})\}\,\mathrm{d}t \geqslant \mu\int_0^T a(u_{\varepsilon\eta})\,\mathrm{d}t + g\int_0^T j_\varepsilon(u_{\varepsilon\eta})\,\mathrm{d}t.\end{aligned} \tag{3.65}$$

Because of (3.62), we can select a sequence, again denoted by $u_{\varepsilon\eta}$, such that

$$\begin{aligned}&u_{\varepsilon\eta}\to u \text{ weakly star in } L^\infty(0,T;H) \text{ and weakly in } L^2(0,T;V),\\ &u'_{\varepsilon\eta}\to u' \text{ weakly in } L^2(0,T;V').\end{aligned} \tag{3.66}$$

It follows then from (3.65) and (3.66) that

$$\begin{aligned}\int_0^T \{&(v', v-u) + \mu a(u,v) - b(u,v,u) + gj(v) - (f, v-u)\}\,\mathrm{d}t\\ &\geqslant \liminf \mu\int_0^T a(u_{\varepsilon\eta})\,\mathrm{d}t + \liminf g\int_0^T j_\varepsilon(u_{\varepsilon\eta})\,\mathrm{d}t.\end{aligned} \tag{3.67}$$

But since the function $v\to\int_0^T a(v)\,\mathrm{d}t$ is lower semi-continuous for $L^2(0,T;V)$ with the weak topology, we have:

$$\liminf \int_0^T a(u_{\varepsilon\eta})\,\mathrm{d}t \geqslant \int_0^T a(u)\,\mathrm{d}t. \tag{3.68}$$

We prove that

$$\liminf \int_0^T j_\varepsilon(u_{\varepsilon\eta})\,\mathrm{d}t \geqslant \int_0^T j(u)\,\mathrm{d}t. \tag{3.69}$$

We have

$$\int_0^T j(u)\,\mathrm{d}t \leqslant \left(\int_Q D_{\mathrm{II}}(u)^{(1+\varepsilon)/2}\,\mathrm{d}x\,\mathrm{d}t\right)^{1/(1+\varepsilon)}\left(\int_Q \mathrm{d}x\,\mathrm{d}t\right)^{\varepsilon/(1+\varepsilon)}$$

from which

$$\int_0^T j_\varepsilon(u_{\varepsilon\eta})\,dt \geqslant c_\varepsilon\left(\int_0^T j(u_{\varepsilon\eta})\,dt\right)^{1+\varepsilon},$$
$$c_\varepsilon = |Q|^{-\varepsilon}, \qquad |Q| = \text{measure of } Q.$$

$$\liminf \int_0^T j_\varepsilon(u_{\varepsilon\eta})\,dt \geqslant \liminf \int_0^T j(u_{\varepsilon\eta})\,dt \tag{3.70}$$

and, since the function $v \to \int_0^T j(v)\,dt$ is continuous and convex on $L^2(0,T;V)$, it is lower semi-continuous for the weak topology on $L^2(0,T;V)$, therefore

$$\liminf \int_0^T j(u_{\varepsilon\eta})\,dt \geqslant \int_0^T j(u)\,dt \tag{3.71}$$

which, together with (3.70), gives (3.69).

From (3.67)–(3.69), it follows that u satisfies (3.28). □

3.4. Proof of Theorem 3.1

3.4.1. Existence Proof

In the case "$n=2$", the introduction of the regularization $\eta((u,v))_s$ (cf. (3.34)) is unnecessary because $s=1$, therefore $V_s=V$ and the term $\mu a(u,v)$ is "as powerful" as $((u,v))_s$.

Thus, we solve (3.32) *directly* with (instead of (3.33)):

$$u_\varepsilon(0) = u^0. \tag{3.72}$$

The method of Section 3.3 proves the existence of a solution u_ε of (3.32), (3.72) with

$$u_\varepsilon \text{ remains in a bounded set of } L^2(0,T;V), \tag{3.73}$$

$$u'_\varepsilon \text{ remains in a bounded set of } L^2(0,T;V'). \tag{3.74}$$

The *essential* point which permits us to obtain existence of a "strong" solution of (3.19) (instead of the "weak" formulation (3.28)), and which permits us to prove uniqueness, is that estimates (3.73), (3.74) hold in the two spaces $L^2(0,T;V)$ and $L^2(0,T;V')$ *which are dual to each other.*

For $v \in L^2(0,T;V)$, we introduce (compare with (3.63)):

$$\begin{aligned} Z_\varepsilon = \int_0^T \{&(u'_\varepsilon, v-u_\varepsilon) + \mu a(u_\varepsilon, v-u_\varepsilon) + b(u_\varepsilon, v-u_\varepsilon) \\ &+ g j_\varepsilon(v) - g j_\varepsilon(u_\varepsilon) - (f, v-u_\varepsilon)\}\,dt. \end{aligned} \tag{3.75}$$

Using (3.32), we see that

$$Z_\varepsilon = g\int_0^T \{j_\varepsilon(v) - j_\varepsilon(u_\varepsilon) - (j'_\varepsilon(u_\varepsilon), v-u_\varepsilon)\}\,dt \geqslant 0. \tag{3.76}$$

Therefore

$$\begin{aligned}(3.77)\qquad &\int_0^T \{(u_\varepsilon', v)+\mu a(u_\varepsilon, v)+b(u_\varepsilon, u_\varepsilon, v)+g j_\varepsilon(v)-(f, v-u_\varepsilon)\}\,\mathrm{d}t\\ &\geqslant \int_0^T \{(u_\varepsilon', u_\varepsilon)+\mu a(u_\varepsilon)+g j_\varepsilon(u_\varepsilon)\}\,\mathrm{d}t\\ &=\tfrac{1}{2}|u_\varepsilon(T)|^2-\tfrac{1}{2}|u^0|^2+\mu\int_0^T a(u_\varepsilon)\,\mathrm{d}t+g\int_0^T j_\varepsilon(u_\varepsilon)\,\mathrm{d}t\,.\end{aligned}$$

According to (3.73), (3.74), we can assume that a subsequence has been selected from u_ε, again denoted by u_ε, such that

$$(3.78)\qquad u_\varepsilon \to u \text{ (resp. } u_\varepsilon' \to u') \text{ weakly in } L^2(0,T;V) \text{ (resp. weakly in } L^2(0,T;V')).$$

From (3.78), it follows that $u_\varepsilon(T)\to u(T)$ weakly in H; then every term on the right hand side of (3.77) is lower semi-continuous for the convergence defined by (3.78), from which[20]

$$\begin{aligned}&\int_0^T \{(u', v)+\mu a(u, v)+b(u, u, v)+g j(v)-(f, v-u)\}\,\mathrm{d}t\\ &\geqslant \tfrac{1}{2}\liminf|u_\varepsilon(T)|^2-\tfrac{1}{2}|u^0|^2+\mu\liminf\int_0^T a(u_\varepsilon)\,\mathrm{d}t\\ &\quad+g\liminf\int_0^T j_\varepsilon(u_\varepsilon)\,\mathrm{d}t\\ &\geqslant \tfrac{1}{2}|u(T)|^2-\tfrac{1}{2}|u^0|^2+\mu\int_0^T a(u)\,\mathrm{d}t+g\int_0^T j(u)\,\mathrm{d}t\\ &=\int_0^T \{(u', u)+\mu a(u)+g j(u)\}\,\mathrm{d}t\end{aligned}$$

and, consequently,

$$\begin{aligned}(3.79)\qquad &\int_0^T \{(u', v-u)+\mu a(u, v-u)+b(u, u, v)+g j(v)-g j(u)\ -(f, v-u)\}\,\mathrm{d}t\geqslant 0\,,\\ &\forall v\in L^2(0,T;V)\,.\end{aligned}$$

We now prove that if u satisfies (3.79), *then u satisfies* (3.19). For this purpose, let w be an arbitrary fixed element of V, and let, for the moment, t_0 be fixed arbitrarily in $]0,T[$; we introduce

$$\mathcal{O}_j =]t_0-1/j,\, t_0+1/j[\, \subset\,]0,T[$$

for sufficiently large j, and we define by

$$(3.80)\qquad v(t)=\begin{cases} w & \text{if } t\in\mathcal{O}_j\\ u(t) & \text{if } t\notin\mathcal{O}_j, \qquad t\in[0,T]\,.\end{cases}$$

For v defined by (3.80), (3.79) reduces to

$$(3.81)\qquad \int_{\mathcal{O}_j}\{(u', w-u)+\mu a(u, w-u)+b(u, u, w)+g j(w)-g j(u)-(f, w-u)\}\,\mathrm{d}t\geqslant 0\,.$$

---

[20] As for (3.58), we see that

$$\int_0^T b(u_\varepsilon, u_\varepsilon, v)\,\mathrm{d}t \to \int_0^T b(u, u, v)\,\mathrm{d}t\,.$$

But

$$b(u,u,w)=(\tilde{u},w)\,,\qquad \tilde{u}\in L^2(0,T;V')$$

and, dividing (3.81) by $|\mathcal{O}_j|$ = measure of $\mathcal{O}_j$, we have:

$$\begin{aligned}(3.82)\qquad &|\mathcal{O}_j|^{-1}\int_{\mathcal{O}_j}(u'+\mu Au+\tilde{u}-f,w)\,dt\\ &\quad -|\mathcal{O}_j|^{-1}\int_{\mathcal{O}_j}\{(u',u)+\mu a(u)+gj(u)-(f,u)\}\,dt+gj(w)\geqslant 0\,.\end{aligned}$$

But according to the Lebesgue theorem on the differentiation of set functions (cf. for example, Dunford-Schwartz [1], III.12.9), we have:

$$\begin{aligned}&|\mathcal{O}_j|^{-1}\int_{\mathcal{O}_j}(u'+\mu Au+\tilde{u}-f)\,dt\to u'(t_0)+\mu Au(t_0)+\tilde{u}(t_0)\\ &\qquad -f(t_0)\quad\text{on } V' \text{ for }\quad t_0\in E_1,\quad \text{measure}\quad (E_1)=0\end{aligned}$$

and

$$\begin{aligned}&|\mathcal{O}_j|^{-1}\int_{\mathcal{O}_j}\{(u',u)+\mu a(u)+gj(u)-(f,u)\}\,dt\to(u'(t_0),u(t_0))\\ &\qquad +\mu a(u(t_0))+gj(u(t_0))-(f(t_0),u(t_0))\end{aligned}$$

for $t_0\notin E_2$, measure $(E_2)=0$.

Then, for $t_0\notin E_1\cup E_2$, we can pass to the limit in (3.82) and find

$$\begin{aligned}&(u'(t_0)+Au(t_0)+\tilde{u}(t_0)-f(t_0),w)-(u'(t_0),u(t_0))-\mu a(u(t_0))\\ &\qquad -(f(t_0),u(t_0))+gj(w)-gj(u(t_0))\geqslant 0\end{aligned}$$

i.e. (3.19). □

3.4.2. Uniqueness Proof

Let u and u_* be two possible solutions of (3.19) which satisfy (3.23), (3.24), (3.25) (and the analogous conditions for u_*). Taking $v=u_*(t)$, which is permissible, (resp. $v=u(t)$) in (3.19) (resp. in the analogous inequality for u), we find by addition (and setting $U=u-u_*$):

$$(3.83)\qquad -(U',U)-\mu a(U)-b(u,u,U)+b(u_*,u_*,U)\geqslant 0$$

from which

$$(2.84)\qquad \frac{1}{2}\frac{d}{dt}|U(t)|^2+\mu\alpha\|U(t)\|^2\leqslant b(u-U,u-U,U)-b(u,u,U)=-b(U,u,U)\,.$$

But according to (3.13)

$$|b(U(t),u(t),U(t))|\leqslant c\|U(t)\|\,|U(t)|\,\|u(t)\|\leqslant\mu\alpha\|U(t)\|^2+c'\|u(t)\|^2|U(t)|^2\,.$$

If we set

$$(3.85)\qquad m(t)=\|u(t)\|^2\,,$$

we then have

$$\frac{1}{2}\frac{\mathrm{d}}{\mathrm{d}t}|U(t)|^2 \leqslant c'm(t)|U(t)|^2$$

from which

$$|U(t)|^2 \leqslant 2c' \int_0^t m(\sigma)|U(\sigma)|^2 \,\mathrm{d}\sigma .$$

Since $m \in L^1(0, T)$, we conclude that $U=0$. ▯

Remark 3.7 (Proof of Remark 3.2). If $n=2$, every function u which satisfies (3.29) and (3.28), satisfies (3.19) (and therefore, when $n=2$, (3.28), (3.29) have a unique solution).

Indeed, let us set

$$W_1 = \{v| \quad v \in L^2(0, T; V), \quad v' \in L^2(0, T; V'), \quad v(0)=0\} . \tag{3.86}$$

The form $v \to$(right hand side of (3.28)) is (for $n=2$) continuous on W in the topology induced by W_1, and, since W is dense in W_1, we see that (3.28) is true $\forall v \in W_1$.

Let us now take an arbitrary w in W_1. Since $u \in W_1$, we can choose in (3.28)

$$v = v_\theta = (1-\theta)u + \theta w , \quad \theta \in]0,1[. \tag{3.87}$$

We note that

$$b(u, u_\theta, u) = \theta b(u, w, u)$$

$$j(v_\theta) \leqslant (1-\theta) j(u) + \theta j(w) .$$

Therefore, (3.28) yields

$$\theta \int_0^T \{(v'_\theta, w-u) + \mu a(u, w-u) - b(u, w, u) + gj(w) - gj(u) - (f, w-u)\} \,\mathrm{d}t \geqslant 0 . \tag{3.88}$$

Dividing by θ and then letting $\theta \to 0$, we deduce from (3.88) that

$$\int_0^T \{(u', w-u) + \mu a(u, w-u) - b(u, w, u) + gj(w) - gj(u) - (f, w-u)\} \,\mathrm{d}t \geqslant 0 \tag{3.89}$$

which implies (since W_1 is dense in $L^2(0, T; V)$) that (3.79) holds whence the result follows. ▯

4. A Regularity Theorem in Two Dimensions

Theorem 4.1. *We assume that* $n=2$ *and that* f *and* u^0 *are given with*

$$f \in L^2(0,T;V'), \quad f' \in L^2(0,T;V'), \quad f(0) \in H, \tag{4.1}$$

$$u^0 \in V \quad \text{and} \quad u^0 \in (H^2(\Omega))^2. \tag{4.2}$$

Then there exists one and only one function u *such that*

$$u \in L^2(0,T;V) \text{ [21]}, \tag{4.3}$$

$$u' \in L^2(0,T;V) \cap L^\infty(0,T;H), \tag{4.4}$$

and satisfying (3.19) *and* (3.25).

Proof. We start out with a solution u_ε of (3.32), (3.72).
The theorem then is the result of the supplementary estimates:

$$\|u_\varepsilon'\|_{L^2(0,T;V)} + \|u_\varepsilon'\|_{L^\infty(0,T;H)} \leqslant C. \tag{4.5}$$

To obtain (4.5), we prove analogous estimates for the solutions u_m of

$$(u_m', w_j) + \mu a(u_m, w_j) + b(u_m, u_m, w_j) + g(j_\varepsilon'(u_m), w_j) = (f, w_j), \qquad 1 \leqslant j \leqslant m, \tag{4.6}$$

$$u_m(0) = u_m^0, u_m^0 \to u^0 \quad \text{in} \quad V \cap (H^2(\Omega))^2. \tag{4.7}$$

It follows from (4.6) for $t=0$ that

$$|u_m'(0)|^2 = -\mu a(u_m^0, u_m'(0)) - b(u_m^0, u_m^0, u_m'(0)) - g(j_\varepsilon'(u_m^0), u_m'(0)) + (f(0), u_m'(0)). \tag{4.8}$$

Because of (4.2), $|a(u_m^0, u_m'(0))| \leqslant c|u_m'(0)|$.
Similarly, using in particular (4.1):

$$|g(j_\varepsilon'(u_m^0), u_m'(0)) + (f(0), u_m'(0))| \leqslant c|u_m'(0)|,$$

so that (4.8) gives

$$|u_m'(0)|^2 \leqslant c|u_m'(0)| + |b(u_m^0, u_m^0, u_m'(0))|. \tag{4.9}$$

But in two dimensions, $H^2(\Omega) \subset L^\infty(\Omega)$, therefore

$$|b(u_m^0, u_m^0, u_m'(0))| \leqslant c\|u_m^0\|_{(L^\infty(\Omega))^2} \|u_m^0\| \, |u_m'(0)| \leqslant c|u_m'(0)|$$

and (4.9) gives

$$|u_m'(0)| \leqslant C. \tag{4.10}$$

---

[21] Actually, due to (4.4), u is continuous from $[0,T] \to V$.

We now differentiate (4.6) with respect to t (actually, this is a formal procedure and is justified by taking difference quotients which approximate the derivative with respect to t). It follows that:

$$(4.11)\qquad \begin{aligned}&(u''_m, w_j)+\mu a(u'_m, w_j)+b(u'_m, u_m, w_j)+b(u_m, u'_m, w_j) \;+g((j'_\varepsilon(u_m)), w_j)=(f', w_j)\,,\\ &\qquad\qquad 1\leqslant j\leqslant m\,.\end{aligned}$$

But, as we already saw (Chap. III)

$$(j'_\varepsilon(u_m)', u'_m)\geqslant 0$$ [22]

and, consequently, it follows from (4.11) that

$$(4.12)\qquad (u''_m, u'_m)+\mu a(u'_m)+b(u'_m, u_m, u'_m)+b(u_m, u'_m, u'_m)\leqslant (f', u'_m)\,.$$

But

$$b(u_m, u'_m, u'_m)=0\,,$$

$$|b(u'_m, u_m, u'_m)|=|-b(u'_m, u'_m, u_m)|\leqslant$$

according to (3.13)

$$\begin{aligned}&\leqslant c\,\|u'_m\|^{3/2}\,|u'_m|^{1/2}\,\|u_m\|^{1/2}\,|u_m|^{1/2}\\ &\leqslant \tfrac{1}{2}\mu\alpha\,\|u'_m\|^2+c\,|u'_m|^2\,\|u_m\|^2\,|u_m|^2\,.\end{aligned}$$

Since we know that $|u_m(t)|\leqslant c,$ we have

$$(4.13)\qquad |b(u'_m, u_m, u'_m)|\leqslant \tfrac{1}{4}\mu\alpha\,\|u'_m\|^2+c\,|u'_m|^2\,\|u_m\|^2\,.$$

Thus, it follows from (4.12) that

$$\begin{aligned}\frac{1}{2}\frac{\mathrm{d}}{\mathrm{d}t}|u'_m(t)|^2+\mu\alpha\,\|u'_m(t)\|^2 &\leqslant \tfrac{1}{4}\mu\alpha\,\|u'_m(t)\|^2+c\,|u'_m(t)|^2\,\|u_m(t)\|^2\\ &\quad+\|f'(t)\|_*\,\|u'_m(t)\|\\ &\leqslant \tfrac{1}{2}\mu\alpha\,\|u'_m(t)\|^2+c\,|u'_m(t)|^2\,\|u_m(t)\|^2+c\,\|f'(t)\|_*^2\end{aligned}$$

from which, due to (4.10):

$$(4.14)\qquad \begin{aligned}|u'_m(t)|^2+\mu\alpha\int_0^t\|u'_m(\sigma)\|^2\,\mathrm{d}\sigma &\leqslant c+c\int_0^t\|f'(\sigma)\|_*^2\,\mathrm{d}\sigma\\ &\quad+c\int_0^t|u'_m(\sigma)|^2\,\|u_m(\sigma)\|^2\,\mathrm{d}\sigma\,.\end{aligned}$$

---

[22] Actually, the *discretized analogue* of this inequality is correct, as a consequence of the monotonicity of j'_ε.

According to Gronwall's inequality, we deduce from (4.14) that[23]

$$|u_m'(t)|^2 \leqslant c \exp(\textstyle\int_0^t \|u_m(\sigma)\|^2 \, d\sigma), \qquad t \leqslant T$$

and since

$$\textstyle\int_0^T \|u_m(\sigma)\|^2 \, d\sigma \leqslant C$$

we have:

$$|u_m'(t)| \leqslant c. \tag{4.15}$$

But then (4.14) gives

$$\textstyle\int_0^T \|u_m'(\sigma)\|^2 \, d\sigma \leqslant c \tag{4.16}$$

from which the desired estimates follow. ▯

5. Newtonian Fluids as Limits of Bingham Fluids

5.1. Statement of the Result

Throughout this section, we assume that

$$n = 2. \tag{5.1}$$

By u_g, we denote *the* solution given in Theorem 3.1; in the following we intend to let g vary (and, in particular, to let it tend toward zero) while the other data stay fixed.

Therefore

$$u_g \in L^2(0,T;V), \qquad \partial u_g/\partial t \in L^2(0,T;V'), \qquad u_g(0) = u^0 \tag{5.2}$$

and

$$(u_g'(t), v - u_g(t)) + \mu a(u_g(t), v - u_g(t)) + b(u_g(t), u_g(t), v) + gj(v) - gj(u_g(t)) \geqslant (f(t), v - u_g(t)) \qquad \forall v \in V. \tag{5.3}$$

Furthermore, let u be the solution of the *Navier-Stokes equation*, then:

$$u \in L^2(0,T;V), \qquad u' \in L^2(0,T;V'), \qquad u(0) = u^0, \tag{5.4}$$

and

$$(u'(t), v) + \mu a(u(t), v) + b(u(t), u(t), v) = (f(t), v) \qquad \forall v \in V. \tag{5.5}$$

---

[23] We recall once more that the c's denote *different* constants.

We now prove[24]

Theorem 5.1. *When* $g\to 0$, *we have:*

$$(5.6)\qquad \begin{aligned} &u_g\to u \quad \textit{weakly in} \quad L^2(0,T;V)\,,\\ &u_g'\to u' \quad \textit{weakly in} \quad L^2(0,T;V')\,. \end{aligned}$$

5.2. Proof of Theorem 5.1

From the proof of Theorems 3.1 and 3.2, it follows that

$$(5.7)\qquad \|u_g\|_{L^2(0,T;V)}+\|u_g'\|_{L^2(0,T;V)}\leqslant c \quad \textit{for all } g>0,\ g \textit{ bounded}.$$

Consequently, when $g\to 0$, we can select from the u_g a sequence, again denoted by u_g, such that

$$(5.8)\qquad \begin{aligned} &u_g\to w \ \text{ weakly in } L^2(0,T;V)\,,\\ &u_g'\to w' \ \text{ weakly in } L^2(0,T;V')\,,\\ &u_{ig}\ (i\text{-th component of } u_g)\to w_{ig} \text{ strongly in } L^2(Q) \text{ and a.e. }^{25}\,,\\ &u_{ig}u_{jg}\to w_iw_j \ \text{ in } L^2(0,T;L^2(\Omega))=L^2(Q)\ ^{26}\,. \end{aligned}$$

Then

$$(5.9)\qquad b(u_g,u_g,v)=-b(u_g,v,u_g)\to -b(w,v,w) \quad \text{weakly in } L^2(0,T), \quad \forall v\in V.$$

If we take in (5.3): $v=v(t)$ a.e., $v\in L^2(0,T;V)$, we have:

$$\begin{aligned} &\int_0^T\{(u_g',v)+\mu a(u_g,v)-b(u_g,v,u_g)+gj(v)-gj(u_g)-(f,v-u_g)\}\,dt\\ &\qquad\geqslant \int_0^T\{(u_g',u_g)+\mu a(u_g)\}\,dt=\tfrac12|u_g(T)|^2-\tfrac12|u^0|^2+\mu\int_0^T a(u_g)\,dt\,. \end{aligned}$$

Applying (5.8), (5.9), it follows, since $\int_0^T gj(u_g)\,dt\to 0$:

$$\begin{aligned} &\int_0^T\{(w',v)+\mu a(w,v)-b(w,v,w)-(f,v-w)\}\,dt\\ &\qquad\geqslant \liminf \tfrac12|u_g(T)|^2-\tfrac12|u^0|^2+\mu\liminf\int_0^T a(u_g)\,dt\\ &\qquad\geqslant \tfrac12|w(T)|^2-\tfrac12|u^0|^2+\mu\int_0^T a(w)\,dt\\ &\qquad= \int_0^T(w',w)\,dt+\mu\int_0^T a(w)\,dt \end{aligned}$$

and therefore

$$(5.10)\qquad \int_0^T\{(w',v-w)+\mu a(w,v-w)-b(w,v,w)-(f,v-w)\}\,dt\geqslant 0 \quad \forall v\in L^2(0,T;V).$$

---

[24] As a complement to Remark 5.3 below, we give a general result on the dependence of u_g on g.

[25] Cf. (3.54).

[26] Cf. (3.56).

From this, it follows (as at the end of Section 3.4.1) that almost everwhere on $[0,T]$:

$$(w'(t),v-w(t))+\mu a(w(t),v-w(t))-b(w(t),v,w(t)) \\ -(f,v-w(t))\geq 0 \quad \forall v\in V. \tag{5.11}$$

If we take in (5.11): $v=w(t)\pm\psi$, $\psi\in V$, it follows that

$$(w'(t),\psi)+\mu a(w(t),\psi)-b(w(t),\psi,w(t))-(f(t),\psi)=0 \quad \forall\psi\in V \tag{5.12}$$

which (because of the uniqueness of the solution of (5.4), (5.5)) implies that

$$w=u,$$

from which the Theorem follows. □

Remark 5.1. We can make use of the proof of Theorem 5.1 to show the *existence* of a solution u of (5.4), (5.5) (as for *uniqueness* of the solution of (5.4), (5.5), the proof is analogous to the proof of Section 3.4.2), but this is somewhat artificial since the preceding proofs make use of, *among others*, the techniques applied in the Navier-Stokes equations. □

Remark 5.2. Under the conditions of Theorem 4.1, we can obtain the supplementary result:

$$u'_g\to u' \text{ weakly in } L^2(0,T;V) \text{ and weakly star in } L^\infty(0,T;H). \tag{5.13}$$ □

Remark 5.3[27]. We now prove

Theorem 5.2. For $g_1,g_2\in[0,g_0]$, $g_0>0$ arbitrary and finite, there exists a $c(g_0)=c$ such that

$$\|u_{g_1}-u_{g_2}\|_{L^\infty(0,T;H)}+\|u_{g_1}-u_{g_2}\|_{L^2(0,T;V)}\leq c|g_1-g_2|. \tag{5.14}$$

Proof. We take $v=u_{g_2}$ (resp. $v=u_{g_1}$) in the inequality corresponding to u_{g_1} (resp. u_{g_2}); setting $u_{g_1}-u_{g_2}=w$ and $a(w,w)=a(w)$, it follows by addition:

$$-(w',w)-\mu a(w)+b(u_{g_1},u_{g_1},u_{g_2})+b(u_{g_2},u_{g_2},u_{g_1}) \\ +(g_1-g_2)(j(u_{g_1})-j(u_{g_2}))\geq 0$$

or again

$$(w',w)+\mu a(w)\leq|g_1-g_2|\,|j(u_{g_1})-j(u_{g_2})|+b(w,w,u_{g_1}). \tag{5.15}$$

But

$$|b(w,w,u_{g_1})|\leq c_1\|u_{g_1}\|\,\|w\|\,|w|\leq\tfrac{1}{2}\mu a(w)+c_1\|u_{g_1}\|^2|w|^2$$

[27] This Remark can be passed over.

so that (5.15) gives

$$\frac{d}{dt}|w(t)|^2+\mu a(w(t))\leqslant 2|g_1-g_2||j(u_{g_1})-j(u_{g_2})|+2c_1\|u_{g_1}\|^2|w|^2 . \tag{5.16}$$

But

$$|j(u_{g_1})-j(u_{g_2})|\leqslant j(u_{g_1}-u_{g_2})=j(w)\leqslant \text{(by Cauchy-Schwarz)}\leqslant c_2 a(w)^{1/2},$$

so that

$$2|g_1-g_2||j(u_{g_1})-j(u_{g_2})|\leqslant \tfrac{1}{2}\mu a(w)+c_2|g_1-g_2|^2$$

and we deduce from (5.16) that

$$\frac{d}{dt}|w(t)|^2+\tfrac{1}{2}\mu a(w(t))\leqslant 2c_1\|u_{g_1}\|^2|w|^2+c_2|g_1-g_2|^2$$

and therefore for $t\leqslant T$:

$$|w(t)|^2+\tfrac{1}{2}\mu\int_0^t a(w(\sigma))\,d\sigma\leqslant c_2 T|g_1-g_2|^2+2c_1\int_0^t\|u_{g_1}\|^2|w(\sigma)|^2\,d\sigma . \tag{5.17}$$

In particular, we conclude from this by Gronwall's inequality that

$$\begin{aligned}|w(t)|^2&\leqslant c_2 T|g_1-g_2|^2\exp\Big(2c_1\int_0^t\|u_{g_1}(\sigma)\|^2\,d\sigma\Big)\\&\leqslant c_2 T|g_1-g_2|^2\exp\Big(2c_1\int_0^T\|u_{g_1}\|^2\,d\sigma\Big).\end{aligned} \tag{5.18}$$

But for $g_1\in[0,g_0]$, u_{g_1} remains in a bounded set of $L^2(0,T:V)$, from which it follows that

$$|w(t)|\leqslant c_3|g_1-g_2|, \quad t\in[0,T]. \tag{5.19}$$

Applying (5.19) on the right hand side of (5.18), we can deduce that

$$\|w\|_{L^2(0,T;V)}\leqslant c_4|g_1-g_2|$$

whence (5.14). □

Remark 5.4. *The case* "$g\to+\infty$". We now prove

$$\text{when}\quad g\to\infty, \quad u_g\to 0 \quad \text{weakly in } L^2(0,T:V). \tag{5.20}$$

Indeed, we can first make part of (5.7) more precise by observing that

$$\|u_g\|_{L^2(0,T;V)}\leqslant c \quad \text{when} \quad g\to\infty . \tag{5.21}$$

(On the other hand, *it is not true* that $\|u_g'\|_{L^2(0,T;V)}\leqslant c$ when $g\to\infty$, for in the proof of Theorem 3, the estimate of the u_ε' *involves* g).

If in (5.3) we set $v=0$, we further obtain;

$$g\int_0^T j(u_g)\,dt+\mu\int_0^T a(u_g)\,dt+\tfrac{1}{2}|u_g(T)|^2\leqslant\int_0^T(f,u_g)\,dt+\tfrac{1}{2}|u^0|^2 \tag{5.22}$$

from which it follows that

$$\int_0^T j(u_g)\,dt\leqslant c/g\,. \tag{5.23}$$

Due to (5.21), we can select a sequence, again denoted by u_g, such that

$$u_g\to w \quad \text{weakly in } L^2(0,T:V)\,. \tag{5.24}$$

Since the function $v\to\int_0^T j(v)\,dt$ is lower semi-continuous for the weak topology of $L^2(0,T:V)$, we have

$$\liminf\int_0^T j(u_g)\,dt\geqslant\int_0^T j(w)\,dt$$

which, together with (5.23), proves that

$$\int_0^T j(w)\,dt=0$$

therefore $j(w)=0$ almost everywhere, and therefore $w=0$, from which the result follows. ▯

Obviously, from the point of view of mechanics, result (5.20) is clear, namely that in the limit (for $g\to\infty$) the fluid behaves like a rigid body.

Moreover, we can supplement (5.20) under the assumptions:

$$f=\{f_1,f_2\}\,,\qquad f_i\in L^\infty(0,T;L^2(\Omega))\,,\qquad i=1,2;\qquad u^0=0\,. \tag{5.25}$$

We will prove that then we have

$$u_g=0 \quad \text{for} \quad g\geqslant g_c=\text{suitable yield limit}\,. \tag{5.26}$$

Indeed, because of uniqueness, it is sufficient to prove that, for g sufficiently large, the zero function is a solution of (5.3), i.e.

$$gj(v)\geqslant(f(t),v) \qquad \forall v\in V\,. \tag{5.27}$$

But, due to (5.25),

$$|(f(t),v)|\leqslant c_1(\|v_1\|_{L^2(\Omega)}+\|v_2\|_{L^2(\Omega)})$$

and according to an inequality due to L. Nirenberg [1], extended by M. J. Strauss [1],

$$\|v_1\|_{L^2(\Omega)}+\|v_2\|_{L^2(\Omega)}\leqslant c_2 j(v)\,,$$

therefore

$$|(f(t), v)| \leqslant c_3 j(v)$$

so that (5.27) does hold for sufficiently large g. ▯

Open problems. 1) For the function $g \to u_g$ which is Lipshitz continuous on every compact set of g's and with values in $L^\infty(0,T;H) \cap L^2(0,T;V)$, *calculate* $\partial u_g / \partial g$ almost everywhere in g;

2) can one prove—which agrees with mechanical intuition and is confirmed by the numerical results of M. Fortin [1]—that *the region where* $D_{ij}(u_g) = 0$ *increases with* g?

6. Stationary Problems

6.1. Statement of Results

For reasons of future convenience, we introduce a parameter $\lambda \geqslant 0$ in the statements below.

Theorem 6.1. *Let* f *be given in* V'. *There exists* $u \in V$ *such that*

$$\mu a(u, v-u) + \lambda b(u,u,v) + gj(v) - gj(u) \geqslant (f, v-u) \quad \forall v \in V_\sigma, \tag{6.1}$$

where

$$\sigma = \max(1, n/2 - 1) \quad \text{(therefore } \sigma = 1 \text{ if } n \leqslant 4). \tag{6.2}$$

We note that this statement has a meaning, because $u, v, w \to b(u,v,w)$ is continuous on $V \times V \times V_\sigma$ (apply Hölder's inequality and Sobolov's imbedding theorem which imply that $u \in L^{q_1}(\Omega)$, $1/q_1 = \frac{1}{2} - 1/n$ if $n > 2$, and $w \in L^{q_2}(\Omega)$, $1/q_2 = \frac{1}{2} - \sigma/n$ if $n > 2$). ▯

We can also study the behavior of the problem when $g \to 0$:

Theorem 6.2. *We assume* $n \leqslant 4$. *We can find a family* u_g *of solutions of* (6.1) *such that, when* $g \to 0$, *we have*

$$u_g \to u \quad \text{weakly in } V, \tag{6.3}$$

where u *is a solution of the stationary problem for Newtonian fluids*:

$$\begin{aligned} &\mu a(u,v) + \lambda b(u,u,v) = (f,v) \quad \forall v \in V, \\ &u \in V. \end{aligned} \tag{6.4}$$

▯

Remark 6.1. We assume $n \leqslant 4$. Then $b(u,v,w)$ is continuous on $V \times V \times V$ and

(6.5) $$b(u,v,w) \leqslant c_1 a(u)^{1/2} a(v)^{1/2} a(w)^{1/2}.$$

Furthermore, for $f \in V'$, we set

(6.6) $$[f]_* = \sup(|(f,v)|/a(v)^{1/2}), \qquad v \in V.$$

Then:

(6.7) if $\mu^2 > \lambda c [f]_*$, then the solution $u(=u_g)$ of (6.1) is unique, and, similarly, the solution u of the stationary Navier-Stokes equations is unique, and we obtain (6.3) without having to select a subsequence.

Indeed, if u (resp. u_*) is a solution of (6.1), and if we take $v = u_*$ (resp. u) in (6.1) (resp. the analogous inequality for u_*), which is permissible when $b \leqslant 4$, then we obtain by addition (if we set $w = u - u_*$):

$$-\mu a(w) + \lambda b(u,u,u_*) + \lambda b(u_*,u_*,u) \geqslant 0$$

therefore

(6.8) $$\mu a(w) \leqslant \lambda b(w,w,u) \leqslant \lambda c_1 a(w) a(u)^{1/2}.$$

But taking $v = 0$ in (6.1), we find that

$$\mu a(u) + g j(u) \leqslant (f,u) \leqslant [f]_* a(u)^{1/2}$$

from which follows in particular

$$\mu a(u)^{1/2} \leqslant [f]_*$$

so that (6.8) gives

$$\mu a(w) \leqslant \lambda c_1 \mu^{-1} [f]_* a(w)$$

which implies "$w = 0$" if we have (6.7).

(This remark is analogous to Remark 7.6, Chap. 1 of Lions [1]. □

Remark 6.2. If $\lambda = 0$, we have existence *and uniqueness* of the solution u in V of

(6.9) $$\mu a(u, v-u) + g j(v) - g j(u) \geqslant (f, v-u) \qquad \forall v \in V.$$

We will return to this problem in Section 9. □

6.2. Proof

Proof of Theorem 6.1. We only give the outline of the proofs.
We start with the *bi-regularized*[28] equation

$$(6.10)\qquad \mu a(u_{\varepsilon\eta}, v) + \eta((u_{\varepsilon\eta}, v))_\sigma + \lambda b(u_{\varepsilon\eta}, u_{\varepsilon\eta}, v) + g(j'_\varepsilon(u_{\varepsilon\eta}), v) = (f, v) \qquad \forall v \in V_\sigma,$$

j_ε being defined as in Section 3.
We prove *existence* of $u_{\varepsilon\eta}$ by monotonicity methods (cf. Lions [1], Chap. 2). Moreover, we prove that

$$(6.11)\qquad \|u_{\varepsilon\eta}\| \leqslant C \quad \text{(a constant independent of } \varepsilon, \eta, \lambda, g),$$

$$(6.12)\qquad \eta^{1/2} \|u_{\varepsilon\eta}\|_\sigma \leqslant C.$$

We pass to the limit in ε, η by a procedure analogous to the one used in the proof of Theorem 3.2: we introduce

$$(6.13)\qquad \begin{aligned} Y_{\varepsilon\eta} = \mu a(u_{\varepsilon\eta}, v - u_{\varepsilon\eta}) + \lambda b(u_{\varepsilon\eta}, u_{\varepsilon\eta}, v - u_{\varepsilon\eta}) + \eta((u_{\varepsilon\eta}, v - u_{\varepsilon\eta}))_\sigma \\ + g j_\varepsilon(v) - g j_\varepsilon(u_{\varepsilon\eta}) - (f, v - u_{\varepsilon\eta}). \end{aligned}$$

Applying (6.10), we see that

$$Y_{\varepsilon\eta} = g[j_\varepsilon(v) - j_\varepsilon(u_{\varepsilon\eta}) - (j'_\varepsilon(u_{\varepsilon\eta}), v - u_{\varepsilon\eta})] \geqslant 0$$

from which

$$(6.14)\qquad \begin{aligned} \mu a(u_{\varepsilon\eta}, v) + \eta((u_{\varepsilon\eta}, v))_\sigma + \lambda b(u_{\varepsilon\eta}, u_{\varepsilon\eta}, v) + g j_\varepsilon(v) - (f, v - u_{\varepsilon\eta}) \\ \geqslant \mu a(u_{\varepsilon\eta}) + g j_\varepsilon(u_{\varepsilon\eta}). \end{aligned}$$

According to (6.11), we can select a subsequence, again denoted by $u_{\varepsilon\eta}$, such that

$$(6.15)\qquad u_{\varepsilon\eta} \to u \quad \text{weakly in } V.$$

The left hand side of (6.14) then converges toward[29]

$$(6.16)\qquad \mu a(u, v) + \lambda b(u, u, v) + g j(v) - (f, v - u), \qquad v \in V_\sigma.$$

Moreover,

$$\liminf a(u_{\varepsilon\eta}) \geqslant a(u)$$

[28] The term $\mu((u_{\varepsilon\eta}, v))$ is unnecessary if $n \leqslant 4$.

[29] One uses a compactness argument to prove that $b(u_{\varepsilon\eta}, u_{\varepsilon\eta}, v) \to b(u, u, v)$.

and, by an analogous argument to the one which led to (3.71), we have:

$$\liminf j_\varepsilon(u_{\varepsilon\eta}) \geqslant j(u) .$$

From this it follows, with (6.16) and (6.14), that u satisfies (6.1). ☐

Proof of Theorem 6.2. The proof of Theorem 6.1 shows the existence of a solution $u_g \in V$ of

$$\mu a(u_g, v-u_g) + \lambda b(u_g, u_g, v) + gj(v) - gj(u_g) \geqslant (f, v-u_g) \qquad \forall v \in V \tag{6.17}$$

and

$$\|u\| \leqslant c \quad \text{(a constant, independent of } g, g \geqslant 0 \text{ arbitrary)}. \tag{6.18}$$

Then we can select a sequence, again denoted by u_g such that (6.3) holds, from which Theorem 6.2 follows provided that we show that u is a solution of (6.4).

But it follows from (6.17) that

$$\mu a(u_g, v) + \lambda b(u_g, u_g, v) + gj(v) - (f, v-u_g) \geqslant \mu a(u_g) ; \tag{6.19}$$

when $g \to 0$, the left hand side of (6.19) converges toward

$$\mu a(u, v) + \lambda b(u, u, v) - (f, v-u)$$

and

$$\liminf \mu a(u_g) \geqslant \mu a(u) ;$$

therefore

$$\mu a(u, v-u) + \lambda b(u, u, v) - (f, v-u) \geqslant 0 . \tag{6.20}$$

If we take—which is permissible—$v = u + w$, $w \in V$, we see that u satisfies (6.4). ☐

Remark 6.3. One can verify by an argument analogous to the one of Remark 5.3 that

$$u \to 0 \quad \text{weakly in } V \text{ when } \quad g \to +\infty . \tag{6.21}$$

(whatever the solution u_g of (6.17)). ☐

7. Exterior Problem

7.1. Formulation of the Problem as a Variational Inequality

We use the spaces V, V_s as in Section 3 (the definitions given in that section being valid whether Ω is bounded *or not*).

We introduce the manifold $\mathscr{U}_{\mathrm{ad}}$ of functions v in a function class, which is to be specified, that have divergence zero and are such that

$$v=0 \quad \text{on } \Gamma, \quad v=\{U,0,0\} \quad \text{at infinity}, \tag{7.1}$$

the second condition (7.1) being taken in a suitable sense which we will also specify.

Formally, the problem is then to find $u=u(t)$ with

$$u(t)\in\mathscr{U}_{\mathrm{ad}}, \tag{7.2}$$

$$\begin{aligned}\partial u(t)/\partial t,\ v-u(t)+\mu a(u(t),\ v-u(t))+b(u(t),u(t),\ v-u(t))+gj(v)-gj(u(t))\\ \geqslant(f(t),\ v-u(t)) \qquad \forall v\in\mathscr{U}_{\mathrm{ad}},\end{aligned} \tag{7.3}$$

$$u(0)=u_0, \quad u_0\in\mathscr{U}_{\mathrm{ad}}. \tag{7.4}$$

If we introduce

$$w(t)=u(t)-u_0, \tag{7.5}$$

we then seek

$$w(t)\in V \tag{7.6}$$

(which, once u_0 is given, specifies the sense in which to interpret the condition "$u(t)=\{U,0,0\}$ at infinity" with

$$\begin{aligned}(w'(t),\ v-w(t))+\mu a(w(t),\ v-w(t))+b(w(t),\ w(t),\ v-w(t))+c(w(t),\ v-w(t))\\ +gj(v+u_0)-gj(w(t)+u_0)\geqslant(\tilde f(t),\ v-w(t)) \qquad \forall v\in V,\end{aligned} \tag{7.7}$$

where

$$c(w,v)=b(u_0,w,v)+b(w,u_0,v), \tag{7.8}$$

$$(\tilde f(t),v)=(f(t),v)-a(u_0,v)-b(u_0,u_0,v), \tag{7.9}$$

and with $w(0)=0$.

We assume that

$$\partial u_0/\partial x_i\in(L^\infty(\Omega))^3, \quad i=1,2,3 \quad (\text{or } (L^\infty(\Omega))^2 \text{ in two dimensions}). \tag{7.10}$$

Then

$$(7.11)\qquad |c(v,v)| \leqslant c_0 |v|^2 . \quad \square$$

We can now pose the problem precisely by distinguishing two cases according as the dimension n is 2 or 3:

1) in the case $n=2$, we seek w which satisfies (7.7) and

$$(7.12)\qquad w \in L^2(0,T;V), \qquad \partial w/\partial t \in L^2(0,T;V'),$$

$$(7.13)\qquad w(0)=0;$$

2) in the case $n=3$, we seek w which satisfies

$$(7.14)\qquad w \in L^2(0,T;V) \cap L^\infty(0,T;H), \qquad \partial w/\partial t \in L^2(0,T;V_s'),$$

(7.13) and the "weak" inequality

$$(7.15)\qquad \begin{aligned} &\textstyle\int_0^T [(v', u-w) + \mu a(w, v-w) - b(w,v,w) + c(w, v-w) \\ &\qquad + gj(v+u_0) - gj(w+u_0) - (\tilde{f}, v-w)]\, dt \geqslant 0 \end{aligned}$$

$\forall v$ such that

$$(7.16)\qquad v \in L^2(0,T;V_s), \qquad v' \in L^2(0,T;H), \qquad v(0)=0. \quad \square$$

7.2. Results

We have the following results:

(7.17) for f given in $L^2(0,T;V')$ and $u^0 \in V$ with (7.10), there exists a *unique* solution u of (7.7), (7.12), (7.13);

(7.18) for f given in $L^2(0,T;V')$ and $u_0 \in V$ with (7.10), there exists a solution w of (7.13), (7.14), (7.15) (the case $n=3$). $\square$

Remark 7.1. The problem of uniqueness in (7.18) is open. $\square$

Remark 7.2. Similarly, we can obtain results analogous to those of Sections 4 and 5. $\square$

For the proof of results (7.17), (7.18), we proceed as in the preceding sections. We limit ourselves to the essential steps for the case (7.17); for the case (7.18), we introduce a supplementary regularizing term as in the preceding sections.

Thus, we begin with the solution of

$$(7.19)\qquad (w_\varepsilon', v) + \mu a(w_\varepsilon, v) + b(w_\varepsilon, w_\varepsilon, v) + c(w_\varepsilon, v) + g(j_\varepsilon'(w_\varepsilon + u_0), v) = (\tilde{f}, v)$$

with

$$(7.20)\qquad w_\varepsilon(0)=0.$$

For this purpose, we use the Galerkin method but with an arbitrary "base" $w_1, \ldots, w_n, \ldots$ for V [30]. There are some differences in the a priori estimates. We take $V = w_\varepsilon$ in (7.19). We no longer have $a(v,v) \geqslant \alpha \|v\|^2$, $\alpha > 0$, because Ω is not bounded, but instead

$$a(v,v) + |v|^2 \geqslant \alpha \|v\|^2, \quad \alpha > 0, \quad v \in V. \tag{7.21}$$

Moreover, we make use of (7.11) and the fact that

$$\begin{aligned}(j'_\varepsilon(w_\varepsilon + u_0), w_\varepsilon) &= (j'_\varepsilon(w_\varepsilon + u_0), w_\varepsilon + u_0) - (j'_\varepsilon(w_\varepsilon + u_0), u_0) \\ &\geqslant -(j'_\varepsilon(w_\varepsilon + u_0), u_0) \geqslant -c_1(1 + \|w_\varepsilon(t)\|),\end{aligned}$$

whence, due to Gronwall's inequality[31],

$$|w_\varepsilon(t)| \leqslant c, \tag{7.22}$$

$$\int_0^T \|w_\varepsilon\|^2 \, d\sigma \leqslant c. \tag{7.23}$$

Now there is a difference in the estimate of w'_ε. We apply the Fourier transformation in t [32] to obtain the following:

(7.24) if $D_t^{1/4-\beta} w_{\varepsilon m}$ = restriction to $(0,T)$ of the inverse Fourier transform in τ of $|\tau|^{1/4-\beta} \hat{w}_{\varepsilon m}$, where $\hat{w}_{\varepsilon m}$ is the Fourier transform in t of $\tilde{w}_{\varepsilon m}$ (extension of $w_{\varepsilon m}$ as 0 outside of $(0,T)$);

then

(7.25) $\forall \beta > 0$, $D_t^{1/4-\beta} w_{\varepsilon m}$ remains in a bounded set of $L^2(0,T;H)$.

From this we conclude, applying Theorem 5.2 of Lions [1], Chap. 1, that we can select a sequence $w_{\varepsilon\mu}$ such that

$$\begin{aligned}&w_{\varepsilon\mu} \to w_\varepsilon \quad \text{weakly star in } L^2(0,T;H) \text{ and weakly in } L^2(0,T;V), \\ &w_{\varepsilon i\mu} w_{\varepsilon j\mu} \to w_{\varepsilon i} w_{\varepsilon j} \quad \text{weakly in } L^2(Q)\end{aligned}$$ [33]

from which it follows that w_ε satisfies (7.19), (7.20) and that the estimates

$$\|w_\varepsilon\|_{L^2(0,T;V)} + \|w_\varepsilon\|_{L^\infty(0,T;H)} + \|D_t^{1/4-\beta} w_\varepsilon\|_{L^2(0,T;H)} \leqslant c \tag{7.26}$$

hold.

We finally pass to the limit in ε as in Section 3. ▯

---

[30] In an open unbounded set, we can no longer use the bases of eigenfunctions. Here, we could make use of the method proposed in Lions-Strauss [1], p. 62.

[31] These estimates are first established for the approximation $w_{\varepsilon m}$, then we pass to the limit in m.

[32] Cf. Lions [1], Chap. 1, Sec. 6.5. The *same* proof is valid in the present situation.

[33] We apply Theorem 5.2, loc. cit. above, *On every compact set of* $\bar{\Omega}$.

Remark 7.2. Another procedure for the proof consists in introducing the open *bounded sets*

(7.27) $$\Omega_M = \Omega \cap \{x \mid |x| < M\}.$$

Then one solves the analogous problems on Ω_M (applying Sec. 3 with fixed M) and after that, one lets M tend toward $+\infty$. □

8. Laminar Flow in a Cylindrical Pipe

8.1. Recapitulation of the Equations

Let Ω be a cross section of the pipe; we assume that Ω is an open bounded set of $\mathbb{R}^2$ with a regular boundary Γ. We want to find the velocity field $u = u(x)$ (u = scalar) satisfying

(8.1) $$\sigma^D_{31,1} + \sigma^D_{32,2} = -c \quad \text{in } \Omega$$ [34]

(8.2) $$\sigma^D_{3i} = g\frac{D_{3i}}{[(D_{13})^2 + (D_{23})^2]^{1/2}} + 2\mu D_{3i}, \qquad i = 1,2 \quad \text{in } \Omega$$

and the boundary condition

(8.3) $$u = 0 \quad \text{on } \Gamma.$$

We recall that in (8.2)

(8.4) $$D_{3i} = \tfrac{1}{2} u_{,i}, \qquad i = 1,2.$$

We now give a variational formulation which will lead to the precise formulation of the problem.

8.2. Variational Formulation

Theorem 8.1. *If a velocity field u is a "strong" solution of* (8.1), (8.2), (8.3), *then u satisfies the variational inequality*

(8.5) $$\mu a(u, v-u) + gj(v) - gj(u) \geq (c, v-u) \qquad \forall v \in H^1_0(\Omega)$$

[34] c = drop in pressure per unit length.

where

$$(8.6)\qquad a(u,v)=\int_\Omega \operatorname{grad} u\,.\operatorname{grad} v\,\mathrm{d}x\,,$$

$$(8.7)\qquad j(v)=\int_\Omega |\operatorname{grad} v|\,\mathrm{d}x\,,$$

$$(8.8)\qquad (f,v)=\int_\Omega f v\,\mathrm{d}x\,.$$

Proof. Multiplying (8.1) by $v-u$, it follows, after integration by parts, that

$$(8.9)\qquad \int_\Omega \sigma_{3i}^D (v-u)_{,i}\,\mathrm{d}x-(c,v-u)=0$$

and, applying (8.2), we deduce

$$(8.10)\qquad \mu\int_\Omega u_{,i}(v-u)_{,i}\,\mathrm{d}x+g\int_\Omega \frac{u_{,i}(v-u)_{,i}}{(u_{,i}u_{,i})^{1/2}}\,\mathrm{d}x=(c,v-u)\,.$$

Since

$$u_{,i}v_{,i}\leqslant (u_{,i}u_{,i})^{1/2}(v_{,i}v_{,i})^{1/2}$$

we conclude (8.5). ☐

The **precise** formulation of the problem of laminar flow in a cylindrical pipe is then: *To find the solution* $u\in H_0^1(\Omega)$ *of* (8.5).

Since the fact that the right hand side of (8.5) contains a *constant* c does not play an essential role, we can clarify the exposition by introducing a slightly more general problem:

$$(8.11)\qquad \begin{aligned}&\textit{for a given } f^2 \textit{ in } L^2(\Omega), \textit{ to find a solution } u\in H_0^1(\Omega) \text{ of}\\ &\mu a(u,v-u)+gj(v)-gj(u)\geqslant (f,v-u)\qquad \forall v\in H_0^1(\Omega)\,.\end{aligned}$$ ☐

Remark 8.1. The formulation makes sense—and Theorem 8.2 below is still valid—if we assume more generally that

$$(8.12)\qquad f\in H^{-1}(\Omega)\,.$$

However, the assumption "$f\in L^2(\Omega)$" will play a role in Theorem 8.3. ☐

Problem (8.11) is equivalent to minimizing the functional

$$(8.13)\qquad J_g(v)=\tfrac{1}{2}\mu a(v)+gj(v)-(f,v)$$ [35]

where

$$(8.14)\qquad a(v)=a(v,v)\,.$$

[35] We introduce the index g, because later on we will let g vary, everything else remaining the same.

Therefore, according to Section I.3, we have:

Theorem 8.2. *Problem* (8.11) (*or the equivalent problem* (8.13)) *has a unique solution* u_g.

We now examine the properties of u_g and of the function $g \to u_g$.

8.3. Properties of the Solution

Theorem 8.3. *The function* $g = u_g$ *is Lipshitz continuous from* $g \geqslant 0 \to H_0^1(\Omega)$. *Furthermore,*

(8.15) $u_g = 0$ *for* $g \geqslant g_c =$ *yield limit dependent on* Ω *and* f,

and

(8.16) *the function* $g \to a(u_g)$ (resp. $g \to j(u_g)$) *is continuous and decreasing (zero for* $g \geqslant g_c$*).*

Proof. If we take $v = 0$ and then $v = 2u = 2u_g$ in (8.11), we find

(8.17) $$\mu a(u_g) + g j(u_g) = (f, u_g)$$

so that

(8.18) $$J_g(u_g) = -\tfrac{1}{2}\mu a(u_g)\,.$$

It follows from (8.17) that (and this is valid when $f \in H^{-1}(\Omega)$)

(8.19) $$\|u_g\| \leqslant c \quad (\|\ \| = \text{norm in } H_0^1(\Omega)) \text{ for } g \geqslant 0.$$

Now, we prove the *strong* continuity of $g \to u_g$ and the (Lipshitz) property:

(8.20) $$a(u_g - u_{g0})^{1/2} \leqslant \mu^{-1} \quad (\text{measure } \Omega)^{1/2} |g - g_0|\,.$$

Indeed, taking $v = u_{g0}$ (resp. u_g) in inequality (8.11) for u_g (resp. for u_{g0}) and adding, it follows that:

(8.21) $$-\mu a(u_g - u_{g0}) - (g - g_0)(j(u_g) - j(u_{g0})) \geqslant 0\,.$$

But

(8.22) $$\begin{aligned} |j(v) - j(w)| &= |\textstyle\int_\Omega (|\text{grad}\, v| - |\text{grad}\, w|)\,dx| \leqslant j(v - w) \\ &\leqslant (\text{Cauchy-Schwarz}) \leqslant (\text{measure } \Omega)^{1/2}\, a(v - w)^{1/2} \end{aligned}$$

so that (8.21) gives

(8.23) $$\mu a(u_g - u_{g_0}) \leqslant (\text{measure } \Omega)^{1/2} |g - g_0| \, a(u_g - u_{g_0})^{1/2}$$

from which (8.20) follows.

We now prove (8.15). According to the theorem of Nirenberg and M. Strauss [1], we have:

(8.24) $$|(f, v)| \leqslant \|f\|_{L^\infty(\Omega)} \|v\|_{L^1(\Omega)} \leqslant c_1 \|f\|_{L^2(\Omega)} j(v) .$$

Then, $u = 0$ is a solution of (8.11) if $gj(v) \geqslant (f, v) \; \forall v$, which is true according to (8.24), as long as

(8.25) $$g \geqslant c_1 \|f\|_{L^2(\Omega)} .$$

It remains to prove properties (8.16). We have

$$J_g(v) \leqslant J\,(v) \qquad \forall v \quad \text{if} \quad g \leqslant h ,$$

therefore

$$\inf J_g(v) = -\tfrac{1}{2}\mu a(u_g) \leqslant \inf J_h(v) = -\tfrac{1}{2}\mu a(u_h)$$

from which the decreasing character of $g \to a(u_g)$ follows.

Moreover, (8.21) proves that

$$(g - g_0)(j(u_g) - j(u_{g_0})) \leqslant 0$$

from which the decreasing character of $g \to j(u_g)$ follows. ∎

We now prove

Theorem 8.4. *If we assume that*

(8.26) $$f \geqslant 0 \quad \text{a.e.} \quad in\ \Omega$$

then

(8.27) $$u_g \geqslant 0 \quad \text{a.e.} \quad in\ \Omega .$$

Proof. In (8.11), (writing u instead of u_g), we take $v = u^+$; then $v - u = u^-$ and it follows that

(8.28) $$\mu a(u, u^-) + g(j(u^+) - j(u)) \geqslant (f, u^-)$$

but $a(u, u^-) = -a(u^-)$ and, since

(8.29) $$j(u) = j(u^+) + j(u^-)$$

it follows that

(8.30) $$\mu a(u^-)+gj(u^-)+(f,u^-)\leqslant 0\,.$$

Since $f\geqslant 0$, we have: $(f,u^-)\geqslant 0$ and thus (8.30) implies that $a(u^-)=0$, therefore

$$u^-=0\,.\quad \square$$

We will now give *a comparison property of the solutions* for two domains $\Omega_1,\Omega_2,\bar{\Omega}_1\subset\Omega_2$; let u_i be the solution of

(8.31) $$\begin{aligned}&\mu a_{\Omega_i}(u_i, v-u_i)+g(j_{\Omega_i}(v)-j_{\Omega_i}(u_i))\geqslant(f_i, v-u_i)_{\Omega_i},\\ &\forall v\in H_0^1(\Omega_i)\,,\qquad i=1,2\,,\qquad (u_i\in H_0^1(\Omega_i))\end{aligned}$$

where

$$\begin{aligned}&a_{\Omega_i}(u,v)=\textstyle\int_{\Omega_i}\operatorname{grad}u\,.\operatorname{grad}v\,\mathrm{d}x\,,\\ &j_{\Omega_i}(v)=\textstyle\int_{\Omega_i}|\operatorname{grad}v|\,\mathrm{d}x\,,\\ &(f,v)_{\Omega_i}=\textstyle\int_{\Omega_i}fv\,\mathrm{d}x\,.\end{aligned}$$

Theorem 8.5. *We assume that* $\bar{\Omega}_1\subset\Omega_2$ *and that*

(8.32) $$f_2\geqslant 0\quad\text{a.e.}\quad in\ \Omega_2\,,\qquad f_2\geqslant f_1\quad\text{a.e.}\quad in\ \Omega_1\,.$$

Then

(8.33) $$u_1<u_2\quad\text{a.e.}\quad\text{in }\Omega_1\,.$$

Proof. In (8.31), we take for $i=1$:

$$v=u_1-(u_2-u_1)^-$$

(this is permissible, since on $\Gamma_1=$boundary of Ω_1, we have according to (8.32): $u_1=0$, $u_2\geqslant 0$ and Theorem 8.4, therefore $(u_2-u_1)^-=0$ and therefore $v=0$ on Γ_1); it follows that

(8.34) $$\begin{aligned}&\mu a_{\Omega_1}(u_1,-(u_2-u_1)^-)+g(j_{\Omega_1}(u_1-(u_2-u_1)^-)-j_{\Omega_1}(u_1))\\ &\qquad\qquad\qquad\qquad\geqslant-(f_1,(u_2-u_1)^-)_{\Omega_1}\,.\end{aligned}$$

Now, let $u_1=$extension of u_1 to Ω_2 as 0 outside of Ω_1, then $u_1\in H_0^1(\Omega_2)$ and let

(8.35) $$v=u_2+(u_2-\tilde{u}_1)^-\,;$$

we have: $v\in H_0^1(\Omega_2)$ and therefore we can take this element v in (8.31) for $i=2$. Since $v=u_2$ in $\Omega_2-\Omega_1$, it follows that:

(8.36) $$\mu a_{\Omega_1}(u_2,(u_2-u_1)^-)+g(j_{\Omega_1}(u_1+(u_2-u_1)^-)-j_{\Omega_1}(u_2))\geqslant(f_2,(u_2-u_1)^-)_{\Omega_1}\,.$$

But we observe that

$$j_{\Omega_1}(u_1-(u_2-u_1)^-)+j_{\Omega_1}(u_1+(u_2-u_1)^-)-j_{\Omega_1}(u_1)-j_{\Omega_1}(u_2)=0$$

so that (8.34) and (8.36) imply:

$$\mu a_{\Omega_1}(u_2-u_1,(u_2-u_1)^-)\geqslant(f_2-f_1,(u_2-u_1)^-)_{\Omega_1}$$

from which

$$\mu a_{\Omega_1}((u_2-u_1)^-)+(f_2-f_1,(u_2-u_1)^-)_{\Omega_1}\leqslant 0\,. \tag{8.37}$$

But according to the second condition of (8.32), we have

$$(f_2-f_1,(u_2-u_1)^-)_{\Omega_1}\geqslant 0$$

so that (8.37) implies that $a_{\Omega_1}((u_2-u_1)^-)=0$, therefore that $(u_2-u_1)^-=0$ in Ω_1 (because $(u_2-u_1)^-\in H_0^1(\Omega_1)$). ▯

9. Interpretation of Inequalities with Multipliers

We will prove the following results:

Theorem 9.1. *In the context of Theorem 3.1, we denote the solution of the problem by u_g. Then there exist functions $m^g=\{m_{ij}^g\}$ and a distribution p^g on Q such that*

$$m_{ij}^g\in L^\infty(Q)\quad m_{ij}^g=m_{ji}^g\,,\qquad \forall i,j\,,\qquad m_{kk}^g=0\,, \tag{9.1}$$

$$m_{ij}^g m_{ij}^g\leqslant 1\qquad \text{a.e.}\quad in\ Q\,, \tag{9.2}$$

$$m_{ij}^g D_{ij}(u_g)=(D_{ij}(u_g)D_{ij}(u_g))^{1/2}\quad \text{a.e.}\quad in\ Q\,, \tag{9.3}$$

$$\partial u_{gi}/\partial t-\mu\Delta u_{gi}+u_{gj}\partial u_{gi}/\partial x_j u_{gi}-g\sqrt{2}\,\partial m_{ij}^g/\partial x_j=f_i-\partial p^g/\partial x_i\,,\qquad i=1,2\,. \tag{9.4}$$

Conversely, let u_g, m_{ij}^g, p^g be given with

$$u_g\in L^2(0,T;V)\,,\qquad u_g'\in L^2(0,T;V')\,, \tag{9.5}$$

$$u_g(0)=u_0\,, \tag{9.6}$$

and the relations (9.1), ..., (9.4). Then u_g is the solution of the problem stated in Theorem 3.1.

The following Theorem proves that the "multipliers" m_{ij}^g depend continuously on g in a suitable topology.

Theorem 9.2. *If* $g \to g_0$ $(g_0 \geqslant 0)$, *one can choose the multipliers* m_{ij}^g *in such a way that*

$$m_{ij}^{g_0} \to m_{ij}^g \quad \text{weakly star in } L^\infty(Q) \text{ [36]} \tag{9.7}$$

where the $m_{ij}^{g_0}$ *are multipliers corresponding to the solution* u_{g_0} (and thus have properties analogous to those given in the preceding Theorem 9.1). □

First, we prove the converse property in the statement of Theorem 9.1. From (9.4), it follows that

$$\begin{aligned}(u'(t), v-u(t)) + \mu a(u(t), v-u(t)) + b(u(t), u(t), v-u(t))\\ + gj(v) - gj(u(t)) - (f(t), v-u(t)) = g[j(v) - j(u)\\ - 2(m_{ij}^g, D_{ij}(v) - D_{ij}(u))] = X\,;\end{aligned}$$

but, according to (9.3), $j(u) = 2(m_{ij}^g, D_{ij}(u))$ and according to (9.2)

$$2(m_{ij}^g, D_{ij}(v)) \leqslant 2\int_\Omega (D_{ij}(v)\, D_{ij}(v))^{1/2}\, dx = j(v)$$

therefore $X \geqslant 0$, from which the result follows. □

We now prove the *existence* of the multipliers m_{ij}^g.

In order to simplify the writing, we begin by omitting the index g. We introduce

$$B(u, v) \in V\,,$$

defined by

$$b(u, v, w) = (B(u, v), w)\,, \qquad u, v, w \in V\,, \tag{9.8}$$

and we set

$$F = \partial u/\partial t - \mu \Delta u + B(u, u) - f\,; \tag{9.9}$$

F is an element of $L^2(0, T; V')$ and the variational inequality can be written

$$(F(t), v) + gj(v) - [(F(t), u(t)) + gj(u(t))] \geqslant 0 \qquad \forall v \in V\,. \tag{9.10}$$

If we replace v by $\pm \lambda v$, $\lambda > 0$, (9.10) is equivalent to

$$\lambda[\pm (F(t), v) + gj(v)] - [(F(t), u(t)) + gj(u(t))] \geqslant 0\,; \tag{9.11}$$

taking $v = v(t)$ in (9.11) where $t \to v(t)$ is in $L^2(0, T; V)$, it follows that

$$\begin{aligned}&\lambda[\pm \textstyle\int_0^T (F, v)\, dt + g\int_0^T j(v)\, dt] - [\int_0^T ((F, u) + gj(u))\, dt] \geqslant 0\\ &\forall v \in L^2(0, T; V) \quad \text{and} \quad \forall \lambda \geqslant 0\,.\end{aligned} \tag{9.12}$$

---

[36] $m_{ij}^{g_0}$ if $g_0 = 0$.

From this it follows that

(9.13) $$\left|\int_0^T (F,v)\,\mathrm{d}t\right| \leqslant g\int_0^T j(v)\,\mathrm{d}t \qquad \forall v\in L^2(0,T;V)$$

and

(9.14) $$\int_0^T ((F,u)+gj(u))\,\mathrm{d}t = 0\,.$$

Indeed, we conclude from (9.12) that $\int_0^T ((F,u)+gj(u))\,\mathrm{d}t\leqslant 0$ and since, according to (9.13), $\int_0^T ((F,u)+gj(u))\,\mathrm{d}t\geqslant 0$, (9.14) follows.

We now apply the Hahn-Banach theorem. We introduce the space

$$\Phi = \{\varphi| \quad \varphi=\{\varphi_{ij}\}, \quad \varphi_{ij}=\varphi_{ji}, \quad \varphi_{ij}\in L^1(Q)\}$$

with the norm

(9.15) $$\|\varphi\|_\Phi = \int_Q (\varphi_{ij}\varphi_{ij})^{1/2}\,\mathrm{d}x\,\mathrm{d}t\,.$$

We consider the mapping

(9.16) $$v \xrightarrow{\pi} D_{ij}(v)=\tfrac{1}{2}(v_{i,j}+v_{j,i})$$

of $L^2(0,T;V)\to\Phi$. The inequality (9.13) is equivalent to

(9.17) $$\left|\int_0^T (F,v)\,\mathrm{d}t\right| \leqslant g\sqrt{2}\,\|\pi v\|_\Phi\,.$$

Consequently, according to the Hahn-Banach theorem, there exists

$$m^g\in\Phi' = \{\varphi\,| \quad \varphi=\{\varphi_{ij}\}, \quad \varphi_{ij}=\varphi_{ji}, \quad \varphi_{ij}\in L^\infty(Q)\}\,,$$

such that

(9.18) $$\int_0^T (F,v)\,\mathrm{d}t = -g\sqrt{2}\int_Q m_{ij}^g D_{ij}(v)\,\mathrm{d}x\,\mathrm{d}t\,,$$

(We can even take $\{m_{ij}^g\}$ in Φ' such that $m_{kk}^g=0$ because $D_{kk}(v)=0$—which we will do) and such that

(9.19) $$\|m^g\|_\Phi\leqslant 1$$

where Φ' is given the norm dual to (9.15).

But (9.19) is equivalent to (9.2), and (9.18) is equivalent to (9.4) (according to the definition of F) and (9.14) is equivalent to (using (9.18))

(9.20) $$\int_Q m_{ij}^g D_{ij}(u)\,\mathrm{d}x\,\mathrm{d}t = \int_Q (D_{ij}(u)D_{ij}(u))^{1/2}\,\mathrm{d}x\,\mathrm{d}t\,;$$

since we have (9.2), (9.20) implies (9.3). □

Remark 9.1. We set

$$\sigma_{ij}^g = -p^g \delta_{ij} + 2\mu D_{ij}(u) + g\sqrt{2}\, m_{ij}^g;$$

then we see immediately that (9.2), (9.3) are equivalent to: $\{\sigma_{ij}^g\}$ and $\{D_{ij}(u)\}$ are related by the constituent law for the Bingham fluid.

Furthermore, (9.14) is then equivalent to

$$\partial u_{gi}/\partial t + u_{gj} u_{gi,j} = \sigma_{ij,j} + f_i,$$

which establishes that σ_{ij}^g is the stress field solution of the problem.

Moreover, this constitutes a *rigorous interpretation* of the solution $u(x,t)$ of the variational problem (2.7). (We recall that a formal interpretation had been given in Remark 2.2.) □

Proof of Theorem 9.2. From Theorem 5.2, we know that when $g \to g_0$, we have:

(9.21) $$u_g \to u_{g_0} \quad \text{in} \quad L^2(0,T;V) \cap L^\infty(0,T;H).$$

Furthermore, from the analogue to Theorem 5.1 (with $g \to g_0$ instead of $g \to 0$), we have:

(9.22) $$\partial u_g/\partial t \to \partial u_{g_0}/\partial t \quad \text{weakly in} \quad L^2(0,T;V').$$

According to (9.2), we can select a subsequence, again denoted by m_{ij}^g such that

(9.23) $$m_{ij}^g \to \mu_{ij} \quad \text{weakly star in } L^\infty(Q)$$

and

(9.24) $$\mu_{ij}\mu_{ij} \leqslant 1 \quad \text{a.e. in } Q.$$

It follows from (9.4) that

(9.25) $$(u_g', v) + \mu a(u_g, v) + b(u_g, u_g, v) - (f, v) = -2g(m_{ij}^g, D_{ij}(v))$$

and we can pass to the limit in g in (9.25); we obtain

(9.26) $$(u_{g_0}', v) + \mu a(u_{g_0}, v) + b(u_{g_0}, u_{g_0}, v) - (f, v) = -2g_0(\mu_{ij}, D_{ij}(v)).$$

But u_{g_0} is the solution of the variational inequality corresponding to g_0, therefore

$$(u_{g_0}', v - u_{g_0}) + \mu a(u_{g_0}, v - u_{g_0}) + b(u_{g_0}, u_{g_0}, v - u_{g_0}) - (f, v - u_{g_0}) + g_0 j(v) - g_0 j(u_{g_0}) \geqslant 0$$

from which, comparing with (9.26)

$$g_0\{j(v) - j(u_{g_0}) - 2(\mu_{ij}, D_{ij}(v - u_{g_0}))\} \geqslant 0 \qquad \forall v$$

therefore

$$J(v)-2(\mu_{ij}, D_{ij}(v))-[j(u_{go})-2(\mu_{ij}, D_{ij}(u_{go}))]\geqslant 0, \tag{9.27}$$

therefore, taking $v=0$:

$$j(u_{go})-2(\mu_{ij}, D_{ij}(u_{go}))\leqslant 0.$$

But according to (9.24) $j(u_{go})-2(\mu_{ij}, D_{ij}(u_{go}))\geqslant 0$, therefore

$$j(u_{go})-2(\mu_{ij}, D_{ij}(u_{go}))=0$$

and therefore

$$\mu_{ij} D_{ij}(u_{go})=(D_{ij}(u_{go}) D_{ij}(u_{go}))^{1/2} \quad \text{a.e. in } Q. \tag{9.28}$$

Thus we can take $m_{ij}^{go}=\mu_{ij}$. ▯

Remark 9.2. We have analogous results for the *stationary* inequalities discussed in this chapter. ▯

10. Comments

The equations of hydrodynamics for Newtonian viscous incompressible fluids have given rise to numerous investigations.

For the "mathematical" point of view, one should consult J. Leray [1–3], O. A. Ladyzenskaya [1], J. Serrin [1], R. Finn [1], Lions-Prodi [1], Lions [1].

For the "mechanical" point of view, one can find in R. Barker [1] a very complete study of explicit solutions as well as references to earlier work.

The theory presented here, in Sections 1 to 7, appears to be new (we published a note on this subject in the Contes Rendus, Duvaut-Lions [4].

From the mechanical point of view, we deal with a generalization of the usual problems for a non-Newtonian fluid; the Bingham fluid is retained, because it is the simplest one which leads to variational inequalities. The corresponding problems for other types of non-Newtonian fluids probably are not beyond reach for the methods of this book, but we did not try to treat them here.

From the mathematical point of view, we deal with a (strict) generalization of the usual problems. Naturally, the question then arises to possibly extend all known results for the Navier-Stokes equations to the inequalities pertaining to Bingham fluids. This is an enormous undertaking; several investigations are in progress on this subject.

We did not take up the following problems:

1) study of solutions of the inequalities which are *local* in t and *strong*;
2) behavior of the solution when $t\to+\infty$;

3) behavior of the solution when $\mu \to 0$; theory of the boundary layer;

4) investigation of solutions which are periodic or almost-periodic (in t);

5) possible extension of the work of C. Foias and G. Prodi [1,2] concerning the Navier-Stokes equations;

6) adaptation of the theory of turbulence.

The study of numerical approximation of the solution is given in Fortin [1], D. Begis [1], R. Glowinski [1] and in the already mentioned work by Glowinski-Lions-Trémolières [1].

Chapter VII

Maxwell's Equations. Antenna Problems

This chapter assumes familiarity with Sections I.1 to I.3.

1. Introduction

The phenomena of electromagnetism can lead to variational inequalities, either with Maxwell's equations in a polarizable medium or, when the phenomena are coupled, with the mechanics of Bingham fluids yielding the magnetodynamics of Bingham fluids. This second possibility is treated in Duvaut-Lions [7]. The present chapter is devoted to unilateral problems connected with Maxwell's equations. Since we cannot refer to a unified exposition of the classical phenomena of electromagnetism, we will set up equations for the phenomena in Section 2 and develop a mathematical theory for the classical solutions in Sections 4–6. This theory is indispensable for the subsequent treatment of Maxwell's equations in a polarizable medium. Section 3 is a presentation of the physical problems under consideration.

2. The Laws of Electromagnetism

The laws of electromagnetism derive from

i) physical concepts (electric charge, current density) which we introduce here without experimental explanations; for such, we refer to the classical works (Panofsky-Phillips [1], G. Bruhat [1];

ii) universal laws (conservation of electric charge, Faraday's law) which we will state and translate into mathematical form (G. Germain [3]);

iii) constituent laws, characteristic for each medium.

The equations which result from the totality of these laws are the *Maxwell equations*, if the medium considered is at rest. In the case where it is a continuous medium in motion, they are the equations *of magnetodynamics of fluids.*

2.1. Physical Quantities

We introduce the following concepts:

—*Electric charge*: it will be represented by the scalar q that will be, according to the case, a volume, surface or line density of electric charge; without further specification, we will deal with volume density.

—*Electric current*: this is a vector $\mathbf{J}$ of $\mathbb{R}^3$ that measures a flux of electric charges: the flux of electric charges across a surface element dS in the sense of the unit normal $\mathbf{n}$ to dS, given by the scalar product $\mathbf{J}.\mathbf{n}dS$. The vector $\mathbf{J}$ will also be called the current density vector.

—*Magnetic induction*: this is a vector $\mathbf{B}$ of $\mathbb{R}^3$, that occurs in Faraday's law.

—*Electric field*: this is a vector $\mathbf{H}$ of $\mathbb{R}^3$.

Other physical quantities will be introduced in the course of the exposition.

2.2. Conservation of Electric Charge

For an arbitrary fixed domain $\mathscr{D}$ of $\mathbb{R}^3$ with regular boundary $\partial\mathscr{D}$, the change per unit time of the total electric charge contained in the interior of $\mathscr{D}$ is produced by the flux of charges through $\partial\mathscr{D}$ and by a possible volume addition g of charges per unit of volume and of time; this can be expressed by

$$(2.1)\qquad \frac{\partial}{\partial t}\int_{\mathscr{D}} q\,dx = -\int_{\partial\mathscr{D}} \mathbf{J}.\mathbf{n}\,dS + \int_{\mathscr{D}} g\,dx$$

where $\mathbf{n}$ is the exterior unit normal at the boundary $\partial\mathscr{D}$ of $\mathscr{D}$.

Since this relation is valid for any $\mathscr{D}$, it follows that pointwise

$$(2.2)\qquad \partial q/\partial t + \operatorname{div}\mathbf{J} = g. \quad \square$$

We now introduce a vector $\mathbf{D}$, *the electric induction or potential of charge*, such that

$$(2.3)\qquad q = \operatorname{div}\mathbf{D},$$

and a vector $\mathbf{G}$ such that

$$(2.4)\qquad g = \operatorname{div}\mathbf{G}.$$

Then, Equation (2.2) shows that the vector $\partial\mathbf{D}/\partial\mathbf{t} + \mathbf{J} - \mathbf{G}$ has divergence zero. Thus, there exists at least one vector $\mathbf{H}$ which we call the *magnetic field* such that

$$(2.5)\qquad \partial\mathbf{D}/\partial t + \mathbf{J} - \operatorname{curl}\mathbf{H} = \mathbf{G}.$$

If the volume production of charges g is zero, we can take $\mathbf{G}=0$; then equation (2.5) furnishes

$$(2.6)\qquad \int_{\Sigma}(\partial\mathbf{D}/\partial t + \mathbf{J}).\mathbf{n}\,d\Sigma = \int_{\partial\Sigma}\mathbf{H}.d\mathbf{s},$$

where Σ is a two-dimensional manifold with boundary $\partial\Sigma$ in $\mathbb{R}^3$ (the unit vector **n** to Σ orients the boundary $\partial\Sigma$ whose infinitesimal tangent vector is d**s**).

Relation (2.6) is known as *Ampère's theorem.* □

Remark 2.1. The vector **G** is defined up to a rotational field. *The choice of* **G** *thus directly influences the value of the magnetic field.* □

Remark 2.2. If the vector field **D** is discontinuous across a surface Σ with unit normal **n**, and if $\mathscr{D}$ is a domain which is divided by Σ into two subdomains $\mathscr{D}_1$ and $\mathscr{D}_2$, **n** being directed towards $\mathscr{D}_2$, we have

(2.7) $$\int_{\partial\mathscr{D}_1} \mathbf{D}.\mathbf{n}\,dS = \int_{\mathscr{D}_1} q\,dx\,,$$

(2.8) $$\int_{\partial\mathscr{D}_2} \mathbf{D}.\mathbf{n}\,dS - \int_{\mathscr{D}_2} q\,dx\,,$$

(2.9) $$\int_{\mathscr{D}} q \equiv \int_{\mathscr{D}_1} q\,dx + \int_{\mathscr{D}_2} q\,dx + \int_{\Sigma\cap\mathscr{D}} q\,d\Sigma = \int_{\partial\mathscr{D}} \mathbf{D}.\mathbf{n}\,dS\,.$$

By comparison, we obtain

(2.10) $$\int_{\Sigma\cap\mathscr{D}} q\,d\Sigma = -\int_{\Sigma\cap\mathscr{D}} \mathbf{D}^{(1)}.\mathbf{n}\,d\Sigma + \int_{\Sigma\cap\mathscr{D}} \mathbf{D}^{(2)}.\mathbf{n}\,d\Sigma\,;$$

this relation holds for arbitrary Σ and $\mathscr{D}$.

From this, there follows the pointwise equality on Σ

(2.11) $$(\mathbf{D}^{(2)} - \mathbf{D}^{(1)})\mathbf{n} = q\,.$$

In the above relation, the indices (1) and (2) refer to the regions $\mathscr{D}_1$ and $\mathscr{D}_2$, respectively, and q represents a surface density of electric charges on Σ. Similarly, we specify that in Equation (2.9) the left hand side $\int_{\mathscr{D}} q$ represents the sum of all charges in the interior of $\mathscr{D}$, i.e., the sum of the volume charges in the interior of $\mathscr{D}_1$ and $\mathscr{D}_2$ and of the surface charges on $\Sigma\cap\mathscr{D}$.

An altogether analogous argument, based on the relation

(2.12) $$\int_{\mathscr{D}} \operatorname{curl}\mathbf{V}\,dx = \int_{\partial\mathscr{D}} \mathbf{n}\wedge\mathbf{V}\,dS\,,$$

and carried out starting with Equation (2.5), gives

(2.13) $$n\wedge(\mathbf{H}^{(2)} - \mathbf{H}^{(1)}) = \mathbf{J}\,,$$

where **J** denotes a surface density of current on (Σ_1).

We note that Equations (2.11) and (2.13) are contained in (2.3) and (2.5), if understood in the sense of distributions. □

2.3. Faraday's Law

The derivative with respect to the time of the flux of magnetic induction **B** *across a portion of the fixed surface* Σ, *with the boundary* $\partial\Sigma$, *is the opposite to the circulation of the electric field along the contour* $\partial\Sigma$.

This law can be expressed analytically by

$$\frac{\partial}{\partial t}\int_{\Sigma}\mathbf{B}.\mathbf{n}\,\mathrm{d}\Sigma+\int_{\partial\Sigma}\mathbf{E}.\mathbf{ds}=0. \tag{2.14}$$

From (2.14) it follows that, if Σ is a closed surface $\partial\mathscr{D}$ which is the boundary of an open set $\mathscr{D}$, that

$$\frac{\partial}{\partial t}\int_{\mathscr{D}}\operatorname{div}B\,\mathrm{d}x=0. \tag{2.15}$$

i.e., that pointwise

$$\partial(\operatorname{div}B)/\partial t=0. \tag{2.16}$$

If we assume that, at an instant t_0, the field $\mathbf{B}_0$ is such that $\operatorname{div}\mathbf{B}_0=0$, then at every moment t

$$\operatorname{div}\mathbf{B}=0. \tag{2.17}$$

Equation (2.14) also furnishes, pointwise,

$$\partial B/\partial t+\operatorname{curl}E=0 \tag{2.18}$$

that, strictly, implies (2.16). □

Remark 2.3. If Σ is a possible surface of discontinuity for the fields **E** and **B**, we prove, as we did in Remark 2.2, that

$$(\mathbf{B}^{(2)}-\mathbf{B}^{(1)}).\mathbf{n}=0, \tag{2.19}$$

$$\mathbf{n}\wedge(E^{(2)}-E^{(1)})=0. \quad \square \tag{2.20}$$

Remark 2.4. If we introduce a new frame of reference in uniform translation of velocity $\mathbf{V}_0$ with respect to the first one, and if we define by

$$\left|\begin{array}{ll} q'=q & \mathbf{J}'=\mathbf{J}-q\mathbf{V}_0 \\ \mathbf{D}'=\mathbf{D} & \mathbf{E}'=\mathbf{E}+\mathbf{V}_0\wedge\mathbf{B} \\ \mathbf{B}'=\mathbf{B} & \mathbf{H}'=\mathbf{H}-\mathbf{V}_0\wedge\mathbf{D}, \end{array}\right.$$

the new electromagnetic quantities in this system, it is easy to see that the Maxwell equations (2.5), (2.18), as well as, obviously, (2.3) and (2.17), remain valid in this new frame of reference. These *equations possess Galilean invariance.*

2.4. Recapitulation. Maxwell's equations

This is the name given to the system of the four equations:

(2.21) $\partial \mathbf{D}/\partial t + \mathbf{J} - \operatorname{curl} \mathbf{H} = \mathbf{G},$

(2.22) $\partial \mathbf{B}/\partial t + \operatorname{curl} \mathbf{E} = 0,$

(2.23) $\operatorname{div} \mathbf{D} = q,$

(2.24) $\operatorname{div} \mathbf{B} = 0\,.$

At this stage, Equation (2.23) can be considered as definition of q, the electric charge, and Equation (2.24) is a consequence of (2.22) and of the assumption $\operatorname{div} \mathbf{B} = 0$ at a particular instant.

Moreover, if we take the divergence of the expression occurring in (2.21), we again obtain the equation of conservation of electric charge.

The two essential Maxwell equations therefore are (2.21) and (2.22). If the vector $\mathbf{G}$ is assumed to be given, these equations constitute six scalar relations between fifteen unknown scalars, the components of the vectors $\mathbf{B}, \mathbf{E}, \mathbf{D}, \mathbf{H}, \mathbf{J}$. Thus it is clear, that this group of equations is insufficient to permit prediction of electromagnetic phenomena. This justifies the introduction of laws of a less universal character, called constituent laws, which vary from one continuous medium to another.

2.5. Constituent Laws

We consider two groups:

i) *Proportionality of fields and inductions*: this law can be expressed by two relations,

(2.25) $\mathbf{D} = \varepsilon \mathbf{E},$

(2.26) $\mathbf{B} = \mu \mathbf{H}$

where ε is the dielectric constant of the medium and μ is the magnetic permeability. These scalars are assumed to be independent of the electromagnetic phenomena taking place in the medium. These then are *linear laws.*

ii) *Ohm's law*: if the medium is "*stable*", that is to say, possesses a resistivity σ independent of electromagnetic quantities, Ohm's law furnishes us with the linear relation

(2.27) $\mathbf{J} = \sigma E.$

If, under the influence of the electric field, the medium is "*ionizable*" then its resistivity σ varies enormously as a function of the electric field, and Ohm's law must be replaced by the relations

$$\mathbf{J} = \sigma \mathbf{E} \quad \text{if} \quad |\mathbf{E}| < E_0$$
$$\mathbf{J} = (\sigma + \lambda)\mathbf{E} \quad \text{if} \quad |\mathbf{E}| = E_0, \tag{2.28}$$

where E_0 is a positive constant, the threshold of ionization, and where λ is a suitable scalar, positive or zero.

Such behavior occurs in gases which are subjected to strong electric fields. It gives rise to the phenomena of the electric arc or of antenna breakdown. ▯

Remark 2.5. The mathematical law (2.28) does not describe the experimentally observed phenomenon entirely exactly. Indeed, if E and J represent the lengths of the vectors E and J, respectively, (2.28) gives a graph of the function $J = J(E)$, represented in Fig. 21, whereas the experimentally measured graph rather resembles that in Fig. 22. All the same, we will retain law (2.28), for the true graph (Fig. 22) shows a phenomenon of *delay of ionization* which in fact corresponds to a *physical instability.*

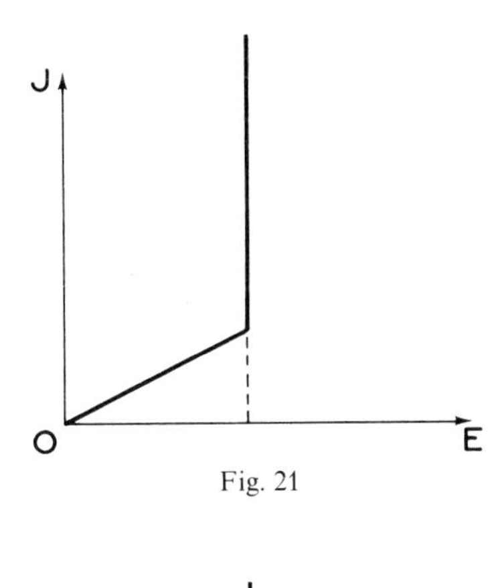

Fig. 21

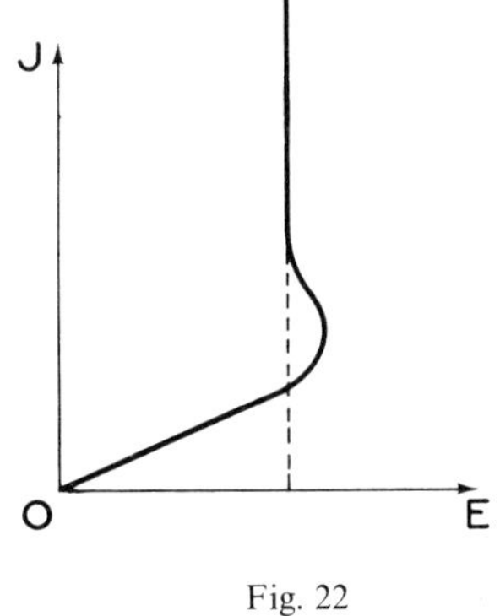

Fig. 22

Remark 2.6. Equations (2.25) and (2.26) *do not possess Galilean invariance.* Therefore it is indispensable, in the general case, to apply them only in a fixed reference frame, or, at least, always in the same reference frame.

In Duvaut-Lions [7], for the particular case of magnetohydrodynamics, these laws are applied in local reference frames tied to the moving fluid. It is then necessary to make simplifying assumptions beforehand that make these relations Galilean. □

Remark 2.7. The constituent laws represent nine supplementary relations between the fifteen unknown scalars. Thus, with Maxwell's laws, we have as many equations as unknowns.

3. Physical Problems to be Considered

3.1. Stable Medium with Supraconductive Boundary

Let Ω be a domain in $\mathbb{R}^3$ with a regular bounded boundary Γ. This open set Ω may or may not be bounded. We want to find the vector fields $\mathbf{B}, \mathbf{D}, \mathbf{J}$ that satisfy

$$\text{(3.1)} \qquad \begin{aligned} &\partial\mathbf{D}/\partial t + \mathbf{J} - \operatorname{curl}(\hat{\mu}\mathbf{B}) = \mathbf{G}_1 \quad \text{in } \Omega \\ &\partial\mathbf{D}/\partial t + \operatorname{curl}(\hat{\varepsilon}\mathbf{D}) = \mathbf{G}_2 \quad \text{in } \Omega \end{aligned}$$

where the right hand sides $\mathbf{G}_1$ and $\mathbf{G}_2$ satisfy

$$\text{(3.2)} \qquad \operatorname{div} \mathbf{G}_2 = 0\,, \qquad \mathbf{G}_2 . \mathbf{n} = 0\,,$$

(we symmetricized the right hand sides of (2.21) and (2.22)), and where we set

$$\text{(3.3)} \qquad \hat{\mu} = 1/\mu\,, \qquad \hat{\varepsilon} = 1/\varepsilon\,.$$

These quantities $\hat{\varepsilon}$ and $\hat{\mu}$ are strictly positive and remain bounded; they may depend on x, in particular be piecewise constant.

We have the initial conditions

$$\text{(3.4)} \qquad \mathbf{B}(x,t)|_{t=0} = \mathbf{B}_0(x)\,, \qquad \mathbf{D}(x,t)|_{t=0} = \mathbf{D}_0(x)\,,$$

and we recall that $\operatorname{div} \mathbf{D} = q$, $\operatorname{div} \mathbf{B} = 0$.

(3.1) and (3.4) are general equations valid for all the problems we consider. The assumption of a "stable medium" is equivalent to *Ohm's law*

$$\text{(3.5)} \qquad \mathbf{J} = \sigma \mathbf{E} = \sigma \varepsilon \mathbf{D} \quad \text{in } \Omega$$

and the supraconductive boundary imposes the conditions on Γ

$$\text{(3.6)} \qquad \mathbf{B}.\mathbf{n} = 0\,, \quad \mathbf{D} \wedge \mathbf{n} = 0$$

which we assume to be satisfied also by the initial conditions $\mathbf{B}_0$ and $\mathbf{D}_0$; condition (3.6) for $\mathbf{B}$ is then included in (3.1) and (3.2).

The scalar σ, the conductivity of the medium, is positive and bounded; it may depend on x.

3.2. Polarizable Medium with Supra Conductive Boundary

Equations and conditions here are the same as in the problem of Section 3.1, with the exception of (3.5) that has to be replaced by (2.28); this we rewrite in terms of $\mathbf{D}$

$$\begin{aligned}&|\mathbf{D}|<\mathscr{D}_0 \Rightarrow \mathbf{J}=\sigma\hat{\varepsilon}\mathbf{D}\\ &|\mathbf{D}|=\mathscr{D}_0 \Rightarrow \exists\lambda\geqslant 0 \qquad \text{such that } \mathbf{J}=(\sigma\hat{\varepsilon}+\lambda)\mathbf{D}\\ &(\mathscr{D}_0=\hat{\varepsilon}E_0)\,.\end{aligned}\tag{3.7}$$

The scalar E_0 is a positive constant that is called the threshold of breakdown or breakdown voltage. This name is particularly descriptive in the case of a plane condensor which is being charged more and more: between the plates, an electric field $\mathbf{E}$ of growing intensity is created that, on attaining the threshold E_0, ionizes the dielectric between the plates: this abruptly becomes a conductor and allows a current to pass through which discharges the condensor and, generally, destroys it. An analogous phenomenon can occur with antennas: if the electric field generated by the antenna is too intense, the atmospheric gas may become ionized, therefore become a conductor, and a considerable electric current then passes through the antenna which deteriorates: this is the phenomenon of breakdown. □

The conductivity σ which occurs in (3.7) is always very weak, if not zero.

Remark 3.1. In everyone of the above problems, one can replace the boundary conditions (3.6) by

$$\mathbf{D}.\mathbf{n}=0\,,\qquad \mathbf{B}\wedge\mathbf{n}=0\tag{3.8}$$

but the problems to which this leads seem less interesting from a physical point of view. At any rate, all that follows can easily be adapted to the case (3.8).

3.3. Bipolar Antenna

Let the open set Ω be the entire space $\mathbb{R}^3$. We consider two open disjoint sets Ω_1 and Ω_2, contained in $\mathbb{R}^3$ as indicated in Fig. 23, which are the seat of given injections of electricity. Let $g(x,t)$ be the volume density of electric charges injected per unit time in Ω_1 and Ω_2.

The problem thus posed reduces to the problems of Sections 3.1 and 3.2, if we take for Γ the empty set and specify the right hand sides G_1 and G_2 of equations (3.1), in accordance with Remark 2.1. □

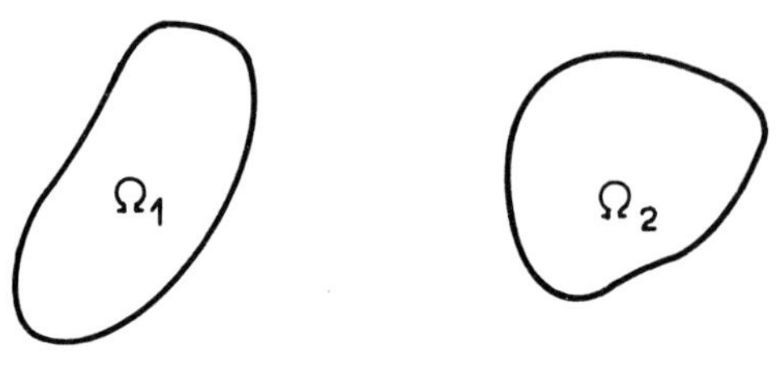

Fig. 23

Determination of G_1 *and* G_2. We take $G_2=0$, which does not cause any difficulty. The function G_1 must be such that

$$\text{div}\,\mathbf{G}_1 = g \tag{3.9}$$

where the function g, the injection of electricity, is given on Ω_1 and Ω_2. We extend g to $\mathbb{R}^3$ by taking $g=0$ on the complement of $\Omega_1 \cup \Omega_2$ and observe that, since the entire space has to remain neutral at every moment,

$$\int_{\mathbb{R}^3} g(x,t)\,\mathrm{d}x = 0\,, \qquad \forall t\,. \tag{3.10}$$

This condition (3.10) is the integrability condition for equation

$$\Delta\Phi + g = 0 \quad \text{in } \mathbb{R}^3\,. \tag{3.11}$$

This induces us to choose, among all functions G_1 which satisfy (3.9), the one whose curl is zero, i.e.,

$$\mathbf{G}_1 = -\,\mathbf{grad}\,\Phi \tag{3.12}$$

where Φ is the solution of (3.11). ▯

The problem of the bipolar antenna is thus a special case of the problem in Section 3.1 if the material is stable, and of Problem 3.2 if the material is polarizable.

3.4. Slotted Antenna. Diffraction of an Electromagnetic Wave by a Supraconductor

The following problem occurs frequently.

Let $\{B^{(1)}, D^{(1)}\}$ be an electro magnetic wave in $\mathbb{R}^3$, i.e., a solution of the Maxwell equations with zero right hand side in a *stable* and generally homogeneous medium. What kind of perturbation does the solution $\{B^{(1)}, D^{(1)}\}$ undergo if one introduces a supraconducting body into $\mathbb{R}^3$? This is the problem of scattering of an electromagnetic wave by a supraconductor. The slotted antenna can be schematized by an analogous phenomenon. If the open set Ω

is the exterior of a body $\mathscr{V} = \tilde{\Omega}$, a bounded supraconductor, let there be given a distribution of dipoles on a curved arc $\widehat{AB}$ in Ω (Fig. 24). In the absence of $\mathscr{V}$, the given distribution of dipoles will radiate an electromagnetic wave $\{B^{(1)}, D^{(1)}\}$ in $\mathbb{R}^3$, which is easily written down, since we deal with a distribution of dipoles.

The presence of a supraconductor perturbs the wave $\{B^{(1)}, D^{(1)}\}$ and we deal with a scattering phenomenon.

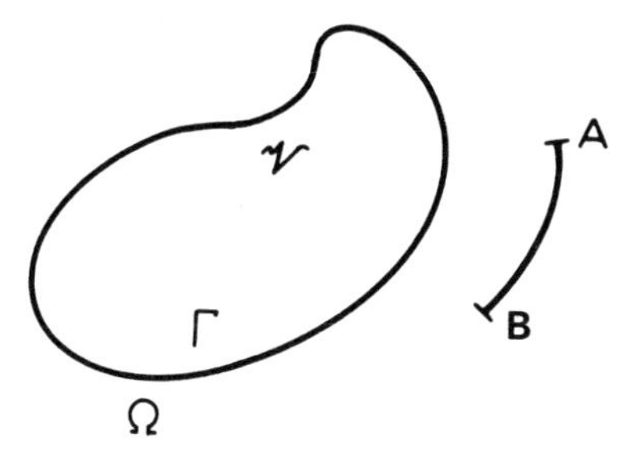

Fig. 24

Formulation of the equations. We set

$$\{\mathbf{B}^{(2)}, \mathbf{D}^{(2)}\} = \{\mathbf{B}^{(1)}, \mathbf{D}^{(1)}\} + \{\mathbf{B}, \mathbf{D}\} \tag{3.13}$$

where $\{\mathbf{B}^{(2)}, \mathbf{D}^{(2)}\}$ is the solution in the presence of a supraconductor. The electromagnetic waves $\{\mathbf{B}^{(1)}, \mathbf{D}^{(1)}\}$ and $\{\mathbf{B}^{(2)}, \mathbf{D}^{(2)}\}$ satisfy the Maxwell equations with zero right hand side, therefore, because of linearity, the same holds for $\{\mathbf{B}, \mathbf{D}\}$, that is to say

$$\begin{aligned} &\partial \mathbf{D}/\partial t + \sigma\hat{\varepsilon}\mathbf{D} - \operatorname{curl}(\hat{\mu}\mathbf{B}) = 0 \\ &\partial \mathbf{B}/\partial t + \operatorname{curl}(\hat{\varepsilon}\mathbf{D}) = 0\,, \end{aligned} \tag{3.14}$$

and at the boundary Γ of the supradonductor

$$\mathbf{B}^{(2)}\mathbf{n} = 0\,, \qquad \mathbf{D}^{(2)} \wedge \mathbf{n} = 0 \tag{3.15}$$

from which it follows that

$$\mathbf{B.n} = -\mathbf{P}^{(1)}\mathbf{n}\,, \qquad \mathbf{D} \wedge \mathbf{n} = -\mathbf{D}^{(1)} \wedge \mathbf{n} \text{ on } \Gamma\,. \tag{3.16}$$

We complete equations (3.14) and (3.16) by the intial data

$$\mathbf{B}(x,t)|_{t=0} = \mathbf{B}_0(x)\,, \qquad \mathbf{D}(x,t)|_{t=0} = \mathbf{D}_0(x) \tag{3.17}$$

which satisfy

$$\operatorname{div} \mathbf{B}_0(x) = 0\,. \quad \square \tag{3.18}$$

3.5. Recapitulation. Unified Formulation of the Problems

Let Ω be an open set of $\mathbb{R}^n$, *bounded* or *not*[1], with *regular* and *bounded*[2] boundary Γ (for example, Γ is a once continuously differentiable, two-dimensional manifold, with Ω lying locally on one side of Γ).

We note: $\tilde{\Omega} = \complement\Omega$ ($\tilde{\Omega} = \mathscr{V}$ in Sec. 3.4).

We prescribe *piecewise constant* functions ε and μ in Ω[3]; more precisely, we assume that

$$\begin{aligned} &\varepsilon = \varepsilon_i, \ \mu = \mu_i \quad \text{in } \Omega_i, \qquad i = 1, \dots, q\ ^4 \\ &\varepsilon_i > 0, \ \mu_i > 0, \end{aligned} \tag{3.19}$$

where we assume that the common boundary (cf. Fig. 25) Σ_{ij} of Ω_i and Ω_j is *regular and bounded*.

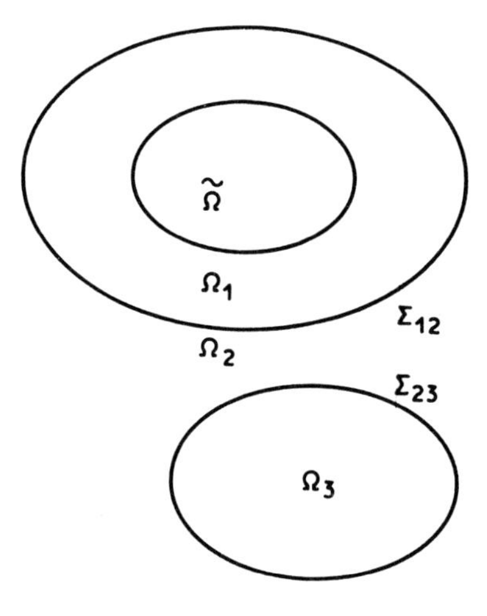

Fig. 25

We then look for the vector fields **B** and **D** such that, with the notation (3.3) ($\hat{\mu} = 1/\mu$, $\hat{\varepsilon} = 1/\varepsilon$):

$$\partial \mathbf{D}/\partial t + \mathbf{J} - \operatorname{curl}(\hat{\mu}\mathbf{B}) = \mathbf{G}_1, \tag{3.20}$$

$$\partial \mathbf{B}/\partial t + \operatorname{curl}(\hat{\varepsilon}\mathbf{D}) = \mathbf{G}_2 \tag{3.21}$$

$$n \wedge \mathbf{D} = 0 \quad \text{on } \Gamma \tag{3.22}$$

$$\mathbf{B}(0) = \mathbf{B}_0, \qquad \mathbf{D}(0) = \mathbf{D}_0 \quad \text{on } \Omega, \tag{3.23}$$

---

[1] Ω is bounded in Problems 3.1 and 3.2, unbounded in the other cases.

[2] Or empty.

[3] One could generalize this to piecewise regular functions. We explicitly assume that ε is constant in the neighborhood of Γ.

[4] $q = 2$ for bipolar antennas.

with, *in stable media*:

$$\mathbf{J}=\sigma\hat{\varepsilon}\mathbf{D} \tag{3.24}$$

and, *in polarizable media*, with:

$$\begin{aligned} &\mathbf{J}=\sigma\hat{\varepsilon}\mathbf{D} \quad \text{when} \quad |\mathbf{D}|<\mathscr{D}_0 \quad \text{(positive constant)} \\ &\mathbf{J}=(\sigma+\lambda)\hat{\varepsilon}\mathbf{D} \quad \text{for a suitable} \quad \lambda\geqslant 0 \text{ when } |\mathbf{D}|=\mathscr{D}_0. \end{aligned} \tag{3.25}$$

Remark 3.2. We omit from the formulation conditions of the type

$$\operatorname{div}\mathbf{B}=0, \qquad n.\mathbf{B}=0 \quad \text{on } \Gamma \qquad \text{etc.}$$

which will appear as properties of the solution of the above problem when the data $\mathbf{G}_i$ and $\mathbf{B}_0, \mathbf{D}_0$ are "suitable". □

Remark 3.3. The conditions at the interfaces Σ_{ij} are *consequences* of the equations when taken in the sense of distributions in Ω (cf. Remark 2.2). □

Orientation. Sections 4 to 7 deal with stable media; in Section 8, polarizable media are discussed. Subsequently, we will discuss *non homogeneous* problems which correspond to slotted antennas.

4. Discussion of Stable Media. First Theorem of Existence and Uniqueness

4.1. Tools from Functional Analysis for the "Weak" Formulation of the Problem

Remark on notation. Vectors are not written in bold face except in the statements of the principal results.

The space $H(\text{curl}; \Omega)$. We define

$$H(\text{curl};\Omega)=\{v| \quad v\in(L^2(\Omega))^3, \quad \text{curl}\, v\in(L^2(\Omega))^3\}. \tag{4.1}$$

With the (graph) norm

$$(\|v\|^2_{L^2(\Omega)^3}+\|\text{curl}\, v\|^2_{L^2(\Omega)^3})^{1/2}, \tag{4.2}$$

the space $H(\text{curl}; \Omega)$ is a Hilbert space.

Lemma 4.1. *We assume that* Ω *has a regular bounded boundary* Γ. *Let* $(C^1_K(\bar{\Omega}))^3$ *be the space of vectors which are once continuously differentiable in* $\bar{\Omega}$ *and have compact support in* $\bar{\Omega}$ (a superfluous condition when Ω is bounded). *Then* $(C^1_K\bar{\Omega}))^3$ *is dense in* $H(\text{curl}: \Omega)$.

Proof. If $v \in H(\text{curl};\Omega)$ and if $\varphi \in \mathscr{D}(\bar{\Omega})$, then $\varphi v \in H(\text{curl};\Omega)$, the operator "curl" being a differential system *of first order*.

Then, let $\psi \in \mathscr{D}(\mathbb{R}^n)$, $\psi = 1$ in a neighborhood of the origin, and let ψ_m be defined by

$$\psi_M(x) = \psi(x/M).$$

If we again denote by ψ_M the restriction of ψ_M to $\bar{\Omega}$, we have:

$$\psi_M v \in H(\text{curl};\Omega) \quad \text{when} \quad v \in H(\text{curl};\Omega)$$

and one can verify without difficulty that

$$\psi_M v \to v \quad \text{in } H(\text{curl};\Omega) \quad \text{when } M \to +\infty.$$

Now, $\forall M$, $\psi_M v$ has compact support in $\bar{\Omega}$. Thus, we only have to approximate (in the sense of $H(\text{curl};\Omega)$) an element v *with compact support in* $\bar{\Omega}$.

We now introduce a finite open covering $\mathcal{O}_1, \ldots, \mathcal{O}_N$ of a neighborhood of Γ in $\mathbb{R}^3$, where each $\mathcal{O}_i$ is a bounded open set with the property[5]:

$$(4.3)\qquad \left|\begin{array}{l} \forall i,\ \forall \varepsilon > 0, \text{ there exists a vector } \lambda_i \in \mathbb{R}^3 \text{ such that} \\ |\lambda_i| \leqslant \varepsilon, \\ \text{and such that the translated open sets } \mathcal{O}_i + \lambda_i \text{ form an open covering} \\ \text{of a neighborhood of } \Gamma \text{ in } \mathbb{R}^3. \end{array}\right.$$

Then, if $\alpha_1 \ldots, \alpha_N$ is a partition of unity of a neighborhood of Γ in $\mathbb{R}^3$ subordinate to $\{\mathcal{O}_i\}$, $\alpha_i \in \mathscr{D}(\mathcal{O}_i)$, we can write

$$(4.4)\qquad v = \textstyle\sum_{i=1}^N \alpha_i v + v_0,$$

$v \in H(\text{curl};\Omega)$, with compact support in Ω.

We can approximate v_0 (by regularization) by elements of $\mathscr{D}(\Omega)$ and thus everything reduces to approximating $\alpha_i v = w$ (i fixed arbitrarily).

We introduce a sequence t_n of vectors of $\mathbb{R}^3$ such that (cf. (4.3))

$$(4.5)\qquad \begin{aligned} &|t_n| \to 0, \quad (n \to \infty) \\ &w_n(x) = w(x - t_n) \in H(\text{curl};\Omega_n), \quad \bar{\Omega} \subset \Omega_n. \end{aligned}$$

Then, if

$$\varphi_n = \text{restriction of } w_n \text{ to } \Omega,$$

we have:

$$\varphi_n \to w \text{ in } H(\text{curl};\Omega) \text{ when } n \to \infty$$

---

[5] This is a variant of a property introduced by Hörmander [1]. We note that such a covering $\{\mathcal{O}_i\}$ exists.

and, consequently, everything *reduces to approximating by a sequence* in $(C_K^1(\bar{\Omega}))^3$

$$\varphi_n = \psi \qquad (n \text{ fixed}).$$

But, (according to (4.5)), ψ is a restriction to Ω of $\Psi \in H(\text{curl}; \Omega')$, $\Omega' \supset \bar{\Omega}$. If we introduce $\Theta \in \mathscr{D}(\Omega')$, $\Theta = 1$ on $\bar{\Omega}$, we have

$$(4.6) \qquad \begin{aligned} &\Theta\Psi \in H(\text{curl}; \mathbb{R}^3) \text{ (extending } \Theta\Psi \text{ as 0 outside of } \Omega') \\ &\Theta\Psi = \psi \text{ on } \Omega . \end{aligned}$$

But (by regularization), there exist $F_j \in \mathscr{D}(\mathbb{R}^3)$, $F_j \to \Theta\Psi$ in $H(\text{curl}; \mathbb{R}^3)$ and, if $f_j =$ restriction of F_j to Ω, it follows that $f_j \to \psi$ in $H(\text{curl}; \Omega)$; since $f_j \in \mathscr{D}(\Omega)$, this concludes the proof of the lemma. □

Remark 4.1. The preceding proof is general and does not make use of the particular structure of the differential system "curl". □

Lemma 4.2. *Assumptions as in Lemma* 4.1. *Let n be the normal to* Γ *directed* (to be definite) *toward the exterior of* Ω. *The mapping*

$$(4.7) \qquad u \to n \wedge u|_\Gamma = n \wedge u$$

of $(C_k^1(\bar{\Omega}))^3 \to (C^1(\Gamma))^3$, *can be extended by continuity to a mapping again denoted by* $u \to n \wedge u$, *which is linear and continuous from* $H(\text{curl}; \Omega) \to (H^{-1/2}(\Gamma))$.

Proof. For $\varphi \in (H^{1/2}(\Gamma))^3$, let $\Phi \in (H^1(\Omega))^3$ such that

$$\Phi|_\Gamma = \varphi ,$$

the mapping $\varphi \to \Phi$ being linear and continuous from $(H^{1/2}(\Gamma))^3 \to (H^1(\Omega))^3$.

For a given u in $H(\text{curl}; \Omega)$, we put

$$(4.8) \qquad \pi(\varphi) = (\text{curl}\, u, \Phi) - (u, \text{curl}\, \Phi)$$

where, generally

$$(4.9) \qquad (f, g) = \int_\Omega f_i g_i \,\mathrm{d}x .$$

The notation in (4.8) is legitimate. Indeed, the right hand side is *independent of the choice* of Φ (provided that $\Phi|_\Gamma = \varphi$); if indeed Ψ is a second vector such that $\Psi \in (H^1(\Omega))^3$, $\Psi|_\Gamma = \varphi$, then

$$\Phi - \Psi = \Theta \in (H_0^1(\Omega)^3$$

and therefore

$$(\text{curl}\, u, \Theta) = (u, \text{curl}\, \Theta).$$

The mapping $\varphi \to \pi(\varphi)$ is linear and continuous on $(H^{1/2}(\Gamma))^3$; accordingly, it has the form

(4.10)
$$\pi(\varphi) = (\sigma_\mu, \varphi)_\Gamma, \quad \sigma_\mu \in (H^{-1/2}(\Gamma))^3,$$
$(\cdot)$ denoting the scalar product of $(H^{-1/2}(\Gamma))^3$ and $(H^{1/2}(\Gamma))^3$ $\mu \to \sigma_u$ being linear and continuous from $H(\mathrm{curl};\Omega) \to H^{-1/2}(\Gamma))^3$.

But if $u \in (C_K^1(\bar{\Omega}))^3$, we have:

$$(4.11) \qquad (\mathrm{curl}\, u, \Phi) - (u, \mathrm{curl}\, \Phi) = \int_\Gamma (n \wedge u) \varphi \, d\Gamma \; (\Phi|_\Gamma = \varphi)$$

therefore

$$\sigma_u = n \wedge u \quad \text{if} \quad u \in (C_K^1(\bar{\Omega}))^3$$

from which the lemma follows. □

Lemma (4.2) allows us to introduce the space

$$(4.12) \qquad H_0(\mathrm{curl};\,\Omega) = \{v| \quad v \in H(\mathrm{curl};\,\Omega), \quad n \wedge v = 0 \text{ on } \Gamma\},$$

which is *closed* in $H(\mathrm{curl};\,\Omega)$.

Lemma 4.3. *Assumptions as in Lemma* 4.1. *Let*

$$(4.13) \qquad X = \text{space of vectors } \varphi \text{ of } (C_K^1(\bar{\Omega}))^3 \text{ such that } n \wedge \varphi = 0 \text{ on } \Gamma.$$

The space X *is dense in* $H_0(\mathrm{curl};\Omega)$.

Proof. For $u \in H_0(\mathrm{curl};\Omega)$, let $\tilde{u}$ = extension of u to $\mathbb{R}^3$ as 0 outside of Ω; then, if $\Phi \in (\mathscr{D}(\mathbb{R}^3))^3$, we have[6]

$$\begin{aligned}(\mathrm{curl}\,\tilde{u}, \Phi)_{\mathbb{R}^3} &= (\tilde{u}, \mathrm{curl}\,\Phi)_{\mathbb{R}^3} = (u, \mathrm{curl}\,\varphi)_\Omega \\ &= (\mathrm{curl}\, u, \varphi)_\Omega - (n \wedge u, \varphi)_\Gamma \\ &= (\mathrm{curl}\, u, \varphi)_\Omega \qquad (\text{since } n \wedge u = 0 \text{ on } \Gamma),\end{aligned}$$

therefore

$$\mathrm{curl}\,(\tilde{u}) = (\mathrm{curl}\,\tilde{u}) \in (L^2(\mathbb{R}^3))^3$$

$\tilde{u} \in H(\mathrm{curl};\mathbb{R}^3)$, $\tilde{u}$ obviously having its support in $\bar{\Omega}$.

We introduce $\{\mathcal{O}_i\}, \{\alpha_i\}$ as in the proof of Lemma 4.1[7].

We can write

$$\tilde{u} = \sum_{i=1}^q \alpha_i \tilde{u} + u_0,$$

[6] The indices denote the domains of integration. By φ, we denote the restriction of Φ to Ω.

[7] The reasoning now is *exactly* as in Hörmander [1], Theorem 1.

$u_0 \in H(\text{curl}; \mathbb{R}^3)$ with compact support in Ω, and everything reduces to approximating $\alpha_i \tilde{u} = w$.

We now apply translations (as in (4.5)); but while we applied *translations "toward the exterior"* of Ω in (4.5), here we move w *"toward the interior"* of Ω: we consider $t_n \in \mathbb{R}^3$ such that

$$(4.14) \qquad \begin{aligned} &|t_n| \to 0\,, \\ &(\mathcal{O}_i \cap \Omega) + t_n \subset \Omega\,; \end{aligned}$$

we define w_n by

$$w_n(x) = w(x - t_n)\,;$$

$w_n \in H(\text{curl}; \mathbb{R}^3)$ and w_n has compact support in Ω.

Thus we approximate w (in the sense of $H(\text{curl}; \Omega)$) by elements of $(\mathcal{D}(\Omega))^3$ by regularization, from which the lemma follows. □

Remark 4.2. In fact, we have proved more than the lemma, namely

$$(4.15) \qquad \mathcal{D}(\Omega)^3 \text{ is dense in } H_0(\text{curl}; \Omega). \quad □$$

4.2. The Operator $\mathcal{A}$. "Weak" Formulation of the Problem

The space $\mathcal{H}$. We set

$$(4.16) \qquad \mathcal{H} = (L^2(\Omega))^6 = (L^2(\Omega))^3 \times (L^2(\Omega))^3\,;$$

if $\Phi = \{\varphi, \psi\} \in \mathcal{H}$ $(\varphi \in (L^2(\Omega))^3,\ \psi \in (L^2(\Omega))^3)$, and if $\Phi_* = \{\varphi_*, \psi_*\} \in \mathcal{H}$, we set

$$(4.17) \qquad (\Phi, \Phi_*)_{\mathcal{H}} = (\hat{\varepsilon}\varphi, \varphi_*) + (\hat{\mu}\psi, \psi_*)\,,$$

(where we use the notation of (4.9)); since the functions $\hat{\varepsilon}$ and $\hat{\mu}$ are bounded in Ω and satisfy

$$(4.18) \qquad \inf \hat{\varepsilon} > 0\,, \qquad \inf \hat{\mu} > 0\,,$$

the scalar product (4.17) is *equivalent* to the "usual" scalar product

$$(\varphi, \varphi_*) + (\psi, \psi_*)\,. \quad □$$

Remark 4.3. Naturally, the choice of the particular scalar product (4.17) is not absolutely indispensable—since the final result (Theorem 4.1) is *independent* of the scalar product on $\mathcal{H}$; however, this choise simplifies the proofs, since, for example, property (4.24) below *does depend* on the scalar product on $\mathcal{H}$. □

The domain $D(\mathscr{A})$. In order to define the unbounded operator $\mathscr{A}$ in $\mathscr{H}$, we first define its domain $D(\mathscr{A})$:

$$\begin{aligned}D(\mathscr{A}) = \{\Phi| \quad &\Phi=\{\varphi,\psi\}\in\mathscr{H}, \quad \operatorname{curl}(\hat{\varepsilon}\varphi)\in(L^2(\Omega))^3,\\ &\operatorname{curl}(\hat{\mu}\psi)\in(L^2(\Omega))^3, \quad n\wedge\varphi=0 \text{ on } \Gamma\}. \quad \square\end{aligned} \tag{4.19}$$

Remark 4.4. Definition (4.19) *has a meaning*; to prove this, it is necessary that $n\wedge\varphi$ has a meaning; now, $\hat{\varepsilon}$ is constant in the neighborhood of Γ; therefore, there exists an open set $\Omega_1\subset\Omega$ (as in Fig. 25) such that $\Gamma\subset$ boundary of Ω_1; then

$$\varphi\in H(\operatorname{curl};\Omega_1) \tag{4.20}$$

and since the boundary $\partial\Omega_1$ of Ω_1 contains Γ, we can (an immediate variant of Lemma 4.2) define

$$n\wedge\varphi\in(H^{-1/2}(\Gamma))^3, \tag{4.21}$$

and

$$\begin{aligned}&\text{the mapping } \varphi\to n\wedge\varphi \text{ is linear and continuous from}\\ &H(\operatorname{curl};\Omega_1)\to(H^{-1/2}(\Gamma))^3. \quad \square\end{aligned} \tag{4.22}$$

The operator $\mathscr{A}$. For $\Phi\in D(\mathscr{A})$, we set

$$\mathscr{A}\Phi=\{-\operatorname{curl}(\hat{\mu}\psi),\operatorname{curl}(\hat{\varepsilon}\varphi)\} \quad (\in\mathscr{H}). \tag{4.23}$$

We have

Lemma 4.4. *The domain* $D(\mathscr{A})$ *is dense in* $\mathscr{H}$ *and* $\mathscr{A}$ *is closed. We have*:

$$\mathscr{A}^*=-\mathscr{A}, \quad D(\mathscr{A}^*)=D(\mathscr{A}). \tag{4.24}$$

Proof. Let us denote by $(\pi\mathscr{D}(\Omega_i))$ the space of Φ such that the restriction Φ^i of Φ to Ω_i is in $\mathscr{D}(\Omega_i)^6$, $i=1,\dots,q$. Evidently, $(\pi\mathscr{D}(\Omega_i))^6$ is dense in $\mathscr{H}$ and contained in $D(\mathscr{A})$, therefore $D(\mathscr{A})$ is dense in $\mathscr{H}$.

Let us prove that $\mathscr{A}$ is *closed*. If $\Phi_j\in D(\mathscr{A})$, $\Phi_j=\{\varphi_j,\psi_j\}\to\Phi$ in $\mathscr{H}$ and $\mathscr{A}\Phi_j\to\Psi$ in $\mathscr{H}$, we have:

$$\varphi_j\to\varphi, \psi_j\to\psi \text{ in } (L^2(\Omega))^3$$

and $\operatorname{curl}(\hat{\mu}\psi_j)$ and $\operatorname{curl}(\hat{\varepsilon}\varphi_j)$ converge in $(L^2(\Omega))^3$; but $\operatorname{curl}(\hat{\mu}\psi_j)\to\operatorname{curl}(\hat{\mu}\psi)$ in $(\mathscr{D}'(\Omega))^3$, $\operatorname{curl}(\hat{\varepsilon}\varphi_j)\to\operatorname{curl}(\hat{\varepsilon}\varphi)$ in $(\mathscr{D}'(\Omega))^3$, therefore

$$\operatorname{curl}(\hat{\varepsilon}\varphi)\in(L^2(\Omega))^3, \quad \operatorname{curl}(\hat{\mu}\psi)\in(L^2(\Omega))^3.$$

Moreover, φ_j^1 (restriction of φ_j to Ω_1)$\to\varphi^1$ in $H(\operatorname{curl}:\Omega_1)$ and therefore (cf. (4.22)) $n\wedge\varphi_j^1(=0)\to n\wedge\varphi^1$ (on Γ) in $(H^{-1/2}(\Gamma))^3$, therefore $n\wedge\varphi^1=0$ on Γ, therefore $\Phi\in D(\mathscr{A})$.

We observe here that we can *describe* $D(\mathscr{A})$ in the following way: if

$$\Phi=\{\varphi,\psi\},\ \Phi^i=\{\varphi^i,\psi^i\} = \text{restriction of } \Phi \text{ to } \Omega_i,$$

then

$$(4.25)\qquad \left|\begin{array}{l} \operatorname{curl}\varphi^i,\ \operatorname{curl}\psi^i \in (L^2(\Omega_i))^3, \\ \hat{\mu}_i n\wedge\psi^i=\hat{\mu}_j n\wedge\psi^j \quad \text{on} \quad \Sigma_{ij}\ (\text{same } n), \\ \hat{\varepsilon}_i n\wedge\varphi^i=\hat{\varepsilon}_j n\wedge\varphi^j \quad \text{on} \quad \Sigma_{ij}, n\wedge\varphi^1=0 \ \text{ on } \Gamma, \end{array}\right.$$

and conversely.

Let then $\Phi_*\in D(\mathscr{A}^*)$, i.e., an element of $\mathscr{H}$ such that the linear form

$$\Phi\to(\mathscr{A}\Phi,\Phi_*)_{\mathscr{H}}$$

is continuous on $D(\mathscr{A})$ for the topology induced by $\mathscr{H}$. Thus, the form $\Phi\to(\mathscr{A}\Phi,\Phi_*)_{\mathscr{H}}$ is continuous, in particular on $(\pi\mathscr{D}(\Omega_i))^6$ for the topology induced by $\mathscr{H}$: but if $\Phi\in(\pi\mathscr{D}(\Omega_i))^6$, we have:

$$\begin{aligned}(\mathscr{A}\Phi,\Phi_*) = &-\textstyle\sum_i(\hat{\mu}_i\psi^i, \operatorname{curl}(\hat{\varepsilon}_i\varphi^i_*))\Omega_i \\ &+\textstyle\sum_i(\hat{\varepsilon}_i\varphi^i, \operatorname{curl}(\hat{\mu}_i\psi^i_*))\Omega_i,\end{aligned}$$

from which it follows that

$$\operatorname{curl}(\hat{\varepsilon}_i\varphi^i_*)\in(L^2(\Omega_i))^3\,,\qquad \operatorname{curl}(\hat{\mu}_i\psi^i_*)\in(L^2(\Omega_i))^3$$

or again

$$(4.26)\qquad \operatorname{curl}\varphi^i_*\in(L^2(\Omega_i))^3\,,\qquad \operatorname{curl}\psi^i_*\in(L^2(\Omega))^3\,.$$

Thus, (Lemma 4.2), we can define $n\wedge\varphi^i_*$, $n\wedge\psi^i_*$ on Σ_{ij} and Γ.

Let us now take $\Phi\in D(\mathscr{A})$ with $\Phi^i\in(C^1_K(\bar{\Omega}_i))^6$, thus satisfying relations (4.25). Then

$$\begin{aligned}(4.27)\qquad (\mathscr{A}\Phi,\Phi_*) &= -\Sigma_i(\operatorname{curl}(\hat{\mu}_i\psi^i),\hat{\varepsilon}_i\varphi^i_*)_{\Omega_i} \\ &\quad + \Sigma_i(\operatorname{curl}(\hat{\varepsilon}_i\varphi^i)\mu_i\psi^i_*)_{\Omega_i} \\ &= -\textstyle\int_{\Sigma_{ij}}(n\wedge\psi^i)\varphi^i_*\hat{\mu}_i\hat{\varepsilon}_i\,d\Sigma_{ij}+\int_{\Sigma_{ij}}(n\wedge\varphi^i)\psi^i_*\hat{\varepsilon}_i\hat{\mu}_i\,d\Sigma_{ij} \\ &\quad -\textstyle\int_\Gamma(n\wedge\psi^1)\varphi^1_*\hat{\varepsilon}_1\hat{\mu}_1\,d\Gamma-(\hat{\mu}\varphi,\operatorname{curl}(\hat{\varepsilon}\varphi_*))+\hat{\varepsilon}\varphi,\operatorname{curl}(\mu\psi_*)).\end{aligned}$$

According to (4.26), the volume integrals in (4.27) are continuous for variable Φ in the topology induced by $\mathscr{H}$ and therefore the same is true for the sum of the surface integrals—thus the integrals:

$$(4.28)\qquad \begin{aligned}&\textstyle\int_{\Sigma_{ij}}((n\wedge\psi^i)\varphi^i_*\hat{\varepsilon}_i\hat{\mu}_i-(n\wedge\psi^j)\varphi^j_*\hat{\varepsilon}_j\hat{\mu}_j)\,\mathrm{d}\Sigma_{ij},\\ &\textstyle\int_{\Sigma_{ij}}((n\wedge\varphi^i)\psi^i_*\hat{\varepsilon}_i\hat{\mu}_i-(n\wedge\varphi^j)\psi^j_*\hat{\varepsilon}_j\hat{\mu}_j)\,\mathrm{d}\Sigma_{ij},\\ &\textstyle\int_\Gamma(n\wedge\psi^1)\varphi^1_*\hat{\varepsilon}_1\hat{\mu}_1\,\mathrm{d}\Gamma\end{aligned}$$

are continuous for the topology induced by $\mathscr{H}$.

Taking account of (4.25), the first integral in (4.28) is equal to

$$-\textstyle\int_{\Sigma_{ij}}\{\hat{\varepsilon}_i n\wedge\varphi^i_*-\hat{\varepsilon}_j n\wedge\varphi^j_*\}\,\mu_j\psi^j\,\mathrm{d}\Sigma_{ij}$$

which is continuous for the topology induced by $\mathscr{H}$ if and only if

$$\hat{\varepsilon}_i n\wedge\varphi^i_*=\hat{\varepsilon}_j n\wedge\varphi^j_*\quad\text{on }\Sigma_{ij}.$$

In the same way, the second integral of (4.28) leads to

$$\hat{\mu}_i n\wedge\psi^i_*=\hat{\mu}_j n\wedge\psi^j_*\quad\text{on }\Sigma_{ij}$$

and the third integral to $n\wedge\varphi^1_*=0$ on Σ.

Therefore, $\Phi_*\in D(\mathscr{A})$ and $\mathscr{A}\Phi_*=-\mathscr{A}\Phi$. Conversely, if $\forall\Phi\in D(\mathscr{A})$

$$(\mathscr{A}\Phi,\Psi)_{\mathscr{H}}=-(\Phi,\mathscr{A}\Psi)_{\mathscr{H}},$$

then $\Psi\in D(\mathscr{A})$, from which (4.24) follows. □

We are now in a position to pose the following problem: we set

$$\mathscr{M}\Phi=\{\sigma\hat{\varepsilon}\varphi,0\}\quad\text{if }\Phi=\{\varphi,\psi\}\tag{4.29}$$

which defines

$$\mathscr{M}\in\mathscr{L}(\mathscr{H};\mathscr{H}).\tag{4.30}$$

We want to find a function $U=\{D,B\}$ such that

$$U\in L^\infty(0,T;\mathscr{H}),\tag{4.31}$$

$$\textstyle\int_0^T[-(U,\partial\Phi/\partial t)_{\mathscr{H}}-(U,\mathscr{A}\Phi)_{\mathscr{H}}+(\mathscr{M}U,\Phi)_{\mathscr{H}}]\,\mathrm{d}t=\int_0^T(G,\Phi)_{\mathscr{H}}\,\mathrm{d}t+(U_0,\Phi(0))_{\mathscr{H}}\tag{4.32}$$

$\forall\Phi$ such that

$$(4.33)\qquad \begin{aligned}&\Phi\in L^2(0,T;D(\mathscr{A})),\\ &\partial\Phi/\partial t\in L^2(0,T;\mathscr{H}),\qquad \Phi(T)=0;\end{aligned}$$

in (4.32), G and U_0 are given by

$$G=\{G_1,G_2\}\in L^2(0,T;\mathscr{H}),\tag{4.34}$$

$$U_0=\{D_0,B_0\}\in\mathscr{H}.\tag{4.35}$$

We now prove

Lemma 4.5. *Problem* (4.31), (4.32) *is a "weak" formulation of the Maxwell problem for "stable" media.*

Proof. Indeed, let us assume that $U=\{D,B\}$ satisfies (3.20) ... (3.24). Then, if we take $\Phi=\{\varphi,\psi\}$ with (4.33) and if we take the scalar products of (3.20) and (3.21) with $\hat{\varepsilon}\varphi$ and $\hat{\mu}\psi$, respectively, we obtain (4.32) after integration by parts. □

Program. We will now (in Sec. 4.3 below) solve problem (4.31), (4.32). Then, in Sections 5 and 6, we will discuss the question of existence of "strong" solutions for the problem. □

4.3. Existence and Uniqueness of the Weak solution

Theorem 4.1. *Problem* (4.31), (4.32) *has a unique solution.*

Existence proof. We observe that the mapping

$$\Phi=\{\varphi,\psi\}\rightarrow\Phi,\ \operatorname{curl}(\hat{\varepsilon}\varphi),\ \operatorname{curl}(\hat{\mu}\psi)$$

permits us to identify $D(\mathscr{A})$ with a closed subspace of $(L^2(\Omega))^{12}$, from which it follows that $D(\mathscr{A})$ is separable. Let then $\Phi_1,\Phi_2,\ldots,\Phi_m,\ldots$ be a "base" for $D(\mathscr{A})$ in the (usual) sense:

$$\begin{aligned}&\forall m,\ \Phi_1,\ldots,\Phi_m \text{ are linearly independent,}\\ &\text{the finite combinations } \textstyle\sum \xi_j\Phi_j,\ \xi_j\in\mathbb{R}, \text{ are dense in } D(\mathscr{A}).\end{aligned} \tag{4.36}$$

We apply the *Galerkin method.* We look for an "approximate" solution U_m of (4.32)

$$U_m(t)\in[\Phi_1,\ldots,\Phi_m]=\text{space spanned by } \Phi_1\ldots\Phi_m;$$

$$(U_m'(t),\Phi_j)_{\mathscr{H}}+(\mathscr{A}U_m(t),\Phi_j)_{\mathscr{H}}+(\mathscr{M}U_m(t),\Phi_j)_{\mathscr{H}}=(G(t),\Phi_j)_{\mathscr{H}},\quad 1\leqslant j\leqslant m, \tag{4.37}$$

$$U_m(0)=U_{0m},\qquad U_{0m}\in[\Phi_1,\ldots,\Phi_m],\qquad U_{0m}\rightarrow U_0\quad\text{in } \mathscr{H}. \tag{4.38}$$

This uniquely defines U_m in $[0,T]$. If

$$U_m(t)=\textstyle\sum_{j=1}^m k_{jm}(t)\Phi_j,$$

we multiply (4.37) by $k_{jm}(t)$ and sum over j; taking account of

$$(\mathscr{A}U_m(t),\ U_m(t))_{\mathscr{H}}=0$$

if follows (according to (4.24)) that

$$(U_m'(t),U_m(t))_{\mathscr{H}}+(\mathscr{M}U_m(t),U_m(t))_{\mathscr{H}}=(G(t),U_m(t))_{\mathscr{H}} \tag{4.39}$$

from which, according to (4.30)

$$(4.40)\qquad \frac{1}{2}\frac{\mathrm{d}}{\mathrm{d}t}\|U_m(t)\|^2_{\mathscr{H}} \leqslant C_j\|U_m(t)\|^2_{\mathscr{H}} + \|G(t)\|_{\mathscr{H}}\,\|U_m(t)\|_{\mathscr{H}}\,.$$

Consequently, since $\|U_{0m}\|_{\mathscr{H}} \leqslant c_2$:

$$\|U_m(t)\|^2_{\mathscr{H}} \leqslant c_3 + c_4\int_0^t \|U_m(\sigma)\|^2_{\mathscr{H}}\,\mathrm{d}\sigma$$

and according to Gronwall's lemma, we have:

$$(4.41)\qquad \|U_m(t)\|_{\mathscr{H}} \leqslant \text{constant, independent of } m.$$

According to (4.41), we can then select a sequence U_μ such that

$$(4.42)\qquad U_\mu \to U \quad \text{weakly star in } L^\infty(0,T;\mathscr{H}).$$

Let then:

$$(4.43)\qquad \begin{aligned} &\xi_j \in C^1([0,T])\,, \qquad \xi_j(T)=0\,,\\ &\textstyle\sum_{j=1}^{m_0} \xi_j \Phi_j = \Psi\,. \end{aligned}$$

We use (4.37) for $m=\mu$; we multiply by ξ_j, $j \leqslant m_0 \leqslant \mu$; summing over j and integrating by parts, we have, taking account of (4.24):

$$(4.44)\qquad \begin{aligned} &\int_0^T [-(U_\mu, \partial\Psi/\partial t)_{\mathscr{H}} - (U_\mu, \mathscr{A}^*\Psi)_{\mathscr{H}} + (\mathscr{M}U_\mu, \Psi)_{\mathscr{H}}]\,\mathrm{d}t\\ &\qquad\qquad = \int_0^T (G,\Psi)_{\mathscr{H}}\,\mathrm{d}t + (U_{0\mu}, \Psi(0))_{\mathscr{H}}\,. \end{aligned}$$

Due to (4.42) and (4.38), we can pass to the limit in (4.44). Thus, we obtain the existence of a U which satisfies (4.31) and (4.32) for *every function* $\Phi=\Psi$ *of the form* (4.43).

But, due to (4.36), if Φ is given with (4.33), we can find a sequence Ψ_k of functions of the form (4.43) such that $\Psi_k \to \Phi$ in $L^2(0,T;D(\mathscr{A}))$,

$$\partial\Psi_k/\partial t \to \partial\Phi/\partial t \quad \text{in} \quad L^2(0,T;\mathscr{H})\,,$$

therefore (4.32) holds $\forall \Phi$ with (4.33). □

Uniqueness Proof. Let us assume that (4.32) holds with $G=0, U_0=0$. Let $\tilde{U}$ = extension of U by 0 for $t<0$; then we take in (4.32)

$$\begin{aligned} &\Phi = \xi\Psi\,,\\ &\xi = \text{restriction to } [0,T] \text{ of } \Xi \in \mathscr{D}(]-\infty, T[)\,. \end{aligned}$$

It follows, in the sense of $\mathscr{D}'(]-\infty,T[)$:

$$\mathrm{d}(\tilde{U},\Psi)_{\mathscr{H}}/\mathrm{d}t - (\tilde{U}, \mathscr{A}\Psi)_{\mathscr{H}} + (\mathscr{M}\tilde{U},\Psi)_{\mathscr{H}} = 0\,.$$

If $\tilde{U}$ denotes the extension by 0 for $t>T$, we obtain on $\mathscr{D}'(\mathbb{R}_t)$ this time:

$$\text{(4.45)} \qquad \mathrm{d}(\tilde{U},\Psi)_{\mathscr{H}}/\mathrm{d}t-(\tilde{U},\mathscr{A}\Psi)_{\mathscr{H}}+(\mathscr{M}\tilde{U},\Psi)_{\mathscr{H}}=c\delta(t-T)\,.$$

If $\rho\in\mathscr{D}(R_t)$, with support in $[0,\varepsilon]$, it follows from (4.45) *by convolution in t* by ρ:

$$\text{(4.46)} \qquad \left(\frac{\mathrm{d}}{\mathrm{d}t}(\tilde{U}*\rho),\Psi\right)_{\mathscr{H}}-(\tilde{U}*\rho,\mathscr{A}\Psi)_{\mathscr{H}}+(\mathscr{M}(\tilde{U}*\rho),\Psi)=c\rho(t-T)$$

from which, since $\rho(t-T)=0$ if $t\leqslant T$:

$$\text{(4.47)} \qquad \left(\frac{\mathrm{d}}{\mathrm{d}t}(\tilde{U}*\rho(t)),\Psi\right)_{\mathscr{H}}-(\tilde{U}*\rho(t),\mathscr{A}\Psi)_{\mathscr{H}}+(\mathscr{M}(\tilde{U}*\rho(t)),\Psi)_{\mathscr{H}}=0\,, \qquad t\leqslant T.$$

But from (4.47) it follows that the form

$$\Psi\to(\tilde{U}*\rho(t),\mathscr{A}\Psi)_{\mathscr{H}}$$

is continuous on $D(\mathscr{A})$ for the topology induced by $\mathscr{H}$. Therefore, $\tilde{U}*\rho(t)\in D(\mathscr{A})$, and we can take $\Psi=\tilde{U}*\rho(t)$ in (4.47); making use of (4.24), it follows that (setting $\tilde{U}*\rho(t)=w(t)$):

$$\frac{1}{2}\frac{\mathrm{d}}{\mathrm{d}t}\|w(t)\|^2_{\mathscr{H}}+(\mathscr{M}w(t),w(t))_{\mathscr{H}}=0\,, \qquad t\leqslant T,$$

$w=0$ if $t\leqslant 0$, from which $w=0$ follows. Therefore

$$\tilde{U}*\rho=0 \quad \text{in} \quad t<T, \qquad \forall\rho$$

therefore $U=0$. □

4.4. Continuous Dependence of the Solution on the Dielectric Constants and on the Magnetic Permeabilities

Let $\{\varepsilon^j,\mu^j\}$ be a sequence of functions such that

$$\text{(4.48)} \qquad \begin{aligned} &\hat{\varepsilon}^j=1/\varepsilon^j, \quad \hat{\mu}^j=1/\mu^j\in \text{bounded set of } L^\infty(\Omega),\\ &\hat{\varepsilon}^j\geqslant c_1>0, \quad \mu^j\geqslant c_2>0 \text{ almost everywhere in } \Omega,\\ &\varepsilon^j,\mu^j \text{ are piecewise constant, } \varepsilon^j=\text{constant in a neighborhood of } \Gamma; \end{aligned}$$

$$\text{(4.49)} \qquad \hat{\varepsilon}^j\to\hat{\varepsilon},\ \hat{\mu}^j\to\hat{\mu} \quad \text{almost everywhere in } \Omega,$$

and let σ^j be such that

$$\text{(4.50)} \qquad \sigma^j\in \text{bounded set of } L^\infty(\Omega). \quad \sigma^j\to\sigma \text{ almost everywhere,}$$

where ε,μ,σ are given as in the preceding sections.

Let $\mathscr{A}^j$ and $\mathscr{M}^j$ be the analogous operators to $\mathscr{A}$ and $\mathscr{M}$ corresponding to $\varepsilon^j, \mu^j, \sigma^j$.

Let U^j be the solution corresponding to the problem which is analogous to (4.31), (4.32). We have:

Theorem 4.2. *Under assumptions* (4.48), (4.49), (4.50), *we have*:

$$U^j \to U \quad \textit{weakly star in} \quad L^\infty(0,T;\mathscr{H}). \tag{4.51}$$

Proof. Let U_m^j be the approximate solution of the problem (analogous to (4.37))

$$\begin{aligned} &\left(\frac{\mathrm{d}}{\mathrm{d}t} U_m^j(t), \Phi_k\right)_{\mathscr{H}} + (\mathscr{A}^j U_m^j(t), \Phi_k)_{\mathscr{H}} + (\mathscr{M}^j U_m^j(t), \Phi_k)_{\mathscr{H}} = (G(t), \Phi_k)_{\mathscr{H}}, \\ &\qquad\qquad\qquad\qquad\qquad\qquad\qquad\qquad\qquad\qquad 1 \leqslant k \leqslant m, \\ &U_m^j(0) = U_{0m}. \end{aligned} \tag{4.52}$$

Then we obtain (here we only use (4.50), because the term in $\mathscr{A}^j$ disappears)

$$\|U_m^j(t)\|_{\mathscr{H}} \leqslant \text{constant, independent of } m \text{ and } j.$$

Thus we obtain

$$U^j \text{ remains in a bounded set of } L^\infty(0,T;\mathscr{H}). \tag{4.53}$$

Therefore, we can select a sequence, again denoted by U^j such that

$$U^j \to U_* \quad \text{weakly star in} \quad L^\infty(0,T;\mathscr{H}). \tag{4.54}$$

Thus, we will have (4.51), if we can prove that U_* is a solution of problem (4.31), (4.32), therefore that $U_* = U$. We now introduce Φ^j with

$$\Phi^j \in L^2(0,T;D(\mathscr{A}^j)), \quad (\Phi^j)' \in L^2(0,T;\mathscr{H}), \quad \Phi^j(T) = 0. \tag{4.55}$$

We have

$$\begin{aligned} &\int_0^T \{U^j, (\Phi^{j\prime}))_{\mathscr{H}} - (U^j, \mathscr{A}^j \Phi^j)_{\mathscr{H}} + (\mathscr{M}^j U^j, \Phi^j)_{\mathscr{H}}\} \,\mathrm{d}t \\ &\qquad = \int_0^T (G, \Phi^j)_{\mathscr{H}} \,\mathrm{d}t + (U_0, \Phi^j(0))_{\mathscr{H}}. \end{aligned} \tag{4.56}$$

Let us accept for the moment

Lemma 4.6. *One can find, for given* Φ *with* (4.33) *and such that*

$$\Phi' \in L^2(0,T;D(\mathscr{A})),$$

a sequence Φ^j *of functions satisfying* (4.55) *and such that*

$$\Phi^j \to \Phi \text{ in } L^2(0,T;\mathscr{H}), \ (\Phi^j)' \to \Phi' \text{ in } L^2(0,T;\mathscr{H}), \tag{4.57}$$

$$\mathscr{A}^j \Phi^j \to \mathscr{A}\Phi \text{ in } L^2(0,T;\mathscr{H}). \tag{4.58}$$

Due to (4.54), (4.57), (4.58), we can pass to the limit in (4.56); thus we find that U_* satisfies (4.32), because the space of the Φ which satisfy the conditions of Lemma 4.6 is dense in the space of the Φ which satisfy (4.33), from which the theorem follows, provided we prove Lemma 4.6. □

We begin by proving another lemma, interesting for its own sake:

Lemma 4.7. *Let f be given in $\mathscr{H}$. For all $\lambda>0$, there exists a unique $U\in D(\mathscr{A})$ such that*

$$(4.59)\qquad (\mathscr{A}+\lambda)U=f,$$

and

$$(4.60)\qquad \|U\|_{\mathscr{H}}\leqslant\lambda^{-1}\|f\|_{\mathscr{H}}.$$

Remark 4.5. Lemma 4.7 expresses that $-\mathscr{A}$ is an *infinitesimal generator* of a semigroup of contractions in $\mathscr{H}$[8] (cf. also Sec. 10). □

Proof of Lemma 4.7 We apply the Galerkin method; let $\Phi_1,\ldots,\Phi_m,\ldots$ be a base for $D(\mathscr{A})$, as in the proof of Theorem 4.1, and let $U_m\in[\Phi_1,\ldots,\Phi_m]$ satisfying

$$(4.61)\qquad ((\mathscr{A}+\lambda)U_m,\Phi_j)_{\mathscr{H}}=(f,\Phi_j)_{\mathscr{H}},\qquad 1\leqslant j\leqslant m;$$

the system of Equations (4.61) (in finite dimension) has a unique solution; if $U_m=\xi_j\Phi_j$, then, multiplying (4.61) by ξ_j and summing over j, it follows, since $(\mathscr{A}U_m,U_m)_{\mathscr{H}}=0$:

$$\lambda\|U_m\|^2_{\mathscr{H}}=(f,U_m)_{\mathscr{H}}$$

from which $\|U_m\|_{\mathscr{H}}\leqslant\lambda^{-1}\|f\|_{\mathscr{H}}$. Therefore, we can select a sequence U_μ such that

$$U_\mu\to U_*\ \text{weakly in}\ \mathscr{H}.$$

Taking (4.61) for $m=\mu$ and with fixed $j\leqslant\mu$, which we write

$$\lambda(U_\mu,\Phi_j)_{\mathscr{H}}-(U_\mu,\mathscr{A}\Phi_j)_{\mathscr{H}}=(f,\Phi_j)_{\mathscr{H}},$$

we obtain in the limit

$$\lambda(U_*,\Phi_j)_{\mathscr{H}}-(U_*,\mathscr{A}\Phi_j)_{\mathscr{H}}=(f,\Phi_j)_{\mathscr{H}}$$

and since this holds $\forall_j$, we have

$$(4.62)\qquad \lambda(U_*,\Phi)_{\mathscr{H}}-(U_*,\mathscr{A}\Phi)_{\mathscr{H}}=(f,\Phi)_{\mathscr{H}},\qquad \forall\Phi\in D(\mathscr{A}).$$

---

[8] Here *the choice of the norm in $\mathscr{H}$* is very important.

From this it follows that the form $\Phi \to (U_*, \mathscr{A}\Phi)_{\mathscr{H}}$ is continuous on $D(\mathscr{A})$ for the topology induced by $\mathscr{H}$, therefore that $U_* \in D(\mathscr{A})$ and that (4.62) can be written

$$((\mathscr{A}+\lambda)U_*, \Phi)_{\mathscr{H}} = (f, \Phi)_{\mathscr{H}}, \qquad \forall \Phi \in D(\mathscr{A});$$

therefore, U_* is a solution of (4.59) and we can take $U = U_*$; since $(\mathscr{A}U, U)_{\mathscr{H}} = 0$, it follows from (4.59) that

$$\lambda \|U\|_{\mathscr{H}}^2 = (f, U)_{\mathscr{H}}$$

from which uniqueness and (4.60) follow. □

Proof of Lemma 4.6 For a given $\Phi = \Phi(t)$, we define $\Phi^j(t)$ (almost everywhere in t) as *the* solution in $D(\mathscr{A}^j)$ of

$$(4.63) \qquad (\mathscr{A}^j + \lambda)\Phi^j(t) = (\mathscr{A}+\lambda)\Phi(t), \qquad (\lambda > 0 \text{ fixed}).$$

According to Lemma 4.7, $\Phi^j(t)$ exists and is unique in $D(\mathscr{A}^j)$. We prove without difficulty that (since $\Phi \in L^2(0, Z; D(\mathscr{A}))$)

$$(4.64) \qquad (\mathscr{A}^j + \lambda)(\Phi^j(t))' = (\mathscr{A}+\lambda)\Phi'(t).$$

Everything then reduces to proving that

$$(4.65) \qquad \Phi^j \to \Phi \text{ weakly in } L^2(0, T; \mathscr{H}).$$

Indeed, if we assume for a moment that (4.65) is true, then (4.63) leads to

$$\int_0^T ((\mathscr{A}^j+\lambda)\Phi^j, \Phi^j)_{\mathscr{H}}\, dt = \lambda \int_0^T \|\Phi^j\|_{\mathscr{H}}^2\, dt = \int_0^T ((\mathscr{A}+\lambda)\Phi, \Phi^j)_{\mathscr{H}}\, dt$$
$$\to \int_0^T ((\mathscr{A}+\lambda)\Phi, \Phi)_{\mathscr{H}}\, dt = \lambda \int_0^T \|\Phi\|_{\mathscr{H}}^2\, dt$$

therefore

$$\int_0^T \|\Phi^j\|_{\mathscr{H}}^2\, dt \to \int_0^T \|\Phi\|_{\mathscr{H}}^2\, dt$$

which, together with (4.65), proves *strong* convergence in $L^2(0, T; \mathscr{H})$. But then $\mathscr{A}^j\Phi^j = (\mathscr{A}+\lambda)\Phi - \lambda\Phi^j \to \mathscr{A}\Phi$ strongly in $L^2(0, T; \mathscr{H})$ and, in the same way, (4.64) will show that $(\Phi^j) \to \Phi'$ strongly in $L^2(0, T; \mathscr{H})$ as well, moreover, that $\mathscr{A}^j(\Phi^j) \to \mathscr{A}\Phi'$ in $L^2(0, T; \mathscr{H})$.

It remains then to prove (4.65). We conclude from (4.63) that

$$\|\Phi^j(t)\|_{\mathscr{H}} \leqslant \lambda^{-1} \|(\mathscr{A}+\lambda)\Phi(t)\|_{\mathscr{H}}$$

therefore that

$$(4.66) \qquad \Phi^j \text{ remains in a bounded set of } L^2(0, T; \mathscr{H}).$$

Then we can select a sequence—again denoted by Φ^j—such that

$$(4.67) \qquad \Phi^j \to \Phi_* \text{ weakly in } L^2(0, T; \mathscr{H}).$$

It remains to prove that $\Phi_* = \Phi$. Now, if $\Theta = \{\theta, \chi\} \in (\mathscr{D}(Q))^6$ where $Q = \Omega \times]0, T[$, $\Phi^j = \{\varphi^j, \psi^j\}$ and, if $(\ , \ )$ denotes the scalar product between $(\mathscr{D}'(Q))^6$ and $(\mathscr{D}(Q))^6$ and between $(\mathscr{D}'(Q))^3$ and $(\mathscr{D}(Q))^3$, we have:

$$(4.68) \qquad \begin{aligned} (\mathscr{A}^j \Phi^j, \Theta) &= -(\operatorname{curl} \hat{\mu}^j \psi^j, \theta) + (\operatorname{curl} \hat{\varepsilon}^j \varphi^j, \chi) \\ &= -(\hat{\mu}^j \psi^j, \operatorname{curl} \theta) + (\hat{\varepsilon}^j \varphi^j, \operatorname{curl} \chi); \end{aligned}$$

but according to (4.68), (4.69) and the Lebesgue theorem, we have:

$$\begin{aligned} &\hat{\mu}^j \operatorname{curl} \theta \to \hat{\mu} \operatorname{curl} \theta \ \text{ in } \ (L^2(Q))^3 \\ &\hat{\varepsilon}^j \operatorname{curl} \chi \to \hat{\varepsilon} \operatorname{curl} \chi \ \text{ in } \ (L^2(Q))^3, \end{aligned}$$

and (4.68) with (4.67) then gives, if $\Phi_* = \{\varphi_*, \psi_*\}$:

$$(\mathscr{A}^j \Phi^j, \Theta) \to -(\psi_*, \hat{\mu} \operatorname{curl} \theta) + (\varphi_*, \hat{\varepsilon} \operatorname{curl} \chi) = (\mathscr{A} \Phi_*, \Theta)$$

and therefore

$$(4.69) \qquad \mathscr{A}^j \Phi^j \to \mathscr{A} \Phi_* \ \text{ in } \ (\mathscr{D}'(Q))^6.$$

Moreover, according to (4.63) and (4.67):

$$\mathscr{A}^j \Phi^j \to (\mathscr{A} + \lambda) \Phi - \lambda \Phi_* \quad \text{weakly in } \ L^2(0, T; \mathscr{H})$$

from which, comparing with (4.69):

$$(4.70) \qquad \mathscr{A}^j \Phi^j \to \mathscr{A} \Phi_* \quad \text{weakly in } \ L^2(0, T; \mathscr{H})$$

and

$$(4.71) \qquad (\mathscr{A} + \lambda) \Phi_* = (\mathscr{A} + \lambda) \Phi.$$

Thus, we would have $\Phi_* = \Phi$, if we prove that $\Phi_* \in L^2(0, T; D(\mathscr{A}))$. Now, according to (4.67), (4.69) and Lemma 4.2, we have:

$$n \wedge \Phi^j \to n \wedge \Phi_* \ \text{ weakly in } \ L^2(0, T; H^{-1/2}(\Gamma))^3);$$

since $n \wedge \Phi^j = 0$, it follows that $n \wedge \Phi_* = 0$, from which the desired result follows. ▯

Application 4.1. We assumed that ε, μ are *piecewise constant* in the various media Ω_i. As a matter of fact, this constitutes an *idealization*, the ε, μ in fact being *continuous in* $\bar{\Omega}$, constant in the "interior" of the Ω_i, and passing "quickly" from one value to another in the neighborhood of the interfaces. Theorem 4.2 shows that the "ideal" problem is an *approximation* to the "real" problem. ▯

Application 4.2. Let

$$\Omega = \bigcup_{i=1}^{q} \Omega_i \bigcup_{i,j} \Sigma_{ij}.$$

If *one* of the domains Ω_{i_0} has a volume tending to 0 (*for fixed constants* $\varepsilon_{i_0}, \mu_{i_0}$), then the corresponding solution converges, always according to Theorem 4.2, toward the solution of the problem for

$$\Omega = \bigcup_{i \neq i_0} \Omega_i \bigcup_{i \neq i_0} \Sigma_{ij}.$$

5. Stable Media. Existence of "Strong" Solutions

5.1. Strong Solution in $\mathbf{D}(\mathscr{A})$

We now prove

Theorem 5.1. *We assume that* G *and* U_0 *are given with*

$$G,\ \partial G/\partial t \in L^2(0,T;\mathscr{H}), \tag{5.1}$$

$$U_0 \in D(\mathscr{A}), \quad (U_0 = \{\mathbf{D}_0, \mathbf{B}_0\}). \tag{5.2}$$

Then there exists one and only one function $U = \{D, B\}$, *which is a solution of*

$$U \in L^\infty(0,T;D(\mathscr{A})), \tag{5.3}$$

$$\partial U/\partial t \in L^\infty(0,T;\mathscr{H}), \tag{5.4}$$

$$\partial \mathbf{D}/\partial t - \operatorname{curl}(\hat{\mu}\mathbf{B}) + \sigma\hat{\varepsilon}\mathbf{D} = \mathbf{G}_1, \tag{5.5}$$

$$\partial \mathbf{B}/\partial t + \operatorname{curl}(\hat{\varepsilon}D) = \mathbf{G}_2, \tag{5.6}$$

$$n \wedge \mathbf{D} = 0 \quad \text{on} \quad \Gamma \times]0,T[, \tag{5.7}$$

$$\mathbf{D}(0) = \mathbf{D}_0, \quad \mathbf{B}(0) = \mathbf{B}_0. \tag{5.8}$$

Proof. 1) We again use the Galerkin method applied in the proof of Theorem 4.1; since $U_0 \in D(\mathscr{A})$, we can take the "base" $\Phi_1, \ldots, \Phi_m, \ldots$ in such a way that

$$U_0 \in [\Phi_1]. \tag{5.9}$$

We take (4.37), (4.38) (with $U_{0m} = U_0$, permissible due to (5.9)). We can differentiate (4.37) with respect to t which gives:

$$(U''_m(t), \Phi_j)_{\mathscr{H}} + (\mathscr{A}\, U'_m(t), \Phi_j)_{\mathscr{H}} + (\mathscr{M}\, U'_m(t), \Phi_j)_{\mathscr{H}} = (G'(t), \Phi_j)_{\mathscr{H}}. \tag{5.10}$$

It follows from (4.37) that

$$(U'_m(0), \Phi_j)_{\mathscr{H}} = (G(0) - \mathscr{A} U_0 - \mathscr{M} U_0, \Phi_j)$$

from which

$$\|U'_m(0)\|_{\mathscr{H}} \leqslant \|G(0) - \mathscr{A} U_0 - \mathscr{M} U_0\|_{\mathscr{H}}. \tag{5.11}$$

Multiplying (5.10) by k'_{jm} and summing over j, it follows that

$$\frac{1}{2}\frac{\mathrm{d}}{\mathrm{d}t}\|U'_m(t)\|^2_{\mathscr{H}} + (\mathscr{M} U'_m(t), U'_m(t))_{\mathscr{H}} = (G'(t), U'_m(t))_{\mathscr{H}},$$

which, together with (5.11), gives

$$\|U'_m(t)\|_{\mathscr{H}} \leqslant \text{constant, independent of } m. \tag{5.12}$$

From this it follows that *the* solution U of the weak problem (given by Theorem 4.1) satisfies (5.4) and that $U(0) = U_0$ from which (5.8) follows.

2) Then we can integrate by parts with respect to t in (4.32), whence

$$\int_0^T [(\partial U/\partial t, \Phi)_{\mathscr{H}} - (U, \mathscr{A}\Phi)_{\mathscr{H}} + (\mathscr{M} U, \Phi)_{\mathscr{H}}]\,\mathrm{d}t = \int_0^T (G, \Phi)_{\mathscr{H}}\,\mathrm{d}t. \tag{5.13}$$

Taking in (5.13)

$$\Phi \in (\mathscr{D}(\Omega \times]0, T[))^6$$

we conclude (5.5) and (5.6). Consequently,

$$\operatorname{curl}(\hat{\mu}B) = \partial D/\partial t + \sigma\hat{\varepsilon}D - G_1 \in L^\infty(0, T; L^2(\Omega)),$$
$$\operatorname{curl}(\hat{\varepsilon}D) = -\partial B/\partial t + G_2 \in L^\infty(0, T; L^2(\Omega)),$$

and consequently, in order to obtain (5.3), it only remains to prove (5.7).

For that purpose, we consider $\Phi = \{\varphi, \psi\}, \Phi$ satisfying (4.33) and having values in $(C^1_K(\bar{\Omega}))^6$ (cf. Lemma 4.1) with, moreover, $n \wedge \varphi = 0$ on $\Gamma \times]0, T[$. Taking the scalar product of (5.5) (resp. (5.6)) with $\hat{\varepsilon}\varphi$ (resp. $\hat{\mu}\psi$), we have

$$(\partial U/\partial t, \Phi)_{\mathscr{H}} + \int_\Gamma \hat{\varepsilon}\hat{\mu}(n \wedge D)\psi\,\mathrm{d}\Gamma - (U, \mathscr{A}\Phi)_{\mathscr{H}} + (\mathscr{M} U, \Phi)_{\mathscr{H}} = (G, \Phi)_{\mathscr{H}}$$

from which, by integration with respect to t and by comparison with (5.13):

$$\int_{\Gamma \times]0,T[} \hat{\varepsilon}\hat{\mu}(n \wedge D)\psi\,\mathrm{d}\Gamma\,\mathrm{d}t = 0$$

from which (5.7) follows. □

5.2. Solution of the Physical Problem

In physical examples, the function G_2 has some properties in addition to (5.1). In order to cleanly separate the occurence of the assumptions, we give two statements where we derive supplementary properties of **B** from properties of the data; first we have

Theorem 5.2. *Under the assumptions of Theorem 5.1 and the additional ones:*

(5.14) $\operatorname{div} G_2 = 0 \quad in \quad \Omega \times]0, T[\,,$

(5.15) $\operatorname{div} B_0 = 0\,,$

we have

(5.16) $\operatorname{div} B = 0 \quad in \quad \Omega \times]0, T[\,.$

Proof. Applying the operator div, in the sense of distributions, to both sides of (5.6), we deduce that (because of div curl $\varphi = 0$ and (5.14)):

(5.17) $\partial(\operatorname{div} B)/\partial t = 0$

from which (5.16) follows, due to (5.15). ▯

Now we prove that if $nG_2 = 0$ and $n.B_0 = 0$ on $\Gamma \times]0, T[$ and Γ, then $n.B = 0$ on $\Gamma \times]0, T[$. For this purpose, however, we need some additional results from Functional Analysis which we will state. ▯

We introduce the space (compare with (4.1))

(5.18) $H(\operatorname{div}; \Omega) = \{v \mid v \in (L^2(\Omega))^3, \quad \operatorname{div} v \in L^2(\Omega)$

which is a Hilbert space for the norm

$$(\|v\|^2_{(L^2(\Omega))^3} + \|\operatorname{Div} v\|^2_{L^2(\Omega)})^{1/2}\,.$$

As in Lemma 4.1 (cf. Remark 4.1), we prove:

Lemma 5.1. *The space $(C^1_K(\bar{\Omega}))^3$ is dense in $H(\operatorname{div}; \Omega)$.*

We prove

Lemma 5.2. *The mapping*

(5.19) $v \to n.v|_\Gamma = n.v$

of $(C^1_K(\bar{\Omega}))^3 \to C^1(\Gamma)$ can be extended by continuity to a linear mapping, again denoted $v \to n.v$, which is continuous from $H(\operatorname{div}; \Omega) \to H^{-1/2}(\Gamma)$.

Proof. The principle is the same as in the proof of Lemma 4.2. For $\varphi \in H^{1/2}(\Gamma)$, we define $\Phi \in H^1(\Omega)$ with

(5.20) $\Phi|_\Gamma = \varphi\,,$

and

(5.21) the mapping $\varphi \to \Phi$ is linear continuous from $H^{1/2}(\Gamma) \to H^1(\Omega)$.

For a given u in $H(\operatorname{div}; \Omega)$, let us put

(5.22) $$\pi(\varphi) = (\operatorname{div} u, \Phi) - (u, \operatorname{div} \Phi)$$

where

$$(f, g) = \int_\Omega f(x) g(x) \, dx;$$

the notation (5.22) is legimate, because the right hand side of (5.22) does not depend on the choice of Φ, provided that (5.20) is satisfied.

The mapping $\varphi \to \pi(\varphi)$ is linear continuous from $H^{1/2}(\Gamma) \to \mathbb{R}$, therefore

(5.23)
$$\pi(\varphi) = (\tau_u, \varphi), \quad \tau_u \in H^{-1/2}(\Gamma),$$
$(\ ,\ )$ denoting the scalar product between $H^{-1/2}(\Gamma)$ and $H^{1/2}(\Gamma)$,
the mapping $u \to \tau_u$ being linear continuous from $H(\operatorname{div}; \Omega) \to H^{-1/2}(\Gamma)$.

But if $u \in (C_K^1(\bar{\Omega}))^3$, then

$$\pi(\varphi) = \int_\Gamma (n \cdot u) \varphi \, d\Gamma$$

from which the lemma follows. □

We observe that, if $G_2 \in L^2(0, T; (L^2(\Omega))^3)$ with (5.14), then

(5.24) $$G_2 \in L^2(0, T; H(\operatorname{div}; \Omega))$$

and therefore, applying Lemma 5.2,

(5.25) $$n \cdot G_2 \in L^2(0, T; H^{-1/2}(\Gamma)).$$

Now we can state

Theorem 5.3. *We make the same assumptions as for Theorem 5.2 and moreover assume that*

(5.26) $$n G_2 = 0 \quad on \quad \Gamma \times]0, T[$$

and

(5.27) $$n B_0 = 0 \quad on\ \Gamma.$$

Then

(5.28) $$n \cdot B = 0 \quad on \quad \Gamma \times]0, T[.$$

Proof. From (5.6), it follows that

$$B(t)+\int_0^t \operatorname{curl}(\hat{\varepsilon}D)(\sigma)\,d\sigma = B_0+\int_0^t G_2(\sigma)\,d\sigma\,. \tag{5.29}$$

Since $\operatorname{curl}(\hat{\varepsilon}D)\in(L^2(\Omega))^3$, and $\operatorname{div}(\operatorname{curl}(\hat{\varepsilon}D))=0$, we have:

$$\operatorname{curl}(\hat{\varepsilon}D)\in L^2(0,T;H(\operatorname{div};\Omega))$$

and consequently, we can apply the operator $v\to n.v$ to both sides of Equation (5.29); using (5.26) and (5.27) we deduce that

$$n.B(t)+\int_0^t n.\operatorname{curl}(\hat{\varepsilon}D)(\sigma)\,d\sigma=0\,.$$

Thus, we will have proved (5.28), if we verify that

$$n.\operatorname{curl}(\hat{\varepsilon}D)(\sigma)=0 \quad (\text{a.e. in } \sigma)\,. \tag{5.30}$$

Now, since $\hat{\varepsilon}$ is constant on Γ:

$$n.\operatorname{curl}(\hat{\varepsilon}D)=\hat{\varepsilon}n.\operatorname{curl} D$$

and generally

(5.31) $n.\operatorname{curl}\psi=$ operator of *tangential* differentiation *on* Γ on the vector $n\wedge\psi$.

Now, $n\wedge D=0$, therefore we have (5.30). ☐

6. Stable Media. Strong Solutions in Sobolev Spaces

6.1. Imbedding Theorem

Theorem 6.1. *Let $\mathcal{O}$ be an open bounded set with regular boundary $\partial\mathcal{O}$* [9]. *Let X be the space defined by*

$$X=\{v \mid v\in(L^2(\mathcal{O}))^3,\ \operatorname{curl} v\in(L^2(\mathcal{O})),\ \operatorname{div} v\in L^2(\mathcal{O}),\ n.v=0 \text{ on } \partial\mathcal{O}\} \tag{6.1}$$

which is a Hilbert space for the norm

$$\|v\|_X=(\|v\|^2_{(L^2(\mathcal{O}))^3}+\|\operatorname{curl} v\|^2_{(L^2(\mathcal{O}))^3}+\|\operatorname{div} v\|^2_{L^2(\mathcal{O})})^{1/2}\,; \tag{6.2}$$

[9] The result is valid whether $\mathcal{O}$ is bounded or not.

then

$$X=(H^1(\mathcal{O}))^3, \tag{6.3}$$

(algebraic and topological identity).

Proof. At the end of this section, we will prove

Lemma 6.1. *The space of vectors* $\varphi\in(C^1(\mathcal{O}))^3$ *such that* $n.\varphi=0$ *on* $\partial\mathcal{O}$ *is dense in* X.

We start from the identity, for $\varphi\in(C^1(\bar{\mathcal{O}}))^3$

$$\int_{\mathcal{O}}\varphi_{i,j}\varphi_{i,j}\,dx=\int_{\mathcal{O}}(|\operatorname{div}\varphi|^2+|\operatorname{curl}\varphi|^2)\,dx+\int_{\mathcal{O}}(n_i\varphi_i\varphi_{j,j}-n_j\varphi_i\varphi_{j,i})\,dS, \tag{6.4}$$

dS = surface element of $\partial\mathcal{O}$.

But if $n.\varphi=0$, the surface integral in (6.4) reduces to

$$-\int_{\partial\mathcal{O}}n_j\varphi_i\varphi_{j,i}\,dS=-\int_{\partial\mathcal{O}}\varphi(n_j\varphi_j)_{,i}\,dS+\int_{\partial\mathcal{O}}\varphi_i\varphi_j n_{j,i}\,dS \tag{6.5}$$

where we continued the function $x\to n(x)$ into a neighborhood of $\partial\mathcal{O}$ as a C^1 function, so that $n_{j,i}$ has a meaning. But, since $n\varphi=0$, the operator $\varphi_i\partial/\partial x_i$ is a tangential differential operator on $\partial\mathcal{O}$ and therefore $\varphi_i(n\varphi)_{,i}=0$ on $\partial\mathcal{O}$.

Thus (6.5) proves that

$$-\int_{\partial\mathcal{O}}n_j\varphi_i\varphi_{j,i}\,dS=\int_{\partial\mathcal{O}}\varphi_i\varphi_j n_{j,i}\,dS \tag{6.6}$$

from which, after substitution in (6.4), it follows that

$$\int_{\mathcal{O}}\varphi_{i,j}\varphi_{i,j}\,dx=\int_{\mathcal{O}}(|\operatorname{div}\varphi|^2+|\operatorname{curl}\varphi|^2)\,dx+\int_{\partial\mathcal{O}}\varphi_i\varphi_j n_{j,i}\,dS. \tag{6.7}$$

But

$$\left|\int_{\partial\mathcal{O}}\varphi_i\varphi_j n_{j,i}\,dS\right|\leqslant C_1\int_{\mathcal{O}}\varphi_i\varphi_i\,dS. \tag{6.8}$$

Moreover, $\forall\varepsilon>0$, there exists a $c(\varepsilon)$ such that if $\psi\in C^1(\bar{\mathcal{O}})$:

$$\int_{\partial\mathcal{O}}\psi^2\,dS\leqslant\varepsilon\int_{\mathcal{O}}\psi_{,i}\psi_{,i}\,dx+c(\varepsilon)\int_{\mathcal{O}}\psi^2\,dx; \tag{6.9}$$

using (6.9) in (6.8), we have, choosing ε so that $c_1\varepsilon=\frac{1}{2}$:

$$\left|\int_{\partial\mathcal{O}}\varphi_j\varphi_j n_{j,i}\,dS\right|\leqslant\tfrac{1}{2}\int_{\mathcal{O}}\varphi_{i,j}\varphi_{i,j}\,dx+c_2\int_{\mathcal{O}}|\varphi|^2\,dx \tag{6.10}$$

which, together with (6.7), gives

$$\int_{\mathcal{O}}\varphi_{i,j}\varphi_{i,j}\,dx\leqslant 2\int_{\mathcal{O}}[|\operatorname{div}\varphi|^2+|\operatorname{curl}\varphi|^2+c_2|\varphi|^2]\,dx \tag{6.11}$$

from which the theorem follows, due to Lemma 6.1. □

Theorem 6.2. *With the conditions of Sections 4 and 5, let Y be the space defined by*

$$Y=\{v \mid v\in(L^2(\Omega))^3,\ \operatorname{div} v=0,\ \operatorname{curl}(\hat{\mu}v)\in(L^2(\Omega))^3,\ n.v=0 \text{ on } \Gamma\}, \tag{6.12}$$

a Hilbert space for the norm

$$\|v\|_Y=(\|v\|^2_{(L^2(\Omega))^3}+\|\operatorname{curl}(\hat{\mu}v)\|^2_{(L^2(\Omega))^3})^{1/2}. \tag{6.13}$$

Then, $\forall v\in Y$, we have:

$$v_i\in(H^1(\Omega_i))^3 \tag{6.14}$$

where v_i is the restriction of v to Ω_i, $\mu=1/\hat{\mu}=$constant in Ω_i [10] *and*

$$\|v_i\|_{(H^1(\Omega_i))^3}\leqslant c\,\|v\|. \tag{6.15}$$

Proof. 1) Let Ω_1 be the open set where $\mu=$value of μ on Γ and θ is a scalar function of $C^1_K(\bar{\Omega}_1)$, $\theta=1$ in the neighborhood of Γ and $\theta=0$ in the neighborhood of $\partial\Omega_1-\Gamma$. For $v\in Y$, the vector θv satisfies:

$$\begin{aligned}&\theta v\in(L^2(\Omega_1))^3\\ &\operatorname{div}(\theta v)\in L^2(\Omega_1),\quad \operatorname{curl}(\theta v)\in(L^2(\Omega_1))^3\\ &n(\theta v)=0 \quad \text{on } \partial\Omega_1\end{aligned} \tag{6.16}$$

and therefore, according to Theorem 6.1, we have:

$$\theta v\in(H^1(\Omega_1))^3. \tag{6.17}$$

2) Setting $\hat{\mu}v=u$, we thus have:

$$\begin{aligned}&u\in(L^2(\Omega))^3,\quad \operatorname{div}(\mu u)=0,\quad \operatorname{curl} u\in(L^2(\Omega))^3\\ &nu=0 \quad \text{on } \Gamma\end{aligned} \tag{6.18}$$

$$u\in(H^1(\mathcal{O}_1))^3,\quad \mathcal{O}_1=\text{neighborhood of } \Gamma \text{ in } \Omega_1. \tag{6.19}$$

Then we can extend u to a vector function w defined on $\mathbb{R}^3$ so that

$$\begin{aligned}&w\in(H^1(\mathcal{O}_2))^3,\quad \mathcal{O}_2=\text{neighborhood of } \Omega \text{ in } \mathbb{R}^3\\ &w=u \quad \text{on } \Omega.\end{aligned} \tag{6.20}$$

If we set

$$f=\operatorname{curl} w \tag{6.21}$$

[10] We recall that μ is constant in the neighborhood of Γ.

and if we define $\Psi=\{\Psi_1,\Psi_2,\Psi_3\}$ by Fourier transformation[11] in the following way

$$(6.22)\qquad \hat{\Psi}_1=\frac{i}{2\pi}\frac{1}{|\xi|^2}(\xi_2\hat{f}_3-\xi_3\hat{f}_2),$$

(and the formulas derived from (6.22) by cyclic permutation), we have:

$$(6.23)\qquad \operatorname{curl}\Psi=f$$

and

$$(6.24)\qquad \partial\Psi/\partial x_i\in(L^2)^3 \quad \text{where} \quad L^2=L^2(\mathbb{R}^3).$$

Then $\operatorname{curl}(w-\Psi)=0$, therefore

$$(6.25)\qquad w=\Psi=\operatorname{grad}.P, \qquad \partial P/\partial x_i\in L^2.$$

If we denote the restriction of Ψ (resp. P) to Ω by ψ (resp. p), it follows from (6.25) that

$$(6.26)\qquad u-\psi=\operatorname{grad}p,$$

from which, since $\operatorname{div}(\mu u)=0$ in Ω:

$$(6.27)\qquad \operatorname{div}(\mu\operatorname{grad}p)+\operatorname{div}(\mu\psi)=0.$$

Denoting the restriction of p (resp. ψ) to Ω_i by p_i (resp. ψ_i), we have

$$(6.28)\qquad \operatorname{div}(\mu_i\operatorname{grad}p_i)+\operatorname{div}(\mu_i\psi_i)=0 \quad \text{in } \Omega_i.$$

Moreover, the p_i satisfy the following *boundary and transmission conditions*: since $n.u=0$ on Γ, we have:

$$(6.29)\qquad \partial p/\partial n=-n.\psi \quad \text{on } \Gamma,$$

and since $\operatorname{div}(\mu u)=0$ on Ω, we have, at the interface Σ_{ij}, the common boundary of Ω_i and Ω_j:

$$\mu_i n.u_i=\mu_j n.u_j$$

from which

$$(6.30)\qquad \mu_i\partial p/\partial n-\mu_j\partial p/\partial n=(\mu_j-\mu_i)n\psi.$$

---

[11] The notation is that of Sec. I.3.2.

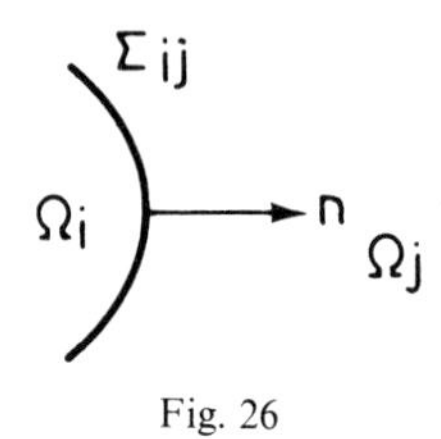

Fig. 26

But since according to (6.24)

(6.31) $$n\psi \in H^{1/2}(\Gamma) \quad \text{and} \quad \in H^{1/2}(\Sigma_{ij}),$$

it follows from (6.28), (6.29), (6.30) and *the regularity of the solutions of the transmission problems* that

(6.32) $$\partial^2 p_i/\partial x_j \partial x_k \in L^2(\Omega_i) \qquad \forall j,k$$

from which the result follows because of (6.26) and $v_i = \mu_i u_i$. □

Proof of Lemma 6.1. It is equivalent to prove that the space of vectors φ of $(H^2(\mathcal{O}))^3$, such that $n\varphi = 0$ on $\partial\mathcal{O}$, is dense in X.

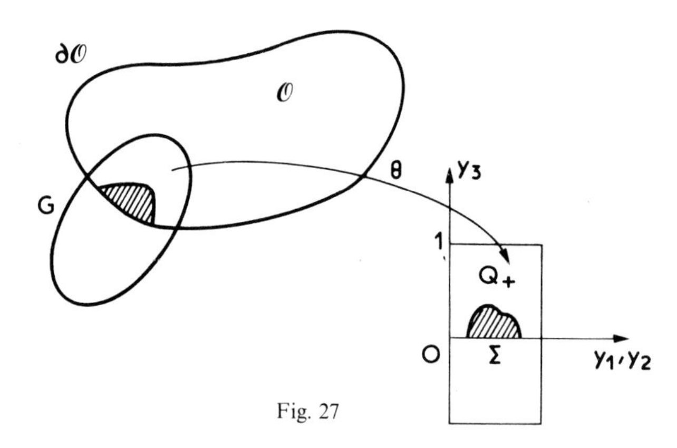

Fig. 27

Since $v \in X$ and $\psi \in \mathscr{D}(\mathcal{O})$ implies $v\psi \in X$, the property is in fact local and reduces to the following situation. We consider (cf. Fig. 27) a "local map", i.e. an open set G and a mapping θ of $G \to Q$,

$$Q = \Sigma \times]-1,1[\,,$$
$$\Sigma =]0,1[\times]0,1[\,,$$

once continuously differentiable as well as its inverse from $G \to Q$ and $Q \to G$, mapping $\mathcal{O} \cap G \to Q_+ = \Sigma \times]-1,1[$ and $\partial\mathcal{O} \cap G \to \Sigma$.

We can assume that the direction normal to $\partial\mathcal{O}\cap G$ becomes the direction of y_3. Let the P_i $(1\leqslant i\leqslant 4)$ be the images under θ of the differential operators curl (three components) and div; then

(6.33) $$P_i\varphi = p^i_{jk}\,\partial\varphi_j/\partial x_k$$

and we can assume that

(6.34) $$p_{33}=1\,.$$

We start from a vector function v of X with support as indicated (cross hatched) in Fig. 27 (in $G\cap\mathcal{O}$) and consider its image w under θ. Then $w\in Y$, Y being defined by

$$Y=\{\varphi \mid \varphi\in(L^2(Q_+))^3,\ P_i\varphi\in L^2(Q_+),\ \varphi_3=0 \text{ on } \Sigma\}$$ [12]

(since v tangent to $\partial\mathcal{O}$ becomes w tangent to Σ).

Moreover, w has a support, as indicated by cross hatching in Fig. 27, i.e., *it is zero* in the neighborhood of $y_3=1$ and of $\partial\Sigma\times]0,1[$; one says that w has the *property* (S).

Thus the problem is the following: approximate a $w\in Y$, having the property (S) (in the sense of the topology of Y) by elements of $\mathcal{Y}$:

$$\mathcal{Y}=\{\varphi \mid \varphi\in(H^1(Q_+))^3,\ \varphi_3=0 \text{ on } \Sigma,\ \varphi \text{ has the property } (S)\}\,.$$

We use a *tangential regularization* in y_1, y_2; introducing the regularizing sequence $\rho^m=\rho^m(y_1,y_2)$, i.e.

$$\rho^m\in\mathcal{D}(\mathbb{R}^2_{y_1y_2}),\qquad \rho^m\geqslant 0,\qquad \int\rho^m(y_1,y_2)\,dy_1\,dy_2=1\,,$$

ρ^m with support in $y_1^2+y_2^2\leqslant 1/m$, we introduce

(6.35) $$\psi^m=w*\rho^m$$

(i.e. $\psi^m(y)=\int w(y_1-\lambda_1, y_2-\lambda_2, y_3)\rho^m(\lambda_1,\lambda_2)\,d\lambda_1\,d\lambda_2$); then

(6.36) $$\psi_i^m\in L^2(0,1;H^1(\Sigma))$$

and ψ^m has the property (S) and satisfies $\psi_3^m=0$ on Σ; moreover, according to the *Lemma of Friedrichs* (cf. K. O. Friedrichs [2]):

$$\psi^m\to w \text{ in } Y\,.$$

According to (6.34), we have

(6.37) $$\partial\psi_3^m/\partial y_3\in L^2(Q_+)$$

---

[12] A Hilbert space for the norm
$$\left(\|\varphi\|^2_{(L^2(Q_+))^3}+\sum\|P_i\varphi\|^2_{L^2(Q_+)}\right)^{1/2}.$$

which, together with (6.36), proves that

$$\psi_3^m \in H^1(Q_+) . \tag{6.38}$$

Fixing m, we are thus led to the following: let ψ be given with

$$\begin{aligned} &\psi_i \in L^2(0,1; H^1(\Sigma)), \quad i=1,2, \\ &\psi_3 \in H^1(Q_+), \quad \psi_3 = 0 \quad \text{on } \Sigma \\ &\rho \text{ has the property } (S). \end{aligned} \tag{6.39}$$

We have to approximate ψ in the sense of Y by elements φ of $\mathscr{Y}$. We can take:

$$\varphi_3 = \psi_3 .$$

If we set $\hat{\psi} = \{\psi_1, \psi_2\}$, we have:

$$P_i \psi = Q_i \hat{\psi} + p_{3k}^i \partial \psi_3 / \partial y_k \tag{6.40}$$

and consequently, it is sufficient to approximate $\hat{\psi}$ by elements $\hat{\varphi}$ in the sense:

$$\begin{aligned} &Q_i \hat{\varphi} \to Q_i \hat{\psi} \quad \text{in } (L^2(Q_+))^2 \\ &\hat{\varphi} \to \psi \quad \text{in } (L^2(Q_+))^2, \quad \hat{\varphi} \in (H^1(Q_+))^2, \end{aligned} \tag{6.41}$$

where the functions $\hat{\varphi}$ are zero in the neighborhood of $y_3 = 1$ and of $\partial\Sigma \times]0,1[$, *but without boundary conditions.* This is possible according to Lemma 4.1 and Remark 4.1. ▯

6.2. *B* as Part of a Sobolev Space

Theorem 6.3. *We make the same assumptions as in Theorem 5.3 and, in addition, assume that the boundary Γ and the "interfaces" Σ_{ij} are twice continuously differentiable. Then, denoting by B^i the restriction of B to Ω_i,* we have

$$B^i \in L^\infty(0, T; (H^1(\Omega_i))^3) . \tag{6.42}$$

Proof. According to Theorems 5.1, 5.2, 5.3, we have, in the notation of (6.12):

$$B \in L^\infty(0, T; Y) \tag{6.43}$$

so that (6.42) is a consequence of Theorem 6.2. ▯

Remark 6.1. The result (6.42) permits us to apply *Sobolev's imbedding theorem* (Sobolev [1]): $H^1(\Omega_i) \subset L^6(\Omega_i)$, therefore

$$B^i \in L^\infty(0, T; (L^6(\Omega_i)^3) . \tag{6.44}$$

This result is essential for the discussion of the magneto-hydrodynamic equations (Duvaut-Lions [7]). □

6.3. D as Part of a Sobolev Space

Theorem 6.4. *We make the same assumptions as for Theorem* 6.3 *and assume in addition that*

(6.45) $\sigma\hat{\varepsilon} = \text{constant}$ *in* Ω (*for example* $\sigma = 0$),

(6.46) $\operatorname{div} G_1 = 0$,

(6.47) $\operatorname{div} D_0 = 0$.

Then, if D^i denotes the restriction of D to Ω_i, we have

(6.48) $D^i \in L^\infty(0, T; (H^1(\Omega_i))^3)$.

Proof. Applying the operator div to both sides of (5.5), it follows (due to (6.45)):

$$\partial(\operatorname{div} D)/\partial t + \sigma\hat{\varepsilon} \operatorname{div} D = \operatorname{div} G_1,$$

from which

(6.49) $\operatorname{div} D = 0$.

From this, (6.48) follows by arguments analogous to those of Sections 6.1 and 6.2. For all details, we refer to C. Bardos [1]. □

7. Slotted Antennas. Non-Homogeneous Problems[13]

7.1. Statement of the Problem (cf. Sec. 3.4)

The problem is the following: we wish to find vectors **D** and **B** such that[14]

(7.1) $\partial \mathbf{D}/\partial t - \operatorname{curl}(\hat{\mu}\mathbf{B}) + \sigma\varepsilon\mathbf{D} = \mathbf{G}_1$, $(\operatorname{div} G_1 = 0)$,

(7.2) $\partial \mathbf{B}/\partial t + \operatorname{curl}(\hat{\varepsilon}\mathbf{D}) = \mathbf{G}_2$, $(\operatorname{div} G_2 = 0)$,

(7.3) $n.\mathbf{B} = g$, $n \wedge \mathbf{D} = h$ on $\Gamma \times]0, T[$,

---

[13] This section can be passed over. It is not necessary for the understanding of the rest of the chapter.

[14] We slightly generalize the situation of Section 3.4.

(7.4) $\operatorname{div} B=0$

(7.5) $\mathbf{B}(0)=\mathbf{B}_0, \quad \mathbf{D}(0)=\mathbf{D}_0 \qquad (\operatorname{div}\mathbf{B}_0=0, \ \operatorname{div}\mathbf{D}_0=0)$.

Remark 7.1. The problem is called "non homogeneous" because of the boundary conditions (7.3). □

Remark 7.2. In the context of the problem in Section 3.4, we have

(7.6) $g=-n.\mathbf{B}^{(1)} \quad h=+n\wedge\mathbf{D}^{(1)}$,

(7.7) $G_1=0$,

(7.8) $G_2=0$. □

7.2. Statement of the Result

We set generally

(7.9) $\varphi\tau=\varphi-n(n\varphi) \quad (\text{on } \Gamma)$.

The operator $\varphi\to n.\operatorname{curl}\varphi|_\Gamma$ can then be written

(7.10) $n.\operatorname{curl}\varphi|_\Gamma=Q\varphi_\tau$,
$Q=$differential operator of the first order tangential to Γ.

We now derive *necessary* conditions (formal ones for the moment) on the data in order that problem (7.1) ... (7.5) has a solution.
We note that we can replace $n\wedge D=h$ by an (equivalent) condition

(7.11) $D_\tau=h_* \quad \text{on} \quad \Gamma\times]0,T[$.

It follows from (7.2) that

$$\partial(nB)/\partial t+n\operatorname{curl}(\hat{\varepsilon}D)=nG_2 \quad \text{on } \Gamma$$

from which, taking account of (7.10),

$$\partial(nB)/\partial t+\hat{\varepsilon}QD\tau=nG_2$$

from which

(7.12) $\partial g/\partial t+\hat{\varepsilon}Qh_*=nG_2 \quad \text{on} \quad \Gamma\times]0,T[$.

Since $nD_\tau=0$, it follows from (7.11) that

(7.13) $nh_*=0 \quad \text{on} \quad \Gamma\times]0,T[$.

Moreover, if $\operatorname{div} B=0$, then $\int_{\partial\Omega} nB \mathrm{d}(\partial\Omega)=0$, from which

$$\int_\Gamma g \,\mathrm{d}S=0. \tag{7.14}$$

Finally, $nB(0)=g(0)$ yields

$$nB_0=g(0). \quad \square \tag{7.15}$$

We will prove, always formally, that *if* $\{\mathbf{D},\mathbf{B}\}$ *is a solution of* (7.1), (7.2) *with*

$$n \wedge D=h \quad \text{on} \quad \Gamma\times\,]0,T[\,, \tag{7.16}$$

and (7.5), and if (7.12) ... (7.15) hold, then[15] we have (7.3), (7.4), (7.5).

Indeed, (7.2) implies

$$\partial(nB)/\partial t+\hat{\varepsilon}Qh_*=nG_2$$

from which, according to (7.12)

$$\partial(nB)/\partial t=\partial g/\partial t$$

and since we have (7.15), it follows that $n\,.\,B=g$. $\square$

Let us assume for a moment that there exists a vector P such that

$$\begin{aligned} &P, \partial P/\partial t \in L^2(0,T;(H^1(\Omega))^3)\,,\\ &\operatorname{div} P=0\\ &P_\tau=h_* \quad \text{on} \quad \Gamma\times\,]0,T[\,. \end{aligned} \tag{7.17}$$

If we introduce

$$D^*=D-P\,, \tag{7.18}$$

then

$$\begin{aligned} &\partial D^*/\partial t-\operatorname{curl}(\hat{\mu}B)+\sigma\hat{\varepsilon}D^*=G_1-\partial P/\partial t-\sigma\hat{\varepsilon}P=G_1\,,\\ &\partial B/\partial t+\operatorname{curl}(\hat{\varepsilon}D)=G_2-\operatorname{curl}(\hat{\varepsilon}P)=G_2\\ &n\wedge D=0 \quad \text{on} \quad \Gamma\times\,]0,T[\,,\\ &D^*(0)=D_0-P(0)\,, \qquad B(0)=B_0\,, \end{aligned} \tag{7.19}$$

and problem (7.1) ... (7.5) is *equivalent* to (7.19) which is a *homogeneous* problem.

We then say that *problem* (7.1) ... (7.5) *has a (unique) "weak" solution if problem* (7.19) *has (in the sense of Lemma* 4.5) *a (unique) "weak" solution.*

---

[15] We assume that $\operatorname{div} G_2=0$, $\operatorname{div} B_0=0$.

With this understanding, we will prove in Section (7.3) below

Theorem 7.1. *We assume that*

(7.20) $$G_1, G_2 \in L^2(0,T;(L^2(\Omega))^3),$$

(7.21) $$h_*, \partial h_*/\partial t \in L^2(0,T;(H^{1/2}(\Gamma))^3),$$

(7.22) $$g \in L^2(0,T;H^{-1/2}(\Gamma))$$

and that conditions (7.12), (7.13), (7.14) *hold. Then problem* (7.1) ... (7.5) *has a unique "weak" solution.*

7.3. Proof of Theorem 7.1

According to what we saw in Section 7.2, all comes back to proving

Lemma 7.1. *With the assumptions for Theorem* 7.1, *we can find a P satisfying* (7.17).

Proof. We introduce the space Z defined by

(7.23) $$Z = \{\varphi \mid \varphi \in (H^1(\Omega))^3, \ \varphi = 0 \ \text{ on } \Gamma, \ \operatorname{div} \varphi = 0\},$$

which is closed in $(H^1(\Omega))^3$. The "trace" $\varphi|_\Gamma$ of φ on Γ satisfies

(7.24) $$\varphi = \mathscr{Z}$$

where

(7.25) $$\mathscr{Z} = \{\psi \mid \psi \in (H^{1/2}(\Gamma))^3, \ \textstyle\int_\Gamma n\psi \, d\Gamma = 0\}$$

(closed sub-space of $(H^{1/2}(\Gamma))^3$).

According to Cattabriga [1], for example, one can find a mapping

$$\psi \to \mathscr{R}\psi$$

which is linear continuous from $\mathscr{Z} \to Z$ such that

(7.26) $$\mathscr{R}\psi|_\Gamma = \psi.$$

We now define

(7.27) $$P(t) = \mathscr{R}h_*(t), \qquad t \in [0,T].$$

According to (7.13), h_* and $\partial h_*/\partial t$ are elements of $L^2(0,T;\mathscr{Z})$ and therefore, since $\mathscr{R}\in\mathscr{L}(\mathscr{Z};Z)$, we have:

$$P, \partial P/\partial t \in L^2(0,T;Z)$$

which gives the first two conditions of (7.17).

Then we have $P|_\Gamma = h_*$ (according to (7.26)), from which

$$P_\tau = P - n(nP) = h_* - n(nh_*) = h_*$$

and thus all conditions of (7.17) are satisfied. ☐

Remark 7.3. One certainly could derive necessary and sufficient conditions for obtaining *strong* solutions, or also, on the other hand, more general solutions (by transposition); but there appears to be nothing in the literature...

For the systematic study of non homogeneous problems for parabolic systems and for certain systems which are hyperbolic (or well posed in the sense of Petrowsky)—but not containing the case of Maxwell's equations—cf. Lions-Magenes [1]. ☐

8. Polarizable Media

8.1. Existence and Uniqueness Result for a Variational Inequality Associated with the Operators of Maxwell

The notation is that of Section 4.

We give $G=\{G_1,G_2\}$ and $U_0=\{D_0,B_0\}$ with (as in Theorem 5.1)

$$G, \partial G/\partial t \in L^2(0,T;\mathscr{H}), \tag{8.1}$$

$$U_0 \in D(\mathscr{A}). \tag{8.2}$$

We define the set

$$K=\{\varphi \mid \varphi\in(L^2(\Omega))^3,\ \hat{\varepsilon}|\varphi(x)|\leqslant \mathscr{D}_0 \text{ a.e. in } \Omega\} \tag{8.3}$$

where

(8.4) $\mathscr{D}_0$ is a function >0, piecewise constant and, more precisely: $\mathscr{D}_0=$constant in the regions where $\hat{\varepsilon}$ and $\hat{\mu}$ are constant. ☐

Remark 8.1. An interesting case in physics is (cf. Fig. 28)

$$\mathscr{D}_0 = \left| \begin{array}{l} \text{constant } d_0>0 \text{ in } \Omega_3, \\ +\infty \text{ in } \Omega_1\cup\Omega_2, \end{array}\right. \tag{8.5}$$

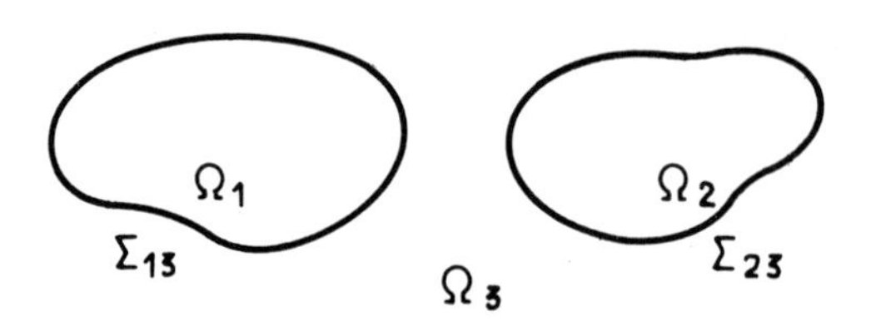

Fig. 28

the definition of K then being equivalent to

(8.6) $$K=\{\varphi \mid \varphi\in(L^2(\Omega))^3,\ \hat{\varepsilon}_3|\varphi(x)|\leqslant d_0 \text{ a.e. in } \Omega_3,\ \varphi \text{ arbitrary in } \Omega_1\cup\Omega_2\},$$

(in this case, $\Omega=\mathbb{R}^3$). □

We have

(8.7) K is a *convex closed* set of $(L^2(\Omega))^3$. □

In Section 8.3 below, we will prove

Theorem 8.1. *We assume G and U_0 to be given with* (8.1), (8.2) *and that*

(8.8) $$\mathbf{D}_0\in K$$

and[16]

(8.9) $$\begin{aligned}&\operatorname{curl}\operatorname{curl}(\hat{\varepsilon}\mathbf{D}_0)\in(L^2(\Omega))^3, \quad n\wedge\operatorname{curl}(\hat{\varepsilon}\mathbf{D}_0)=0 \quad on\ \Gamma,\\ &n\wedge\mathbf{B}_0=0 \quad on\ \Gamma, \quad \operatorname{curl}\operatorname{curl}(\hat{\mu}\mathbf{B}_0)\in(L^2(\Omega))^3,\\ &n\wedge\operatorname{curl}(\hat{\mu}\mathbf{B}_0)=0 \quad on\ \Gamma.\end{aligned}$$

Then there exists one and only one pair of vectors $\{\mathbf{D},\mathbf{B}\}$ *such that*

(8.10) $$\begin{aligned}&\mathbf{D},\mathbf{B}\in L^\infty(0,T;(L^2(\Omega))^3),\\ &\partial\mathbf{D}/\partial t,\ \partial\mathbf{B}/\partial t\in L^\infty(0,T;(L^2(\Omega))^3),\end{aligned}$$

(8.11) $$\mathbf{D}(t)\in K, \quad t\in[0,T],$$

(8.12) $$\operatorname{curl}(\hat{\varepsilon}\mathbf{D})\in L^\infty(0,T;(L^2(\Omega))^3),$$

(8.13) $$n\wedge\mathbf{D}=0 \quad on \quad \Gamma\times]0,T[,$$

(8.14) $$\left|\begin{aligned}&(\partial\mathbf{D}(t)/\partial t,\varepsilon(\varphi-\mathbf{D}(t)))+(\sigma\varepsilon\mathbf{D}(t),\varepsilon(\varphi-\mathbf{D}(t)))-(\hat{\mu}\mathbf{B}(t),\operatorname{curl}(\hat{\varepsilon}(\varphi-\mathbf{D}(t)))\\ &\qquad\qquad\geqslant(\mathbf{G}_1(t),\hat{\varepsilon}(\varphi-\mathbf{D}(t)))\ ^{17}\\ &\forall\varphi\in K, \quad such\ that \quad \operatorname{curl}\hat{\varepsilon}\varphi\in(L^2(\Omega))^3 \quad and \quad n\wedge\varphi=0 \quad on\ \Gamma,\end{aligned}\right.$$

[16] These conditions can be avoided if one considers "weak" solutions of problems in polarizable media.

[17] In this expression, $(f,g)=\int_\Omega f_i g_i\,dx$, $i=1,2,3$.

(8.15) $\partial \mathbf{B}/\partial t + \operatorname{curl}(\hat{\varepsilon}\mathbf{D}) = \mathbf{G}_2$,

(8.16) $\mathbf{D}(0) = \mathbf{D}_0$, $\mathbf{B}(0) = \mathbf{B}_0$. □

Remark 8.2. If $\operatorname{div} \mathbf{G}_2 = 0$ and $n\mathbf{B}_0 = 0$, it *follows* from (8.15) that

(8.17) $\operatorname{div} \mathbf{B} = 0$

and

(8.18) $n \cdot \mathbf{B} = 0$ on Γ. □

Before proving Theorem 8.1, we show that this Theorem *solves the problems for polarizable media.*

8.2. Interpretation of the Variational Inequality. Solution of the Problems for Polarizable Media

The interpretation which we give is partly formal. If we assume that $\operatorname{curl}(\hat{\mu}B) \in (L^2(\Omega))^3$, then (8.14) can be written

(8.19) $(F(t), \hat{\varepsilon}(\varphi - D(t))) \geqslant 0 \quad \forall \varphi \in K$,

where

(8.20) $F(t) = \partial D(t)/\partial t + \sigma\hat{\varepsilon}D(t) - \operatorname{curl}(\hat{\mu}B(t)) - G_1(t)$.

But (8.19) is equivalent to a *pointwise* condition

$$F(x,t) \cdot \hat{\varepsilon}(\varphi(x) - D(x,t)) \geqslant 0 \quad \forall \varphi \in K, \quad \text{a.e. in } \Omega,$$

or again:

(8.21) $F(x,t) \cdot (k - \hat{\varepsilon}D(x,t)) \geqslant 0 \quad \forall k \in \mathbb{R}^3$ with $|k| \leqslant \mathscr{D}_{0i}$ if $x \in \Omega_i$ [18].

We now have to distinguish *two cases*:

i) if $\hat{\varepsilon}|D(x,t)| < \mathscr{D}_0$, then

$$F(x,t) = 0;$$

ii) if $\varepsilon|D(x,t) = \mathscr{D}_{0i}$, then there exist $\lambda_i(x,t)$ such that

$$\lambda_i(x,t) \geqslant 0, \quad F(x,t) = -\lambda_i(x,t)\hat{\varepsilon}D(x,t)$$

[18] $\mathscr{D}_0 = \mathscr{D}_{0i}$ in Ω_i.

or again

$$\begin{aligned}(8.22)\qquad & \partial D(x,t)/\partial t+(\sigma+\lambda_i(x,t))\hat{\varepsilon}D(x,t)-\operatorname{curl}\hat{\mu}B(x,t)=G_1(x,t)\,,\\ & \lambda_i(x,t)\geqslant 0\,.\end{aligned}$$

This covers the problems for polarizable media.

8.3. Proof of Theorem 8.1

8.3.1. Existence Proof

We will use

1) *a regularization* in order to arrive at operators of parabolic type;
2) *a penalization* in order to arrive at *equations*.

We introduce the space $\mathscr{V}$ defined by

$$\begin{aligned}(8.23)\qquad \mathscr{V}=\{\Phi\mid\ \Phi=\{\varphi,\psi\},\ \varphi\in(L^2(\Omega))^3,\ \psi\in(L^2(\Omega))^3,\\ \operatorname{curl}(\hat{\varepsilon}\varphi)\in(L^2(\Omega))^3,\ \operatorname{curl}(\hat{\mu}\psi)\in(L^2(\Omega))^3,\ n\wedge\varphi=0\ \text{ on } \Gamma\}\,;\end{aligned}$$

for $\Phi=\mathscr{V}$, $\Phi_*=\{\varphi_*,\psi_*\}\in\mathscr{V}$, we set

$$(8.24)\qquad ((\Phi,\Phi_*))=(\operatorname{curl}(\hat{\varepsilon}\varphi),\operatorname{curl}(\hat{\varepsilon}\varphi_*))+(\operatorname{curl}(\hat{\mu}\psi),\operatorname{curl}(\mu\psi_*))$$

and we observe that $\mathscr{V}$ is a Hilbert space for the norm

$$(8.25)\qquad \|\Phi\|_{\mathscr{V}}=(\|\varphi\|^2_{(L^2(\Omega))^3}+\|\psi\|^2_{(L^2(\Omega))^3}+((\Phi,\Phi)))^{1/2}\,.$$

We now introduce a penalization operator β of $(L^2(\Omega))^3\to(L^2(\Omega))^3$, attached to K; then

$$\begin{aligned}(8.26)\qquad & \beta \text{ is bounded, monotone and Lipschitz continuous for}\\ & (L^2(\Omega))^3\to(L^2(\Omega))^3\,,\\ & \beta(\varphi)=0\Leftrightarrow\varphi\in K\,.\end{aligned}$$

For example, we could take

$$(8.27)\qquad \beta(\varphi)(x)\ \left|\begin{array}{l}=\varphi_i(x)-\mathscr{D}_{0i}\dfrac{\varphi_i(x)}{\hat{\varepsilon}_i|\varphi_i(x)|}\quad\text{if}\quad x\in\Omega_i,\quad \hat{\varepsilon}_i|\varphi_i(x)|\geqslant\mathscr{D}_{0i}\\ =0\quad\text{if}\quad x\in\Omega_i,\quad \hat{\varepsilon}_i|\varphi_i(x)|\leqslant\mathscr{D}_{0i}\,.\quad\square\end{array}\right.$$

Regularized and penalized equation. For η and $\lambda>0$, we want to find solutions

$$(8.29)\qquad D=D_{\eta\lambda}\,,\qquad B=B_{\eta\lambda}\,,\qquad U=U_{\eta\lambda}=\{D_{\eta\lambda},B_{\eta\lambda}\}\,,$$

of

$$(8.30)\quad \left|\begin{array}{l}(D',\hat{\varepsilon}\varphi)+(B',\mu\psi)+((\sigma\hat{\varepsilon}D),\hat{\varepsilon}\varphi)-(\hat{\mu}B,\operatorname{curl}(\hat{\varepsilon}\varphi))\\ \qquad +(\operatorname{curl}(\hat{\varepsilon}D),\hat{\mu}\psi)+\eta((U,\Phi))+\lambda^{-1}(\beta(D),\varphi)=(G,\Phi)_{\mathscr{H}}\\ \forall\Phi=\{\varphi,\psi\}\in\mathscr{V}\quad\text{with}\end{array}\right.$$

$$(8.31)\qquad D(0)=D_0,\qquad B(0)=B_0.$$

In (8.30), we set: $G=\{G_1,G_2\}$.

Problem (8.30), (8.31) has a unique solution (special case of Theorem 1,2, Chap. 2 of Lions [1]) which satisfies

$$(8.32)\qquad U\in L^2(0,T;\mathscr{V})\cap L^\infty(0,T;\mathscr{H}).$$

If we let "$t=0$" in (8.30), we find (observing that $\beta(D_0)=0$ since $D_0\in K$):

$$\begin{aligned}&(D'(0),\hat{\varepsilon}\varphi(0))+(B'(0),\hat{\mu}\psi(0))+(\sigma\hat{\varepsilon}D_0,\hat{\varepsilon}\varphi(0))-(\operatorname{curl}(\hat{\mu}B_0),\hat{\varepsilon}\varphi(0))\\ &\qquad +(\operatorname{curl}(\hat{\varepsilon}D_0),\hat{\mu}\psi(0)+\eta(\operatorname{curl}(\hat{\varepsilon}D_0),\operatorname{curl}(\hat{\varepsilon}\varphi(0))\\ &\qquad +\eta(\operatorname{curl}(\hat{\mu}B_0),\operatorname{curl}(\hat{\mu}\psi(0))=(G(0),\Phi(0))_{\mathscr{H}}\end{aligned}$$

from which it follows, due to (8.9), that

$$(8.33)\qquad \begin{aligned}&D'(0)=G_1(0)-\sigma\hat{\varepsilon}D_0+\operatorname{curl}(\mu B_0)-\eta\operatorname{curl}(\operatorname{curl}(\hat{\varepsilon}D_0)),\\ &B'(0)=G_2(0)-\operatorname{curl}(\hat{\varepsilon}D_0)-\eta\operatorname{curl}(\operatorname{curl}(\hat{\mu}B_0)).\end{aligned}$$

Therefore

$$(8.34)\qquad \{D'(0),B'(0)\}=\{D'_{\eta\lambda}(0),B'_{\eta\lambda}(0)\}\quad\text{remains in a bounded set of }\mathscr{H}\quad\text{when }\eta\text{ and }\lambda\to 0.$$

If (8.30) is to hold in t [19], we obtain:

$$(8.35)\qquad \begin{aligned}&(D'',\hat{\varepsilon}\varphi)+(B'',\mu\psi)+(\sigma\hat{\varepsilon}D',\hat{\varepsilon}\varphi)-(\hat{\mu}B',\operatorname{curl}(\varepsilon\varphi))\\ &\qquad +(\operatorname{curl}(\hat{\varepsilon}D'),\hat{\mu}\psi)+\eta((U',\Phi))+\lambda^{-1}((\beta D))',\varphi)=(G',\Phi)_{\mathscr{H}};\end{aligned}$$

In (8.35), we take $\varphi=D'$, $\varphi=B'$; it follows[20] that

$$(8.36)\qquad (U'',U')_{\mathscr{H}}+(\sigma\hat{\varepsilon}D',\hat{\varepsilon}D')+\eta\|U'\|^2+\lambda^{-1}((\beta(D))',D')=(G',U')_{\mathscr{H}}.$$

But (this can always be justified by the method of difference quotients)

$$((\beta(D))',D')=\lim h^{-2}(\beta(D(t+h))-\beta(D(t)),D(t+h)-D(t))\geqslant 0$$

[19] Which can be justified by the method of difference quotients.

[20] $\|U'\|^2=((U',U'))$.

(because of the monotonicity of β) and therefore (8.36) gives[21]

$$\frac{1}{2}\frac{d}{dt}\|U'(t)\|^2_{\mathscr{H}}+\eta\|U'(t)\|^2\leqslant c_1\|U'(t)\|^2_{\mathscr{H}}+c_2\|G'(t)\|^2_{\mathscr{H}}. \tag{8.37}$$

From (8.37) and (8.34) it follows, with Gronwall's inequality:

(8.38) $U'_{\eta\lambda}$ remains in a bounded set of $L^\infty(0,T;\mathscr{H})$

(8.39) $\eta^{1/2}U'_{\eta\lambda}$ remains in a bounded set of $L^2(0,T;\mathscr{V})$.

Since $U_{\eta\lambda}(0)=U_0=\{D_0,B_0\}$, it follows from (8.38)[22] that

(8.40) $U_{\eta\lambda}$ remains in a bounded set of $L^\infty(0,T;\mathscr{H})$. □

Passage to the limit in η and λ. According to (8.38) and (8.40), we can select a sequence, again denoted by $U_{\eta\lambda}$, such that, when $\lambda\to 0$, $\eta\to 0$:

(8.41) $U_{\eta\lambda}$(resp. $U'_{\eta\lambda}$)$\to U$ (resp. U') weakly star in $L^\infty(0,T;\mathscr{H})$.

We now prove that $\forall\Phi=\{\varphi,\psi\}$ given with

(8.42) $\Phi\in L^2(0,T;\mathscr{V})$, $\varphi(t)\in K$ a.e.,

we have

$$\begin{aligned}\int_0^T\{(U',\Phi-U)_{\mathscr{H}}+(\sigma\hat{\varepsilon}D,\hat{\varepsilon}(\varphi-D))-(\hat{\mu}B,\operatorname{curl}\hat{\varepsilon}\varphi)&\\ +(\hat{\varepsilon}D,\operatorname{curl}\hat{\mu}\psi)-(G,\Phi-U)\}\,dt\geqslant 0.&\end{aligned} \tag{8.43}$$

Indeed, replacing $\Phi=\{\varphi,\psi\}$ by $\Phi-U_{\eta\lambda}$ in (8.30), we have

$$\begin{aligned}(U'_{\eta\lambda},\Phi-U_{\eta\lambda})_{\mathscr{H}}+(\sigma\hat{\varepsilon}D_{\eta\lambda},\hat{\varepsilon}(\varphi-D_{\eta\lambda}))-(\hat{\mu}B_{\eta\lambda},\operatorname{curl}\hat{\varepsilon}(\varphi-D_{\eta\lambda}))&\\ +(\operatorname{curl}\hat{\varepsilon}D_{\eta\lambda},\hat{\mu}(\psi-B_{\eta\lambda}))+\eta((U_{\eta\lambda},\Phi))-(G,\Phi-U_{\eta\lambda})_{\mathscr{H}}&\\ =\eta\|U_{\eta\lambda}\|^2-\lambda^{-1}(\beta(D_{\eta\lambda}),\varphi-D_{\eta\lambda}).&\end{aligned} \tag{8.44}$$

But since $\varphi=\varphi(t)\in K$, we have $\beta(\varphi)=0$ and the right hand side of (8.44) equals

$$\eta\|U_{\eta\lambda}\|^2+\lambda^{-1}(\beta(\varphi)-\beta(D_{\eta\lambda}),\varphi-D_{\eta\lambda})\geqslant 0.$$

Observing that, obviously,

$$(\hat{\mu}B_{\eta\lambda},\operatorname{curl}\hat{\varepsilon}D_{\eta\lambda})-(\operatorname{curl}\hat{\varepsilon}D_{\eta\lambda},\hat{\mu}B_{\eta\lambda})=0,$$

---

[21] We note that $|(\sigma\hat{\varepsilon}D',\hat{\varepsilon}D')|\leqslant c\|U'(t)\|^2_{\mathscr{H}}$.

[22] Alternatively, one can obtain (8.40) directly starting from (8.30).

it follows from (8.44) that

$$\int_0^T \{(U'_{\eta\lambda}, \Phi - U_{\eta\lambda})_{\mathscr{H}} + (\sigma\hat{\varepsilon}D_{\eta\lambda}, \hat{\varepsilon}(\varphi - D_{\eta\lambda})) - (\hat{\mu}D_{\eta\lambda}, \operatorname{curl}(\hat{\varepsilon}\varphi)) + (\hat{\varepsilon}D_{\eta\lambda}, \operatorname{curl}(\hat{\mu}\psi)) + \eta((U_{\eta\lambda}, \Phi)) - (G, \Phi - U_{\eta\lambda})_{\mathscr{H}}\} \, dt \geqslant 0 . \tag{8.45}$$

Consequently,

$$\begin{aligned} \int_0^T \{(U'_{\eta\lambda}, \Phi)_{\mathscr{H}} + (\sigma\hat{\varepsilon}D_{\eta\lambda}, \hat{\varepsilon}\varphi) - (\hat{\mu}B_{\eta\lambda}, \operatorname{curl}(\hat{\varepsilon}\varphi)) \\ + (\hat{\varepsilon}D_{\eta\lambda}, \operatorname{curl}(\hat{\mu}\psi)) + \eta((U_{\eta\lambda}, \Phi)) - (G, \Phi - U_{\eta\lambda})_{\mathscr{H}}\} \, dt \\ \geqslant \tfrac{1}{2}\|U_{\eta\lambda}(T)\|^2_{\mathscr{H}} - \tfrac{1}{2}\|U_0\|^2_{\mathscr{H}} + \int_0^T (\sigma\hat{\varepsilon}D_{\eta\lambda}, \hat{\varepsilon}D_{\eta\lambda}) \, dt . \end{aligned} \tag{8.46}$$

Using (8.41) and (8.39), we see that the left hand side of (8.46) converges toward

$$\int_0^T \{(U', \Phi)_{\mathscr{H}} + (\sigma\hat{\varepsilon}D, \hat{\varepsilon}\varphi) - (\hat{\mu}B, \operatorname{curl}\hat{\varepsilon}\varphi) + (\hat{\varepsilon}D, \operatorname{curl}\hat{\mu}\psi) - (G, \Phi - U)_{\mathscr{H}}\} \, dt$$

and that the limit inferior of the right hand side is

$$\geqslant \tfrac{1}{2}\|U(T)\|^2_{\mathscr{H}} - \tfrac{1}{2}\|U_0\|^2_{\mathscr{H}} + \int_0^T (\sigma\hat{\varepsilon}D, \hat{\varepsilon}D) \, dt = \int_0^T \{(U', U)_{\mathscr{H}} + (\sigma\hat{\varepsilon}D, \hat{\varepsilon}D)\} \, dt ,$$

from which (8.43) follows. □

We now use the fact that, in (8.43), *there are no restrictions on the component* ψ *of* Φ. Thus, if we replace ψ by $k\psi$, $k \in \mathbb{R}$, φ being unchanged, we still have (8.43) and, letting $k \to \pm\infty$, it follows that *the coefficient of* ψ *is zero*, i.e.,

$$\int_0^T \{(B', \hat{\mu}\psi) + (\hat{\varepsilon}D, \operatorname{curl}(\hat{\mu}\psi)) - (G_2, \hat{\mu}\psi)\} \, dt = 0. \tag{8.47}$$

It follows that (8.15) holds, therefore $\operatorname{curl}(\hat{\varepsilon}D) = G_2 - \partial B/\partial t$ satisfies (8.12). Then, taking the scalar product of (8.15) with $\hat{\mu}\psi$ and integrating, it follows by comparison with (8.47) that

$$\int_{\Gamma \times (0,T)} \hat{\varepsilon}\hat{\mu}(n \wedge D)\psi \, d\Gamma \, dt = 0$$

from which (8.13) follows. □

On the other hand, from (8.15) it follows that

$$\int_0^T \{B', \hat{\mu}(\psi - B)) + (\operatorname{curl}(\hat{\varepsilon}D), \hat{\mu}(\psi - B)) - (G_2, \hat{\mu}(\psi - B))\} \, dt = 0$$

so that (8.43) can be written

$$\begin{aligned} \int_0^T \{(D', \hat{\varepsilon}(\varphi - D)) + (\sigma\hat{\varepsilon}D, \hat{\varepsilon}(\varphi - D)) - (\hat{\mu}B, \operatorname{curl}\hat{\varepsilon}(\varphi - D)) \\ - (G_1, \hat{\varepsilon}(\varphi - D))\} \, dt \geqslant 0 . \end{aligned} \tag{8.48}$$

But from this, as we already saw, we can pass to a *local condition in t*, i.e. to (8.14). □

Thus, it only remains to prove (8.11) to obtain existence in Theorem 8.1. From (8.11) it follows (taking $\varphi = D_{\eta\lambda}$, $\psi = B_{\eta\lambda}$) that

(8.49) $$\lambda^{-1} \int_0^T (\beta(D_{\eta\lambda}), D_{\eta\lambda}) \mathrm{d}t \leqslant c\,.$$

Moreover, according to (8.40) and the definition of β, $\beta(D_{\eta\lambda})$ remains in a bounded set of $L^\infty(0, T; (L^2(\Omega))^3)$ and therefore we can assume, possibly by selection of a sub-sequence, that

(8.50) $$\beta(D_{\eta\lambda}) \to \chi \quad \text{weakly star in } L^\infty(0, T; (L^2(\Omega))^3)\,.$$

From (8.30) it follows that

$$\partial D_{\eta\lambda}/\partial t + \sigma \hat{\varepsilon} D_{\eta\lambda} - \operatorname{curl}(\hat{\mu} B_{\eta\lambda}) + \eta \operatorname{curl}\operatorname{curl}(\hat{\varepsilon} D_{\eta\lambda}) + \lambda^{-1}(D_{\eta\lambda}) = G_1$$

from which (with (8.38), (8.39), (8.40)) it follows that

$$\beta(D_{\eta\lambda}) \to 0 \quad \text{in} \quad (\mathscr{D}'(\Omega \times]0, T[)) \quad \text{for example};$$

comparing with (8.50) we have

(8.51) $$\beta(D_{\eta\lambda}) \to 0 \quad \text{weakly star in } L^\infty(0, T; (L^2(\Omega))^3)\,.$$

If φ is an *arbitrary* function of $L^2(0, T; (L^2(\Omega))^3)$, we have:

$$\int_0^T (\beta(\varphi) - \beta(D_{\eta\lambda}), \varphi - D_{\eta\lambda}) \mathrm{d}t \geqslant 0\,.$$

Using (8.49) and (8.51), it follows that

(8.52) $$\int_0^T (\beta(\varphi), \varphi - D) \mathrm{d}t \geqslant 0\,.$$

In (8.52) we take

$$\varphi = D + s\theta\,, \qquad s > 0\,, \qquad \theta \in L^2(0, T; (L^2(\Omega))^3) \quad \text{arbitrary}\,.$$

It follows that

$$s \int_0^T (\beta(D + s\theta), \theta) \mathrm{d}t \geqslant 0$$

therefore

$$\int_0^T (\beta(D + s\theta), \theta) \mathrm{d}t \geqslant 0$$

therefore, letting $s \to 0$,

$$\int_0^T (\beta(D), \theta) \mathrm{d}t \geqslant 0 \qquad \forall \theta$$

therefore

$$\beta(D)=0$$

from which (8.11) follows. □

8.3.2. Uniqueness Proof

Let $\{D,B\}$, $\{D_*,B_*\}$ be two possible solutions of the problem. We set:

$$u=D-D_*, \quad v=B-B_*. \tag{8.53}$$

In (8.41) we take (resp. in the analogous inequality for D_*,B_*)

$$\varphi=D_*(\text{resp. } \varphi=D).$$

Adding, we get:

$$-(u',\hat{\varepsilon}u)-(\sigma\hat{\varepsilon}u,\hat{\varepsilon}u)+(\hat{\mu}v,\operatorname{curl}\hat{\varepsilon}u)\geqslant 0. \tag{8.54}$$

But from (8.15) and the analogous equation for D_*,B_*, it follows that

$$v'+\operatorname{curl}(\hat{\varepsilon}u)=0. \tag{8.55}$$

Therefore

$$(v',\hat{\mu}v)+(\operatorname{curl}(\hat{\varepsilon}u),\hat{\mu}v)=0$$

and (8.54) is equivalent to

$$-(u',\hat{\varepsilon}u)-(v',\hat{\mu}v)-(\sigma\hat{\varepsilon}u,\hat{\varepsilon}u)\geqslant 0$$

from which, in particular if $W=\{u,v\}$, it follows that:

$$-\frac{\mathrm{d}}{\mathrm{d}t}\|W(t)\|^2_{\mathscr{H}}\geqslant 0$$

and since $W(0)=0$, we have $W=0$. □

9. Stable Media as Limits of Polarizable Media

9.1. Statement of the Result

We now prove that when "the polarization increases indefinitely", then the solution of the corresponding problem converges toward the solution of the usual problem (for a stable medium). Precisely:

Theorem 9.1. *We assume that*

(9.1) $$\mathscr{D}_0 \to +\infty$$

(i.e. inf $\mathscr{D}_0 \to +\infty$). *Let* $U^{\mathscr{D}_0} = \{D^{\mathscr{D}_0}, B^{\mathscr{D}_0}\}$ (resp. $U = \{D, B\}$) *be the solution of the polarization problem* (Theorem 8.1) (resp. *of the usual problem, Theorem* 5.1). *Then*

(9.2) $$D^{\mathscr{D}_0} \to D, \quad \partial D^{\mathscr{D}_0}/\partial t \to \partial D/\partial t, \quad B^{\mathscr{D}_0} \to B, \quad \partial B^{\mathscr{D}_0}/\partial t \to \partial B/\partial t$$
weakly star in $L^\infty(0, T; (L^2(\Omega))^3)$,

(9.3) $$\operatorname{curl}(\hat{\varepsilon} D^{\mathscr{D}_0}) \to \operatorname{curl}(\hat{\varepsilon} D) \quad \textit{weakly star in } L^\infty(0, T; (L^2(\Omega))^3).$$

9.2. Proof of Theorem 9.1

The proof of Theorem 8.1 shows that

(9.4) $D^{\mathscr{D}_0}, \partial D^{\mathscr{D}_0}/\partial t, B^{\mathscr{D}_0}, \partial B^{\mathscr{D}_0}/\partial t$ remain in a *bounded set* of $L^\infty(0, T; (L^2(\Omega))^3)$ when $\mathscr{D}_0 \to +\infty$.

Equation (8.15) then gives:

$$\operatorname{curl}(\varepsilon D^{\mathscr{D}_0}) = G_2 - \partial B^{\mathscr{D}_0}/\partial t$$

remains in a bounded set of $L^\infty(0, T; (L^2(\Omega))^3)$.

We can then select a sequence, again denoted by $D^{\mathscr{D}_0}, B^{\mathscr{D}_0}$ such that we have (9.2), (9.3), *but where we still have to prove that* D, B *is a solution of the ordinary problem*; for that, it only remains to prove that

(9.5) $$\partial D/\partial t + \sigma \hat{\varepsilon} D - \operatorname{curl}(\hat{\mu} B) = G_1.$$

Let φ be given with

(9.6) $$\varphi \in (L^2(\Omega))^3, \quad \operatorname{curl}(\hat{\varepsilon}\varphi) \in (L^2(\Omega))^3, \quad n \wedge \varphi = 0 \quad \text{on } \Gamma,$$
$$\varphi \in (L^\infty(\Omega))^3$$

and let

(9.7) $$K^{\mathscr{D}_0} = \text{convex set } K \text{ given by (8.3)}.$$

Due to the fact that $\varphi \in (L^\infty(\Omega))^3$, we have:

(9.8) $$\varphi \in K^{\mathscr{D}_0} \quad \text{for sufficiently large } \mathscr{D}_0$$

and therefore (8.14) gives

$$(\partial D^{\mathcal{D}_0}/\partial t, \hat{\varepsilon}(\varphi - D^{\mathcal{D}_0})) + (\sigma\hat{\varepsilon}D^{\mathcal{D}_0}, \hat{\varepsilon}(\varphi - D^{\mathcal{D}_0})) - (\hat{\mu}B^{\mathcal{D}_0}, \operatorname{curl}\hat{\varepsilon}(\varphi - D^{\mathcal{D}_0})) \geqslant (G_1, \hat{\varepsilon}(\varphi - D^{\mathcal{D}_0}))$$

from which, if we now take $\varphi = \varphi(t)$ with

$$\|\varphi(t)\|_{(L^\infty(\Omega))^3} \leqslant \text{constant}, \quad t \in [0, T]: \tag{9.9}$$

$$\begin{aligned} \int_0^T \{ &(\partial D^{\mathcal{D}_0}/\partial t, \hat{\varepsilon}\varphi) + (\sigma\hat{\varepsilon}D^{\mathcal{D}_0}, \hat{\varepsilon}\varphi) - (\hat{\mu}B^{\mathcal{D}_0}, \operatorname{curl}(\hat{\varepsilon}\varphi)) \\ &+ (\hat{\mu}B^{\mathcal{D}_0}, \operatorname{curl}(\hat{\varepsilon}D^{\mathcal{D}_0})) - (G_1, \hat{\varepsilon}(\varphi - D^{\mathcal{D}_0}))\}\, dt \\ &\geqslant \int_0^T [(\partial D^{\mathcal{D}_0}/\partial t, \hat{\varepsilon}D^{\mathcal{D}_0}) + (\sigma\hat{\varepsilon}D^{\mathcal{D}_0}, \hat{\varepsilon}D^{\mathcal{D}_0})]\, dt. \end{aligned} \tag{9.10}$$

But from (8.15)

$$\partial B^{\mathcal{D}_0}/\partial t + \operatorname{curl}(\hat{\varepsilon}D^{\mathcal{D}_0}) = G_2 \tag{9.11}$$

so that (9.10) can be written

$$\begin{aligned} \int_0^T \{ &(\partial D^{\mathcal{D}_0}/\partial t, \hat{\varepsilon}\varphi) + (\sigma\hat{\varepsilon}D^{\mathcal{D}_0}, \hat{\varepsilon}\varphi) - (\hat{\mu}B^{\mathcal{D}_0}, \operatorname{curl}(\hat{\varepsilon}\varphi)) \\ &+ (\hat{\mu}B^{\mathcal{D}_0}, G_2) - (G_1, \hat{\varepsilon}(\varphi - D^{\mathcal{D}_0}))\}\, dt \\ &\geqslant \int_0^T \{(\partial U^{\mathcal{D}_0}/\partial t, U^{\mathcal{D}_0})_{\mathcal{H}} + (\sigma\hat{\varepsilon}D^{\mathcal{D}_0}, \hat{\varepsilon}D^{\mathcal{D}_0})\}\, dt \\ &- \tfrac{1}{2}\|U^{\mathcal{D}_0}(T)\|_{\mathcal{H}}^2 - \tfrac{1}{2}\|U_0\|_{\mathcal{H}}^2 + \int_0^T (\sigma\hat{\varepsilon}D^{\mathcal{D}_0}, \hat{\varepsilon}D^{\mathcal{D}_0})\, dt \end{aligned} \tag{9.12}$$

and we can pass to the limit (resp. limit inferior) on the left hand (resp. right hand) side of (9.12); we get:

$$\begin{aligned} \int_0^T \{ &(\partial D/\partial t, \hat{\varepsilon}\varphi) + (\sigma\hat{\varepsilon}D, \hat{\varepsilon}\varphi) - (\hat{\mu}B, \operatorname{curl}(\hat{\varepsilon}\varphi)) + (\hat{\mu}B, G_2) - (G_1, \hat{\varepsilon}(\varphi - D))\}\, dt \\ &\geqslant \int_0^T (\partial U/\partial t, U)_{\mathcal{H}} + (\sigma\hat{\varepsilon}D, \hat{\varepsilon}D)\}\, dt. \end{aligned} \tag{9.13}$$

But on the other hand, in the limit (9.11) gives

$$\partial B/\partial t + \operatorname{curl}(\hat{\varepsilon}D) = G_2 \tag{9.14}$$

and taking (9.14) into account in (9.13), we find

$$\begin{aligned} \int_0^T \{ &(\partial D/\partial t, \hat{\varepsilon}(\varphi - D) + (\sigma\hat{\varepsilon}D, \hat{\varepsilon}(\varphi - D)) - (\hat{\mu}B, \operatorname{curl}(\hat{\varepsilon}(\varphi - D))) \\ &- (G_1, \hat{\varepsilon}(\varphi - D))\}\, dt \geqslant 0. \end{aligned} \tag{9.15}$$

From this we pass to a *pointwise* condition and consequently to

$$\begin{aligned} &(\partial D/\partial t, \hat{\varepsilon}(\varphi - D)) + (\sigma\hat{\varepsilon}D, \hat{\varepsilon}(\varphi - D)) - (\hat{\mu}B, \operatorname{curl}(\hat{\varepsilon}(\varphi - D))) \\ &- G_1, \hat{\varepsilon}(\varphi - D)) \geqslant 0 \quad \text{a.e. in } t, \end{aligned} \tag{9.16}$$

$\forall \varphi$ satisfying (9.6).

But (proceeding analogously to Section 6) we can prove that the space of the φ which satisfy (9.6) is *dense* in the space of the functions φ such that

$$\varphi \in (L^2(\Omega))^3, \quad \operatorname{curl}(\hat{\varepsilon}\varphi) \in (L^2(\Omega))^3, \quad n \wedge \varphi = 0 \quad \text{on } \Gamma. \tag{9.17}$$

Then (9.16) holds $\forall \varphi$ with (9.17). Thus we can replace φ by $D \pm \varphi$ in (9.16) from which

$$(\partial D/\partial t, \hat{\varepsilon}\varphi) + (\sigma\hat{\varepsilon}D, \hat{\varepsilon}\varphi) - (\hat{\mu}B, \operatorname{curl}(\hat{\varepsilon}\varphi)) - (G_1, \hat{\varepsilon}\varphi) = 0$$

$\forall \varphi$ with (9.6) from which the result follows. □

10. Various Additions

Remark 10.1. Lemma (4.7) combined with the Hille-Yosida theorem (cf. Hille-Phillips [1], Yosida [1]) proves that

Theorem 10.1. *The operator* $-\mathscr{A}$ *is an infinitesimal generator of a semigroup* $t \to G(t)$, *continuous from* $t \geqslant 0 \to \mathscr{L}(\mathscr{H}; \mathscr{H})$.

In other words, if U_0 is given in $\mathscr{H}$ (resp. $D(\mathscr{A})$), there exists one and only one function $t \to U(t)$ which is the solution of

$$\partial U/\partial t + \mathscr{A}U = 0, \tag{10.1}$$

$$U(0) = U_0, \tag{10.2}$$

weak (resp. strong) solution of (10.1), such that $t \to U(t)$ is continuous (resp. once continuously differentiable) from $t \geqslant 0 \to \mathscr{H}$, and given by

$$U(t) = G(t)\,U_0. \tag{10.3}$$

If $U_0 \in D(\mathscr{A})$, then $U(t) \in D(\mathscr{A})$ $\forall t \geqslant 0$. □

Moreover—since here the choice of the norm for $\mathscr{H}$ is important—$G(t)$ is a *contraction operator*, i.e.

$$\|G(t)\|_{\mathscr{L}(\mathscr{H};\, \mathscr{H})} \leqslant 1, \quad t \geqslant 0. \quad \square \tag{10.4}$$

11. Comments

In this chapter, we discussed two types of media: "stable" and "ionizable", characterized, respectively, by Ohm's law (2.27) and the law (2.28). In practice, intermediate laws will frequently be of the form

$$J = \Phi(E).$$

When Φ is a multi-valued mapping of maximal monotone graph, the results obtained here should lend themselves to generalization. However, the physical measurements corresponding to these phenomena seem to involve phenomena of delay as indicated in Remark 2.5; then the function Φ is no longer monotone. The corresponding problems have not been investigated from a mathematical point of view. Nevertheless, the case where Φ is Lipshitz continuous can be treated. □

In all the problems touched upon here, we never took into account the fact that the speed of light is large (with respect to L/T, where L is a representative length), which implies that periodic stationary phenomena are reached at the end of a short time, when the excitation itself is sinusoidal. The mathematical justification for this situation leads to the following problem:

We go back to the example of Section 3.1 with

$$G_1(x,t) = Y(t)\, G_1(x)\, e^{i\omega t}$$

$$G_2(x,t) = Y(t)\, G_2(x)\, e^{i\omega t}$$

where $Y(t)$ is zero for $t<0$ and $+1$ for $t>0$, where $i=\sqrt{-1}$, where ω is a positive constant and $G_1(x)$ and $G_2(x)$ are given functions of x alone.

To be proved is that, for sufficiently large t, the solution (B, D) is close to an expression of the form.

$$B(x,t) = B^*(x)\, \mathrm{e}^{i\omega t}, \qquad D(x,t) = D^*(x)\, \mathrm{e}^{i\omega t}.$$

One can pose the same problem starting from the situation in Section 3.2 with

$$g(x,t) = g^*(x)\, \mathrm{e}^{i\omega t}$$

and starting from Section 3.3 with

$$B^{(1)} . n = Y(t)\, b^*(x)\, \mathrm{e}^{i\omega t},$$

$$D^{(1)} . n = Y(t)\, d^*(x)\, \mathrm{e}^{i\omega t}. \quad □$$

The imbedding theorems of Section 6 reproduce results of C. Goulaouic and B. Hanouzet [1], J. Gobert [3] and G. Schmidt [1].

Bibliography

Annin, B. D.
[1] Existence and uniqueness of the solution of the elastic-plastic torsion problem for a cylindrical bar of oval cross section. P. M. M. of Appl. Math. Mech. **29**, 1038–1047 (1965).

Artola, M.
[1] Sur les perturbations des équations d'évolution. Application à des problèmes de retard. Ann. Ecole Nat. Sup. Mec. Nantes **2**, 137–253 (1969).

Baiocchi, C.
[1] C. R. Acad. Sci. Paris **273**, 1215–1217 (1971)

Balaban, M. M., Green, A. E., Naghdi, P. M.
[1] Acceleration waves in elastic-plastic materials. Internat. J. Engrg. Sci. **8**, 315–335 (1970).

Bardos, C.
[1] Lecture Notes, University of Paris-Nord 1973.

Begis, D.
[1] Thesis of 3rd cycle, Paris 1972.

Bellman, R., Cooke, K.
[1] Differential-Difference Equations. New York: Acad. Press 1963.

Berker, R.
[1] Mouvement d'un fluide visqueux incompressible. Handbuch der Physik **VIII**, 2 (1963).

Beurling, A., Deny, J.
[1] Espaces de Dirichlet. I. Le cas élémentaire. Acta Math. **99**, 203–224 (1958).

Bihovski, E. B.
[1] Solution des problèmes mixtes pour les équations de Maxwell dans le cas de frontières supraconductrices (in Russian). Vestnik Leningrad. Univ. **13**, 50–65 (1957).

Bihovski, E. B., Smirnov, N. V.
[1] Sur les décompositions orthogonales des espaces de fonctions vecteurs... Trudy Mat. Inst. Steklov **LIX**, 5–33 (1960).

Biroli, M.
[1] Sulla perturbazione delle disequazioni d'evoluzione paraboliche. Ann. Scuola Norm. Sup. Pisa (3) **25**, 1–24 (1971)

Bourbaki, N.
[1] Espaces vectoriels topologiques. Chap. 3 and 4. Paris: Hermann 1955.

Bourgat, J. F.
[1] Thesis of 3rd cycle, Paris 1971

Brézis, H.
[1] Equations et inéquations non linéaires dans les espaces vectoriels en dualité. Ann. Inst. Fourier Grenoble **18**, 115–175 (1968).
[2] Inéquations variationnelles. J. Math. Pures Appl. (1971).
[3] Results not published.

Brézis, H., Lions, J. L.
[1] Sur certains problèmes unilatéraux hyperboliques. C. R. Acad. Sci. Paris **264**, 928–931 (1967).

Brézis, H., Sibony, M.
[1] Arch. Rational Mech. Anal. (1971).

Brézis, H., Stampacchia, G.
[1] Sur la régularité de la solution d'inéquations elliptiques. Bull. Soc. Math. France **96**, 153–180 (1968).

Browder, F.
[1] Non linear elliptic boundary value problems. Bull. Amer. Math. Soc. **69**, 862–674 (1963).

Bruhat, G.
[1] Electricité. Masson.

Brun, L.
[1] Méthodes énergétiques dans les systèmes évolutifs linéaires. I. Séparation des énergies. II. Théorèmes d'unicité. J. Mécanique **8**, 125–192 (1969).

Bui, H. D., Dangvan, K.
[1] Sur le problème aux limites en vitesse des contraintes du solide élasto-plastique. Internat. J. Solids and Structures **6**, 183–193 (1970).

Cabannes, H.
[1] Magnétodynamique des fluides. «Les cours de Sorbonne» C. D. U. Paris 1969.

Casal, P.
[1] Capillarité interne en mécanique des milieux continus. C. R. Acad. Sci. **256**, 29th April 1963.

Cattabriga, L.
[1] Su un problema al conterno relativo al sistema di equazioni di Stokes. Rend. Sem. Mat. Univ. Padova **31**, 1–33 (1961).

Céa, J., Glowinski, R.
[1] To appear.

Céa, J., Glowinski, R., Nédélec, J.
[1] To appear.

Coleman, B. D., Noll, W.
[1] Material symmetry and thermodynamic inequalities in finite elastic deformations. Arch. Rational Mech. Anal. (Berlin) **15**, n^0 2, 87–111 (1964).

Comincioli, V.
[1] Un risultato relativo a disequazioni variazionali d'evoluzione per operatori del primo ordine in t con termini di ritardo. Ann. Mat. Pura Appl. (4) **88**, 357–378 (1971).
[2] Disequaziono variazionali d'evoluzione per operatori del 2^0 ordine in t con termini di ritardo. (French Summary) Boll. Un. Mat. Ital. **4**, 273–289 (1971).
[3] Publications du Laboratoire de Calcul de l'Université de Pavie 1971.

Cooke: Cf. Bellman and Cooke.

Courjaret, B.
[1] To appear.

Coutris, N.
[1] Flexion élastoplastique d'une plaque. C. R. Acad. Sci. Paris **270**, 1377–1380 (1970).

Cowling, T. G.
[1] Magneto hydrodynamics. Interscience. Tracts n^0 4 (1957).

Critescu, N.
[1] Dynamic plasticity. Amsterdam: North Holland 1967.

Dafermos, C. M.
[1] An abstract Volterra equation with applications to linear-visco-elasticity. To appear in: J. Differential Equations.
[2] On the existence and the asymptotic stability of solutions of the equations of linear thermoelasticity. Arch. Rational Mech. Anal. **29**, 241–271 (1968)

Dangvan: Cf. Bui and Dangvan.

Day, W. A.
[1] Time reversal and the symmetry of the relaxation function of a linear viscoelastic material. Arch. Rational Mech. Anal. **40** (3), 155–159 (1971).

Deny: Cf. Beurling and Deny.

Dinga
[1] Sur la monotonie d'après Minty-Browder de l'opérateur de la théorie de plasticité. C. R. Acad. Sci. **269**, 535–538 (1969).

Distefano, J. N.
[1] On a class of Volterra integral equations ... Univ. of Southern Calif. **243**, Jan. 1968.

Dunford, N., Schwartz, J. T. S.
[1] Linear operators. Part I, Interscience Publ. (1958).

Duvaut, G.
[1] Application du principe de l'indifférence matérielle à un milieu élastique matériellement polarisé. C. R. Acad. Sci. Paris **258**, 3631–3634 (1964).
[2] Lois de comportement pour un milieu isotrope matériellement polarisé de degré deux. C. R. Acad. Sci. Paris **261**, 3178–3179 (1965).
[3] Problème de Signorini en viscoélasticité linéaire. C. R. Acad. Sci. Paris **268**, 1044–1046 (1969).
[4] Problèmes unilatéraux en mécanique des milieux continus. International Congress of Mathematicians, Nice 1970.

Duvaut, G., Lions, J. L.
[1] Sur de nouveaux problèmes d'inéquations variationnelles posés par la Mécanique. Le cas stationnaire. C. R. Acad. Sci. Paris **269**, 510–513 (1969).
[2] Sur de nouveaux problèmes d'inéquations variationnelles posés par la Mécanique. Le cas d'évolution. C. R. Acad. Sci. Paris **269**, 570–572 (1969).
[3] Nouvelles inéquations variationnelles rencontrées en thermique et en thermo-élasticité. C. R. Acad. Sci. Paris **269**, 1198–1201 (1969).
[4] Ecoulement d'un fluide rigide viscoplastique incompressible. C. R. Acad. Sci. Paris **270**, 58–61 (1970).
[5] Sur les équations de Maxwell des milieux polarisables et sur la magnéto-dynamique des fluides de Bingham. C. R. Acad. Sci. Paris **270**, 1600–1603 (1970).
[6] Elasticité avec frottement. J. Mécanique **10**, N° 3, 409–420 (1971).
[7] Inéquations en thermo-élasticité et magnéto-hydrodynamique. Arch. Rational Mech. Anal. **46,** 241–279 (1972).
[8] Transfert de chaleur dans un fluide de Bingham... J. Functional Analysis **11,** 93–110 (1972).

Eringen, A. C., Suhubi
[1] Non linear theory of micro-elastic solids. Internat. J. Engrg. Sci. **2**, 4, 389–404 (1964).

Fichera, G.
[1] Problemi elastostatici con vincoli unilaterali il problema die Signorini con ambigue condizioni al contorno. Mem. Accad. Naz. Lincei **8** (7), 91–140 (1964).

Finn, R.
[1] On the exterior stationary problem for the Navier Stokes equations and associated perturbation problems. Arch. Rational Mech. Anal. **19**, 363–406 (1965).

Flügge, W.
[1] Viscoelasticity. Blaisdell Publishing Company 1967.

Foias, C., Prodi, G.
[1] Sur le comportement global des solutions non stationnaires des équations de Navier Stokes en dimension 2. Rend. Sem. Mat. Univ. Padova **XXXIX**, 1–34 (1967).
[2] Sur les solutions statistiques des équations de Navier Stokes (To appear).

Fortin, M.
[1] Résolution numérique d'écoulements newtoniens et non newtoniens. Thesis, Paris 1962.

Frémond, M.
[1] Solide posé sur un sol élastique. C.R. Acad. Sci. Paris **271**, 508–510 (1970).
[2] Thesis, Paris 1971.

Fredrickson, A.G.
[1] Principles and applications of rheology. Englewood Cliffs, N.J.: Prentice-Hall Inc. 1964.

Freudenthal, A.M., Geiringer, H.
[1] The mathematical theories of the inelastic continuum. Handbuch der Physik – Encyclopedia of Physics, **VII**, Elasticity and Plasticity. Berlin-Göttingen-Heidelberg: Springer 1958.

Friedrichs, K.O.
[1] Differential forms on Riemannian manifolds. Comm. Pure Appl. Math. **8**, 551–590 (1955).
[2] Symmetric hyperbolic linear differential equations. Comm. Pure Appl. Math. **7**, 345–392 (1954).

Fusciardi, A., Mosco, U., Scarpini, F., Schiaffino, A.
[1] A dual method for the numerical solution of some variational inequalities. J. Math. Anal. Appl. **40**, 471–493 (1972).

Geiringer: Cf. Freudenthal and Geiringer.

Germain, P.
[1] Mécanique des milieux continus. Masson 1962.
[2] Cours de mécanique des solides (1964–1965). Faculté des Sciences de Paris.
[3] Théorie des ondes de chocs en dynamique des gaz et en magnétodynamique des fluides. Cours à la Faculté des Sciences de Paris. Département de Mécanique (1962–1963).

Glowinski, R., Lions, J.L., Trémolières, R.
[1] Approximation numérique des solutions des inéquations en mécanique et en physique. Paris: Dunod 2 Vol. 1975, 1976. Cf. Céa-Glowinski, Céa-Glowinski-Nédélec.

Gobert, J.
[1] Une inéquation fondamentale de la théorie de l'élasticité. Bull. Soc. Roy. Sci. Liège, n°s 3 and 4 (1962).
[2] Opérateurs matriciels de dérivation elliptiques et problèmes aux limites. Mém. Soc. Roy. Sci. Liège (6), 7–143 (1961).
[3] Sur une inégalité de coercivité. J. Math. Anal. Appl. **35** (1971).

Goulaouic, C., Hanouzet, B.
[1] Un résultat de régularité pour les solutions d'un système d'équations différentielles (to appear).

Goursat, M.
[1] Thesis of 3rd cycle, Paris 1971.

Green, A. E.
[1] On Reissner's theory of bending of elastic plates. Quart. App. Math. **7** (1949).

Green, A.E., Rivlin, R.S.
[1] Multipolar continum mechanics: functional theory I. Proc. Roy. Soc. Ser. A **284** (1965).

Green, A.E., Zerna, W.
[1] Theoretical elasticity. Oxford: Clarendon Press 1968.

Green, A.E.: Cf. Balaban, Green, Naghdi.

Haar, A., von Karman, Th.
[1] Zur Theorie der Spannungszustände in plastischen und sandartigen Medien. Nachr. Akad. Wiss. Göttingen Math.-Phys. Kl. II, 204–218 (1909).

Halanay, A.
[1] Differential equations, stability, oscillations, time lags. New York: Academic Press 1966.

Hanouzet: Cf. Goulaouic and Hanouzet.

Haugazeau, Y.
[1] Thesis, Paris 1968.

Hayart, R.
[1] Extension des formules de Murnaghan relatives au solide en phase d'élasticité finie, au cas de couples superficiels. C.R. Acad. Sci. **258**, 3rd Feb. 1964.

Hencky, H.
[1] Z. Angew. Math. Phys. **4**, 323 (1924).
[2] Über die Berücksichtigung der Schubverzerrung in ebenen Platten. Ing. Archiv, XVI Band (1947).

Herakovich, C.T., Hodge, P.G.
[1] Elastic-plastic torsion of hollow bars by quadratic programming. Int. J. Mech. Sciences **11**, 11, 53–63 (1969), Pergamon Press.

Hill, R.
[1] Mathematical theory of plasticity. Oxford: University Press 1950.
[2] Quart. J. Mech. Appl. Math. **1**, 18 (1948).
[3] J. Appl. Mech. **17**, 64 (1950).
[4] Philosophical Magazine **42**, 868 (1951).

Hille, E., Phillips, R.S.
[1] Functional Analysis and Semi Groups. Amer. Math. Soc. Coll. Pub. **XXXI** (1957).

Hodge, P.G.
[1] Elastic-plastic torsion as a problem in non linear programming. Int. J. Solids and Structures **3**, 989–999 (1967). Cf. Herakovich and Hodge, Prager and Hodge.

Hörmander, L.
[1] Definitions of maximal differential operators. Ark. Mat. **46**, 501–504 (1958).

Houpeurt, A.
[1] Eléments de mécanique des fluides dans les milieux poreux. I.F.P. (Technip), Paris (1957).

John, F.
[1] Plane strain problem for a perfectly elastic material of harmonic type. Comm. Pure Appl. Math. **XIII**, 239–296 (1960).
[2] Plane elastic waves of finite amplitude. Hadamard material and harmonic materials. Comm. Pure Appl. Math. **XIX**, 309–341 (1966).

Karman: Cf. Haar and von Karman.

Koiter, W.I.
[1] General theorems for elastic plastic solids. Progress in solid mechanics, pp. 165–221. Amsterdam: North Holland 1960.

Krasnoel'skii, M.A.
[1] Topological methods in the theory of non linear integral equations. Pergamon Press 1964 (Translation from the Russian, 1956).

Ladyzenskaya, O.A.
[1] La théorie mathématique des fluides visqueux incompressibles. Moscow 1961 (Translation from the English). New York: Gordon-Breach, 1963.

Ladyzenskaya, O.A., Solonnikov, V.A.
[1] Résolution de certains problèmes non stationnaires de la magnéto-hydrodynamique des fluides visqueux incompressibles (in Russian). Trudy Mat. Inst. Steklov **LIX**, 115–173 (1960).

Lanchon, H.
[1] Solution du problème de torsion élastoplastique d'une barre cylindrique de section quelconque. C.R. Acad. Sci. Paris **269**, 791–794 (1969).
[2] Sur la solution du problème de torsion élastoplastique d'une barre cylindrique de section multiconnexe. C.R. Acad. Sci. Paris **271**, 1137–1140 (1970).
[3] Problème d'élastoplasticité statique pour un matériau régi par la loi de Hencky. C.R. Acad. Sci. Paris **271**, 888–891 (1970).
[4] Thesis, Paris. J. Mécanique, 1972.

Lanchon, H., Duvaut, G.
[1] Sur la solution du problème de torsion élastoplastique d'une barre cylindrique de section quelconque. C.R. Acad. Sci. Paris **264**, 520–523 (1967).

Landau, L., Lifshitz, E.
[1] Théorie de l'élasticité, VII. Editions MIR, Moscow 1967.
[2] Course of theoretical Physics. Fluid Mechanics. Pergamon Press 1959.

Léonard, P.
[1] Problèmes aux limites pour les opérateurs matriciels de dérivation hyperboliques des premier et second ordres. **XI**, 7–128 (1965).

Leray, J.
[1] Etude de diverses équations intégrales non linéaires et de quelques problèmes que pose l'Hydrodynamique. J. Math. Pures Appl. **XII**, 1–82 (1933).
[2] Essai sur le mouvement plan d'un liquide visqueux que limitent des parois. J. Math. Pures Appl. **XIII**, 331–418 (1934).
[3] Sur le mouvement d'un liquide visqueux emplissant l'espace. Acta Math. **63**, 193–248 (1934).

Lévy, M.
[1] Mémoire sur les équations générales des mouvements intérieurs des corps solides ductiles au-delà des limites élastiques. C.R. Acad. Sci. Paris **70**, 1323–1325 (1870).
[2] Mémoire sur les équations des corps solides ductiles au-delà de la limite élastique. J. Math. Pures Appl. **16**, 369–372 (1871).

Lifshitz: Cf. Landau and Lifshitz.

Lin, T.H.
[1] Theory of inelastic structures. John Wiley 1968.

Lions, J.L.
[1] Quelques méthodes de résolution des problèmes aux limites non linéaires. Dunod, Gauthier-Villars 1969.
[2] Sur le contrôle optimal des systèmes gouvernés par des équations aux dérivées partielles. Dunod, Gauthier-Villars 1968.
[3] Les problèmes aux limites en théorie des distributions. Acta Math. **94**, 13–153 (1955).
[4] Sur un nouveau type de problème non linéaire pour opérateurs hyperboliques du deuxième ordre. Sem. J. Leray. Collège de France, 17–33, 1965–66, II.
[5] Singular perturbations and singular layers in variational inequalities. Symp. Non Linear Func. Analysis, April 1971; and C.R. Acad. Sci. Paris (1971).
[6] Cours Faculté des Sciences de Paris 1971.
[7] Inéquations variationnelles d'évolution. International Congress of Mathématicians, Nice 1970.

Lions, J.L., Magenes, E.
[1] Problèmes aux limites non homogènes et applications. Paris: Dunod 1968 (Vol. 1 and 2); 1970 (Vol. 3). English Translation: Non-homogeneous boundary value problems and applications (Grundlehren math. Wiss.). Berlin-Heidelberg-New York: Springer 1972 (Vol. 181 and 182); 1973 (Vol. 183).

Lions, J.L., Peetre, J.
[1] Sur une classe d'espaces d'interpolation. Inst. Hautes Etudes **19**, 5–68, Paris 1964.

Lions, J.L., Prodi, G.
[1] Un théorème d'existence et unicité dans les équations de Navier Stokes en dimension 2. C.R. Acad. Sci. Paris, 248 (1959).

Lions, J.L., Stampacchia, G.
[1] Variational inequalities. Comm. Pure Appl. Math. **XX**, 493–519 (1967).

Lions, J.L., Strauss, W.
[1] Some non linear evolution equations. Bull. Soc. Math. France **93**, 43–96 (1965).

Lions, J.L.: Cf. Brézis, Duvaut, Glowinski.

Love, A.E.H.
[1] A treatise on the mathematical theory of elasticity. Dover 1944.

Magenes, E., Stampacchia, G.
[1] I problema al contorno per le equazioni differenziali di tipo ellittico. Ann. Scuola Norm. Sup. Pisa, Ser. III, **XII**, fasc. III, 247–358 (1958).

Magenes, E.: Cf. Lions, Magenes.

Mandel, G.
[1] Cours de mécanique des milieux continus, Vol. 1, Mécanique des fluides, Vol. 2, Mécanique des solides. Gauthier-Villars 1966.
[2] Séminaire de plasticité. Publ. Sci. Tech. Ministère de l'air. N. T. 116 (1962).

Marocco
[1] Thesis of 3rd cycle, Paris 1970.

Miasnikov: Cf. Mosolov and Miasnikov

Mindlin, R. D., Tiersten, H. F.
[1] Effects of couple-stress in linear elasticity. Arch. Rational Mech. Anal. **11,** n⁰ 5, 415–448 (1962).

Minty, G.
[1] Monotone (non linear) operators in Hilbert space. Duke Math. J. **29,** 341–346 (1962).

Mooney, M.
[1] J. Appl. Phys. **11**, 528 (1940).

Moreau, J. J.
[1] Fonctionnelles convexes. Collège de France (1966–1967).
[2] La notion de surpotentiel et les liaisons unilatérales en élastostatique. C. R. Acad. Sci. Paris. Séance du 16-12-68, **257**, 954–957.
[3] Sur la naissance de la cavitation dans une conduite. C. R. Acad. Sci. Paris **259**, 3948–3951 (1964).

Mosco: Cf. Fusciardi. Mosco, Scarpini, Schiaffino.

Mosolov, P. P., Miasnikov, V. P.
[1] Variational methods in the theory of the fluidity of a viscous plastic medium. PMM **29**, 468–492 (1965).

Müller, C.
[1] Foundations of the mathematical theory of electromagnetic waves. (Grundlehren math. Wiss., Vol. 155). Berlin-Heidelberg-New York: Springer 1969.

Muskat
[1] The flow of homogeneous fluid through porous media. MacGraw-Hill 1937.
[2] Multiphase flow through porous media. MacGraw-Hill.

Naghdi: Cf. Balaban, Green, Naghdi.

Nayrolles, B.
[1] Essai de la théorie fonctionnelle des structures rigides plastiques parfaites. J. Mécanique **IX**, n⁰ 3, 491–506 (1970).
[2] Quelques applications variationnelles de la théorie des fonctions duales à la mécanique des solides. J. Mécanique **10**, N⁰ 2, 263–289 (1971).

Necas, J.
[1] Les méthodes directes dans la théorie des équations elliptiques. Prague: Acad. Tchécoslovaque des Sciences 1967.

Nédélec, J. C.
[1] Sur des inéquations variationnelles. Boll. Un. Mat. Ital. **4**, 762–774 (1971).

Nirenberg, L.
[1] Private communication, April 1971.

Nitsche, J. C. C.
[1] Variational problems with inequalities as boundary conditions or: How to fashion a cheap hat for Giacometti's brother. Arch. Rational Mech. Anal. **35,** n⁰ 2, 83–113 (1969).

Noll, W., Truesdell, C.
[1] The non linear field theory of mechanics. Handbuch der Physik, Vol. III/3. Berlin-Heidelberg-New York: Springer 1965. Cf. Coleman and Noll.

Panofsky, W., Phillips, M.
[1] Classical electricity and magnetism. Addison-Wesley 1962.

Peetre, J.
[1] Espaces d'interpolation et théorème de Sobolev. Ann. Inst. Fourier, **16**, 279–317 (1966). Cf. Lions and Peetre.

Phillips, M.: Cf. Panofsky and Phillips.

Phillips, R. S.: Cf. Hille and Phillips.

Piau, M.
[1] Conduction de la chaleur et propagation des ondes dans les milieus élasto-plastiques. C. R. Acad. Sci. Paris **271**, 1133–1136 (1970).

Prager, W., Hodge, P. G.
[1] Theory of perfectly plastic solids. Wiley 1961.

Prager, W.
[1] Introduction to mechanics of continua. Ginn and Company 1961.
[2] Problèmes de plasticité théorique. Dunod 1958.
[3] On ideal locking materials. Transaction of the Society of Rheology **1**, 169–175 (1957).

Prodi: Cf. Foias and Prodi, Lions and Prodi.

Reissner, E.
[1] On the theory of bending of elastic plates. J. Math. and Phys. **23** (1944).
[2] The effect of transverse shear deformation on the bending of elastic plates. J. Appl. Mech., June 1945.

Reuss, A.
[1] Z. Angew. Math. Phys. **10**, 266 (1930).

Rivlin: Cf. Green and Rivlin.

Rocard, Y.
[1] Thermodynamique, 2nd ed. Masson et Cie 1967.

Rockafellar, T.
[1] Duality and stability in extremum problems involving convex functions. Pacific J. Math. **21**, 167–187 (1967).
[2] Integrals which are convex functionals. Pacific J. Math. **24**, n° 3 (1968).

de Saint-Venant, M.
[1] Sur l'établissement des équations des mouvements intérieurs opérés dans les corps ductiles au-delà des limites d'élasticité. C. R. Acad. Sci. **70**, 473–480 (1870).
[2] Sur les équations du mouvement intérieur des solides ductiles. J. Math. Pures Appl. **16**, 373–382 (1871).

Sanchez-Palencia, E.
[1] Sur l'existence et l'unicité des solutions de certains problèmes aux limites posés par la magnéto-hydrodynamique. Thèse Faculté des Sciences de Paris, Département de Mécanique (1969).

Sander, G.
[1] Application de la méthode des éléments finis à la flexion des plaques. Publication de l'Université de Liège, n° 15 (1969).

Scarpini: Cf. Fusciardi. Mosco, Scarpini, Schiaffino.

Schatzman, M.
[1] Thesis of 3rd cycle, Paris 1971.

Schiaffino: Cf. Fusciardi. Mosco, Scarpini, Schiaffino.

Schmidt, G.
[1] Spectral and scattering theory for Maxwell's equations in an exterior domain. Arch. Rational Mech. Anal. **28**, 284–322 (1968).

Schwartz, J. T. S.: Cf. Dunford and Schwartz.

Schwartz, L.
[1] Théorie des distributions, I, II, 2nd Ed. 1957. Paris: Hermann 1950–1951.
[2] Distributions à valeurs vectorielles, I, II. Ann. Inst. Fourier **7**, 1–141 (1957); **8**, 1–209 (1958).

Sedov, L. I.
[1] Introduction to the mechanics of a continuous medium. Addison-Wesley 1965.

Serrin, J.
[1] The initial value problem for the Navier Stokes equations. In: Non Linear problems, ed. by R. E. Langer, pp. 69–98 (1963).

Sewell, M. J.
[1] On dual approximation principles and optimization in continuum mechanics. Philos. Trans. Roy. Soc. London Ser. A **265**, 319–351 (1969–1970).

Sibony, M.
[1] Une méthode itérative pour les inéquations variationnelles non linéaires. Pub. IRIA (1968). Cf. Brézis and Sibony.

Signorini, A.
[1] Sopra alcune questioni di elastostatica. Atti della Soc. Ital. per il Progresso della Scienze (1933).
[2] Questioni di elastostatica linearizzata e semilinearizzata. Rend. Mat. e Appl. **XVIII** (1959).

Smirnov: Cf. Bihovski and Smirnov.

Sobolev, S. L.
[1] Applications de l'analyse fonctionnelle aux équations de la physique mathématique. Leningrad 1950.

Solomon, L.
[1] Elasticitate Liniara. Editura Academieri Republicii Socialiste Romania.

Solonnikov, V. A.
[1] Sur certains problèmes aux limites stationnaires de la magnéto-hydrodynamique (In Russian). Trudy Mat. Inst. Steklov **LIX,** 174–187 (1960).

Solonnikov, V. A.: Cf. Ladyzenskaya and Solonnikov.

Stampacchia: Cf. Brézis and Stampacchia, Lions and Stampacchia.

Strauss, M. J.
[1] Variations of Korn's and Sobolev's inequalities. Berkeley Symposium, 1971.

Strauss, W.: Cf. Lions and Strauss.

Suhubi: Cf. Eringen and Suhubi.

Teman, R.
[1] Solutions généralisées d'équations non linéaires non uniformément elliptiques. Arch. Rational Mech. Anal. (1971).
[2] Cours Fac. Sci., Orsay (1971).

Tiersten: Cf. Mindlin and Tiersten.

Timoshenko, S., Woinowski-Krieger, S.
[1] Theory of plates and shells, 2nd Ed. McGraw-Hill 1959.

Ting, T. W.
[1] Elastic plastic torsion problem III. Arch. Rational Mech. Anal. **34,** 228–243 (1969).
[2] Elastic plastic torsion of convex cylindrical bars. J. Math. Mech., n° 19, 531–551 (1969).
[3] Elastic plastic torsion of simply connected cylindrical bars. Indiana University Math. J. **20,** n° 11, 1047–1076 (1971).

Tonti, E.
[1] On the formal structure of continuum mechanics, part I: deformation theory. Meccanica **V,** n° 1 (1970).

Toupin, R. A.
[1] Theories of elasticity with couple stress. Arch. Rational Mech. Anal. **17,** n° 2, 85–112.

Trémolières: Cf. Glowinski-Lions.

Tresca, H.
[1] Comptes rendus de l'Académie des Sciences **59,** 754 (1864); **70,** 27 (1870); **18,** 733 (1868); **20** (1872).

Truesdell, C., Toupin, R.
[1] The classical field theory. Handbuch der Physik, Vol. III/1. Berlin-Göttingen-Heidelberg: Springer 1960. Cf. Noll and Truesdell.

Viaud, D.
[1] Publication IRIA (1971).

Washizu, K.
[1] Variational methods in elasticity and plasticity. Pergamon Press 1968.

Woinowski-Krieger: Cf. Timoshenko and Woinowski-Krieger.

Yosida, K.
[1] Functional Analysis, 4th Ed. (Grundlehren math. Wiss., Vol. 123). Berlin-Heidelberg-New York: Springer 1974.

Zerna: Cf. Green and Zerna.

Additional Bibliography and Comments

1. Comments

Chapters 1 and *2*. Problems of climatization and semi-permeable boundary can be solved by non linear semi-group methods (H. Brezis [1]). In cases where boundary conditions depend on time, we are led to problems with varying convex sets for which we refer to Attouch [1], Damlamian [1] and C. Picard [1].

Chapter 2. Results on elasticity problems in an infinite medium have been treated by M. Boucher [1] in the case of a semi-infinite space.

Elastic materials with non quadratic strain energy have been studied by F. Lene [1].

Chapter 4. Plates in the non linear theory of Von Karman with homogeneous lateral boundary conditions were treated by the authors in G. Duvaut and J. L. Lions [1]. This study was completed by M. Potier [1] [2] who also studied some stability and bifurcation problems.

Some problems of unilateral stability and bifurcation were investigated by Cl. Do in [1] and lead to eigenvalue problem for inequalities. For those problems we refer also to J. P. Dias [1], J. P. Dias and V. Hernandez [1], Beirao da Veiga [1], Beirao da Veiga and J.P. Dias [1].

Chapter 5. Some problems of plasticity connected with variable convex sets have been studied by J. J. Moreau [1]. We refer in particular to the paper by C. Johnson [1], which contains a new a priori estimate.

Chapter 6. Incompressible fluids with a constituent relation of the form

$$\sigma_{ij} + p\delta_{ij} \in \{\partial\theta(\mathbb{D})\}_{ij}$$

(where θ is a convex function of the rate of strain tensor) which in some sense constitute the wider generalization of Bingham fluids, were studied by D. Cioranescu [1].

It was observed by C. Baiocchi [1] that free boundary problems arising in infiltration in dams of non rectangular shape can be transformed into *quasi* variational inequalities (Q.V.I.); Q.V.I. were introduced by Bensous-San-Lions [1], Bensoussan-Goursat-Lions [1] for solving problems arising in *impulse control*; a number of other free boundary problems were solved by reduction to variational inequalities on Q.V.I.; we refer to a book in preparation by the authors—(Duvaut-Lions [2]).

2. Bibliography

Attouch, H.
[1] Thesis Paris, 1975.

Baiocchi, C.
[1] C. R. Acad. Sci. Paris **278,** 1975; lecture at the I.C.M. Vancouver, 1974.

Bensoussan, A., Lions, J. L.
[1] Nouvelle formulation des problèmes de contrôle impulsionnel et applications. C. R. Acad. Sci. Paris **276**, 1189–1192; 1333–1338 (1973); **278**, 675–679, 747–751 (1974).

Bensoussan, A., Goursat, M. and Lions, J. L.
[1] C. R. Acad. Sci. Paris **273,** 1279–1284 (1973).

Boucher, M.
Thesis of 3rd Cycle, Paris-University 1972.

Beirão da Veiga, H.
[1] Differentiability for Green's operators ... J. Functional Analysis 1976.

Beirão da Veiga, H., Dias, J. P.
[1] Sur la subjectivité de certains operateurs non lineaires ..., Boll. Un. Mat. I. **10,** 52–59 (1974).

Cionarescu, D.
[1] Remarks on non newtonian fluids. To appear 1975.

Damlanian, A.
[1] Thesis Paris 1975.

Dias, J. P.
[1] Variational inequalities and eigenvalue problems ... A. J. Math. 1976.

Dias, J. P., Hernandez, J.
[1] A Sturm Liouville theorem ... Proc. Amer. Math. Soc. (1976).
[2] Bifurcation a l'infini ... J. Math. Pures Appl. (1976).

Do Cl.
[1] Problèmes de valeurs propres pour une inéquation variationnelle sur un cône et application au flambement unilatéral d'une plaque mince. C. R. Acad. Sci. Paris A **280,** 45–48 (1975).

Duvaut, G., Lions J. L.
[1] Problèmes unilatéraux dans la théorie de la flexion forte des plaques, I) Le cas stationnaire, II) Le cas d'évolution. J. Mécanique **13**, N° 1, June 1974, **13**, N° 1, March 1974.
[2] Sur les problèmes à frontière libre de la physi-mathématique. To appear.

Johnson, C.
[1] Plasticity and finite elements. To appear.

Lene, F.
Material with non quadratic strain energy. J. Mécanique **13**, N° 3, September 1974.
Moreau, J. J.
[1] CIME Symposium at Bressanone. June 1973, Ed. Cremonese, 1974, p. 173–322.

Picard, C.
[1] Exposé Séminaire Brézis-Lions, 1974/1975.

Potier, M.
[1] Third Cycle thesis. Paris-University 1973.
[2] C. R. Acad. Sci. Paris (to be published)

Subject Index

Die Grundlehren der mathematischen Wissenschaften in Einzeldarstellungen mit besonderer Berücksichtigung der Anwendungsgebiete

Eine Auswahl

---

172. Bühlmann: Mathematical Methods in Risk Theory
173. Maeda/Maeda: Theory of Symmetric Lattices
174. Stiefel/Scheifele: Linear and Regular Celestial Mechanics. Perturbed Two-body Motion—Numerical Methods—Canonical Theory
175. Larsen: An Introduction to the Theory of Multipliers
176. Grauert/Remmert: Analytische Stellenalgebren
177. Flügge: Practical Quantum Mechanics I
178. Flügge: Practical Quantum Mechanics II
179. Giraud: Cohomologie non abélienne
180. Landkof: Foundations of Modern Potential Theory
181. Lions/Magenes: Non-Homogeneous Boundary Value Problems and Applications I
182. Lions/Magenes: Non-Homogeneous Boundary Value Problems and Applications II
183. Lions/Magenes: Non-Homogeneous Boundary Value Problems and Applications III
184. Rosenblatt: Markov Processes. Structure and Asymptotic Behavior
185. Rubinowicz: Sommerfeldsche Polynommethode
186. Handbook for Automatic Computation. Vol. 2. Wilkinson/Reinsch: Linear Algebra
187. Siegel/Moser: Lectures on Celestial Mechanics
188. Warner: Harmonic Analysis on Semi-Simple Lie Groups I
189. Warner: Harmonic Analysis on Semi-Simple Lie Groups II
190. Faith: Algebra: Rings, Modules, and Categories I
192. Mal'cev: Algebraic Systems
193. Pólya/Szegö: Problems and Theorems in Analysis I
194. Igusa: Theta Functions
195. Berberian: Baer *-Rings
196. Athreya/Ney: Branching Processes
197. Benz: Vorlesungen über Geometrie der Algebren
198. Gaal: Linear Analysis and Representation Theory
199. Nitsche: Vorlesungen über Minimalflächen
200. Dold: Lectures on Algebraic Topology
201. Beck: Continuous Flows in the Plane
202. Schmetterer: Introduction to Mathematical Statistics
203. Schoeneberg: Elliptic Modular Functions
204. Popov: Hyperstability of Control Systems
205. Nikol'skii: Approximation of Functions of Several Variables and Imbedding Theorems
206. André: Homologie des Algèbres Commutatives
207. Donoghue: Monotone Matrix Functions and Analytic Continuation
208. Lacey: The Isometric Theory of Classical Banach Spaces
209. Ringel: Map Color Theorem
210. Gihman/Skorohod: The Theory of Stochastic Processes I
211. Comfort/Negrepontis: The Theory of Ultrafilters
212. Switzer: Algebraic Topology—Homotopy and Homology
213. Shafarevich: Basic Algebraic Geometry
214. van der Waerden: Group Theory and Quatum Mechanics
215. Schaefer: Banach Lattices and Positive Operators
216. Pólya/Szegö: Problems and Theorems in Analysis II
217. Stenström: Rings of Quotients
218. Gihman/Skorohod: The Theory of Stochastic Processes II